Advanced Textbooks in Control and Signal Processing

Springer
London
Berlin
Heidelberg
New York
Barcelona
Hong Kong
Milan
Paris
Santa Clara
Singapore
Tokyo

Series Editors

Professor Michael J. Grimble, Professor of Industrial Systems and Director
Professor Michael A. Johnson, Professor of Control Systems and Deputy Director
Industrial Control Centre, Department of Electronic and Electrical Engineering,
University of Strathclyde, Graham Hills Building, 50 George Street, Glasgow G1 1QE, U.K.

Other titles published in this series:

Genetic Algorithms: Concepts and Designs
K.F. Man, K.S. Tang and S. Kwong

Model Predictive Control
E. F. Camacho and C. Bordons

Introduction to Optimal Estimation
E.W. Kamen and J. Su

Neural Networks for Modelling and Control of Dynamic Systems
M. Nørgaard, O. Ravn, L.K. Hansen and N.K. Poulsen
Publication Due March 2000

Modelling and Control of Robot Manipulators (2nd Edition)
L. Sciavicco and B. Siciliano
Publication Due March 2000

Darrell Williamson

Discrete-time Signal Processing

An Algebraic Approach

With 44 Figures

Springer

Darrell Williamson, PhD

Faculty of Engineering and Information Technology, Australian National University, Canberra, ACT 0200. Australia

ISBN 1-85233-161-5 Springer-Verlag Berlin Heidelberg New York

British Library Cataloguing in Publication Data
Williamson, Darrell, 1948-
Discrete-time signal processing. - (Advanced textbooks in control and signal processing)
1.Signal processing 2.Discrete-time systems
I.Title
621.3'822
ISBN 1852331615

Library of Congress Cataloging-in-Publication Data
A catalog record for this book is available from the Library of Congress

Typesetting: Camera ready by author
Printed and bound by Athenæum Press Ltd., Gateshead, Tyne & Wear
69/3830-543210 Printed on acid-free paper SPIN 10715982

To Jan, Naomi, Pia, Leigh and Zoë

Series Editors' Foreword

The topics of control engineering and signal processing continue to flourish and develop. In common with general scientific investigation, new ideas, concepts and interpretations emerge quite spontaneously and these are then discussed, used, discarded or subsumed into the prevailing subject paradigm. Sometimes these innovative concepts coalesce into a new sub-discipline within the broad subject tapestry of control and signal processing. This preliminary battle between old and new usually takes place at conferences, through the internet and in the journals of the discipline. After a little more maturity has been acquired by the new concepts then archival publication as a scientific or engineering monograph may occur.

The applications of signal processing techniques have grown and grown. They now cover the wide range from the statistical properties of signals and data through to the hardware problems of communications in all its diverse aspects. Supporting this range of applications is a body of theory, analysis and techniques which is equally broad. Darrell Williamson has faced the difficult task of organising this material by adopting an algebraic approach. This uses general mathematical and systems ideas and results to form a firm foundation for the discrete signal processing paradigm. Although this may require some extra concentration and involvement by the student or researcher, the rewards are a clarity of presentation and deeper insight into the power of individual results. An additional benefit is that the algebraic language used is the natural language of computing tools like MATLAB and its simulation facility, SIMULINK. Thus, the step from analysis to demonstration and illustrative simulation is a shorter one.

The special bonus in the book is a rare chapter on finite wordlength considerations. Of course, Darrell Williamson is the ideal author for such a contribution having previously written a specialist text on the subject. This chapter is where design meets the implications of real-world implementation problems and it is a very appropriate way to conclude the first signal processing volume in the Advanced Textbooks in Control and Signal Processing Series.

M.J. Grimble and M.A. Johnson
Industrial Control Centre
Glasgow, Scotland, U.K.
July, 1999

Preface

This text provides an *algebraic approach* to both the analysis of discrete time signals, and the analysis and design of *discrete time signal processing* algorithms. The material is presented with the use of algebraically based software design packages (such as MATLAB) in mind, and is written for students in their third and fourth years of an undergraduate engineering program.

The text assumes that the reader has a working knowledge of complex numbers, and has completed an introductory course in vector and matrix analysis. This background is sufficient to cover the material in Chapters 1 and 2 which in itself could be used as a introductory course in difference equations. The material on digital filters presented in Chapter 3 is developed from first principles. However the subsequent material on analog filters assumes that the reader has completed an introductory course on differential equations. The study of signal processing in Chapter 4 which begins with the concepts of norm, inner product and orthogonality would benefit from an earlier course in analysis. Given this background, the coverage of the discrete Fourier transform, and the least squares estimation estimation algorithms is then appropriate. The final chapter on the finite wordlength implementation of a digital filter requires no further prerequisite material although the final two sections of this chapter assumes an advanced understanding in linear algebra.

Chapter 1 begins with a discussion of a number of applications in digital signal processing: market analysis, numerical integration, digital filter design, radar processing and speech compression. As well as providing motivation for subsequent developments, these applications are useful as illustrations of particular results. This introductory chapter provides some fundamental properties of both discrete and continuous time signal including a brief consideration of what is meant by signal processing, and concludes with an overview of hardware methods for digital-to-analog and analog-to-digital conversion.

Chapter 2 is concerned with the analysis of digital signals that can be represented in the form of the solution of a linear time invariant difference equation. The chapter begins with a detailed treatment of first and second order linear time invariant difference equations, and introduces the concepts and significance of order, time invariance and linearity. The chapter then proceeds directly to the consideration of a general linear time invariant difference equation. In particular, it is shown that any linear time invariant difference equation can be expressed as a system of first order linear time invariant difference equations known as the state space representation. This representation is most appropriate for the solution of difference equations via MATLAB.

Chapter 3 introduces the fundamental properties of both recursive (or IIR) and nonrecursive (or FIR) digital filters. All-pass, linear phase, lowpass, bandpass, bandstop and highpass filters are characterized. The first section of this Chapter considers some basic principles for the design of FIR digital filters. Conditions to guarantee a linear phase response are developed, and windowing methods are then introduced to help in matching desired magnitude characteristics. The second section of the chapter then focuses on the design of recursive digital filters, and begins with a brief coverage of analog filters which includes the design of Butterworth, Chebyshev and Elliptic analog low pass filters. Based on this introduction, the design of a digital filter as an approximation of an analog filter is then developed. Finally, the design of a lowpass, bandpass, bandstop or highpass digital filter as a transformation of a given lowpass digital filter is completed. The transformations from an analog filter to a digital filter are all expressed algebraically as a transformation of a state space representation of an analog filter to a state space representation of a digital filter, and as a consequence, can be implemented using MATLAB. Likewise, the transformations from a given lowpass digital filter to another lowpass, or a highpass, or a bandpass, or a bandstop digital filter are all expressed algebraically as a transformation of a state space representation of the given digital filter to a state space representation of the required digital filter, and so, can also be implemented using MATLAB.

Chapter 4 begins by characterizing signals in terms of their size (or norm). An inner product is defined between two signals which then provides a basis for characterizing orthogonal signals, and a method for establishing bounds on the output signal and the state component signals of a digital filter in terms of a bound on the input signal. In particular, bounds are expressed in terms of the solution of a linear algebraic equation which is readily solvable using MATLAB. An important representation of both periodic and finite length signals in terms of orthogonal signals known as the Discrete Fourier Series representation is then developed. The advantage of such a representation in applications of signal processing becomes evident following the development of the fast Fourier transform algorithm. Finally, the theory of least squares estimation is developed and applied to problems in filter design, and recursive and nonrecursive signal estimation. These algorithms are expressed in terms of algebraic equations which can be implemented using MATLAB.

Chapter 5 considers the design of a recursive filter structure for the finite wordlength implementation of a digital filter which takes into account the effects of input, coefficient and signal quantization errors. After covering the basic properties of both fixed and floating point arithmetic representations, this chapter considers the implication of the presence of arithmetic quantizations on both the accuracy and the speed of the implementation of a digital filter. Performance is considered in terms of arithmetic error and overflow, and in this regard, the structure of the FWL implementation is shown to be significant. Both low complexity and optimal filter structures are considered. Algebraic methods are developed for selecting the optimal filter structure which are readily implemented using MATLAB.

For the most part, the material is developed from first principles, and is based on lecture notes delivered to undergraduate engineering students at the Australian National University over a number of years. The material has appeared in different ways in various texts including those listed below. In such cases, no explicit credit is assigned. However, when results are not fully developed, an explicit reference is given. As has already been stated, the presentation in this text has a strong algebraic flavour. Frequency domain characteristics are developed from first principles, but there are no chapters on either the z-transform or the Laplace transform since such material is not needed. Instead a focus on linear algebraic concepts enables the reader to better integrate the analysis, design and implementation of digital signal processing algorithms with numerical simulation and design packages.

Supporting material in vector and matrix analysis can be found in one or more of the following texts:

- M. Marcus and H. Minc, *A Survey of Matrix Theory and Matrix Inequalities*, Allyn and Bacon, Boston, 1964.
- R. Bellman, *Introduction to Matrix Analysis*, (2nd ed.), McGraw-Hill, New York, 1970.
- S. Barnett, *Matrix Methods for Engineers and Scientists*, McGraw-Hill, London, 1979.
- B. Noble and J.W. Daniel, *Applied Linear Algebra*, (3rd ed.), Prentice-Hall, Englewood Cliffs, N.J., 1988.
- G. Strang, *Linear Algebra and Its Applications*, Brace Jovanovich, San Diego, 1988.
- R.A. Horn and C.R. Johnson, *Topics in Matrix Analysis*, Cambridge University Press, Cambridge, 1991

Likewise, one or more of the following texts can be consulted for supporting material in signal analysis, signal processing and filter design:

- L.R. Rabiner and B. Gold, *Theory and Applications of Digital Signal Processing*, Prentice-Hall, Englewood Cliffs, N.J., 1975.
- A.V. Oppenheim and R.V. Schafer, *Digital Signal Processing*, Prentice-Hall, Englewood Cliffs, N.J., 1975.
- S.M. Bozic, *Digital and Kalman Filtering*, Edward Arnold, 1979.

- H.Y.F Lam, *Analog and Digital Filters*, Prentice-Hall, Englewood Cliffs, N.J., 1979.
- R.A. Gabel and R.A. Roberts, *Signals and Linear Systems*, Wiley, 1980.
- W.D. Stanley, G.R. Dougherty and R. Dougherty, *Digital Signal Processing*, Reston Publ.,1984.
- N.S. Jayant and P. Noll, *Digital Coding of Waveforms*, Prentice-Hall, Englewood Cliffs, N.J., 1984.
- J.C. Cluley, *Transducers for Microprocessors*, MacMillan, London 1985.
- T.W. Parks and C.S. Burrus, *Digital Filter Design*, Wiley, 1987.
- M.F. Hordeski, *Transducers for Automation*, Van Nostrand, Reinholt, N.Y., 1987.
- J. R. Johnson, *Introduction to Digital Signal Processing*, Prentice-Hall, Englewood Cliffs, N.J., 1989.
- S.S. Soliman and M.D. Srinath, *Continuous and Discrete Signals and Systems*, Prentice-Hall, 1990.
- H. Kwakernaak and R. Sivan, *Modern Signals and Systems*, Prentice-Hall, Englewood Cliffs, N.J., 1991
- D. Williamson, *Digital Control and Implementation*, Prentice-Hall, Englewood Cliffs, N.J., 1991.
- R.E. Ziemer, W.H. Tranter and D.R. Fannin, *Signals and Systems: Continuous and Discrete*, MacMillan, London, 1993.
- C-T Chen, *System and Signal Analysis*, Saunders College Publ., 1994.
- A.V. Oppenheim and A.S. Wilsky, *Signals and Systems*, Prentice-Hall, Englewood Cliffs, N.J., 1997.
- E.W. Kamen and B.S. Heck, *Fundamentals of Signals and Systems*, Prentice-Hall, Englewood Cliffs, N.J., 1997.
- P. Lapsley, J. Bier, A. Shoham and E.A. Lee, *DSP Processor Fundamentals: Architecture and Features*, IEEE Press, Inc., 1997.

Contents

1. Introduction

A *discrete* time signal is one that has a value only for a finite or infinite number of time instants whereas a *continuous* time signal has a value for every (real) time instant. For example, consider:

- The sound pressure wave of a speech signal
- The electrical signal that is the output of a transducer that is used to measure the sound pressure wave of a speech signal
- The sequence of numbers that is obtained from connecting an analog to digital converter (ADC) to the electrical signal which is the output of a transducer that is used to measure the sound pressure wave of a speech signal

The first two examples are *continuous time* or *analog* signals (of pressure measured in pascals and voltage measured in volts respectively) which vary continuously with time. The third example is a *discrete time* or *digital* signal having no units. In the physical sense, the first two signals are different, but in a mathematical sense, we treat them as being equivalent in that they are both continuous functions of time. On the other hand, the third example has a mathematically representation which is expressed in terms of a sequence of numbers, and so is different from either of the first two signals.

Other examples of physical signals are:

- Position, velocity and acceleration of an aircraft
- Acoustic pressure waves in seismic or sonar applications
- Video signals produced by a television camera
- The number of sales of a brand of motor car per month in a particular city
- The daily currency exchange rates
- The daily value of a particular stock on the stock market

The first three example here are inherently continuous time signals although one usually deals with discrete time versions of these signals which are obtained by viewing, or recording, these signals at discrete instants of time. On the other hand, the last three examples may be thought of as inherently discrete time since the signal data is only available at particular times of the day, week or month.

As well as understanding the mathematical tools which are needed for signal analysis, one must also be concerned about the accuracy of the transducer that records the value of pressure, and the accuracy of the ADC that samples the output of the transducer. In any analysis, both the transducer and ADC errors should be taken into account. It is important to be able to analyse what (if any) information is lost during the sampling process. However, the more samples that are made available, the more data that must be stored and processed which in turn means the more memory and computer processing power that is required.

In this text, we are interested in the representation of signals. However, we are also interested in finding algorithms that can be applied to the signals to estimate particular properties of the signal. In engineering terms, this is referred to as "processing" the signal. Signal processing either seeks to obtain estimates of some particular properties of a signal, or to reduce the effects of some form of error (or distortion) that has inadvertently been added to the signal. For example, we may need to:

- Remove noise from a speech signal that has been added as a result of a faulty telephone connection
- Design a signal processing algorithm which enables a computer to recognise the voice of a person from a given set of possible speakers
- Predict the number of orders a company will have for a particular product, or
- Estimate the velocity of an aircraft based on measurements of position.

Practical implementations of signal processing algorithms need to be concerned with numerical accuracy, complexity and speed of implementation. Linear algebra is a mathematical tool that is useful for both the analysis of signals, and the implementation of signal processing algorithms. Linear algebra enables a large amount of data to be handled in a compact notation. This not only helps to simplify the presentation of results, but also enables one to focus more on the concepts rather than on the details of how to handle any particular set of data.

1.1 Digital and Analog Signals

A constant n-dimensional vector $\boldsymbol{v}^T = [v_1 \ \ v_2 \ \ \cdots \ \ v_n]$ where all the components v_k are real can be used (for example) to define the location of a point P in real n-dimensional space. The location of a point P which moves with time t can be represented at each time instant t by an $n+1$ dimensional vector $\boldsymbol{v}_t$ where

$$\boldsymbol{v}_t^T = [v_1 \ \ v_2 \ \ \cdots \ \ v_n \ \ t] \tag{1.1}$$

However, it is usually more convenient to represent $\boldsymbol{v}_t$ as an n-dimensional *time varying* vector $\boldsymbol{v}(t)$ parametrized by t; that is

$$\boldsymbol{v}^T(t) = [v_1(t) \quad v_2(t) \quad \cdots \quad v_n(t)] \tag{1.2}$$

Each component $v_k(t)$ of $\boldsymbol{v}(t)$ in (1.2) is assumed to be a real (or more generally complex) number for each value of the single *independent* time variable t, and the function $v_k(\cdot)$ is a *function* of t. For simplicity, we shall henceforth refer to the *function* $\boldsymbol{v}(\cdot)$ by $\boldsymbol{v}$, and each component function $v_k(\cdot)$ by v_k. The value of the function $\boldsymbol{v}$ at time t_k is then $\boldsymbol{v}(t_k)$.

In engineering applications, a function of time is more commonly referred to as a *(time) signal.* The signal v_k is referred to as a *scalar* valued signal, and the signal $\boldsymbol{v}$ in (1.2) is referred to as a *vector* valued signal. An example of a scalar signal v_k is the acoustic pressure of a speech signal as a function of time. A vector valued signal $\boldsymbol{v}$ in (1.2) could be the acoustic pressure of n speech signals which have been simultaneously recorded as functions of time. More generally, signals can be represented as mathematical functions of more than one independent variable. For example, a *photograph* can be represented by the greyness (or colour) level of *two* independent spatial variables x and y. A video image can be represented by *three* independent variables (two spatial variables and one time variable). In this text however, we only consider signals involving a single independent variable of time.

The *domain* of a real scalar signal v_k is a subset $\mathcal{T}$ of the real line $\mathcal{R}$ where

$$\mathcal{R} \triangleq \{t \ : \ -\infty < t < \infty\}$$

Throughout the text, we use the symbol "$\triangleq$" to mean "defined by" which is to be distinguished from the symbol "$=$" which has the meaning "equal to". We also use the notation $\mathcal{T} \subset \mathcal{R}$ to mean "the set $\mathcal{T}$ is a subset of the set $\mathcal{R}$". Hence, the domain $\mathcal{T} \subset \mathcal{R}$ of the real signal v_k is a subset $\mathcal{T}$ of the real line $\mathcal{R}$. The domain $\mathcal{T} \subset \mathcal{R}$ of the signal v_k may either be a *discrete* subset of $\mathcal{R}$, or a *continuous* subset of $\mathcal{R}$. We also define the following subsets of the real numbers $\mathcal{R}$ and integers $\mathcal{Z}$:

$$\begin{aligned} \mathcal{R}_+ &\triangleq \{r \in \mathcal{R} : r \geq 0\} \ ; \quad \mathcal{R}_- \triangleq \{r \in \mathcal{R} : r \leq 0\} \\ \mathcal{Z}_+ &\triangleq \{n \in \mathcal{Z} : n \geq 0\} \ ; \quad \mathcal{Z}_- \triangleq \{n \in \mathcal{Z} : n \leq 0\} \end{aligned} \tag{1.3}$$

For example, $\mathcal{T} = \mathcal{Z}$ is a discrete subset of $\mathcal{R}$, and $\mathcal{T} = \mathcal{R}_+$ is a continuous subset of $\mathcal{R}$.

A signal whose domain is a discrete set is called a discrete (time) signal, while a signal whose domain is a continuous set is called a continuous (time) signal. A discrete time signal is also called a digital signal, and a continuous time signal is also called an analog signal .

The *range* of a signal v is the set $\mathcal{S}$ of possible values that $v(t)$ may assume as t varies over the domain of v. Thus we write

$$v : \mathcal{T} \to \mathcal{S} \tag{1.4}$$

If $\mathcal{S} \subset \mathcal{R}$, so that $v(t) \in \mathcal{R}$ is real, then v is called a *real* signal whereas if $v(t) \in \mathcal{C}$ where $\mathcal{C}$ is the set of *complex* numbers, then v is called a *complex* signal. For a real vector valued signal $\boldsymbol{v}$ in (1.2), we write $\boldsymbol{v}(t) \in \mathcal{R}^n$, and for a complex vector valued signal $\boldsymbol{v}$, we write $\boldsymbol{v}(t) \in \mathcal{C}^n$.

The *magnitude* of a scalar α is denoted by $|\alpha|$. If α is *real* scalar, then $|\alpha|$ denotes the *absolute value* of α, whereas if

$$\alpha = \alpha_1 + j\alpha_2 \quad ; \; j^2 = -1 \tag{1.5}$$

is a *complex* number (with α_1, α_2 both real), then

$$|\alpha| = \sqrt{\alpha_1^2 + \alpha_2^2} = \sqrt{\alpha\bar{\alpha}} \tag{1.6}$$

where

$$\bar{\alpha} \triangleq \alpha_1 - j\alpha_2 \tag{1.7}$$

defines the *complex conjugate* of the scalar α.

Infinite Length Signals

A digital signal $s : \mathcal{Z}_+ \rightarrow \mathcal{R}$ corresponds to an infinite sequence of numbers arranged in order as follows:

$$s(0), s(1), s(2), s(3), \; \ldots \; , s(p), s(p+1), \; \ldots$$

Some examples are:

$$s_1(n) = 2^{-n} \quad ; \quad s_2(n) = (\frac{n+1}{n})^n \quad ; \quad s_3(n) = \sum_{k=1}^{n}(1+k)^{-1} \tag{1.8}$$

A signal $s : \mathcal{Z}_+ \rightarrow \mathcal{R}$ is said to converge to $\bar{s}$ (or to have the limit $\bar{s}$), and written

$$\lim_{n\to\infty} s(n) = \bar{s}$$

if for any positive real number ε, an integer N can be found such that

$$|\bar{s} - s(n)| < \varepsilon$$

for all $n \geq N$. If the signal s does not converge, it is said to diverge.

With reference to the signals defined in (1.8),

$$\lim_{n\to\infty} 2^{-n} = 0 \quad ; \quad \lim_{n\to\infty} (\frac{n+1}{n})^n = e = 2.718...$$

and so both s_1 and s_2 are *converging* signals. However

$$\lim_{n\to\infty}\sum_{k=0}^{n}(1+k)^{-1}$$

is infinite, and so s_3 is a *diverging* signal.

In signal processing applications, certain signals arise as the result of summing a number of terms of another signal. Specifically, if the signal s is defined by

$$s(k+1) = s(k) + z(k) \quad ; \quad s(0) = 0 \tag{1.9}$$

then $s(p)$ is given by

$$s(p) \triangleq \sum_{n=1}^{p-1} z(n) \tag{1.10}$$

The infinite series $\sum_{n=1}^{\infty} z(n)$ is convergent if the signal s defined by (1.9) is convergent; otherwise, the infinite series is said to be divergent. If the signal s converges to $\bar{s}$, then $\bar{s}$ is called the sum of the series and written

$$\bar{s} \triangleq \sum_{n=1}^{\infty} z(n)$$

We have the following result.

Theorem 1.1.1.

- *If the sum $\sum_{n=1}^{\infty} z(n)$ is infinite, then $\sum_{n=1}^{p} z(n)$ diverges.*
- *If the sum $\sum_{n=1}^{\infty} |z(n)|$ is convergent, then $\sum_{n=1}^{\infty} z(n)$ converges.*
- *If $|z(n)| \leq w(n)$ for all n, and $\sum_{n=1}^{\infty} w(n)$ converges, then $\sum_{n=1}^{\infty} |z(n)|$ converges.*
- *If $z(n) \geq w(n) \geq 0$ for all n, and $\sum_{n=1}^{\infty} w(n)$ diverges, then $\sum_{n=1}^{\infty} z(n)$ diverges.*
- *The harmonic series of order p*

$$\sum_{n=1}^{\infty} \frac{1}{n^p} = 1 + \frac{1}{2^p} + \frac{1}{3^p} + \ \ldots$$

  converges for $p > 1$ and diverges for $p \leq 1$.
- *The geometric series*

$$\sum_{n=0}^{\infty} ar^n$$

  converges for $|r| < 1$ and diverges for $|r| > 1$. In particular, for $r \neq 1$

$$s(p) \triangleq \sum_{n=0}^{p} ar^n = \frac{a(1-r^{p+1})}{1-r}$$

and for $r = 1$, $s(p) = a(p+1)$. *When* $|r| < 1$

$$\lim_{p\to\infty} s(p) \triangleq \bar{s} = \frac{a}{1-r}$$

- *(Ratio Test) Suppose* $z(n) \neq 0$ *for all* n, *and*

$$\lim_{n\to\infty} \left| \frac{z(n+1)}{z(n)} \right| = L$$

Then $\sum_{n=1}^{\infty} |z(n)|$ *is convergent if* $L < 1$, $\sum_{n=1}^{\infty} z(n)$ *is divergent if* $L > 1$, *and no conclusion can be made if* $L = 1$.
- *(Alternating Series) The series* $\sum_{n=0}^{\infty} (-1)^{n+1} z(n)$ *for* $z(n) \geq 0$ *is convergent if either* $z(n+1) \leq z(n)$ *for all* n, *or* $\lim_{n\to\infty} z(n) = 0$.
- *(Root Test) Suppose*

$$\lim_{n\to\infty} |z(n)|^{\frac{1}{n}} = R$$

Then $\sum_{n=1}^{\infty} |z(n)|$ *is convergent if* $R < 1$, $\sum_{n=1}^{\infty} z(n)$ *is divergent if* $R > 1$, *and no conclusion can be made if* $R = 1$.

Exponential Signal: An *exponential* signal v is defined by

$$v(t) = \varrho exp\{\alpha t\} \quad ; \quad t \in \mathcal{T} \tag{1.11}$$

where (in general) ϱ and α are complex numbers, and

$$exp\{\alpha t\} \triangleq 1 + \alpha t + \frac{\alpha^2 t^2}{2!} + \cdots + \frac{\alpha^n t^n}{n!} + \cdots = \sum_{k=0}^{\infty} \frac{\alpha^k t^k}{k!}$$

where by definition $0! \triangleq 1$.

The ratio of the $(n+1)$th term $z(n+1)$ to the nth term $z(n)$ of the series for $exp\{\alpha t\}$ is given by

$$\frac{z(n+1)}{z(n)} = \frac{\alpha^{n+1} t^{n+1}}{(n+1)!} \cdot \frac{n!}{\alpha^n t^n} = \frac{\alpha t}{n+1}$$

Hence by the *ratio test* in Theorem 1.1.1, the series for $exp\{\alpha t\}$ is *convergent* for all α and t.

Consider an *analog* real exponential signal v as defined by $v(t) = 4exp\{-t\}$ over the continuous domain $\mathcal{T} = \{t \in \mathcal{R} \ : \ -1 \leq t \leq 3\}$. Then a *digital* exponential signal u is given by *sampling* v every (say) 0.5 seconds; that is

$$u(k) \triangleq v(t = 0.5k) = 4a^k \ ; \ a = exp\{-0.5\}, \ k \in \mathcal{T}$$

Note that a discrete signal is only a sequence of numbers.

A *real* exponential signal v is defined by (1.11) when $\varrho = \varrho_0$ and $\alpha = \alpha_0$ are both real numbers. A real exponential signal v increases in magnitude as t increases when $\alpha > 0$, while v decreases in magnitude as t increases for $\alpha < 0$. When $\varrho = \varrho_0$ is real, and $\alpha = j\omega_0$ ($j^2 = -1$) is *purely imaginary*, then the resulting *complex signal* v in (1.11) is given by

$$v(t) = \varrho_0 exp\{j\omega_0 t\} = \varrho_0(\cos\omega_0 t + j\sin\omega_0 t)$$

More generally, when both ϱ and α are complex; that is

$$\varrho = \varrho_0 exp\{j\phi\} \quad ; \quad \alpha = \alpha_0 + j\omega_0$$

where $\{\varrho_0, \phi, \alpha_0, \omega_0\}$ are all real numbers, then

$$v(t) = \varrho_0 exp\{\alpha_0 t\}[\cos(\omega_0 t + \phi) + j\sin(\omega_0 t + \phi)]$$

This complex signal v can also be written in the form

$$v = v_R + jv_I \quad ; \quad j^2 = -1$$

where

$$v_R(t) = \varrho_0 exp\{\alpha_0 t\}\cos(\omega_0 t + \phi) \quad ; \quad v_I(t) = \varrho_0 exp\{\alpha_0 t\}\sin(\omega_0 t + \phi)$$

are both *real* signals.

The *analog* complex exponential signal

$$v(t) = 2exp\{\frac{j\pi}{4}\}exp\{(-1 + j3)t\}$$

can also be equivalently written in the form

$$v(t) = 2exp\{-t\}[\cos(3t + \frac{\pi}{4}) + j\sin(3t + \frac{\pi}{4})]$$

A *digital* complex exponential signal u can then be defined by *sampling* the analog signal v every (say) 0.5 seconds according to

$$\begin{aligned} u(k) &\triangleq v(t = 0.5k) \\ &= 2a^k\left[\cos\left(\frac{3k}{2} + \frac{\pi}{4}\right) + j\sin\left(\frac{3k}{2} + \frac{\pi}{4}\right)\right] \quad ; \quad a = exp\{-0.5\} \end{aligned}$$

Rectangular Pulse Signal: An *analog rectangular pulse* signal ϱ_Δ is defined by

$$\varrho_\Delta(t) \triangleq \begin{cases} \Delta^{-1} \; ; \; 0 \le t \le \Delta, \;\; t \in \mathcal{R} \\ 0 \; ; \; \text{otherwise} \end{cases} \tag{1.12}$$

A *digital* pulse signal u can then be defined by *sampling* the analog signal ϱ_Δ for $\Delta = 2$ every (say) 0.5 seconds according to

$$u(k) = \varrho_\Delta(t = 0.5k) = \begin{cases} 0.5\ ; & 0 \le k \le 4 \\ 0\ ; & \text{otherwise} \end{cases}$$

Triangular Pulse Signal: An *analog triangular pulse* τ_Δ is defined by

$$\tau_\Delta(t) \triangleq \begin{cases} 4\Delta^{-2}t & ;\ 0 \le t \le 0.5\Delta \\ -4\Delta^{-2}t + 4\Delta^{-1} & ;\ 0.5\Delta \le t \le \Delta \quad t \in \mathcal{R} \\ 0 & ;\ \text{otherwise} \end{cases} \tag{1.13}$$

A *digital* triangular pulse signal u can be defined by sampling the analog signal τ_Δ every (say) 0.5 seconds. For example, when $\Delta = 1$, we have

$$u(k) \triangleq \tau_1(t = 0.5k) = \begin{cases} 2k & ;\ 0 \le k \le 1 \\ -2k + 4 & ;\ 1 \le k \le 2 \\ 0 & ;\ \text{otherwise} \end{cases}$$

Unit Step Signal: The *analog unit step* signal s is defined by

$$s(t) \triangleq \begin{cases} 1\ ;\ t \ge 0, & t \in \mathcal{R} \\ 0\ ;\ \text{otherwise} \end{cases} \tag{1.14}$$

A *digital* unit step signal u can be obtained by *sampling* s every (say) 0.5 seconds to give

$$u(k) = \begin{cases} 1\ ;\ k \ge 0 \\ 0\ ;\ \text{otherwise} \end{cases}$$

Unit Ramp Signal: An *analog ramp* signal r is defined by

$$r(t) \triangleq \begin{cases} t\ ;\ t \ge 0, & t \in \mathcal{T} \\ 0\ ;\ \text{otherwise} \end{cases} \tag{1.15}$$

A *digital* ramp signal u can then be obtained by *sampling* v every (say) 0.5 seconds; that is,

$$u(k) = \begin{cases} 0.5k \ ; \ k \geq 0 \\ 0 \ ; \ \text{otherwise} \end{cases}$$

Discrete Unit Impulse Signal: The discrete unit impulse signal δ is defined by

$$\delta(n) \triangleq \begin{cases} 1 \ ; \ n = 0 \\ 0 \ ; \ n \neq 0 \end{cases} \tag{1.16}$$

There is no analog signal d which when sampled at any sampling rate T^{-1} results in the discrete unit impulse signal.

We now provide details about the hardware characteristics of both analog-to-digital and a digital-to-analog conversion. Since most analog-to-digital converters use a digital-to-analog converter, we first discuss some methods for digital-to-analog conversion.

1.2 Digital-to-Analog Conversion

The basic configuration for a *digital-to-analog converter* (DAC) is illustrated in Figure 1.1. A digital interface converts the parallel logic inputs to suitable control levels which operate a set of electronic switches. A precision ladder network then provides either a weighted sum of currents (I_{out}) or a voltage (V_{out}) as an output when referenced against a stable precision voltage source (V_{ref}).

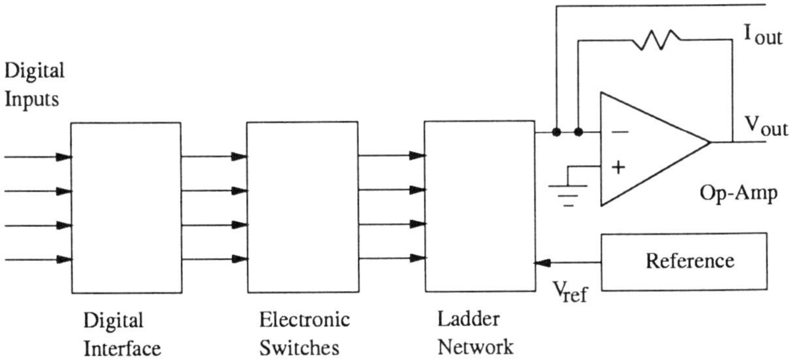

Fig. 1.1. Digital-to-analog converter

One means of achieving a binary weighted combination of currents I_{out} (known as the *Weighted Current Method*) is illustrated in Figure 1.2, where if a voltage output V_{out} is required, the output current I_{out} drives a current-to-voltage circuit. However, some loss in conversion speed results due to the bandwidth limitation of the operational amplifier (Op-Amp). For a voltage output, the high speed associated with current conversion can be preserved by replacing the Op-Amp with a single resistor. But in this case, only limited full-scale output voltage is possible due to the limited positive voltage swing of the collectors of the transistors.

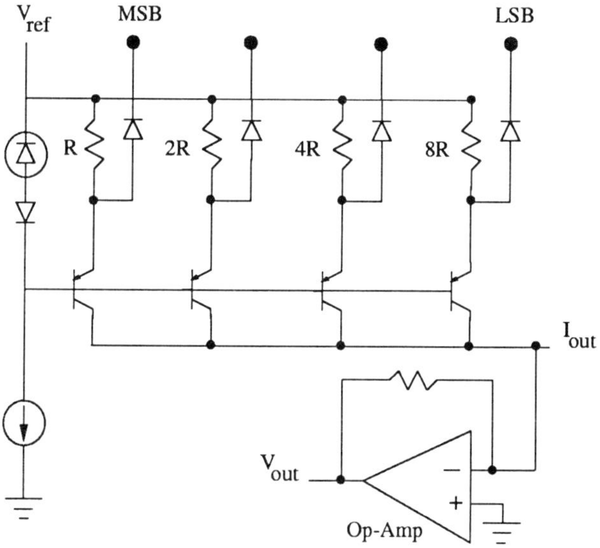

Fig. 1.2. Ladder network for bipolar digital-to-analog conversion

The binary weights in Figure 1.2 are achieved by the emitter resistors with values of R, $2R$, $4R$ and $8R$ (for a 4 bit conversion). The most significant bit (MSB) drives the emitter with resistance R. The current source transistors are turned on and off by the applied logic inputs which are connected via diodes to the emitters. A stable reference voltage V_{ref} which is compensated to allow for variations in the transistor base to emitter voltage with temperature is applied to the base of all transistors so as to set up constant emitter currents. Depending on the logic level at each diode input, the current flows either through the diode or through the transistor. The weighted currents are summed to give I_{out} at the common collector of the transistors. The weighted current method has the *advantages* of simplicity and high speed. The *disadvantage* of using a current output DAC of this type is the wide range of matched resistance values (i.e. R, $2R$, $4R$, ...) are required for conversion to a large wordlength. Some resistors must then necessarily have large resistance values, and this decreases the conversion speed for integrated circuits.

Large wordlength converters with *high* speed conversion can be made by using several groups of 4 bit current output DAC's of the type illustrated in Figure 1.2, and dividing down the output current of each group. This principle is illustrated in Figure 1.3 for a 12 bit conversion. The current dividers at nodes **3** and **2** must reduce the current outputs of each 4 bit weighted source to 1/16 of their original values. This method also enables *bipolar* operation (i.e. both positive and negative voltage swings) for a DAC by connecting a current source with a current equal to the MSB weight to the output of all other weighted sources. This offsets the output of the converter by one half the full scale voltage swing, and is known as *offset binary coding*.

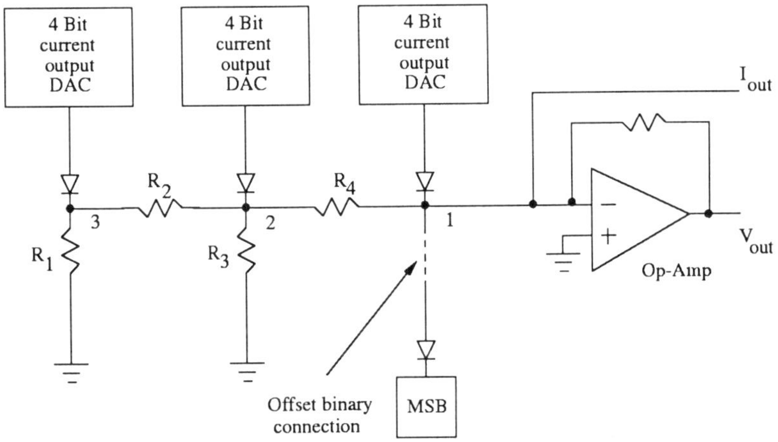

Fig. 1.3. High speed-low complexity circuit for digital-to-analog conversion

Another common method of digital-to-analog conversion (known as the *R-2R Ladder Method*) is to use a network of resistors with series value R and shunt value 2R (referred to as an R-2R ladder) as illustrated in Figure 1.4 for a 4 bit conversion. Depending on the logic level, the bottom of the shunt resistors are switched either to ground or to the voltage reference source V_{ref}. The operation of the ladder network is based on the binary division of current as it flows down the ladder. The output can either be a current I_{out} or a voltage V_{out} at the output of the Op-Amp.

The *advantage* of the *R-2R ladder method* is that resistances of R and 2R are relatively easy to match and to compensate for temperature variations (compared with R, 2R, 4R, 8R, ... in the weighted current source method). Furthermore the resistor values can be kept low thereby ensuring high speed. A *disadvantage* of the method is that the R-2R ladder network requires *two* resistors per binary bit whereas the *weighted current method* as illustrated in Figure 1.2 requires only one. The number of resistors is an important consideration for an integrated circuit realization.

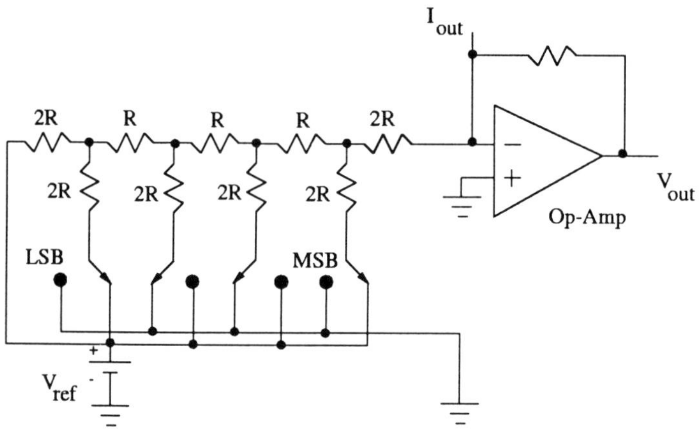

Fig. 1.4. The R-2R ladder method for digital-to-analog conversion

1.3 Analog-to-Digital Conversion

Methods for designing an *analog-to-digital converter* (ADC) can be based on the principles of *servo operation*, *dual slope integration*, *successive approximation*, or *parallel operation.* All these methods are relatively simple and provide good accuracy. Their main difference lies in the time required for conversion.

In the method based on *servo operation* the conversion time is relatively *slow* and in the worst case is proportional to the voltage level to be converted. Conversion commences by gating on the clock, and the digital counter counts the clock pulses. As it counts, it changes the output of the DAC which is compared to the analog input (V_i) via a comparator. When the DAC output exceeds the analog input, the comparator output changes logic levels and inhibits the clock pulses. The conversion time can be reduced by using an up-down counter which counts either up or down from the *previously* converted value rather than resetting to zero for each new conversion.

The method known as the *dual slope integration operation* offers high accuracy and linearity, and has excellent noise rejection characteristics due to the integration operation. The dual slope method is commonly used in digital voltmeters. However, as with the servo operation, a relatively long time for conversion is often necessary. Conversion commences by switching the analog input onto an integrator input. The output of the integrator then ramps up from some negative reference voltage $-V_{\text{ref}}$ until the comparator output changes logic levels. This output (by way of control logic) gates the clock to the counter which counts up to a number N_1 in time t_1. The counter is set to zero and reference input is switched to the integrator input. The integrator then integrates the reference back down to the comparator threshold which stops the counter at a value N_2 after time t_2. The time t_c required for conversion is given by $t_c = t_1 + t_2$, and the analog input voltage V_i is then given by

$$V_i = \frac{N_2}{N_1} \times V_{\text{ref}}$$

The *accuracy* of this approach depends on the stability of the clock frequency and the integrating capacitor as well as on the accuracy of the reference voltage V_{ref}. The *resolution* of the conversion is limited by the analog resolution of the comparator.

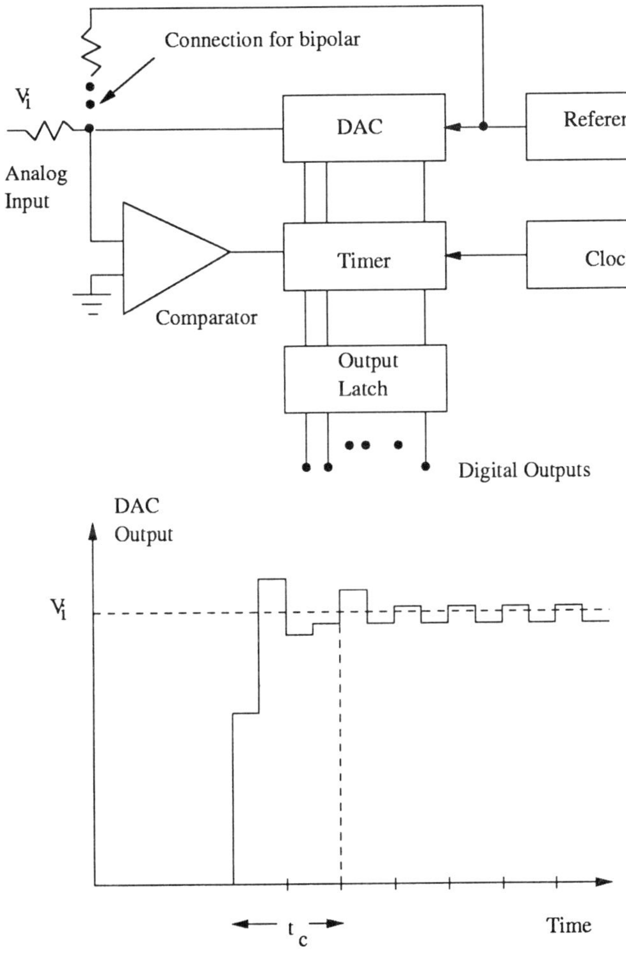

Fig. 1.5. Analog-to-digital conversion based on successive approximation

Another approach for analog to digital conversion known as *successive approximation operation* which has good accuracy, high resolution, and high speed can be realized using the network illustrated in Figure 1.5. Conversion times of the order of 100 nanoseconds per bit are possible. Conversion commences by setting the MSB of the DAC high and comparing with the analog input. If it is smaller than the input, the MSB is left high, otherwise it is turned off. The next most significant bit is then set high and procedure is repeated down to the LSB at which time the output register contains the converted result. The DAC output is also illustrated in Figure 1.5. By connecting the reference source to the comparator, bipolar input levels ($\pm 0.5 V_{\text{ref}}$) are permitted.

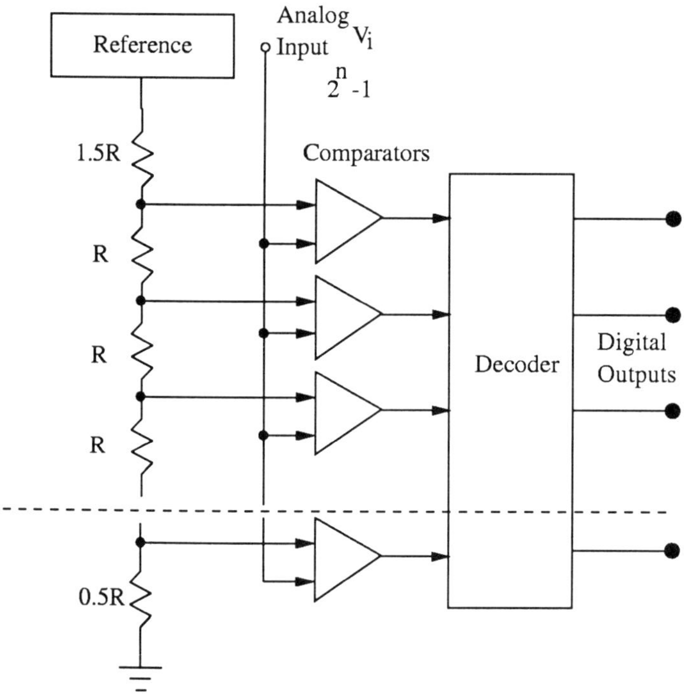

Fig. 1.6. Analog-to-digital conversion based on parallel operation

Another method known as *parallel operation* results in extremely fast conversion in which the order of 4 nanoseconds for 4 bits are possible. However, the number of bits is limited (say 4) due to the large number ($2^n - 1$ for n bits) of comparators required. A schematic which illustrates this technique of operation is illustrated in Figure 1.6. Conversion is accomplished in a single switching time of a single comparator (plus decoder propagation time). The output decoder is necessary since the comparator outputs are not in binary code. Conversion which results in a large number of bits delivered at high speed can be realized using a *hybrid* approach in which a fast parallel conversion stage is followed by a fast DAC whose output is *subtracted* from the analog input voltage. The difference signal is then amplified and converted using another parallel conversion stage.

Companding

The digital output of an ADC can only assume a finite number of possible values. In particular, for an n bit output there are 2^n possibilities. The quantization characteristic in Figure 1.7 which illustrates a uniform relationship between the analog input x and the digital output $Q[x]$ provides a *uniform resolution* between the quantization levels for all $|x| \leq 1$. For $|x| > 1$, the ADC saturates.

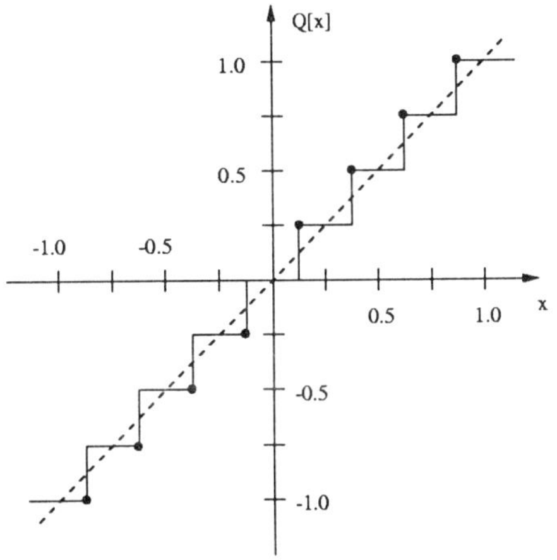

Fig. 1.7. Uniform ADC/DAC quantization

The dynamic range for ADC and DAC operation can be increased without increasing the wordlength if *nonuniform* quantization is used. Normally with nonuniform quantization as illustrated in Figure 1.8, the *resolution* is smaller for low level analog signals and correspondingly larger for large analog signals. This characteristic is called *compression*. If a compression characteristic is used in the analog-to-digital conversion stage, then an 'inverse' characteristic (called *expansion*) should be used in the digital-to-analog stage. The combined effect of compression and expansion is referred to as *companding*. The net effect of companding is that, for a given dynamic range, the quantization distortion is more favourable for small signals. In other words, companding permits a larger dynamic range for the combined ADC/DAC operation at the expense of lower resolution for large signals.

In telephone systems, the companding characteristics that are used are the μ-law in the USA, Canada and Japan, and the A-law in Europe. Both

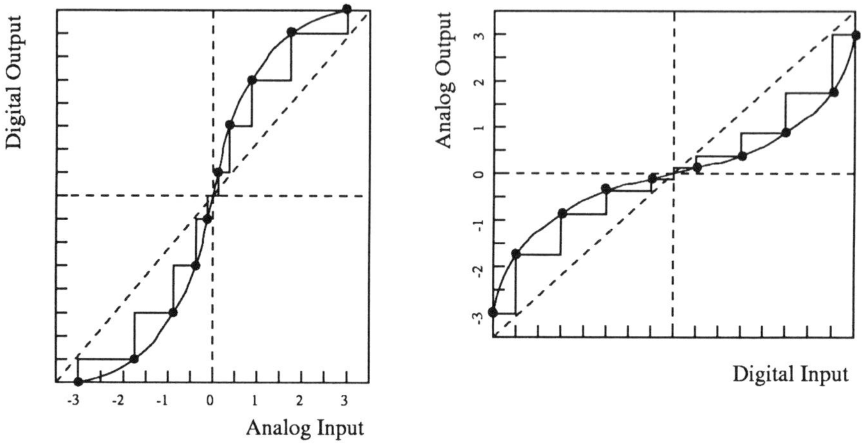

Fig. 1.8. Nonuniform ADC/DAC quantization

the μ-law and the A-law represent piecewise linear approximation to the logarithmic compression curves $F_\mu(x)$ and $F_A(x)$ where

$$F_\mu(x) = \text{sgn}(x)\,\frac{\log_e(1+\mu|x|)}{\log_e(1+\mu)} \quad ; \quad 0 \le |x| \le 1$$

and

$$F_A(x) = \begin{cases} \text{sgn}(x)\dfrac{1+\log_e(A|x|)}{1+\log_e A} & ; A^{-1} \le |x| \le 1 \\ \text{sgn}(x)\dfrac{A|x|}{1+\log_e A} & ; 0 \le |x| \le A^{-1} \end{cases}$$

Companding has the effect of providing more bits for smaller signals at the expense of larger errors when the signal is large.

1.4 Signal Processing

In many applications, it is more convenient to process a continuous signal v using a digital filter. For example, instead of having to keep changing values of the resistance R and/or capacitance C in an *analog* filter, the parameters in a *digital* filter can be easily changed via a computer software instruction. However this convenience can only be achieved at the expense of more assosciated hardware. In particular, processing a continuous signal v using a digital filter involves three steps:

- Convert the analog signal v to a digital signal v_s where $v_s(k) = v(t_k)$ by *sampling* v using a *digital acquisition system* which operates at the discrete sampling times t_k. Often a uniform sampling period is used which

corresponds to $t_k = kT_s + t_0$ for some integer k, and uniform sampling period T_s (or a uniform sampling rate $f_s = T_s^{-1}$).
- Filter the resulting discrete signal v_s by means of a digital filter to produce a discrete output signal $\hat{y}_s$, and finally
- Construct a continuous signal $\hat{y}$ from the discrete filter output signal $\hat{y}_s$ using a *signal reconstruction system*.

One way to achieve signal reconstruction is to use a sample and hold in which a piecewise constant analog signal $\hat{y}$ is constructed from the digital signal $\hat{y}_s$ by

$$\hat{y}(t) = \hat{y}_s(t_k) \quad ; \quad t_k \leq t < t_{k+1}$$

This mathematical statement can be realized in practice using a circuit known as a *sample and hold*.

Aliasing

When uniform sampling is implemented, one decision that needs to be made when constructing a digital processing system is to select the *sampling period* T_s, or equivalently, the *sampling fequency* $f_s = T_s^{-1}$. One must be aware that the process of sampling a continuous signal can introduce a phenomenon known as *aliasing*. Aliasing is said to occur when a high frequency *continuous* frequency component is transformed to a low frequency *discrete* frequency component.

In order to illustrate aliasing, consider the discrete signal u_n which is obtained by sampling the continuous time signal v_n defined by

$$v_n(t) = \sin 2\pi(f_0 + 0.5 f_s n)t \tag{1.17}$$

for any integer n with a uniform sampling frequency $f_s = T_s^{-1}$. Then for all *even* integers n

$$u_n(k) \triangleq v_n(kT_s) = \sin 2\pi(F_0 + 0.5n)k = \sin 2\pi F_0 k \quad ; \quad F_0 \triangleq \frac{f_0}{f_s}$$

That is, for all even integers n, the sampled values $\{u_n(k)\}$ cannot be used to differentiate between any of the analog signals v_n.

For example, suppose the continuous signal $v(t) = \sin(6\pi t - \theta)$ is sampled at the sampling frequency $f_s = T_s^{-1}$. Then

$$u(k) \triangleq v(kT_s) = \sin(6\pi k T_s - \theta)$$

Now if T_s is selected such that

$$P \leq 6T_s < P + 1 \tag{1.18}$$

for some *integer* P, then with $0 < \delta \triangleq 6T_s - P < 1$, we have

$$\sin(6\pi k T_s - \theta) = \sin(kP\pi + k\delta\pi - \theta) = (-1)^{kP}\sin(k\delta\pi - \theta)$$

We therefore conclude that if the sampling frequency $f_s = T_s^{-1}$ is too low (equivalently, the sampling period T_s is too high), then the continuous 3Hz sinusoidal component when sampled will appear like sampled values of a "low" frequency continuous f_0Hz sinusoidal component where

$$2\pi f_0 T_s = \delta\pi \quad \text{or} \quad f_0 = 0.5T_s^{-1}\delta$$

In particular, it follows from (1.18), that the *largest sampling period* $(T_s)_{min}$ is given by

$$6(T_s)_{min} < 1$$

That is, the *uniform sampling period* T_s should be selected such that

$$T_s < T_{min} = \frac{1}{6} \text{ sec} \tag{1.19}$$

which means that the *minimum* sampling frequency should be $(f_s)_{min} = 6$Hz. In any particular application, an iterative process may be required before the final sampling frequency is determined. Often, in order to achieve the required performance objectives, it is necessary to select the sampling rate as much as an order of magnitude greater than the minimum rate that would be required based only on aliasing considerations.

Hardware Considerations

The physical analog signal v_p which is to be filtered is first converted to an electrical signal v_e by means of a *transducer*. Proper conditioning of the signal v_e can be achieved by using a signal amplifier to provide additional gain, electrical isolation and/or impedance matching. Since the output signal of a transducer usually contains high frequency noise, the transducer output (after amplification) would generally be lowpass filtered using an analog filter; otherwise the *aliasing effect* that is associated with the sampling operation will introduce low frequency noise onto the sampled signal.

Transducers can be classified in terms of their basic mode of operation. A measured change in the physical variable may either cause a direct change in a circuit component (such as a resistance, capacitance, or inductance) or, alternatively, part of the energy (such as mechanical or thermal) of the physical variable may be converted into electrical energy. Various types of transducers exist for measurements of linear and rotary motion, force, pressure, vibration, fluid flow, light intensity, temperature, humidity, acidity, and acoustic energy.

Once the transducer output v_e is low-pass filtered, the resulting continuous signal v is sent to an *analog multiplexer* (MUX) if many signals are to be sampled. The MUX performs a time division multiplexing operation between each of its input signals which are to be sampled so that each input is sequentially connected to the output of the multiplexer for a specified period of time. The output of the MUX then is sent to a *sample and hold* (S&H) circuit which samples the output of the multiplexer at specified times and holds the sampled voltage level at its output until the *analog-to-digital converter* (ADC) performs the conversion operation.

The *timing and control* of this overall process is accomplished by a *programmable timer* which controls the MUX, the S&H, and the ADC. This timer may either be controlled by a digital counter at a fixed cycle rate or by the outputs of a data processor. One possibility is to have an *analog multiplexer* (MUX) which has several input channels which can be switched in turn into a single output channel. This option reduces the overall hardware cost and complexity since only one ADC is necessary and only a relatively simple hardware computer interface is required.

The time needed for analog-to-digital conversion (referred to as the *aperture time* t_a) depends on both the method of operation of the ADC, and the wordlength of the output digital word. For a *fixed* aperture time t_a, the amplitude uncertainty ($\triangle A$) in sampling the signal v at time t_1 may be approximated by

$$\frac{\triangle A}{t_a} \simeq \frac{dv(t)}{dt}\Big|_{t=t_1}$$

The *maximum* amplitude uncertainty is obtained when the rate of change of the analog signal v is maximum. For a sinusoidal signal $v(t) = A_0 \sin 2\pi f_0 t$ of frequency f_0Hz, we have $\dot{v}(t) = 2\pi A_0 \cos 2\pi f_0 t$, and so

$$(\Delta A)_{max} = 2\pi f_0 t_a$$

Conversion for a unity (i.e $A_0 = 1$) amplitude signal to B bit accuracy (i.e. such that $|\triangle A| \leq 2^{-B}$) therefore requires an aperture time t_a in nanoseconds (ns) given by

$$t_a \leq \frac{2^{-(B+1)}10^9}{\pi f_0} \quad \text{(ns)}$$

The higher the required resolution (i.e. the larger the ADC wordlength B), the faster the ADC is required to operate.

The problem assosciated with a changing signal while conversion takes place can be overcome if a S&H circuit is used in conjunction with the ADC. However, a S&H device is *not* always necessary. In particular, any change that occurs during signal conversion in the absence of a S&H may actually be beneficial since a more accurate sampled value with respect to the time at the *end of conversion* is produced.

If a single S&H must be *shared* by all the analog input signals to the MUX, then the information *cannot* be acquired simultaneously from each of the channels. If this arrangement is unsuitable, then a separate S&H should be used with each analog signal. In this case, once the programmable timer sends the signal to all sample and holds, all input signals are sampled at the *same* instant. These signals are then held constant until the ADC converts each signal in turn.

We now describe some applications of signal processing ideas.

An Example in Marketing

Suppose two shops supply all sales of bicycles sold in a certain town. Over time, consumers will tend to switch from one shop to the next for various reasons such as cost or after sales service. At the end of one particular month, suppose shop #1 and shop #2 have fractions $x_1(0)$ and $x_2(0)$ respectively of the total market; that is

$$x_1(0) + x_2(0) = 1 \tag{1.20}$$

Also, let $x_1(k)$ and $x_2(k)$ define the respective fractions of the total market at the *end* of the kth month, so that also

$$x_1(k) + x_2(k) = 1; \; k = 1, 2, \cdots \tag{1.21}$$

Now assume that in each month, shop #1 *retains* a fraction a_{11} of its own customers, and *attracts* a fraction a_{12} of the customers of shop #2. That is, at the end of month $k+1$, the total fraction $x_1(k+1)$ of customers for shop #1 is given by

$$x_1(k+1) = a_{11}x_1(k) + a_{12}x_2(k)$$

Similarly, for shop #2, we have

$$x_2(k+1) = a_{21}x_1(k) + a_{22}x_2(k) \tag{1.22}$$

The definitions of the coefficients a_{ij} (which are fractions of particular total numbers of customers) result in the following properties:

Property 1:

$$0 \leq a_{ij} \leq 1 \text{ for all } i, j \; . \tag{1.23}$$

Property 2:

$$a_{11} + a_{21} = 1 \quad ; \quad a_{12} + a_{22} = 1 \tag{1.24}$$

More generally, suppose we have n shops in a town which supply all sales of bicycles. Assume in the ith month that shop #i retains a fraction a_{ii} of its own customers, and attracts a fraction a_{ij} of the customers of shop #j. Then following similar arguments to those above, we conclude that the fraction of total sales $x_p(k+1)$ made by shop #p for $p = 1, 2, ..., n$ after month $k+1$ is given by

$$x_p(k+1) = a_{p1}x_1(k) + a_{p2}x_2(k) + \; \ldots \; + a_{pn}x_n(k) \tag{1.25}$$

where

$$x_1(k) + x_2(k) + \ \ldots \ + x_n(k) = 1 \ ; \ k = 1, 2, \ \ldots$$

and where (based on the definition of a_{ij})

$$0 \leq a_{ij} \leq 1 \text{ for all i, j} \tag{1.26}$$
$$\sum_{i=1}^{n} a_{ij} = 1 \text{ for all } j$$

The following questions then arise:

- Do $\{x_1(k), x_2(k), \ \ldots \ x_n(k)\}$ converge to some limiting values $\{\overline{x}_1(k), \overline{x}_2(k), \ \ldots \ \overline{x}_n(k)\}$ as $k \to \infty$?
- If the limits exist, then how does one calculate the limits ? In particular, do these limits depend on the initial values $\{x_1(0), x_2(0), \ \ldots \ x_n(0)\}$?

An Example in Numerical Integration

A problem that arises in numerical analysis is to determine a numerical approximation for the integral

$$\gamma(t_N, t_0) \triangleq \int_{t_0}^{t_N} z(\sigma) d\sigma \tag{1.27}$$

of a given function z. Of course, if the function z is available in analytical form, then an analytical evaluation may be possible. This form is generally preferred since an explicit expression for the function γ as a function of the limits $\{t_0, t_N\}$ of integration may then be determined.

However is many cases, an analytical solution is not possible, and so a numerical approach must be pursued. To this end, define

$$y(k) \triangleq \int_{t_0}^{t_k} z(\sigma) d\sigma \ ; \ 0 \leq k \leq N \tag{1.28}$$

Then from (1.27), we have that $y(k+1)$ for $k = 0, 1, 2, ..., N-1$ is given by

$$y(k+1) = \int_{t_0}^{t_k} z(\sigma) d\sigma + \int_{t_k}^{t_{k+1}} z(\sigma) d\sigma = y(k) + \int_{t_k}^{t_{k+1}} z(\sigma) d\sigma \tag{1.29}$$

where

$$y(0) = 0 \ ; \ y(N) = \gamma(t_N, t_0)$$

Provided the interval $t_k \leq t \leq t_{k+1}$ is small enough, we can consider a numerical approximation of the integral in (1.29). For example, one can use the *rectangular* (or *backward Euler*) rule for approximation whereby

$$\int_{t_k}^{t_{k+1}} z(\sigma) d\sigma \simeq u(k)[t_{k+1} - t_k] \ ; \ u(k) \triangleq z(t_k) \tag{1.30}$$

If a *uniform integration step size* is selected, so that $T = t_{k+1} - t_k$ is *constant* for all k, then from (1.29), we have that $y(k)$ in (1.28) is approximately given by $y_1(k)$ where y_1 is given by

$$y_1(k+1) = y_1(k) + Tu(k) \; ; \; y_1(0) = 0 \tag{1.31}$$

If instead we use the *forward Euler* integral approximation with $T = t_{k+1} - t_k$ constant, we have

$$\int_{t_k}^{t_{k+1}} z(\sigma)d\sigma \simeq u(k+1)T \; ; \; u(k+1) \triangleq z(t_{k+1}) \tag{1.32}$$

which implies that $y(k)$ in (1.28) is approximately given by $y_2(k)$ where y_2 is given by

$$y_2(k+1) = y_2(k) + Tu(k+1) \; ; \; y_2(0) = 0 \tag{1.33}$$

Alternatively, if we use the so-called *trapezoidal rule* for numerical integration, then

$$\int_{t_k}^{t_{k+1}} z(\sigma)d\sigma \simeq \frac{[u(k+1) + u(k)]T}{2} \; ; \; u(k) \triangleq z(t_k) \tag{1.34}$$

which implies that $y(k)$ in (1.28) is approximately given by $y_3(k)$ where y_3 is given by

$$y_3(k+1) = y_3(k) + \frac{T[u(k+1) + u(k)]}{2} \; ; \; y_3(0) = 0 \tag{1.35}$$

Simpson's 1/3 rule for numerical approximation leads to the approximation $y_4(k)$ of $y(k)$ where y_4 is given by

$$y_4(k+1) = y_4(k) + \frac{T[u(k+1) + 4u(k) + u(k-1)]}{3} \; ; \; y_4(0) = 0 \tag{1.36}$$

The following questions arise:

- How does the accuracy of the approximations $\{y_p; \; 1 \leq p \leq 4\}$ of y depend on the integration step size T and the properties of the function z ?
- How can one measure the accuracy of the approximations, or equivalently, how can one measure the size of the error between the approximating signal and the actual signal y ?
- Are there better ways to the above methods for approximating the integral?

An Example in Filter Design

Suppose a digital signal u can be expressed in terms of a sum of sinusoidal signals in the form

$$u(k) = \alpha_1 \sin[k\omega_1 + \theta_1] + \alpha_2 \sin[k\omega_2 + \theta_2]$$

where the signal component at frequency ω_1 is regarded as the signal of interest, but the component at frequency ω_2 is the undesirable (or interference) component. Suppose after appropriate filtering, the filtered output y is given by

$$y(k) = \alpha_1 A(\omega_1)\sin[(k-\beta_1)\omega_1 + \theta_1] + \alpha_2 A(\omega_2)\sin[(k-\beta_2)\omega_2 + \theta_2]$$

where

$$A(\omega_n) = \sum_{m=1}^{M} b_m \sin m\omega_m \tag{1.37}$$

If the aim of filter design is to use the filter so as to remove the interfering signal of frequency ω_2, but retain the component of frequency ω_1, then the ideal filter design with respect to the characteristic of the gain (or attenuation) function A is to have

$$A(\omega_1) = 1 \quad ; \quad A(\omega_2) = 0 \tag{1.38}$$

Given this performance specification, we then have from (1.37) and (1.38) with (say) $M = 3$ that the coefficients $\{b_1, b_2, b_3\}$ must satisfy

$$\begin{aligned} A(\omega_1) &= b_1 \sin\omega_1 + b_2 \sin 2\omega_1 + b_3 \sin 3\omega_1 = 1 \\ A(\omega_2) &= b_1 \sin\omega_2 + b_2 \sin 2\omega_2 + b_3 \sin 3\omega_2 = 0 \end{aligned} \tag{1.39}$$

The first question that arises is then: Can one always find three coefficients $\{b_1, b_2, b_3\}$ which satisfy these two equations ? Generally, the answer is in the affirmative.

Suppose in another application, we have a signal u

$$u(k) = \sum_{n=1}^{5} \alpha_n \sin[k\omega_n + \theta_n]$$

consisting of five frequency components $\{\omega_1, \omega_2, \omega_3, \omega_4, \omega_5\}$ of which the four components with frequencies $\{\omega_2, \omega_3, \omega_4, \omega_5\}$ represent interference signals. Now the ideal solution is to have

$$\begin{aligned} A(\omega_1) &= b_1 \sin\omega_1 + b_2 \sin 2\omega_1 + b_3 \sin 3\omega_1 = 1 \\ A(\omega_2) &= b_1 \sin\omega_2 + b_2 \sin 2\omega_2 + b_3 \sin 3\omega_2 = 0 \\ A(\omega_3) &= b_1 \sin\omega_3 + b_2 \sin 2\omega_3 + b_3 \sin 3\omega_3 = 0 \\ A(\omega_4) &= b_1 \sin\omega_4 + b_2 \sin 2\omega_4 + b_3 \sin 3\omega_4 = 0 \\ A(\omega_5) &= b_1 \sin\omega_5 + b_2 \sin 2\omega_5 + b_3 \sin 3\omega_5 = 0 \end{aligned} \tag{1.40}$$

We can once again ask a similar question: Are there three coefficients $\{b_1, b_2, b_3\}$ which satisfy these five equations ? In this case, since there are more equations than unknown coefficients, the answer is generally in the negative.

However when there are more equations than unknown coefficients $\{b_m\}$, we can instead seek a solution which makes the gain terms $\{A(\omega_n); n = 2, 3, 4, 5\}$ as "small as possible" within some well-defined sense.

An Example in Recursive Estimation

Consider an experiment in which N students measure the same given constant voltage which is nominally of value v_0 volts. Unfortunately, the available voltmeters are known to be inaccurate so that the kth measurement $z(k)$ is given by

$$z(k) = v_0 + \varepsilon(k) \quad ; \quad 1 \leq k \leq N \tag{1.41}$$

where $\varepsilon(k)$ represents the measurement error. The problem is to use all the students' measurements to come up with a better estimate for the true value v_0.

The measurement representation (or model) (1.41) can also be equivalently expressed in the form

$$\begin{aligned} y(k+1) &= y(k) \quad ; \quad y(0) = v_0 \\ z(k) &= y(k) + \varepsilon(k) \end{aligned} \tag{1.42}$$

for $1 \leq k \leq N$ where $y(0) = v_0$ is unknown and is to be estimated based on the N measurements $z(k)$.

After m students have made m independent measurements $\{z(1), z(2), \ldots, z(m)\}$ with m different voltmeters, one *estimate* $\hat{y}(m)$ of the true voltage y is given by averaging the m measurements; that is

$$\hat{y}(m) = \frac{1}{m} \sum_{k=1}^{m} z(k) \tag{1.43}$$

and after measurement $m+1$, the new estimate $\hat{y}(m+1)$ is given by

$$\hat{y}(m+1) = \frac{1}{m+1} \sum_{k=1}^{m+1} z(k) \tag{1.44}$$

The estimate $\hat{y}(m+1)$ can be arranged as

$$\hat{y}(m+1) = \frac{m}{m+1} \left\{ \frac{1}{m} \sum_{k=1}^{m} z(k) \right\} + \frac{z(m+1)}{m+1}$$

That is, using (1.43)

$$\hat{y}(m+1) = \frac{m}{m+1} \hat{y}(m) + \frac{z(m+1)}{m+1} \quad ; \quad 1 \leq m \leq N-1 \tag{1.45}$$

This relationship expresses the estimate $\hat{y}(m)$ based on m measurements in terms of the updated estimate $\hat{y}(m+1)$ once measurement $z(m+1)$ is made.

A further rearrangement of (1.45) is also possible; that is, since

$$\frac{m}{m+1} = 1 - \frac{1}{m+1}$$

we have that

$$\hat{y}(m+1) = \hat{y}(m) + \frac{1}{m+1}[z(m+1) - \hat{y}(m)] \quad ; \quad 1 \leq m \leq N-1 \quad (1.46)$$

When expressed in this form, we see that the new estimate $\hat{y}(m+1)$ is equal to the old estimate $\hat{y}(m)$ plus an *error term* which is given by a *gain* ($= (m+1)^{-1}$) times the error between the new measurement $z(m+1)$ and the old estimate $\hat{y}(m)$.

In summary, the problem of estimating the constant voltage v_0 by averaging the N measurements $\{z(1), z(2), \ldots, z(N)\}$ can be expressed in terms of the *signal model*:

$$\begin{aligned} y(k+1) &= y(k) \quad ; \quad y(0) = v_0 \\ z(k) &= y(k) + \varepsilon(k) \end{aligned} \quad (1.47)$$

and the *signal estimator*:

$$\begin{aligned} \hat{y}(k+1) &= \hat{y}(k) + g(k)[z(k+1) - \hat{y}(k)] \\ g(k) &\triangleq \frac{1}{k+1} \end{aligned} \quad (1.48)$$

which has the form

$$\text{NEW}_{\text{estimate}} = \text{OLD}_{\text{estimate}} + \text{gain} \times [\text{error in OLD}_{\text{estimate}}]$$

which turns out to be common in many signal processing applications.

The question that arises is: Are there other methods (besides averaging) for processing the N measurements which take into account the characteristics of the measurement errors ? If such methods exist, what are the corresponding expression for the gain $g(k)$ in such cases ?

An Example in Radar Processing

Suppose an aircraft is moving with constant (but unknown) velocity V. Then the velocity $x_2(k) = V$ at time $t = kT$ where k is an integer and T is a constant period is given by

$$x_2(k+1) = x_2(k) \; ; \; x_2(0) = V$$

If $x_1(k)$ denotes the position of the aircraft at time $t = kT$, then we also have that

$$x_1(k+1) = x_1(k) + Tx_2(k) \ ; \ x_1(0) = D$$

where D denotes the position at time $t = 0$ as measured with respect to some reference position.

Now consider the situation where a radar beam is used to determine the range and velocity of the aircraft from the radar transmitter. The information that is required to determine the range is the value of the time interval t_{travel} between the transmission of the radar pulse and the reception of the pulse that is reflected from the object and received back at the transmitter. The ideal *envelopes* of the transmitted and received pulse signals are illustrated respectively in Figure 1.9 (a) and Figure 1.9 (b), while the typical shape of the envelope of an actual received pulse is illustrated in Figure 1.9 (c). In practice, both the received and transmitted pulses are pulsed high frequency signals. A typical transmitted pulse is illustrated in Figure 1.9 (d).

The received pulse is distorted as a result of various disturbances that occur during the transmission of the pulse, and this causes an error between the actual two-way travel time t_{travel} and the measured two-way travel time $\hat{t}_{travel}$. The actual range r and an measured range z are then given by

$$r = \frac{c.\ t_{travel}}{2} \ ; \ z = \frac{c.\ \hat{t}_{travel}}{2} \tag{1.49}$$

where c is the speed of pulse propogation

As one might expect, the measured range z based on only one measurement $\hat{t}_{travel}$ can result in large errors. Therefore in an attempt to reduce this error, a periodic sequence of pulses is transmitted with one pulse every T seconds. This enables a sequence of measurements $\{z(0), z(1), z(2), \ldots z(m)\}$ of the range to be made. The aim is then to use this data in order to predict both the position and velocity of the aircraft. To this end, we define the following quantities:

- The measurement, $z(k)$, of the range of the object based on the travel time $\hat{t}_{travel}(k)$ of the kth transmitted pulse (which is received before sending the $k+1$st pulse).
- The estimate $\hat{x}_1(k)$ of the range, and the estimate $\hat{x}_2(k)$ of the velocity of the object after processing $z(k)$.

Given the current estimate $\hat{x}_1(k)$ of position, and the current estimate $\hat{x}_2(k)$ of velocity, the *predicted* range $x_1(k+1|k)$ at time $k+1$ is given by

$$x_1(k+1|k) \triangleq \hat{x}_1(k) + T\hat{x}_2(k)$$

The *estimated* range $\hat{x}_1(k+1)$ at time $k+1$ can be updated from the *predicted range* $x_1(k+1|k)$ using the *predicted range error* $z(k+1) - \hat{x}_1(k+1|k)$ between the measured range and the predicted range according to:

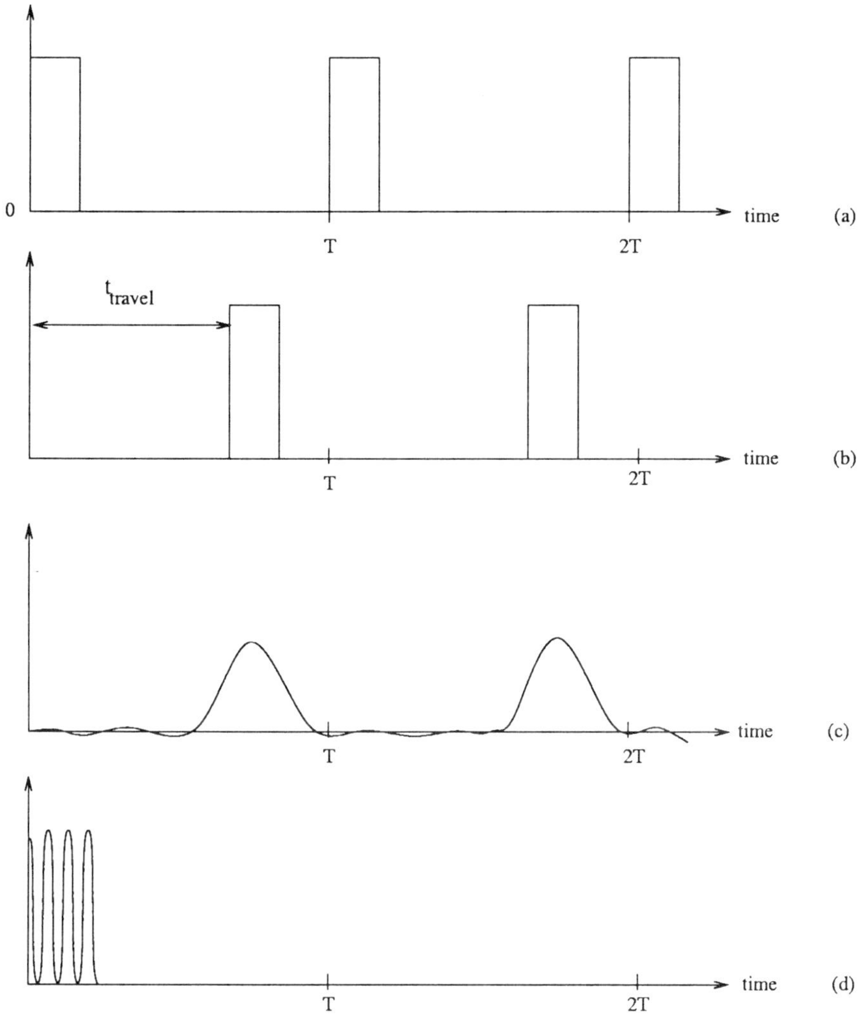

Fig. 1.9. A radar tracking system

$$\hat{x}_1(k+1) = \hat{x}_1(k+1|k) + \alpha[z(k+1) - \hat{x}_1(k+1|k)]$$

where $\alpha > 0$ is a parameter to be selected.

In a similar way, the estimated velocity $\hat{x}_2(k+1)$ at time $k+1$ can be updated from the current velocity $\hat{x}_2(k)$ at time k using the *velocity error* $(1/T)[z(k+1) - \hat{x}_1(k+1|k)]$ according to:

$$\hat{x}_2(k+1) = \hat{x}_2(k) + \frac{\beta}{T}[z(k+1) - \hat{x}_1(k+1|k)]$$

where $\beta > 0$ is another parameter to be selected. The resulting set of equations; namely

$$\begin{aligned}\hat{x}_1(k+1) &= \hat{x}_1(k+1|k) + \alpha[z(k+1) - \hat{x}_1(k+1|k)] \\ \hat{x}_2(k+1) &= \hat{x}_2(k) + \frac{\beta}{T}[z(k+1) - \hat{x}_1(k+1|k)] \\ \hat{x}_1(k+1|k) &= \hat{x}_1(k) + T\hat{x}_2(k)\end{aligned} \tag{1.50}$$

can also be rearranged into the equivalent form

$$\begin{aligned}\hat{x}_1(k+1) &= (1-\alpha)\hat{x}_1(k) + T(1-\alpha)\hat{x}_2(k) + \alpha z(k+1) \\ \hat{x}_2(k+1) &= -\frac{\beta}{T}\hat{x}_1(k) + (1-\beta)\hat{x}_2(k) + \frac{\beta}{T}z(k+1)\end{aligned} \tag{1.51}$$

and describe a signal processing algorithm which in radar applications is known as the *α-β tracker*.

For $\alpha = \beta = 0$, the α-β tracker corresponds to a model of the aircraft moving with constant velocity $\hat{x}_2(k) = \hat{V}$ with initial position $\hat{x}_1(0) = \hat{D}$. When α and/or β are nonzero, then the error between the position measurement $z(k+1)$ and the predicted position measurement $\hat{x}_1(k+1|k)$ can be used to correct for the velocity and position estimates. The input $z(k)$ is obtained from (1.49) based on the estimated travel time $\hat{t}_{travel}(k)$ of the kth pulse, and can be expressed in the form

$$z(k) = r(k) + \varepsilon(k)$$

where $\varepsilon(k)$ represents the measurement error arising as a result of pulse distortion. The best choice of the tracking parameters $\{\alpha, \beta\}$ will depend on knowledge of the characteristics of this error.

An Example in Speech Compression

Consider the problem of the efficient digital storage of speech for use in such applications as voice mail systems and digital dictaphones. Given that a telephone line has a bandwidth of approximately 4 kHz, then in order to satisfy the minimum criterion for speech reproducibility from the sampled values, a sampling rate of at least 8 kHz is required. Hence one possibility is to sample the speech at 8 kHz; that is, 8000 times per second, or every 0.125 milliseconds, using a 16 bit analog to digital converter.

The use of such a sampling rate is equivalent to generating a data rate of N_0 bits per second where $N_0 = 8000 \times 16 = 128,000$ bits/second, or almost 1 million bits every minute. Since this amount of data can present memory difficulties if a lot of speech is to be stored, a better approach to directly storing the sampled speech signal is to apply a speech compression process to the signal *before* storing. This compression process must be reversible in the sense that "near" original speech quality must be able to be reproduced after *decompression*.

A useful technique for speech compression is known as *Linear Predictive Coding* (LPC). The basis of this technique is the assumption that each speech sample is assumed to be dependent on a weighted linear sum of the n previous samples plus an excitation source u. That is, the kth speech sample $y(k)$ taken every 0.125 milliseconds is assumed to be described by

$$y(k) = a_1 y(k-1) + a_2 y(k-2) + \ldots + a_n y(k-n) + u(k) \tag{1.52}$$

(The justification of this representation is based on a linear model of the human vocal tract.)

It turns out that this representation is only a reasonable approximation of the physical speech process over a "short" time frame (usually taken to be somewhere in the range of $10 - 30$ milliseconds). In this way, we represent *each* time frame by the n parameters $\{a_1, a_2, \ldots, a_n\}$ together with two other parameters assosciated with the source signal u. (We omit further description of the source at this time). Speech can then be *reproduced* by recovering these $n+2$ parameters from memory every time frame, and then using these parameters to generate $y(k)$ according to (1.52) every 0.125 milliseconds for the length of the stipulated time frame. The resulting signal y is then sent to a digital to analog converter which is connected to a speaker to produce the sound.

For example, suppose the time frame is 20 milliseconds (or 50 time frames per second), and $n = 12$ so that 14 parameters are used to represent $20/0.125 = 160$ sampled values of speech over each time frame. Also suppose that each of the 14 parameters are represented by 8 bits. Then the number N_1 of bits required to represent the speech signal every second using the LPC approach is given by

$$N_1 = 50 \times 14 \times 8 = 5600$$

When this is compared with the $N_0 = 128,000$ bits per second that is required using direct sampling, we see that a compression ratio of about 23 to 1 is achieved. In practice, by "fine-tuning" the procedure (such as allocating different wordlenths to different parameters), compression ratios as high as 50 to 1 have been found to provide good quality speech reproduction. The key problem is to find an efficient procedure which provides a good estimate of the speech model parameters.

Summary

In this text, we use the term 'signal' to denote both a physical quantity (such as a TV signal) whose value depends upon time, and a mathematical representation of a physical quantity. An analog signal is a signal which has a value for all instants of time whereas a digital signal has a value for only certain instants of time. Some mathematical properties of signals were described in section 1.1.

A digital signal can be obtained from an analog signal by sampling the analog signal at particular instants of time. The sampling is said to be uniform when the time between any two successive time instants is constant; otherwise the sampling is said to be nonuniform. The (nonunique) process of converting an analog signal to a digital signal is known as analog-to-digital conversion. The hardware device which does the conversion is called an analog-to-digital converter (or ADC). The output of an ADC can only represent a finite number of analog values, or quantization levels. The separation between successive digital representations from an ADC is known as the resolution of the ADC. If the resolution is constant (up to some maximum level), then the ADC is said to be uniform; otherwise the ADC is said to be nonuniform.

The reverse process of converting a digital signal to an analog signal (which is also nonunique) is known as digital-to-analog conversion. The hardware device which does the conversion is called an digital-to-analog converter (or DAC). In many cases, a hardware device known as sample and hold is used to produce a 'piecewise constant', or 'staircase like', analog signal from the finite number of digital signals. A uniform DAC is one in which the 'treads' of these steps have equal length; otherwise the DAC is said to be nonuniform.

The basis for two methods of performing digital-to-analog conversion; namely, the weighted current method and the R-2R ladder method were described in section 1.2. Most hardware methods for performing analog-to-digital conversion make use of a DAC. Three common methods of analog-to-digital conversion; namely the servo method of operation, the dual slope integration method, the successive approximation method and the parallel operation method were described in section 1.3.

Signal processing describes the process, or mathematical algorithm, which extracts information from a signal. The final section of this introductory chapter provided a number of applications of signal processing including digital filter design, radar processing for range and velocity estimation, and speech compression. The detailed descriptions of these algorithms will become clearer as the following chapters unfold.

Exercises

1.1 (i) Show that each of the following two series are convergent:

$$(a)\ \cos\theta \triangleq 1 - \frac{\theta^2}{2!} + \frac{\theta^4}{4!} - \ \ldots$$

$$(b)\ \sin\theta \triangleq \theta - \frac{\theta^3}{3!} + \frac{\theta^5}{5!} - \ \ldots$$

(ii) From (i), prove the following:

$$(a)\ exp\{j\theta\} = \cos\theta + j\sin\theta$$

$$(b)\ \cos\theta = 0.5(exp\{j\theta\} + exp\{-j\theta\})$$

$$(c)\ \sin\theta = 0.5(exp\{j\theta\} - exp\{-j\theta\})$$

1.2 (i) Reduce the following complex numbers c to cartesian form $c = x + jy$:

$$(a)\ c = \frac{1-j2}{1+j3}\ \ ;\ \ (b)\ c = (3+j4) + (-2-j3)$$

$$(c)\ c = (2-j)^3\ \ ;\ \ (d)\ c = 4exp\{-j0.6\}$$

$$(e)\ c = (3exp\{j0.1\})^3\ \ ;\ \ (f)\ c = 1.2exp\{-j\pi/4\}$$

$$(g)\ c = 1.2exp\{-j\pi/2\} - 0.8exp\{j0.5\}\ \ ;\ \ (h)\ c = \frac{2exp\{-j\pi/6\}}{1+2j}$$

(ii) Reduce the complex numbers c in (i) to polar form $c = rexp\{j\theta\}$.

1.3 Find all real solutions $\{\alpha, \beta\}$ of the equations:

$$(a)\ (1+j2)\ exp\{j\pi/4\} = \alpha + j\beta$$

$$(b)\ 3 - j4 + 2exp\{-j\pi/3\} = \alpha exp\{j\beta\}$$

1.4 (i) Consider any complex number $h = h_1 + jh_2$ where $\{h_1, h_2\}$ are both real. Show that the real numbers $\{r, \phi\}$ which satisfy the equation

$$rexp\{-j\phi\} = h_1 + jh_2$$

are *not* unique.

(ii) Define

$$\theta_1 \triangleq \tan^{-1}\left(\frac{h_2}{h_1}\right)\ \ ;\ \ -\frac{\pi}{2} \le \theta_1 < \frac{\pi}{2}$$

and show that

$$rexp\{-j\phi\} = |r|exp\{-j\psi_1\}$$

where $|r| = \sqrt{h_1^2 + h_2^2} \geq 0$ and

$$\psi_1 = \begin{cases} \theta_1 & ; \ h_1 \geq 0 \\ \theta_1 + \pi & ; \ h_1 < 0, h_2 \geq 0 \\ \theta_1 - \pi & ; \ h_1 < 0, h_2 \leq 0 \end{cases} \tag{1.53}$$

(iii) Suppose

$$rexp\{-j\phi\} = \frac{h_1 + jh_2}{m_1 + jm_2} \quad ; \quad m = m_1 + jm_2 \neq 0$$

where θ_1 is given in (ii), and θ_2 is defined by

$$\theta_2 \triangleq \tan^{-1}\left(\frac{m_2}{m_1}\right) \quad ; \quad -\frac{\pi}{2} \leq \theta_2 < \frac{\pi}{2} \tag{1.54}$$

Find $|r| > 0$ and ψ such that

$$rexp\{-j\phi\} = |r|exp\{-j\psi\}$$

1.5 Evaluate the limits as $t \to \infty$ (if they exist) of the following functions:

$$(a)\ f(t) = \frac{\sin t}{t} \ ; \ (b)\ f(t) = \frac{2t-4}{3t+7} \ ; \ (c)\ f(t) = \log\log 2t$$

$$(d)\ f(t) = \frac{\exp\{t\} - 3}{5t} \ ; \ (e)\ f(t) = \frac{1 - 2\cos t}{3t^2}$$

1.6 Define

$$S = \sum_{n=n_0}^{n_1} ar^n$$

(i) Show that: for $r \neq 1$ and $\{n_0 = 0, n_1 = N-1\}$ that

$$S = \frac{a(1-r^N)}{1-r}$$

and hence: for $|r| < 1$ and $\{n_0 = 0, n_1 = \infty\}$ that

$$S = \frac{a}{1-r}$$

(ii) Using (i), evaluate S when $\{n_0 = k, n_1 = \infty\}$ for any integer $k \geq 0$.

(iii) Under what conditions does the summation S when $\{n_0 = -\infty, n_1 = 0\}$ exist ? What is the value when it does ?

(iv) Under what conditions does the summation S when $\{n_0 = -\infty, n_1 = \infty\}$ exist ? What is the value when it does exist ?

(v) Using (i), show that: for $|r| < 1$

$$\sum_{n=0}^{\infty} anr^n = \frac{ar}{(1-r)^2}$$

(vi) Using (i)-(ii), find an expression for:

$$\sum_{n=0}^{N-1} an^2 r^n \; ; \; |r| < 1$$

(vii) Describe a general method for evaluating

$$\sum_{n=0}^{N-1} an^p r^n \; ; \; |r| < 1$$

for any integer $p \geq 0$.

1.7 (i) Show that

$$\sum_{k=0}^{N-1} exp\{jk\theta\} = \alpha + j\beta$$

where

$$\alpha = \frac{1 - \cos\theta + \cos(N-1)\theta - \cos N\theta}{2(1-\cos\theta)}$$

$$\beta = \frac{\sin\theta + \sin(N-1)\theta - \sin N\theta}{2(1-\cos\theta)}$$

(ii) Show that

$$\alpha = \frac{\sin(0.5N\theta)\cos[0.5(N-1)\theta]}{\sin 0.5\theta} \; ; \; \beta = \frac{\sin(0.5N\theta)\sin[0.5(N-1)\theta]}{\sin 0.5\theta}$$

1.8 The differential equation

$$\frac{dy(t)}{dt} = z(t) \tag{1.55}$$

can also be equivalently written in the integral form

$$y(t_{k+1}) = y(t_k) + \int_{t_k}^{t_{k+1}} z(\sigma)d\sigma$$

(i) Suppose

$$z(t) = \cos 2\pi t$$

Find the solution $y(t)$ for all $t \geq 0$ given that $y(0) = 0$.

(ii) Show that an approximation $p(k)$ of $y(t)$ at the time instant $t = kT$ (k an integer, T a constant) is given by the solution of the difference equation

$$p(k+1) = p(k) + T[\alpha u(k) + (1-\alpha)u(k+1)]$$

for any $0 \leq \alpha \leq 1$.

(iii) What values of α in (ii) correspond to the backward Euler, forward Euler and trapezoidal methods for approximating an integral ?

(iv) Let $T = 0.2$ secs. Calculate $p(k)$ for $1 \leq k \leq 4$ when $p(0) = 0$ for each value of α in (iii).

(v) Which value of α in (iv) gives the "best" approximation $p(k)$ of $y(0.2k)$ in (i) for $1 \leq k \leq 4$ where "best" is measured in terms of the performance index J where

$$J \triangleq \max_{1 \leq k \leq 4} |p(k) - y(0.2k)|$$

(vi) Repeat (v) when

$$J \triangleq \sqrt{\sum_{k=1}^{4} [p(k) - y(0.2k)]^2}$$

1.9 (i) What is the sampling interval T which corresponds to a sampling frequency f_s when:

$$(a)\ f_s = 5 \text{ kHz} \quad (b)\ f_s = 10 \text{ kHz} \quad (a)\ f_s = 15 \text{ kHz}$$

(ii) Suppose $v(t) = 6\sin(11\pi t + 0.25\pi)$ is an analog signal. Find the three digital signals u which are obtained by sampling v with a sampling frequency f_s given in (i). What are the resulting digital frequencies ?

2. Digital Signals

Given a discrete set $\mathcal{T}$, a real valued digital signal is a function $v : \mathcal{T} \to \mathcal{R}$, and a complex valued signal v is a function $v : \mathcal{T} \to \mathcal{C}$. Without any loss of generality, we assume $\mathcal{T}$ to be a subset of the integers $\mathcal{Z}$. Given one digital signal, then another signal can be defined as the solution of a difference equation in which the given signal is the independent (or input) signal. This Chapter is concerned with the solution of linear difference equations. A difference equation is also one way to represent a digital filter, and so understanding the properties of difference equations is also necessary for the understanding of the properties of digital filters.

2.1 First Order

Consider the digital signal y which is given by the solution of the *first order difference equation*

$$y(k+1) = ay(k) + b_0 u(k) + b_1 u(k+1) \quad ; \quad k \geq k_0 - 1 \tag{2.1}$$

where $\{a,\ b_0,\ b_1\}$ are real constants, k_0 and k_1 are given integers, and k is a variable integer. The signal y is uniquely determined once the *initial condition* $y(k_0 - 1)$ and the input signal u are specified.

Substituting $k = k_0 - 1$ in (2.1), we have

$$y(k_0) = ay(k_0 - 1) + b_0 u(k_0 - 1) + b_1 u(k_0)$$

Then with $k = k_0$, we have

$$y(k_0 + 1) = ay(k_0) + b_0 u(k_0) + b_1 u(k_0 + 1)$$

which after substituting for $y(k_0)$ using the previous equation gives

$$y(k_0 + 1) = a^2 y(k_0 - 1) + ab_0 u(k_0 - 1) + (ab_1 + b_0)u(k_0) + b_1 u(k_0 + 1)$$

Also when $k = k_0 + 1$, we have

$$\begin{aligned}y(k_0+2) &= ay(k_0+1)+b_0u(k_0+1)+b_1u(k_0+2)\\&= a^3y(k_0-1)+a^2b_0u(k_0-1)+(a^2b_1+ab_0)u(k_0)\\&\quad +(ab_1+b_0)u(k_0+1)+b_1u(k_0+2)\end{aligned}$$

Hence in general, we conclude that: for $p \geq 1$

$$y(k_0+p-1) = a^p y(k_0-1) + b_0 \sum_{m=0}^{p-1} a^{p-m-1} u(k_0+m-1) + b_1 \sum_{m=0}^{p-1} a^{p-m-1} u(k_0+m) \tag{2.2}$$

Alternatively, if we let $k = k_0+p-1 \geq k_0$, (2.2) can also be expressed in the form

$$y(k) = a^{k-k_0+1} y(k_0-1) + b_0 \sum_{n=0}^{k-k_0} a^{k-k_0-n} u(k_0+n-1) + b_1 \sum_{n=0}^{k-k_0} a^{k-k_0-n} u(k_0+n) \tag{2.3}$$

Example 2.1.1. Suppose we are given the initial condition $y(-1) = 4$, and the input signal u is defined by $u(n) = 2^{-n}$ for $n \geq -1$. Then from (2.3) for $k \geq 0$, we have with $k_0 = 0$ that

$$\begin{aligned}y(k) &= 4a^{k+1} + b_0 \sum_{n=0}^{k} a^{k-n} 2^{-(n-1)} + b_1 \sum_{n=0}^{k} a^{k-n} 2^{-n}\\&= 4a^{k+1} + (2b_0+b_1) \sum_{n=0}^{k} a^{k-n} 2^{-n}\end{aligned}$$

(Note that the values $y(k)$ for $k \geq 0$ are independent of the values of $u(n)$ for $n < -1$.)

Now for $a \neq 0.5$

$$\sum_{n=0}^{k} (2a)^{-n} = \frac{1-(2a)^{-(k+1)}}{1-(2a)^{-1}}$$

and for $a = 0.5$

$$\sum_{n=0}^{k} (2a)^{-n} = k+1$$

That is, when $a \neq 0.5$

$$y(k) = 4a^{k+1} + (2b_0 + b_1)a^k \left\{ \frac{1 - (2a)^{-(k+1)}}{1 - (2a)^{-1}} \right\}$$

and for $a = 0.5$

$$y(k) = 4(0.5)^{k+1} + (2b_0 + b_1)(0.5)^k (k+1)$$

The advantage of having a *closed form* expression for $y(k)$ in (2.3) is that a particular value (say) $y(10)$ of y can be computed directly without the need to compute all the intermediate values $y(1), y(2), \ldots, y(9)$.

Approximation of Solution of Differential Equation

In Chapter 1, we considered a problem in numerical integration where the aim was to provide a numerical procedure for approximating the integral

$$m(t_N, t_0) = \int_{t_0}^{t_N} z(\sigma) d\sigma$$

of a given signal z.

One approach is based on the *backward Euler* approximation where for $|t_{k+1} - t_k|$ sufficiently small, we have the approximation

$$\int_{t_k}^{t_{k+1}} z(\sigma) d\sigma \simeq z(t_k)[t_{k+1} - t_k] \tag{2.4}$$

Using this approximation, we now derive a first order *difference equation* whose solution provides an approximation to the solution of the first order *differential equation*

$$\dot{x}(t) = \alpha x(t) + \beta v(t) \quad ; \quad t \geq t_0 \tag{2.5}$$

Specifically, write (2.5) in the form of an integral equation

$$x(t) = x(t_0) + \int_{t_0}^{t} [\alpha x(\sigma) + \beta v(\sigma)] d\sigma \tag{2.6}$$

so that with $\{t = t_{k+1}, t_0 = t_k\}$

$$\begin{aligned} x(t_{k+1}) - x(t_k) &= \int_{t_k}^{t_{k+1}} [\alpha x(\sigma) + \beta v(\sigma)] d\sigma \\ &\simeq [\alpha x(t_k) + \beta v(t_k)][t_{k+1} - t_k] \end{aligned}$$

where the approximation follows from (2.4).

In particular with $t_k = kT$ for some fixed constant T and all integers k, we have that an approximation $y_1(k)$ of $x(kT)$ is given by

$$y_1(k+1) - y_1(k) = T[\alpha y_1(k) + \beta u(k)] \quad ; \quad k \geq 0$$

where $u(k) \triangleq v(kT)$. That is

$$y_1(k+1) = (1+\alpha T)y_1(k) + T\beta u(k) \quad ; \quad k \geq 0 \tag{2.7}$$

The *forward Euler* approximation of an integral is given by

$$\int_{t_k}^{t_{k+1}} z(\sigma)d\sigma \simeq z(t_{k+1})[t_{k+1} - t_k] \tag{2.8}$$

Using this approximation, it follows that a first order difference equation whose solution $y_2(k)$ provides an approximation to the solution $x(kT)$ of the first order differential equation (2.5) with $u(k) \triangleq v(kT)$ is given by

$$y_2(k+1) = \frac{1}{1-\alpha T} y_2(k) + \frac{\beta T}{1-\alpha T} u(k+1) \quad ; \quad k \geq 0 \tag{2.9}$$

Another so called *trapezoidal approximation* of an integral (which is like the average of the backward and forward Euler approximations) is given by

$$\int_{t_k}^{t_{k+1}} z(\sigma)d\sigma \simeq 0.5[z(t_{k+1}) + z(t_k)][t_{k+1} - t_k] \tag{2.10}$$

Using this approximation, it follows that a first order difference equation whose solution $y_3(k)$ provides an approximation to the solution $x(kT)$ of the first order differential equation (2.5) with $u(k) \triangleq v(kT)$ is given for $k \geq 0$ by

$$\begin{aligned} y_3(k+1) = {} & \frac{1+0.5\alpha T}{1-0.5\alpha T} y_3(k) + \frac{0.5\beta T}{1-0.5\alpha T} u(k+1) \\ & + \frac{0.5\beta T}{1-0.5\alpha T} u(k) \end{aligned} \tag{2.11}$$

Example 2.1.2. Suppose $\{\alpha = -1, \beta = 2\}$ and $\{x(0) = 0;\ v(t) = 1, t \geq 0\}$. Then the solution x of the differential equation (2.5) is given by

$$x(t) = 2exp\{-t\} + 2 \quad ; \quad t \geq 0$$

which then implies $x(t=1) = 2(1 - e^{-1}) = 1.26$ (to 2 dec. places).

Now suppose $T = 0.2$. Then (to 2 dec. places), we have from (2.7), (2.9) and (2.11) that

$$\begin{aligned} y_1(k+1) &= 0.80y_1(k) + 0.40u(k) \quad ; \quad y_1(0) = 0 \\ y_2(k+1) &= 0.83y_2(k) + 0.33u(k) \quad ; \quad y_2(0) = 0 \\ y_3(k+1) &= 0.82y_3(k) + 0.18u(k+1) + 0.18u(k) \quad ; \quad y_3(0) = 0 \end{aligned}$$

Since $T = 0.2$, the approximations of $x(t = 1)$ are given by $\{y_1(5), y_2(5), y_3(5)\}$.
Now from (2.3) with $k_0 = 1$ and $k = 5$, we have (to 2 dec. places)

$$y_1(5) = 0.40 \sum_{n=0}^{4} (0.80)^{4-n}.1 = 1.34$$

$$y_2(5) = 0.33 \sum_{n=0}^{4} (0.83)^{4-n}.1 = 1.77$$

$$y_1(5) = 0.18 \sum_{n=0}^{4} (0.82)^{4-n}.1 + 0.18 \sum_{n=0}^{4} (0.82)^{4-n}.1 = 1.26$$

Together (2.7), (2.9) and (2.11) provide three *different* ways of approximating the solution of the *same* differential equation (2.5). One issue that arises is: for a given input signal v, parameters $\{\alpha, \beta\}$ and time interval T, which solution y_m provides the most accurate solution for x at the time instants $t = kT$? We will consider this matter in more detail in Chapter 4.

As we later show, the first order differential equation (2.5) in engineering terms can be thought of as a first order *analog* filter. Likewise, the first order difference equations (2.7), (2.9) and (2.11) can each be thought of as first order *digital* filters. The question just posed can therefore be expressed in another equivalent way. Which first order digital filter provides the best approximation of the given first order analog filter ?

Causality

The signal y is referred to as the *output* and the signal u as the *input* of the first order difference equation (2.1). The input u is the *independent* signal whereas the output y is *dependent* on u. From (2.1), we see that the output y at time $k+1$ depends only on the value of the output at time k, and on values of the input u at time k and time $k+1$ (for $b_1 \neq 0$).

Any difference equation whose output at time k depends only on values of the output and input u up to and including time k is said to define a causal system; otherwise the system is said to be noncausal.

Equation (2.1) provides an example of a *causal* system. The system $y(k+1) = ay(k) + u(k+2)$ is an example of a *noncausal* system.

Order

The difference equation (2.1) for $a_1 \neq 0$ is said to be of *first order* since one and only one *initial value* of the output y at time $k = k_0 - 1$ is required to solve (2.1) uniquely for y for all $k \geq k_0$ given the values of the input u.

The (causal) difference equation

$$y(k+1) = u(k) \quad ; \quad k \geq k_0$$

is (formally) said to be of *zero order* since *only* the input signal u (and no initial values of y) is required to determine y.

The (causal) difference equation

$$y(k+1) = ay(k) + bu(k-1) \quad ; \quad k \geq k_0 - 1$$

is of *first order* since only one initial value $y(k_0 - 1)$ of y at time $k = k_0 - 1$ (in addition to knowledge of the input u) is required to solve for y.

To find $y(k_0)$, the (causal) difference equation

$$y(k+1) = ay(k) + y(k-1) + bu(k) \quad ; \quad k \geq k_0 - 1$$

requires knowledge of the initial values $\{y(k_0-1),\ y(k_0-2)\}$ in addition to the input signal u, and so is a *second order* difference equation.

Linearity

The difference equation (2.1) is said to be *linear* since: if $y^{(1)}$ is the output signal that results from input $u^{(1)}$ with initial condition α_1, and $y^{(2)}$ is the output signal that results from input $u^{(2)}$ with initial condition α_2; that is, if

$$\begin{aligned} y^{(1)}(k+1) &= ay^{(1)}(k) + b_0u^{(1)}(k) + b_1u^{(1)}(k+1) \quad ; \quad y^{(1)}(k_0-1) = \alpha_1 \\ y^{(2)}(k+1) &= ay^{(2)}(k) + b_0u^{(2)}(k) + b_1u^{(2)}(k+1) \quad ; \quad y^{(2)}(k_0-1) = \alpha_2 \end{aligned}$$

then when the input signal is $v = c_1u^{(1)} + c_2u^{(2)}$, and the initial condition is $c_1\alpha_1 + c_2\alpha_2$, the output $z = c_1y^{(1)} + c_2y^{(2)}$; that is

$$z(k+1) = az(k) + b_0v(k) + b_1v(k+1) \quad ; \quad z(k_0-1) = c_1\alpha_1 + c_2\alpha_2$$

In other words, in a *linear* system, if one knows the output corresponding to two particular inputs and two particular initial conditions, then for an appropriate initial condition, the output will be given by the linear combination of these inputs.

Time Invariance

The difference equation (2.1) is said to be *time invariant* since: if y is the signal that results from the input u for $k \geq k_0$ with initial condition α; that is,

$$y(k+1) = ay(k) + b_0u(k) + b_1u(k+1) \quad ; \quad k \geq k_0 - 1\ ,\ y(k_0-1) = \alpha$$

then when the input signal v is a *time delayed* version of u; that is, for some k_2

$$v(k) = u(k - k_2)$$

and $y(k_0 + k_2 - 1) = \alpha$, we have that $z(k) \triangleq y(k - k_2)$ for $k \geq k_0 + k_2 - 1$ satisfies the equation

$$z(k+1) = az(k) + b_0 v(k) + b_1 v(k+1) \ ; \ z(k_0 + k_2 - 1) = \alpha$$

In other words, in a *time invariant* system , if one knows the output y as a result of an input u beginning at time $k = k_0$, then one can determine what the output will be to a time delayed version of u applied at time $k = k_0 + k_2$. The time invariance property follows as a consequence of the fact that the coefficients $\{a, b_0, b_1\}$ in (2.1) are constant, and so do not vary with time k.

Example 2.1.3. (i) The following first order difference equations are *causal, time invariant* and *nonlinear*:

$$y(k+1) = ay^2(k) + u(k) \ ; \ k \geq 0$$
$$y(k+1) = y(k)u(k) \ ; \ k \geq 0$$

(ii) The following first order difference equations are *causal, time varying* and *linear*:

$$y(k+1) = 2^k y(k) + bu(k) \ ; \ k \geq 1$$
$$y(k+1) = ay(k) + k^2 u(k) \ ; \ k \geq 1$$

(iii) The following first order difference equations are *causal, time varying* and *nonlinear*:

$$y(k+1) = 2^k y(k) + u^2(k) \ ; \ k \geq 3$$
$$y(k+1) = 0.2y(k) + (k-1)^2 u^3(k) \ ; \ k \geq 3$$

2.1.1 Zero Input and Forced Response

The response y in (2.3) can be decomposed into two parts; namely

$$y = y_i + y_f \tag{2.12}$$

where for $k \geq k_0$

$$y_i(k) \triangleq a^{k-k_0+1} y(k_0 - 1) \tag{2.13}$$
$$y_f(k) \triangleq b_0 \sum_{n=0}^{k-k_0} a^{k-k_0-n} u(k_0 + n - 1) + b_1 \sum_{n=0}^{k-k_0} a^{k-k_0-n} u(k_0 + n)$$

The signal y_i is *independent* of the input u, and depends only on the initial condition $y(k_0 - 1)$. As a consequence, y_i is called the *zero input response* (or *initial condition response*). Conversely, the signal y_f is *independent* of the initial condition $y(k_0 - 1)$, and depends only on the input u. The component y_f is therefore called the *forced response* (or *zero initial condition response*). The sum y in (2.12) of the zero input response and forced response is called the *complete response.*

The *zero input response* y_i can also be rearranged into the form

$$y_i(k) = a^{1-k_0} y(k_0 - 1) \cdot a^k \ ; \ k \geq k_0$$

from which it is clear that: for all initial conditions $y(k_0 - 1)$, $|y_i(k)|$ approaches zero as $k \to \infty$ if and only if $|a| < 1$, while $|y_i(k)|$ approaches infinity as $k \to \infty$ when $|a| > 1$. In the special case when $|a| = 1$, we have that $|y_i(k)| = |y_i(k_0 - 1)|$ is constant for all k. When $a = 1$, $y_i(k)$ has the same sign as $y(k_0 - 1)$, but the sign of $y_i(k)$ oscillates when $a = -1$.

Example 2.1.4. (i) The initial condition response y_i of the difference equation

$$y(k+1) = -0.7y(k) + 3u(k) \quad ; \quad y(2) = 5$$

is given by

$$y_i(k) = 5(-0.7)^{k-2} \quad ; \quad k \geq 2$$

In this form, the initial condition of $y(2) = 5$ is explicitly evident. However this solution can equivalently be expressed in the form

$$y_i(k) = 5(-0.7)^{-2}(-0.7)^k = 10.20(-0.7)^k \quad ; \quad k \geq 2$$

(ii) The initial condition response y_i of the difference equation

$$3y(k+1) = 0.6y(k) + 7u(k+1) - u(k) \quad ; \quad y(4) = -8$$

is given by

$$y_i(k) = (-8)(0.2)^{k-4} \quad ; \quad k \geq 4$$

By definition, the *forced response* y_f in (2.13) depends only on the input signal u. We now consider three particular examples of forced responses when $k_0 = 0$; the *unit impulse response*, the *unit step response*, the *exponential response* and the *sinusoidal response.*

Unit Impulse Response

Consider the forced response y_f in (2.13) when $k_0 = 0$, $y_f(-1) = 0$ and the input signal u is the *digital unit impulse signal* as defined by

$$u(k) \triangleq \begin{cases} 1 \ ; k = 0 \\ 0 \ ; k \neq 0 \end{cases} \tag{2.14}$$

The corresponding forced response (known as the *unit impulse response*) $y_f = h$ is then given from (2.13) by

$$h(k) = \begin{cases} 0 & ; k < 0 \\ b_1 & ; k = 0 \\ b_0 a^{k-1} + b_1 a^k & ; k \geq 1 \end{cases} \tag{2.15}$$

Note that $h(k) \to 0$ as $k \to \infty$ for $|a| < 1$, but $|h(k)| \to \infty$ as $k \to \infty$ for $|a| > 1$. When $a = 1$, $h(k)$ is constant for all $k \geq 1$, but $h(k)$ oscillates between $\pm(b_0 - b_1)$ for $k \geq 2$ when $a = -1$.

The unit impulse response $y_f = h$ can also be found directly from (2.1) with $k_0 = 0$, $h(-1) = 0$ and u given by (2.14). That is

$$\begin{aligned} h(0) &= ah(-1) + b_0 u(-1) + b_1 u(0) = b_1 \\ h(1) &= ah(0) + b_0 u(0) + b_1 u(1) = ab_1 + b_0 \\ h(2) &= ah(1) + b_0 u(1) + b_1 u(2) = ah(1) \end{aligned}$$

and more generally

$$h(k+1) = ah(k) \quad ; \quad k \geq 1$$

That is, the unit impulse response of the first order differende equation (2.1) is given by $\{h(0) = b_1,\ h(1) = ab_1 + b_0\}$ and

$$h(k) = h(1)a^{k-1} \quad ; \quad k \geq 1$$

Example 2.1.5. The unit impulse initial response h of the difference equation

$$y(k+1) = -0.7y(k) + 3u(k)$$

is given by

$$h(k) = \begin{cases} 0 & ; k \leq 0 \\ 3(-0.7)^{k-1} & ; k \geq 1 \end{cases}$$

Note that it is *not* necessary to remember the general form (2.15) of the solution for h in order to obtain this result. Instead from first principles, we have

$$\begin{aligned} h(0) &= -0.7h(-1) + 3u(-1) = 0 \\ h(1) &= -0.7h(0) + 3u(0) = 3 \\ h(2) &= -0.7h(1) + 3u(1) = -0.7h(1) \end{aligned}$$

and in general

$$h(k+1) = -0.7h(k) \quad ; \quad k \geq 1$$

That is, for $k \geq 1$

$$h(k) = (-0.7)^{k-1}h(1) = 3(-0.7)^{k-1}$$

(ii) The unit impulse response h of the difference equation

$$3y(k+1) = 0.6y(k) + 7u(k+1) - u(k)$$

is obtained as follows: with $h(-1) = 0$

$$3h(0) = 0.6h(-1) + 7u(0) - u(-1) = 7 \text{ or } h(0) = 7/3$$

Then

$$3h(1) = 0.6h(0) + 7u(1) - u(0) = 0.6(7/3) - 1 \text{ or } h(1) = 2/15$$

Also

$$3h(k+1) = 0.6h(k) \quad ; \quad k \geq 1$$

That is, $h(k) = 0$ for $k < 0$, $h(0) = 7/3$ and

$$h(k) = \frac{2}{15}(0.2)^{k-1} \quad ; \quad k \geq 1$$

Once again, note that it is not necessary to remember the general form (2.15) of the solution h since h can be derived from first principles.

Unit Step Response

Consider the forced response y_f in (2.13) when $k_0 = 0$, $y_f(-1) = 0$ and the input signal u is the unit step signal given by

$$u(k) = \begin{cases} 0 \; ; \; k < 0 \\ 1 \; ; \; k \geq 0 \end{cases} \tag{2.16}$$

The corresponding forced response (known as the *unit step response*) $y_f = s$ is then given by

$$s(0) = as(-1) + b_0u(-1) + b_1u(0) = b_1$$
$$s(1) = as(0) + b_0u(0) + b_1u(1) = ab_1 + b_0 + b_1$$
$$s(2) = as(1) + b_0u(1) + b_1u(2) = a^2b_1 + a(b_0 + b_1) + (b_0 + b_1)$$

and more generally

$$s(k) = a^k b_1 + (b_0 + b_1)\sum_{n=0}^{k-1} a^n \; ; \; k \geq 1$$

That is, for $k \geq 1$

$$s(k) = \begin{cases} a^k b_1 + (b_0 + b_1)(1-a)^{-1}(1-a^k) \; ; a \neq 1 \\ b_1 + (b_0 + b_1)k \; ; a = 1 \end{cases} \tag{2.17}$$

Note that the unit step response s becomes unbounded (i.e. $|s(k)| \to \infty$) as $k \to \infty$ when $|a| > 1$. For $|a| < 1$, $s(k) \to (b_0 + b_1)(1-a)^{-1}$. When $a = 1$, $s(k) \to \infty$ and when $a = -1$, $s(k)$ for $k \geq 1$ oscillates between b_0 and b_1.

Example 2.1.6. The unit step response s of the difference equation

$$y(k+1) = -0.7y(k) + 3u(k)$$

is given by $s(0) = -0.7s(-1) + 3u(-1) = 0$, and

$$s(k+1) = -0.7s(k) + 3 \;\; ; \;\; k \geq 0$$

That is, $s(1) = 3$ and

$$s(2) = -0.7s(1) + 3 = (-0.7)(3) + 3$$
$$s(3) = -0.7s(2) + 3 = (-0.7)^2(3) + (-0.7)(3) + 3$$

so that in general: for $k \geq 1$

$$s(k) = 3\sum_{n=0}^{k-1}(-0.7)^n = 3\frac{1-(-0.7)^k}{1-(-0.7)} = \frac{30}{17}(1-(-0.7)^k)$$

(ii) The unit step response s of the difference equation

$$3y(k+1) = 0.6y(k) + 7u(k+1) - u(k)$$

is given by $3s(0) = 0.6s(-1) + 7u(0) - u(-1) = 7$ or $s(0) = 7/3$. Then $3s(1) = 0.6s(0) + 7u(1) - u(0) = 0.6s(0) + 6$, and more generally: for $k \geq 0$

$$3s(k+1) = 0.6s(k) + 6 \;\text{ or }\; s(k+1) = 0.2s(k) + 2$$

Hence

$$s(1) = 0.2s(0) + 2 = (0.2)(7/3) + 2$$
$$s(2) = 0.2s(1) + 2 = (0.2)^2(7/3) + (0.2)(2) + 2$$

so that in general: for $k \geq 1$

$$s(k) = (0.2)^k(7/3) + 2\sum_{n=0}^{k-1}(0.2)^n = (0.2)^k(7/3) + \frac{5}{2}(1-(0.2)^k)$$

Exponential Response

Consider the forced response y_f in (2.13) when $k_0 = 0$, $y_f(-1) = 0$ and the input u is a digital exponential signal of the form

$$u(m) = \begin{cases} 0 & ; \; m < 0 \\ z^m & ; \; m \geq 0 \end{cases} \tag{2.18}$$

for some *constant* z. (Recall from Chapter 1 that the digital exponential signal $u(n) \triangleq v(nT)$ where $v(t) = exp\{\alpha t\}$ can be written in the form (2.18) with $z = exp\{\alpha T\}$.)

From (2.13), we then have that $y_f(0) = b_1$, and for $k \geq 1$

$$\begin{aligned} y_f(k) &= b_0 \sum_{n=1}^{k} a^{k-n} z^{n-1} + b_1 \sum_{n=0}^{k} a^{k-n} z^n \\ &= a^k b_1 + (b_0 + b_1 z) \sum_{n=1}^{k} a^{k-n} z^{n-1} \\ &= a^k b_1 + (b_0 + b_1 z) a^{k-1} \sum_{n=1}^{k} (a^{-1} z)^{n-1} \end{aligned}$$

Hence: (i) When $z = a$

$$y_f(k) = a^k b_1 + k(b_0 + b_1 a) a^{k-1} \quad ; \quad k \geq 1 \tag{2.19}$$

For $|a| \geq 1$, we conclude that $|y_f(k)| \to \infty$ as $k \to \infty$, and for $|a| < 1$ that $|y_f(k)| \to 0$ as $k \to \infty$.

(ii) When $z \neq a$, we have

$$y_f(k) = a^k b_1 + (b_0 + b_1 z) a^{k-1} \left\{ \frac{1 - (a^{-1} z)^k}{1 - a^{-1} z} \right\} \quad ; \quad k \geq 1 \tag{2.20}$$

This expression for $z \neq a$ can also be rearranged into the form

$$y_f(k) = a^k b_1 + \left(\frac{b_0 + b_1 z}{z - a} \right) (z^k - a^k) \quad ; \quad k \geq 1 \tag{2.21}$$

For $|a| > 1$, we conclude that $|y_f(k)| \to \infty$ as $k \to \infty$. For $|a| < 1$, we conclude that: in the limit as $k \to \infty$, the forced response y_f approaches the signal $\bar{y}_f$ where

$$\bar{y}_f(k) \triangleq \lim_{k \to \infty} y_f(k) = \left(\frac{b_0 + b_1 z}{z - a} \right) z^k \quad ; \; z \neq a, \; |a| < 1 \tag{2.22}$$

If in addition $|z| < 1$, we conclude from (2.22) that $\bar{y}_f(k) \to 0$, and if $|z| > 1$, then $|\bar{y}_f(k)| \to \infty$. However if $|z| = 1$, then $\bar{y}_f(k)$ will remain nonzero and finite for all k.

Example 2.1.7. (i) The forced response y_f of the difference equation

$$y(k+1) = -0.7y(k) + 3u(k)$$

when $y(-1) = 0$ and u is the exponential signal (2.18) is given by

$$y_f(k) = 3(-0.7)^{k-1} \sum_{n=1}^{k} (-\frac{z}{0.7})^{n-1}$$

That is

$$y_f(k) = \begin{cases} 3k(-0.7)^{k-1} & ; z = -0.7 \\ 3(z+0.7)^{-1}(z^k - (-0.7)^k) & ; z \neq -0.7 \end{cases}$$

When $z = -0.7$, $|y_f(k)| \to 0$ as $k \to \infty$, and

$$\bar{y}_f(k) \triangleq \lim_{k \to 0} y_f(k) = z^k (z+0.7)^{-1} \quad ; \quad z \neq -0.7$$

(ii) The forced response y_f of the difference equation

$$3y(k+1) = 0.6y(k) + 7u(k+1) - u(k)$$

when $y(-1) = 0$ and u is the exponential signal (2.18) is given by

$$y_f(k) = (0.2)^k \left(\frac{7}{3}\right) + (-\frac{1}{3} + \frac{7z}{3})(0.2)^{k-1} \sum_{n=1}^{k} (\frac{z}{0.2})^{n-1}$$

where

$$\sum_{n=1}^{k} (\frac{z}{0.2})^{n-1} = \begin{cases} k & ; z = 0.2 \\ (1-5z)^{-1}(1-(5z)^k) & ; z \neq 0.2 \end{cases}$$

Sinusoidal Response

Consider the forced response y_f in (2.13) when $k_0 = 0$, $y_f(-1) = 0$ and the input signal u is a sinusoidal signal of the form

$$u(k) = \begin{cases} 0 & ; \ k < 0 \\ \sin \omega_0 k & ; \ k \geq 0 \end{cases} \tag{2.23}$$

Then from (2.13), we have that: for $k \geq 1$

$$y_f(k) = b_0 a^k \sum_{n=1}^{k} a^{-n} \sin \omega_0 (n-1) + b_1 a^k \sum_{n=0}^{k} a^{-n} \sin \omega_0 n \tag{2.24}$$

It is not at all obvious as to whether or not there exists a closed form expression for y_f in (2.24), and so we consider another approach.

To begin, let us reconsider the linear time invariant difference equation (2.1) where now the input signal u is *complex*; that is, suppose

$$u(k) = u_1(k) + ju_2(k) \quad ; \quad j^2 = -1 \tag{2.25}$$

where $\{u_1, u_2\}$ are both real signals. It then follows that the resulting output signal y is also complex. In particular, if we let

$$y(k) = y_1(k) + jy_2(k) \quad ; \quad j^2 = -1 \tag{2.26}$$

where $\{y_1, y_2\}$ are both real valued signals, then as a result of the *linearity property*, it follows after substituting (2.25) and (2.26) into (2.1) and equating real and imaginary components that

$$\begin{aligned} y_1(k+1) &= ay_1(k) + b_0u_1(k) + b_1u_1(k+1) \quad ; \quad k \geq -1 \\ y_2(k+1) &= ay_2(k) + b_0u_2(k) + b_1u_2(k+1) \quad ; \quad k \geq -1 \end{aligned}$$

In other words, if $y = y_1 + jy_2$ is the complex output that results from the complex input $u = u_1 + ju_2$ with initial condition $y(-1) = y_1(-1) + jy_2(-1)$, then $y_1 = Re\{y\}$ is the real output that results from the real input signal $u_1 = Re\{u\}$ with initial condition $y_1(-1)$, and $y_2 = Im\{y\}$ is the real output that results from the real input signal $u_2 = Im\{u\}$ with initial condition $y_2(-1)$.

Thus if the input signal u is given by

$$u(m) = z^m \quad ; \quad z = exp\{j\omega_0\} = \cos\omega_0 + j\sin\omega_0, \;\; m \geq 0 \tag{2.27}$$

(which implies $|z| = 1$ for all ω_0), then in (2.18) for $m \geq 0$

$$u(m) = z^m = exp\{j\omega_0 m\} = \cos\omega_0 m + j\sin\omega_0 m \tag{2.28}$$

with

$$Im\{u(m)\} = \sin\omega_0 m \tag{2.29}$$

The corresponding *complete* response y for $k \geq -1$ of (2.1) is then given by

$$y(k) = y_i(k) + y_f(k)$$

where the *initial condition response* y_i is given by (2.13); that is, $y_i(k) = a^{k+1}y(-1)$ for $k \geq -1$, and the *forced response* y_f is given by (2.19) when $z = a$, and (2.20) when $z \neq a$.

Now when z is given by (2.27), we have $|z| = 1$. Then since a is real, the case $z = a$ is only possible if either $\{z = a = 1\}$, or $\{z = a = -1\}$. When $z = 1$, we have $u(m) = 1$, and when $z = -1$, we have $u(m) = (-1)^m$.

For $z \neq a$, the relevant solution y_f when u is given by (2.18) is given by (2.21). That is, for $|a| \neq 1$ when $\{u(-1) = 0;\ u(m) = \sin m\omega_0, m \geq 0\}$, the complete response y is given by $y(0) = ay(-1) + b_1$, and for $k \geq 1$

$$y(k) = a^{k+1}y(-1) + a^k b_1 + Im\left\{rexp\{-j\phi\}(exp\{jk\omega_0\} - a^k)\right\}$$

That is, for $k \geq 1$

$$y(k) = a^{k+1}y(-1) + a^k b_1 + r\sin(k\omega_0 - \phi) + ra^k \sin\phi \tag{2.30}$$

where r and ϕ are given by

$$rexp\{-j\phi\} = H(exp\{j\omega_0\}) \tag{2.31}$$

with the function H defined by

$$H(z) \triangleq \frac{b_0 + b_1 z}{z - a} \tag{2.32}$$

Example 2.1.8. The forced response y_f of the difference equation

$$y(k+1) = -0.7y(k) + 3u(k)$$

when $y(-1) = 0$ and $u(k) = \sin\omega_0 k$ is given by

$$y_f(k) = r\sin(\omega_0 k - \phi) + (-0.7)^k r\sin\phi$$

where

$$rexp\{-j\phi\} = \frac{3}{exp\{j\omega_0\} + 0.7}$$

For example, if $\omega_0 = \pi/4$, then (to 2. dec. places)

$$rexp\{-j\phi\} = \frac{3}{1/\sqrt{2} + j/\sqrt{2} + 0.7} = 1.70 - j0.86$$

(ii) The forced response y_f of the difference equation

$$3y(k+1) = 0.6y(k) + 7u(k+1) - u(k)$$

when $y(-1) = 0$ and $u(k) = \sin\omega_0 k$ is given by

$$y_f(k) = \frac{7}{3}(0.2)^k + r\sin(\omega_0 k - \phi) + r(0.2)^k \sin\phi$$

where

$$rexp\{-j\phi\} = \frac{7exp\{j\omega_0\} - 1}{3exp\{j\omega_0\} - 0.6}$$

2.1.2 Steady State and Transient Response

In (2.12), (2.13) we decomposed the complete response y of the first order difference equation (2.1) into the sum of the *zero input response* y_i and the *forced response* y_f. There is also another decomposition that is sometimes useful. Specifically, one can also express the complete response y in the form

$$y = y_t + y_{ss} \tag{2.33}$$

where y_t is the *transient response* and y_{ss} is the *steady state response*. The *transient response* is defined to be that part of the complete response which tends to zero as time approaches infinity, and the *steady state response* is what remains. Note that in general

$$y_i \neq y_t \quad ; \quad y_f \neq y_{ss}$$

For example, from (2.12) and (2.13), the complete response y of (2.1) when $\{k_0 = 0,\ y(-1) = 2\}$ and u is the digital unit impulse signal (2.16) is given by

$$y(k) = 2a^{k+1} + b_0 a^{k-1} + b_1 a^k \quad ; \quad k \geq 1$$

For $|a| < 1$, the *transient response* $y_t = y$ is the *complete response*, while the *steady state response* $y_{ss} = 0$. For $|a| \geq 1$, the *transient response* $y_t = 0$, while the *steady state response* y_{ss}, which is that component of the complete response y that does not eventually decay to zero, is given by $y_{ss} = y$.

For $|a| < 1$, the complete response can be decomposed as in (2.33) where

$$\begin{aligned} y_t(k) &= a^{k+1} y(-1) + a^k b_1 + r a^k \sin\phi \\ y_{ss}(k) &= r \sin(k\omega_0 - \phi) \end{aligned}$$

where $\{r, \phi\}$ are given by (2.31), (2.30). Hence for $|a| < 1$, a sinusoidal input signal u of the form $u(k) = \sin \omega_0 k$ results in a steady state sinusoidal output signal which differs from the input by a *scaling factor* r and a *phase lag* by $\phi > 0$. Alternatively, if we write

$$\omega_0 k - \phi = \omega_0 (k - \frac{\phi}{\omega_0})$$

then ϕ/ω_0 represents the *time delay*. Plots of r and ϕ as functions of the frequency ω_0 for $\{b_0 = 1, b_1 = 0\}$ and $\{a = 0.2, 0.5, 0.8\}$ are illustrated in Figure 2.1.

First Order Digital Filter

The first order difference equation (2.1) can be used as a *digital filter* to reduce the effect of a high frequency interference signal. Specifically, suppose we receive a signal u where

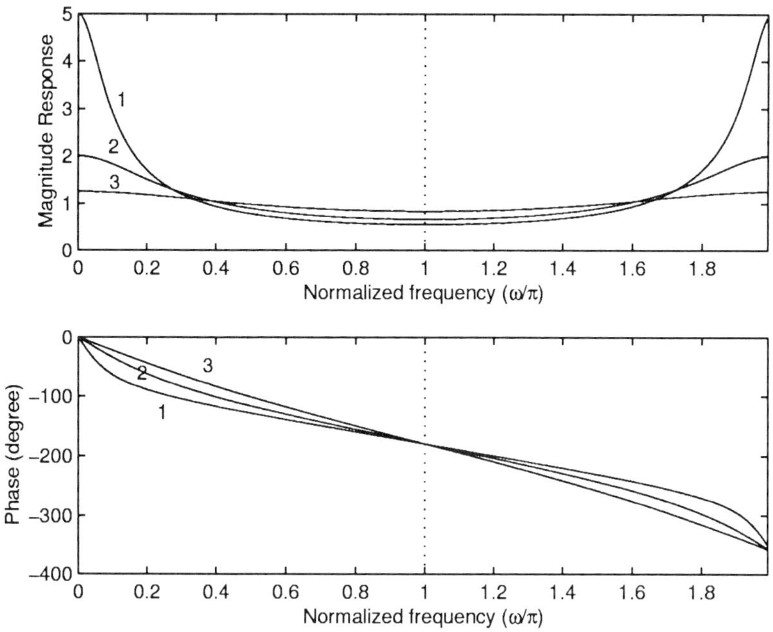

Fig. 2.1. Attenuation and phase characteristics for 1: $a = 0.2$, 2: $a = 0.5$ and 3: $a = 0.8$

$$u(k) = \alpha_0 \sin \omega_0 k + \alpha_1 \sin(\omega_1 k + \theta_1) \quad ; \quad k \geq 0$$

where $\omega_1 > \omega_0$ is the frequency of the interference signal, and the signal component with frequency ω_0 is to be "recovered".

Since (2.1) is linear, it follows that the steady state response $\overline{y}_f$ of (2.1) when $\{b_0 = 1, b_1 = 0\}$ and $0 < a < 1$ to the input signal u is given by

$$\bar{y}_f(k) = r_0\alpha_0 \sin(\omega_0 k - \phi_0) + r_1\alpha_1 \sin(\omega_1 k + \theta_1 - \phi_1)$$

where for $m = 0, 1$

$$r_m exp\{-j\phi\} = \frac{1}{exp\{j\omega_m\} - a}$$

That is

$$r_m = \frac{1}{\sqrt{a^2 - 2a\cos\omega_m + 1}} \quad ; \quad \phi_m = \tan^{-1}\left\{\frac{\sin\omega_m}{\cos\omega_m - a}\right\} \tag{2.34}$$

In particular, suppose $\omega_0 \simeq 0$ and $\omega_1 \simeq \pi$, then

$$\frac{r_0}{r_1} \simeq \frac{1+a}{1-a}$$

which for $a = 0.5$ (say) means that the high frequency signal is attenuated by a factor of 3 with respect to the "desired" low frequency signal. This attenuation factor increases to 9 when $a = 0.8$.

An Inverse Filter

We conclude this section with a simple illustration of what is meant by *inverse filtering*. Specifically, suppose that a digital measurement signal z is received which is known to have been generated as the response of the first order difference equation

$$\begin{aligned} y(k+1) &= ay(k) + b_0 u(k) \ ; \ k \geq 0 \\ z(k) &= y(k) \end{aligned} \tag{2.35}$$

with $b_0 \neq 0$. The aim is to recover the input signal u by processing (or filtering) the measurement signal z.

Observe that if z is the input to the causal filter

$$\begin{aligned} m(k) &= \frac{1}{b_0}[z(k) - az(k-1)] \ ; \ k \geq 1 \\ r(k) &= m(k) \end{aligned} \tag{2.36}$$

with output r, then: for $k \geq 1$, we have

$$r(k) = \frac{1}{b_0}[y(k) - ay(k-1)] = u(k-1)$$

and so the input signal u is "recovered" with one unit time delay. (Note that the *noncausal* filter $r(k) = 1/b_0[z(k+1) - az(k)]$ with input z and output r would result in $r(k) = u(k)$, but since this filter is *noncausal*, it *cannot* be realized in real time.)

In signal processing applications, measurement errors are usually present which means that the input signal u cannot always be perfectly recovered without any errors. Specifically, suppose the measurements z in (2.35) are now described by

$$z(k) = y(k) + w(k) \tag{2.37}$$

where the signal w represents measurement errors. Then the filtering operation (2.36) now leads to

$$r(k) = u(k-1) + \varepsilon(k) \ ; \ k \geq 1 \tag{2.38}$$

where the error ε is given in terms of the measurement error ω in y by

$$\varepsilon(k) = \frac{1}{b_0}[w(k) - aw(k-1)]$$

If information is known about both the measurement errors ω, then information can be inferred about the error ε in the estimate of u.

For example, suppose $|w(k)| \leq \delta$, then

$$|\varepsilon(k)| \leq \left(\frac{1+|a|}{|b_0|}\right)\delta$$

Now suppose that the measurement signal z is generated according to

$$\begin{aligned} y(k+1) &= ay(k) + b_0 u(k) + b_1 u(k+1) \ ; \ k \geq 0 \\ z(k) &= y(k) + w(k) \end{aligned} \tag{2.39}$$

with $b_1 \neq 0$. If z is processed in real time to provide the signal r according to

$$r(k+1) = -\frac{b_0}{b_1} r(k) + \frac{1}{b_1}[z(k+1) - az(k)] \ ; \ k \geq 0 \tag{2.40}$$

then from (2.39)

$$r(k+1) + \frac{b_0}{b_1} r(k) = u(k+1) + \frac{b_0}{b_1} u(k) + \frac{1}{b_1}[w(k+1) - aw(k)]$$

and so the error $\varepsilon(k) \triangleq r(k) - u(k)$ between the signal output r and the desired signal u satisfies the first order difference equation

$$\varepsilon(k+1) = -\frac{b_0}{b_1}\varepsilon(k) + \frac{1}{b_1}[w(k+1) - w(k)]$$

Note that the filter (2.40) must be initialized with some initial condition $r(0)$. This value together with the unknown value $u(0)$ of the desired signal u at $k = 0$, determines the initial error $\varepsilon(0)$ which then in turn determines the propogation of the error signal ε.

When $w(k)$ is a constant, then $\varepsilon(k) \to 0$ as $k \to \infty$ if $|b_0| < |b_1|$. The rate of convergence however decreases as $|b_0|$ approaches $|b_1|$. When $|b_0| > |b_1|$, then for all initial errors $\varepsilon(0)$, $|\varepsilon(k)| \to \infty$ as $k \to \infty$. Hence the signal inversion algorithm (2.39) only applies for $|b_0| < |b_1|$.

2.2 Second Order

Consider the digital signal y which is given for $k \geq k_0 - 2$ by the solution of the difference equation

$$y(k+2) = a_1 y(k+1) + a_0 y(k) + b_0 u(k) + b_1 u(k+1) + b_2 u(k+2) \tag{2.41}$$

where $\{a_1, a_0, b_2, b_1, b_0\}$ are real constants, k_0 is a given integer, and k is a variable integer.

The output signal y is uniquely determined by the initial conditions $\{y(k_0-1),\ y(k_0-2)\}$ and the input signal u. In particular, with $k = k_0 - 2$, we have

$$y(k_0) = a_1 y(k_0-1) + a_0 y(k_0-2) + b_0 u(k_0-2) + b_1 u(k_0-1) + b_2 u(k_0)$$

Then for $k = k_0 - 1$

$$y(k_0+1) = a_1 y(k_0) + a_0 y(k_0-1) + b_0 u(k_0-1) + b_1 u(k_0) + b_2 u(k_0+1)$$

which after substituting for $y(k_0)$ gives

$$\begin{aligned} y(k_0+1) = {} & (a_1^2 + a_0) y(k_0-1) + a_1 a_0 y(k_0-2) + a_1 b_0 u(k_0-2) \\ & + (a_1 b_1 + b_0) u(k_0-1) + (b_1 + a_1 b_2) u(k_0) + b_2 u(k_0+1) \end{aligned}$$

Continuing, for $k = k_0$

$$y(k_0+2) = a_1 y(k_0+1) + a_0 y(k_0) + b_0 u(k_0) + b_1 u(k_0+1) + b_2 u(k_0+2)$$

which after substituting for $y(k_0+1)$ and $y(k_0)$ can also be written in terms of the initial conditions $\{y(k_0-1), y(k_0-2)\}$ and the input values $\{u(k_0-2), u(k_0-1), u(k_0), u(k_0+1), u(k_0+2)\}$. However unlike the case of the first order difference equation (2.1), it is unclear as to how to proceed in order to develop a general expression for $y(k)$ for all $k \geq k_0$. We leave further consideration of this problem until section 2.5.

Causality

Since the output signal y at time $k+2$ depends only on previous values of y, and values of the input u up to time $k+2$, the difference equation (2.41) is said to be *causal*. The difference equation

$$\begin{aligned} y(k+2) = {} & a_1 y(k+1) + a_0 y(k) + b_3 u(k+3) + b_2 u(k+2) \\ & + b_1 u(k+1) + b_0 u(k) \end{aligned}$$

is noncausal unless $b_3 = 0$.

Order

The difference equation (2.41) for $a_0 \neq 0$ is said to be of *second order* since given the input signal u, only *two* initial values $\{y(k_0-2), y(k_0-1)\}$ are necessary and sufficient in order to determine $y(k)$ for $k \geq k_0$. The following three difference equations are respectively of order one, two and three:

$$y(k+2) = a_1 y(k+1) + b_1 u(k+1) \ ; \ a_1 \neq 0$$
$$y(k+2) = a_1 y(k+1) + a_0 y(k) + b_1 u(k+1) + c_1 u(k-1) \ ; \ a_0 \neq 0$$
$$y(k+2) = a_1 y(k+1) + d_1 y(k-1) + b_0 u(k) + c_1 u(k-1) \ ; \ d_1 \neq 0$$

Given that the input u is available, then in the first example, only *one* initial value $\{y(k_0-1)\}$ is required in order to uniquely determine $y(k)$ for $k \geq k_0$. In the second example, *two* initial values $\{y(k_0-2), y(k_0-1)\}$ are required, while in the third example, *three* initial values $\{y(k_0-3), y(k_0-2), y(k_0-1)\}$ are required.

Linearity

The second order difference equation (2.41) is said to be *linear* since: if $y^{(j)}$ for $j = 1, 2$ is the output signal due to an input signal $u^{(j)}$ with initial condition $\{y^{(j)}(k_0-2) = \alpha_j,\ y^{(j)}(k_0-1) = \beta_j\}$ then $z = c_1 y^{(1)} + c_2 y^{(2)}$ is the output signal due to the input signal $c_1 u^{(1)} + c_2 u^{(2)}$ with initial condition $\{z(k_0-2) = c_1\alpha_1 + c_2\alpha_2 \ ; \ z(k_0-1) = c_1\beta_1 + c_2\beta_2\}$.

The following two second order difference equations are both *nonlinear* second order difference equations:

$$y(k+2) = a_1 y(k+1) + a_0 y(k) + d_1 y(k+1) y(k) + u(k) \ ; \ d_1 \neq 0$$
$$y(k+2) = a_0 y(k) + b_1 u^2(k) \ ; \ b_1 \neq 0$$

Time Invariance

The second order linear difference equation (2.41) with output signal y and input signal u is also said to be *time invariant* since

$$z(k) \triangleq y(k-k_2) \ ; \ k \geq k_0 + k_2$$

for $k \geq k_0 + k_2 - 2$ is given by the solution of the second order difference equation

$$z(k+2) = a_1 z(k+1) + a_0 z(k) + b_1 v(k+1) + b_0 v(k)$$

when the input signal v is given by $v(k) \triangleq u(k-k_2)$ and the initial conditions are:

$$z(k_0+k_2-2) = y(k_0-2) \quad ; \quad z(k_0+k_2-1) = y(k_0-1)$$

That is, in a time invariant system, if one knows the response y due to an input signal u, then one can determine what the response z will be due to a *time delayed* version v of u.

The time invariance property is a consequence of the fact that all the coefficients $\{a_1, a_0, b_1, b_0\}$ in (2.41) are *constant*, and therefore do not vary with time. The following two linear second order linear difference equations are both *time varying*:

$$\begin{aligned} y(k+2) &= ky(k+1) + a_0 y(k) + u(k) \\ y(k+2) &= a_0 y(k) + b_1^k u(k+1) \ ; \ b_1 \neq 0 \end{aligned}$$

Coupled First Order Difference Equations

A second order linear difference equation can result from a system of two coupled first order difference equations. Specifically, consider the two equations:

$$\begin{aligned} x_1(k+1) &= f_{11}x_1(k) + f_{12}x_2(k) + g_1 u(k) \\ x_2(k+1) &= f_{21}x_1(k) + f_{22}x_2(k) + g_2 u(k) \end{aligned} \tag{2.42}$$

The first equation can be thought of as a first order difference equation in x_1 with input signal $f_{12}x_2 + g_1 u$, while the second equation is a first order difference equation in x_2 with input signal $f_{21}x_1 + g_2 u$. These equations are said to be *coupled* when either $f_{12} \neq 0$ or $f_{21} \neq 0$ since then the input to one equation depends on the response of the other.

The first equation implies

$$x_1(k+2) = f_{11}x_1(k+1) + f_{12}x_2(k+1) + g_1 u(k+1)$$

and after substituting for $x_2(k+1)$ using the second equation in (2.42), we have

$$\begin{aligned} x_1(k+2) = {} & f_{11}x_1(k+1) + f_{12}\{f_{21}x_1(k) + f_{22}x_2(k) + g_2 u(k)\} \\ & + g_1 u(k+1) \end{aligned}$$

But from the first equation in (2.42), we have

$$f_{12}x_2(k) = x_1(k+1) - f_{11}x_1(k) - g_1 u(k)$$

and so it follows that x_1 satisfies the difference equation

$$\begin{aligned} x_1(k+2) = {} & [f_{11} + f_{22}]x_1(k+1) + [f_{12}f_{21} - f_{11}f_{22}]x_1(k) \\ & + [f_{12}g_2 - f_{22}g_1]u(k) + g_1 u(k+1) \end{aligned} \tag{2.43}$$

Similarly, it follows that the corresponding second order difference equation in x_2 is given by

$$x_2(k+2) = [f_{11} + f_{22}]x_2(k+1) + [f_{12}f_{21} - f_{11}f_{22}]x_2(k) \\ +[f_{21}g_1 - f_{11}g_2]u(k) + g_2u(k+1) \tag{2.44}$$

If $f_{12}f_{21} = f_{11}f_{22}$, then both (2.43) and (2.44) are *first order* difference equations in x_1 and x_2 respectively whose solutions are determined by the input u and the initial conditions $\{x_1(k_0-1), x_2(k_0-1)\}$.

Now suppose $f_{12}f_{21} \neq f_{11}f_{22}$. Then (2.43) and (2.44) are both *second order* difference equations in x_1 and x_2 respectively whose solutions can be found once the initial conditions $\{x_1(k_0-2), x_1(k_0-1)\}$, the initial conditions $\{x_2(k_0-2), x_2(k_0-1)\}$ and the input signal u are given. However from (2.42) with $u(k) = 0$ for $k < k_0$, and $\{x_1(k_0-1), x_2(k_0-1)\}$ given, we have that

$$\begin{bmatrix} x_1(k_0-2) \\ x_2(k_0-2) \end{bmatrix} = \begin{bmatrix} f_{11} & f_{12} \\ f_{21} & f_{22} \end{bmatrix}^{-1} \begin{bmatrix} x_1(k_0-1) \\ x_2(k_0-1) \end{bmatrix}$$

That is, only two *independent* initial conditions $\{x_1(k_0-1), x_2(k_0-1)\}$ are needed to determine the solution of (2.42) given the input signal u.

Example 2.2.1. Consider the system of two difference equations that were presented in Chapter 1 in connection with an application in marketing. Specifically, we proposed a mathematical model that provides the fraction $\{x_1(k), x_2(k)\}$ of customers distributed between two shops at the end of the kth month according to:

$$x_1(k+1) = a_{11}x_1(k) + a_{12}x_2(k) \;\; ; \;\; k \geq 0 \tag{2.45}$$
$$x_2(k+1) = a_{21}x_1(k) + a_{22}x_2(k) \;\; ; \;\; k \geq 0$$

where $a_{ij} \geq 0$ for all i, j, $x_j(0) \geq 0$ for all j, $x_1(0) + x_2(0) = 1$, and

$$a_{11} + a_{21} = 1 \;\; ; \;\; a_{12} + a_{22} = 1 \tag{2.46}$$

If we add both equations together, then

$$x_1(k+1) + x_2(k+1) = (a_{11} + a_{21})x_1(k) + (a_{12} + a_{22})x_2(k) \\ = x_1(k) + x_2(k)$$

where the second equality follows from (2.46). That is, $z(k) \triangleq x_1(k) + x_2(k)$ satisfies the first order linear time invariant difference equation

$$z(k+1) = z(k) \;\; ; \;\; k \geq 1$$

which implies $z(k) = z(0) = 1$ for all $k \geq 1$.

From (2.43), we have

$$x_1(k+2) = (a_{11}+a_{22})x_1(k+1) + (a_{12}a_{21} - a_{22}a_{11})x_1(k) \tag{2.47}$$

where from (2.46)

$$a_{12}a_{21} - a_{22}a_{11} = (1-a_{22})(1-a_{11}) - a_{22}a_{11} = 1 - (a_{11}+a_{22})$$

That is, (2.47) simplifies to

$$\begin{aligned} x_1(k+2) &= \alpha x_1(k+1) + (1-\alpha)x_1(k) \quad ; \quad k \geq 0 \\ x_2(k) &= 1 - x_1(k) \end{aligned} \tag{2.48}$$

where $\alpha \triangleq a_{11} + a_{22}$. From (2.46), we then have that $0 \leq \alpha \leq 2$. Also from (2.45)

$$x_1(1) = a_{11}x_1(0) + a_{12}x_2(0) = a_{11}x_1(0) + a_{12}(1 - x_1(0))$$

or

$$x_1(1) = a_{12} + (a_{11} - a_{12})x_1(0) \tag{2.49}$$

To summarize, we have shown that the system of two first order difference equations (2.45), (2.46) in x_1 and x_2 can be expressed as a second order difference equation in x_1 in (2.48) (or as a second order difference equation in x_2) whose solution is uniquely defined once the initial conditions $\{x_1(0), x_1(1)\}$ are specified. In this particular example, it turns out that $x_1(1)$ in (2.49) is related to $x_1(0)$.

Later in example 2.2.3, we obtain a closed form expression for the solution $x_1(= 1-x_2)$ which gives the value of $x_1(k)$ explicitly for any value of k without the need of calculating any other values $\{x_1(m); m < k\}$.

Approximation of Solution of Differential Equation

In section 2.1, we considered a problem in numerical integration where the aim was to provide a numerical procedure for approximating the solution of the first order differential equation (2.5). We now derive a numerical procedure for approximation the solution of the second order differential equation

$$\ddot{z}(t) = \alpha\dot{z}(t) + \beta z(t) + \gamma v(t) \quad ; \quad t \geq 0 \tag{2.50}$$

Specifically, after integration of both sides of (2.50), we have

$$\begin{aligned} \dot{z}(t) - \dot{z}(t_k) &= \int_{t_k}^{t} [\alpha\dot{z}(\sigma) + \beta z(\sigma) + \gamma v(\sigma)]d\sigma \\ z(t) - z(t_k) &= \int_{t_k}^{t} \dot{z}(\sigma)d\sigma \end{aligned} \tag{2.51}$$

Therefore when $t = t_{k+1}$

$$\dot{z}(t_{k+1}) = \dot{z}(t_k) + \int_{t_k}^{t_{k+1}} [\alpha \dot{z}(\sigma) + \beta z(\sigma) + \gamma v(\sigma)] d\sigma$$
$$z(t_{k+1}) = z(t_k) + \int_{t_k}^{t_{k+1}} \dot{z}(\sigma) d\sigma$$

Using the *backward Euler* approximation, with $t_k = kT$ constant for some fixed T and all integers k, we have that an approximation $x_1(k)$ of $\dot{z}(kT)$ and an approximation $x_2(k)$ of $z(kT)$ are given by

$$x_1(k+1) = x_1(k) + T[\alpha x_1(k) + \beta x_2(k) + \gamma u(k)]$$
$$x_2(k+1) = x_2(k) + T x_1(k)$$

where $u(k) \stackrel{\Delta}{=} v(kT)$. That is

$$x_1(k+1) = (1 + \alpha T) x_1(k) + T\beta x_2(k) + T\gamma u(k) \tag{2.52}$$
$$x_2(k+1) = T x_1(k) + x_2(k)$$

which as in (2.42) is a system of two coupled first order difference equations.

Then with $\{x_1(0) = \dot{z}(0), x_2(0) = z(0)\}$, it follows from (2.43) that the second order difference equation in $y = x_2$ for $k \geq 0$ is given by

$$y(k+2) = [2 + \alpha T] y(k+1) + [\beta T^2 - 1 - \alpha T] y(k) + \gamma T^2 u(k) \tag{2.53}$$

with the initial conditions given by $\{y(0) = z(0), y(1) = T\dot{z}(0) + z(0)\}$.

2.2.1 Zero Input Response

The complete response y in (2.41) can be decomposed into two components as follows:

$$y = y_i + y_f \tag{2.54}$$

where y_i is the *zero input* response and y_f is the *forced* response.

The zero input response y_i is that part of the complete response which is only due to the initial conditions. That is, from (2.41), the zero input response is given by

$$y_i(k+2) = a_1 y_i(k+1) + a_0 y_i(k) \quad ; \quad k \geq k_0 - 2 \tag{2.55}$$

subject to the given initial conditions $\{y_i(k_0 - 2) = y(k_0 - 2)\ ,\ y_i(k_0 - 1) = y(k_0 - 1)\}$.

In order to determine the explicit form of the zero input response, we begin by assuming that

$$y_i(k) = \beta\lambda^{k-k_0+2} \quad ; \quad k \geq k_0 \tag{2.56}$$

is a solution of (2.55) for some parameters $\{\beta, \lambda\}$. Then after substitution of (2.56) into (2.55), we conclude that y_i is in fact a solution of

$$\beta(\lambda^2 - a_1\lambda - a_0)\lambda^{k-k_0+2} = 0 \quad ; \quad k \geq k_0$$

Therefore we conclude that either $\{\beta = 0\}$ or $\{\lambda^2 - a_1\lambda - a_0 = 0\}$.

The first case $\{\beta = 0\}$ implies $y_i(k) = 0$ for all k in (2.56). Therefore if $y_i \neq 0$ in (2.56) is to be a zero input solution, the parameter λ must satisfy the following *algebraic* equation known as the *characteristic equation*:

$$\lambda^2 - a_1\lambda - a_0 = 0 \tag{2.57}$$

This algebraic equation has two (possibly complex) solutions $\{\lambda_1, \lambda_2\}$ given by

$$\lambda_1 = \frac{a_1 + \sqrt{a_1^2 + 4a_0}}{2} \quad ; \quad \lambda_2 = \frac{a_1 - \sqrt{a_1^2 + 4a_0}}{2} \tag{2.58}$$

The roots $\{\lambda_1, \lambda_2\}$ of the characteristic equation (2.57) are called the *characteristic roots.* If λ_1 is *complex*, then since the coefficients $\{a_1,\ a_0\}$ are real, we have that $\lambda_2 = \overline{\lambda_1}$ is the complex conjugate of λ_1.

Example 2.2.2. (i) Consider the difference equation

$$y(k+2) = -0.1y(k+1) + 0.3y(k)$$

Then the characteristic equation is given by

$$\lambda^2 = -0.1\lambda + 0.3$$

and since $\lambda^2 + 0.1\lambda - 0.3 = (\lambda - 0.5)(\lambda + 0.6)$, the characteristic roots are $\lambda_1 = 0.5$ and $\lambda_2 = -0.6$

(ii) Consider the difference equation

$$y(k+2) + y(k+1) + 0.25y(k) = 0$$

Then the characteristic equation is

$$\lambda^2 + \lambda + 0.25 = 0$$

and and since $\lambda^2 + \lambda + 0.25 = (\lambda + 0.5)^2$, the characteristic roots are $\lambda_1 = \lambda_2 = -0.5$.

(iii) Consider the difference equation

$$y(k+2) - y(k+1) + y(k) = 0$$

Then the characteristic equation is

$$\lambda^2 - \lambda + 1 = 0$$

and the characteristic roots are $\lambda_1 = 0.5 + j0.866$, $\lambda_2 = \overline{\lambda_1} = 0.5 - j0.866$.

In order to continue the development of the zero input response in (2.55), we need to consider two separate cases: (i) $a_0 = 0$ and (ii) $a_0 \neq 0$. When $\{a_0 = 0,\ a_1 = 0\}$, both the characteristic roots $\{\lambda_1, \lambda_2\}$ in (2.58) are zero. In this case, the second order difference equation (2.55) for the zero input response y_i reduces to the *zero* order equation

$$y_i(k+2) = 0 \quad ; \quad k \geq k_0 - 2$$

which implies $y_i(k) = 0$ for all $k \geq k_0$.

When $\{a_0 = 0,\ a_1 \neq 0\}$, the characteristic roots in (2.58) are given by $\{\lambda_1 = 0, \lambda_2 = a_1 \neq 0\}$, and (2.55) for the zero input response of (2.55) reduces to the *first order* difference equation

$$y_i(k+2) = a_1 y_i(k+1) \quad ; \quad k \geq k_0 - 2$$

which implies

$$y_i(k) = a_1^{k-k_0+2} y_i(k_0 - 2) \quad ; \quad k \geq k_0 - 2$$

When $a_0 \neq 0$, (2.55) is a *second order* difference equation for all values of a_1. We now consider this case in more detail according to the type of characteristic roots.

Characteristic Equation ; $a_0 \neq 0,\ \lambda_1 \neq \lambda_2 \neq 0$

For $a_0 \neq 0$, the roots $\{\lambda_1, \lambda_2\}$ in (2.58) are real, non equal and non zero when

$$a_1^2 + 4a_0 \neq 0$$

In this case, we have from (2.56), the *zero input* response is of the general form

$$y_i(k) = \beta_1 \lambda_1^{k-k_0+2} + \beta_2 \lambda_2^{k-k_0+2} \quad ; \quad k \geq k_0 - 2 \tag{2.59}$$

The values of the coefficients $\{\beta_1, \beta_2\}$ in (2.59) are obtained from the given initial conditions; that is, from (2.59)

$$y_i(k_0 - 2) = \beta_1 + \beta_2 \quad ; \quad y_i(k_0 - 1) = \beta_1\lambda_1 + \beta_2\lambda_2 \quad ; \quad (\lambda_1 \neq \lambda_2) \tag{2.60}$$

and since $\lambda_1 \neq \lambda_2$, the coefficients $\{\beta_1, \beta_2\}$ are given by

$$\begin{aligned} \beta_1 &= \frac{\lambda_2 y_i(k_0 - 2) - y_i(k_0 - 1)}{\lambda_2 - \lambda_1} \\ \beta_2 &= \frac{\lambda_1 y_i(k_0 - 2) - y_i(k_0 - 1)}{\lambda_1 - \lambda_2} \end{aligned} \tag{2.61}$$

Example 2.2.3. We have shown in example 2.2.1 that the fraction $x_1(k)$ of the total number of customers of shop #1 at the end of the kth month assosciated with the marketing problem defined in Chapter 1 is given by the solution of the second order difference equation

$$x_1(k+2) = \alpha x_1(k+1) + (1-\alpha)x_1(k) \quad ; \quad k \geq 0 \qquad (2.62)$$
$$x_1(1) = a_{12} + (a_{11} - a_{12})x_1(0)$$

for some $0 \leq x_1(0) \leq 1$ and $0 \leq \alpha \leq 2$ where $\alpha \triangleq a_{11} + a_{22}$.

For this difference equation, the characteristic equation is given by

$$\lambda^2 = \alpha\lambda + (1-\alpha) \quad \text{or} \quad (\lambda - 1)(\lambda + 1 - \alpha) = 0$$

which has the characteristic roots

$$\lambda_1 = \alpha - 1 \quad ; \quad \lambda_2 = 1$$

(i) The case $\alpha = 2$ results when $\{a_{11} = 1; a_{12} = 0; a_{22} = 1; a_{21} = 0\}$ which from (2.45), (2.46) then implies

$$x_1(k+1) = x_1(k) \quad ; \quad x_2(k+1) = x_2(k)$$

That is, when $\alpha = 2$, we conclude that $x_1(k) = x_1(0)$ and $x_2(k) = x_2(0)$ for all $k \geq 1$ which means that each shop maintains the same fraction of customers from month to month.

(ii) For $0 \leq \alpha < 2$, we have that $\lambda_1 \neq \lambda_2$, and so from (2.59) the response x_1 is given by

$$x_1(k) = \beta_1(\alpha - 1)^k + \beta_2 \quad ; \quad k \geq 0 \qquad (2.63)$$

Then

$$x_1(0) = \beta_1 + \beta_2 \quad ; \quad x_1(1) = \beta_1(\alpha - 1) + \beta_2$$

which since $\alpha \neq 2$ implies using (2.62) that

$$\beta_1 = \frac{x_1(0) - x_1(1)}{2 - \alpha} = \frac{(1 - a_{11} + a_{12})x_1(0) - a_{12}}{2 - \alpha} = x_1(0) - \frac{a_{12}}{2 - \alpha}$$
$$\beta_2 = x_1(0) - \beta_1$$

However from (2.46), we have

$$a_{11} - a_{12} = a_{22} - a_{21} = (a_{22} + a_{11}) - (a_{21} + a_{11}) = \alpha - 1$$

Therefore

$$\beta_1 = x_1(0) - \frac{a_{12}}{2 - \alpha} \quad ; \quad \beta_2 = \frac{a_{12}}{2 - \alpha}$$

When $\alpha = 0$, $\alpha - 1 = -1$ and then from (2.63)

$$x_1(k) = \beta_1(-1)^k + \beta_2$$

which means that $x_1(k)$ oscillates between $\beta_2 - \beta_1$ and $\beta_2 + \beta_1$.

For $0 < \alpha < 2$, we have $-1 < \alpha - 1 < 1$, and so we conclude from (2.63) that $x_1(k) \to \beta_2$ in the limit as $k \to \infty$. That is, in the limit, shop #1 attracts more customers (i.e. $\beta_2| > 0.5$) if

$$\frac{a_{12}}{2 - \alpha} > 0.5$$

If we assume that there are a total of N possible customers each month, then the total number of customers $S_1(m)$ for shop #1 up to and including the mth month is given by

$$S_1(m) = NT_1(m) \quad ; \quad T_1(m) = \sum_{k=0}^{m} x_1(k)$$

That is

$$\begin{aligned} T_1(m) &= \beta_2(m+1) + \beta_1 \sum_{k=0}^{m} (\alpha - 1)^k \\ &= \beta_2(m+1) + \frac{\beta_1}{2-\alpha}(1 - (\alpha - 1)^{m+1}) \end{aligned}$$

The corresponding number of customers $S_2(m)$ for shop #2 in the same period is given by

$$S_2(m) = NT_2(m) \quad ; \quad T_2(m) = (m+1) - T_1(m)$$

Plots of x_1 and T_1 for $x_1(0) = 0.1$, and $\alpha = 0.1, 1.3, 1.8$ are illustrated in Figure 2.2.

Characteristic Equation; $a_0 \neq 0,\ \lambda_1 = \lambda_2 \neq 0$

For $a_0 \neq 0$, the roots $\{\lambda_1, \lambda_2\}$ in (2.58) are equal and real when

$$a_1^2 + 4a_0 = 0 \ \text{ with } \lambda_1 = \lambda_2 = \frac{a_1}{2}$$

In this case, the general form of the zero input response y_i is given by

$$y_i(k) = [\beta_1 + \beta_2(k - k_0 + 2)]\lambda_1^{k-k_0+2} \quad ; \quad k \geq k_0 - 2 \tag{2.64}$$

To verify this solution, substitute (2.64) into (2.55) to give

$$[p_1 + p_2(k - k_0 + 2)]\lambda_1^{k-k_0+2} = 0 \ ; \ k \geq k_0 - 2$$

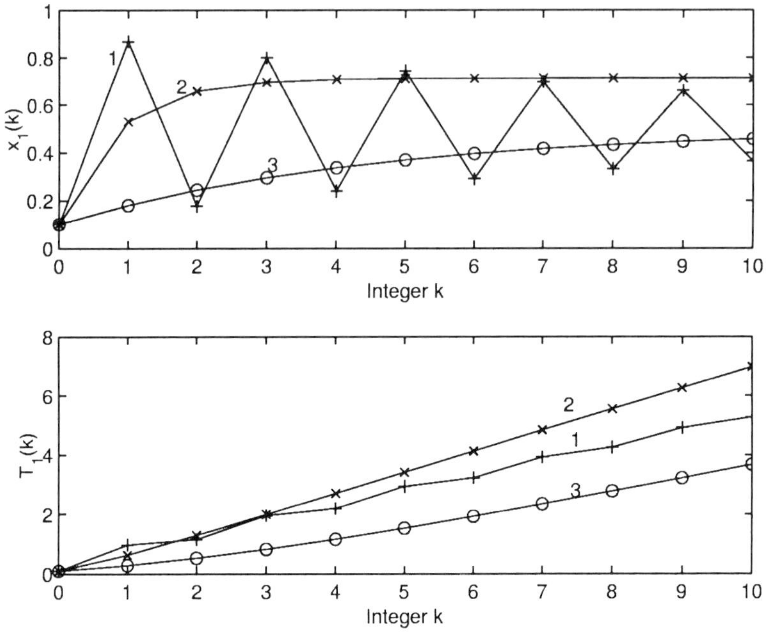

Fig. 2.2. Monthly and total sales for shop #1 for 1: $\alpha = 0.1$, 2: $\alpha = 1.3$ and 3: $\alpha = 1.8$

where

$$p_1 \stackrel{\Delta}{=} \beta_1(\lambda_1^2 - a_1\lambda_1 - a_0) + \beta_2\lambda_1(2\lambda_1 - a_1) \quad ; \quad p_2 \stackrel{\Delta}{=} \beta_2(\lambda_1^2 - a_1\lambda_1 - a_0)$$

It therefore follows that y_i in (2.64) is a solution when $\lambda_1 = \lambda_2$ if and only if

$$p_1 = p_2 = 0$$

for *all* parameters $\{\beta_1,\ \beta_2\}$; that is, since $\lambda_1 \neq 0$, if and only if

$$\lambda_1^2 - a_1\lambda_1 - a_0 = 0 \quad ; \quad 2\lambda_1 - a_1 = 0 \tag{2.65}$$

The first equation is satisfied since this equation is the characteristic equation (2.57). Since $\lambda_1 = \lambda_2 = a_1/2$, the second equation is also satisfied. The form of the solution in (2.64) is therefore established.

The values of the coefficients $\{\beta_1, \beta_2\}$ in (2.64) are obtained from the given initial conditions; that is

$$\beta_1 = y_i(k_0 - 2) \quad ; \quad \beta_2 = \frac{y_i(k_0 - 1)}{\lambda_1} - y_i(k_0 - 2) \tag{2.66}$$

(Note that since $\lambda_1 \neq 0$, β_2 is well-defined.)

Characteristic Equation; $a_0 \neq 0,\ \lambda_1 = \overline{\lambda}_2$ complex

Now consider the case when $\lambda_1 = \overline{\lambda}_2$ is *complex*, or equivalently from (2.58), when $a_1^2 + 4a_0 < 0$. Then

$$\lambda_1 = \overline{\lambda}_2 = \frac{a_1 + j\sqrt{-(a_1^2 + 4a_0)}}{2} \triangleq \lambda_{11} + j\lambda_{12} \quad (\lambda_{12} \neq 0) \tag{2.67}$$

where λ_{11} and λ_{12} are both real. The zero input response y_i is then given by

$$y_i(k) = \beta_1 \lambda_1^{k-k_0+2} + \beta_2 \overline{\lambda}_1^{k-k_0+2} \quad ; \quad k \geq k_0 - 2$$

which implies

$$y_i(k_0 - 2) = \beta_1 + \beta_2 \quad ; \quad y_i(k_0 - 1) = \beta_1 \lambda_1 + \beta_2 \overline{\lambda}_1$$

which gives

$$\beta_1 = \frac{\overline{\lambda}_1 y_i(k_0 - 2) - y_i(k_0 - 1)}{\overline{\lambda}_1 - \lambda_1} \quad ; \quad \beta_2 = \frac{\lambda_1 y_i(k_0 - 2) - y_i(k_0 - 1)}{\lambda_1 - \overline{\lambda}_1}$$

Hence since $\{y_i(k_0-2), y_i(k_0-1)\}$ are both real, and $\overline{\lambda}_1 - \lambda_1 = -j2\lambda_{12} \neq 0$ is purely complex, it follows that

$$\beta_2 = \overline{\beta}_1 \tag{2.68}$$

Therefore

$$y_i(k) = \beta_1 \lambda_1^{k-k_0+2} + \overline{\beta}_1 \overline{\lambda}_1^{k-k_0+2} \quad ; \quad k \geq k_0 - 2 \tag{2.69}$$

If we now define $\{r, \theta\}$ and $\{\varrho, \phi\}$ in terms of λ_1 and β_1 by

$$r exp\{j\theta\} \triangleq \lambda_1 \quad ; \quad \varrho exp\{j\phi\} \triangleq \beta_1 \tag{2.70}$$

we have from (2.69) that

$$y_i(k) = 2\varrho r^{k-k_0+2} \cos[(k - k_0 + 2)\theta + \phi] \quad ; \quad k \geq k_0 - 2 \tag{2.71}$$

Example 2.2.4. Given the initial conditions $\{y(0), y(1)\}$, consider the solution of the the second order difference equation

$$y(k+2) = 2\cos\Omega_0 y(k+1) - y(k)$$

The characteristic equation is given by

$$\lambda^2 = 2\cos\Omega_0 \lambda - 1$$

and so the characteristic roots $\{\lambda_1, \lambda_2\}$ are given by

$$\lambda_1 = \overline{\lambda}_2 = \cos\Omega_0 + j\sin\Omega_0 = exp\{j\Omega_0\}$$

Hence from (2.70) and (2.71) with $k_0 = 2$, we have $\{r = 1,\ \theta = \Omega_0\}$, and so the zero input response y_i is given by

$$y_i(k) = 2\varrho\cos(k\Omega_0 + \phi) \quad ; \quad k \geq 0$$

where the parameters $\{\varrho, \phi\}$ are given from the initial conditions

$$y_i(0) = 2\varrho\cos\phi \quad ; \quad y_i(1) = 2\varrho\cos(\Omega_0 + \phi)$$

Equivalently, the zero input response y_i can be written in the form

$$y_i(k) = \alpha\cos k\Omega_0 + \beta\sin k\Omega_0$$

where

$$y_i(0) = \alpha \quad ; \quad y_i(1) = \alpha\cos\Omega_0 + \beta\sin\Omega_0$$

Example 2.2.5. Consider the second order difference equation

$$y(k+2) = y(k+1) - y(k) + 3u(k) \quad ; \quad k \geq 2$$

when $y(2) = 0.2,\ y(3) = -0.3$. The zero input response y_i is given by

$$y_i(k+2) = y_i(k+1) - y_i(k) \quad ; \quad k \geq 2$$

with $y_i(2) = 0.2,\ y_i(3) = -0.3$.

The characteristic equation is $\lambda^2 = \lambda - 1$, and so the characteristic roots are $\lambda_1 = 0.5 + j0.866 = \overline{\lambda}_2$. Now

$$\lambda_1 = 0.5 + j0.866 = exp\{\frac{j\pi}{3}\}$$

which corresponds in (2.70) to $r = 1$ and $\theta = \pi/3$.

From (2.71), the zero input response is then of the form

$$y_i(k) = 2\varrho\cos[(k-2)\frac{\pi}{3} + \phi] \quad ; \quad k \geq 2$$

Now

$$y_i(2) = 2\varrho\cos\phi = 0.2 \quad ; \quad y_i(3) = 2\varrho\cos\left(\phi + \frac{\pi}{3}\right) = -0.3$$

and so

$$\frac{\cos(\phi + \pi/3)}{\cos\phi} = \frac{-0.3}{0.2} = -1.5$$

which implies $0.5 - 0.5\sqrt{3}\tan\phi = -1.5$. That is,

$$\tan\phi = \frac{4}{\sqrt{3}} \text{ or } \phi = 1.16\ rads = 0.37\pi\ rads$$

Then (to 2 dec. places)

$$2\varrho = \frac{y_i(2)}{\cos 0.37\pi} = 0.50$$

and so finally the zero input response y_i for $k \geq 2$ is given by

$$y_i(k) = 0.50 \cos[(k-2)\frac{\pi}{3} + 0.37\pi]$$

Stability

Let y_i be the zero input response of a difference equation. Then we say that the difference equation is:

1. *Unstable* if for *any* initial conditions $\{y(k_0 - 1), y(k_0)\}$, $|y_i(k)| \to \infty$ as $k \to \infty$.
2. *Stable* if for *all* initial conditions $\{y(k_0 - 1), y(k_0)\}$, $|y_i(k)|$ remains finite as $k \to \infty$.
3. *Asymptotically stable* if for *all* initial conditions $\{y(k_0-1), y(k_0)\}$, $|y_i(k)| \to 0$ as $k \to \infty$

We have the following result.

Theorem 2.2.1. *Consider the algebraic equation*

$$\lambda^2 = a_1\lambda + a_0 \tag{2.72}$$

Then the second order difference equation

$$y(k+2) = a_1 y(k+1) + a_0 y(k) + b_2 u(k+2) + b_1 u(k+1) + b_0 u(k)$$

is:

1. *Unstable if* $|\lambda_1| > 1$ *or* $|\lambda_2| > 1$, *and when* $|\lambda_1| \geq 1$ *if* $\lambda_1 = \lambda_2$.
2. *Stable if* $|\lambda_1| \leq 1$ *and* $|\lambda_2| \leq 1$ *provided when* $\lambda_1 = \lambda_2$, *then* $|\lambda_1| < 1$
3. *Asymptotically stable if* $|\lambda_1| < 1$ *and* $|\lambda_2| < 1$

Furthermore, condition 3 is satisfied if and only if

$$|a_0| < 1 \quad ; \quad |a_1| < 1 - a_0 \tag{2.73}$$

The proof this claim follows from (2.59), (2.64) and (2.69) and is left to an exercise.

The region in the complex plane where the roots $\{\lambda_1,\ \lambda_2\}$ must occur for stability is illustrated in Figure 2.3(a). The region in $\{a_0,\ a_1\}$ parameter space defining the necessary and sufficient conditions (2.73) for asymptotic stability is illustrated in Figure 2.3(b).

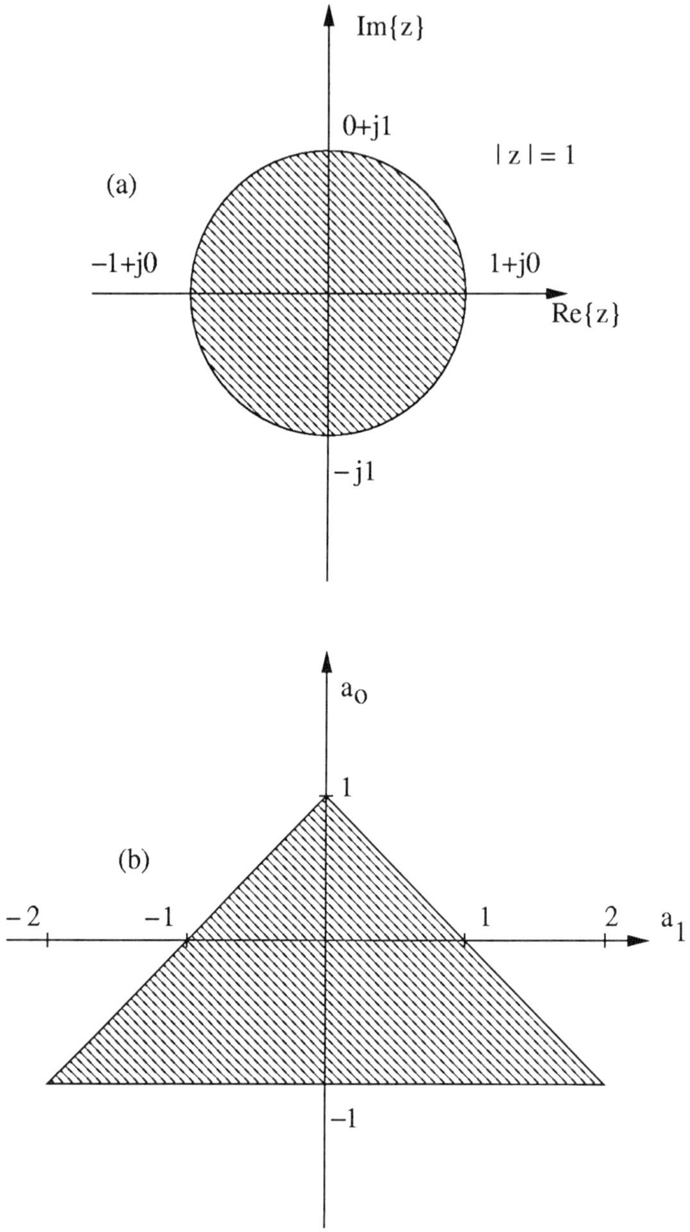

Fig. 2.3. Stability regions for second order difference equation

Example 2.2.6. (i) Consider the second order difference equation

$$2y(k+2) = y(k+1) - 1.6y(k) + u(k+1)$$

The characteristic equation is given by

$$2\lambda^2 = \lambda - 1.6$$

which in (2.72) corresponds to $\{a_1 = 0.5; a_0 = -0.8\}$. Then

$$|a_0| = 0.8 \;\; ; \;\; |a_1| = 0.5 \;\; ; \;\; 1 - a_0 = 1.8$$

and so by (2.73), this difference equation is asymptotically stable. (This conclusion can also be verified by determining the magnitudes of the characteristic roots.)

(ii) For the second order difference equation

$$y(k+2) = \alpha y(k+1) + 0.2y(k) + u(k+1) - 3u(k+1)$$

the characteristic equation is given by

$$\lambda^2 - \alpha\lambda - 0.2 = 0$$

Then $\{a_1 = \alpha, a_0 = 0.2\}$ implies

$$|a_0| = 0.2 \;\; ; \;\; |a_1| = |\alpha| \;\; ; \;\; 1 - a_0 = 0.8$$

and so by (2.73), this difference equation is asymptotically stable if and only if $|\alpha| < 0.8$. (This conclusion can also be verified by determining the magnitudes of the characteristic roots.)

2.2.2 Unit Impulse Response

By definition, a forced response y_f of the second order difference equation (2.41) is the response due to a given input signal u subject to *zero initial conditions*. In this section, we consider a particular forced response known as the *unit impulse response* which is defined when $\{y_f(-2) = y_f(-1) = 0\}$, and the input signal u is a unit impulse applied at $k = 0$; that is,

$$u(k) = \begin{cases} 1 \; ; \; k = 0 \\ 0 \; ; \; k \neq 0 \end{cases} \tag{2.74}$$

(We will see later in Section 2.4 that the importance of the unit impulse response lies in the fact that the forced response due to an *arbitrary* input u can be expressed in terms of a *linear convolution* of the unit impulse response with u.)

The *unit impulse response* $\{h(n) \; ; \; n \geq 0\}$ of the second order difference equation (2.41) is given for $k \geq -2$ by

$$\begin{aligned} h(k+2) &= a_1 h(k+1) + a_0 h(k) + b_0 u(k) \\ &\quad + b_1 u(k+1) + b_2 u(k+2) \end{aligned} \tag{2.75}$$

where $h(-2) = h(-1) = 0$ with u given by (2.74). This implies

$$h(0) = a_1 h(-1) + a_0 h(-2) + b_0 u(-2) + b_1 u(-1) + b_2 u(0) = b_2$$

$$h(1) = a_1 h(0) + a_0 h(-1) + b_0 u(-1) + b_1 u(0) + b_2 u(1) = a_1 b_2 + b_1$$

and

$$\begin{aligned} h(2) &= a_1 h(1) + a_0 h(0) + b_0 u(0) + b_1 u(1) + b_2 u(2) \\ &= a_1(a_1 b_2 + b_1) + a_0 b_2 + b_0 \end{aligned}$$

Then

$$h(3) = a_1 h(2) + a_0 h(1) + b_0 u(1) + b_1 u(2) + b_2 u(3) = a_1 h(2) + a_0 h(1)$$

and generally, since $u(n) = 0$ for all n except $n = 0$,

$$h(k+2) = a_1 h(k+1) + a_0 h(k) \quad ; \quad k \geq 1 \tag{2.76}$$

That is, for $k \geq 3$, the unit impulse response h is a zero input response with respect to a particular set of initial conditions $\{h(1), h(2)\}$ determined by the coefficients of the difference equation.

From section 2.2.1, based on the property of the roots $\{\lambda_1, \; \lambda_2\}$ of the characteristic equation (2.57), we deduce the following results.

Theorem 2.2.2. *The impulse response h of the second order linear time invariant difference equation (2.41) is given by:*

$$h(0) = b_2 \; , \quad h(1) = a_1 b_2 + b_1 \; , \quad h(2) = a_1^2 b_2 + a_1 b_1 + a_0 b_2 + b_0$$
$$h(k+2) = a_1 h(k+1) + a_0 h(k) \quad ; \quad k \geq 1$$

In particular:

(i) For $\lambda_1 \neq \lambda_2 \neq 0$ both real

$$h(n) = \beta_1 \lambda_1^{n-1} + \beta_2 \lambda_2^{n-1} \quad , \quad n \geq 1 \tag{2.77}$$

where

$$\beta_1 = \frac{\lambda_2 h(1) - h(2)}{\lambda_2 - \lambda_1} \quad ; \quad \beta_2 = \frac{\lambda_1 h(1) - h(2)}{\lambda_1 - \lambda_2} \tag{2.78}$$

(ii) For $\lambda_1 = \lambda_2 \neq 0$

$$h(n) = [\beta_1 + \beta_2(n-1)]\lambda_1^{n-1} \quad , \quad n \geq 1 \tag{2.79}$$

where

$$\beta_1 = h(1) \quad ; \quad \beta_2 = \frac{h(2)}{\lambda_1} - \beta_1 \tag{2.80}$$

(iii) For $\lambda_1 = \overline{\lambda}_2 = r_1 exp\{j\theta\}$ *complex*

$$h(n) = 2\varrho r^{n-1}\cos[(n-1)\theta + \phi] \ , \ n \geq 1 \tag{2.81}$$

where

$$2\varrho\cos\phi = h(1) \quad ; \quad 2\varrho r\cos(\theta + \phi) = h(2) \tag{2.82}$$

Example 2.2.7. Consider the second order difference equation

$$y(k+2) = -0.1y(k+1) + 0.3y(k) + u(k+1) - 2u(k) + 4u(k+2) \tag{2.83}$$

The unit impulse response h is given by $h(-2) = h(-1) = 0$ with u given by (2.74); that is

$$\begin{aligned} h(0) &= -0.1h(-1) + 0.3h(-2) + u(-1) - 2u(-2) + 4u(0) = 4 \\ h(1) &= -0.1h(0) + 0.3h(-1) + u(0) - 2u(-1) + 4u(1) = 0.6 \\ h(2) &= -0.1h(1) + 0.3h(0) + u(1) - 2u(0) + 4u(2) = -0.86 \end{aligned}$$

That is, since $u(k) = 0$ for $k \neq 0$, we have

$$\begin{aligned} h(0) = 4 \quad ; \quad h(1) &= 0.6 \quad ; \quad h(2) = -0.86 \qquad (2.84) \\ h(k+2) &= -0.1h(k+1) + 0.3h(k) \quad ; \quad k \geq 1 \end{aligned}$$

Note that $\{h(0), h(1), h(2)\}$ are calculated from first principles without recourse to the general formula in Theorem 2.2.2.

Now the characteristic equation associated with (2.84) is given by

$$\lambda^2 = -0.1\lambda + 0.3$$

and so the characteristic roots are given by $\{\lambda_1 = 0.5 \ ; \ \lambda_2 = -0.6\}$. The zero input response of (2.84) is then of the form

$$h(k) = \beta_1(0.5)^{k-1} + \beta_2(-0.6)^{k-1} \quad ; \quad k \geq 1$$

Now $h(1) = 0.6$ and $h(2) = -0.86$ imply

$$0.6 = \beta_1 + \beta_2 \quad ; \quad -0.86 = \beta_1(0.5) + \beta_2(-0.6)$$

so that (to 2 dec. places) $\{\beta_1 = -0.45; \ \beta_2 = 1.05\}$. Hence the unit impulse response of (2.83) is given by

$$h(0) = 4 \quad ; \quad h(k) = -0.45(0.5)^{k-1} + 1.05(-0.6)^{k-1} \quad , \quad k \geq 1$$

Example 2.2.8. Consider the second order difference equation

$$y(k+2) = 0.5y(k+1) - 0.25y(k) + u(k+1) \tag{2.85}$$

The unit impulse response h is given by $h(-2) = h(-1) = 0$ with u is the unit impulse signal; that is

$$\begin{aligned} h(0) &= 0.5h(-1) - 0.25h(-2) + u(-1) = 0 \\ h(1) &= 0.5h(0) - 0.25h(-1) + u(0) = 1 \\ h(2) &= 0.5h(1) - 0.25h(0) + u(1) = 0.5 \end{aligned}$$

That is, the unit impulse response h is given by

$$\begin{aligned} h(0) = 0 \;\; ; \;\; h(1) = 1.0 \;\; ; \;\; h(2) = 0.5 \\ h(k+2) = 0.5h(k+1) - 0.25h(k) \;\; ; \;\; k \geq 1 \end{aligned} \tag{2.86}$$

The characteristic equation associated with (2.86) is given by

$$\lambda^2 = 0.5\lambda - 0.25$$

and so the characteristic roots $\{\lambda_1, \lambda_2\}$ are

$$\lambda_1 = \overline{\lambda}_2 = \frac{1}{4} + j\frac{\sqrt{3}}{4} = 0.5exp\{j\frac{\pi}{3}\} = 0.5exp\{j1.05\}$$

The zero input response of (2.86) is therefore of the form

$$h(k) = 2\varrho(0.5)^{k-1}\cos[(k-1)\frac{\pi}{3} + \phi] \;\; ; \;\; k \geq 1$$

Now $h(1) = 1$ and $h(2) = 0.5$ imply

$$1 = 2\varrho\cos\phi \;\; ; \;\; 0.5 = 2\varrho(0.5)\cos[\frac{\pi}{3} + \phi]$$

which (to 2 dec. places) gives $\{\varrho = 0.58; \; \phi = -0.53\}$. Hence the unit impulse response of (2.85) is given by

$$h(0) = 0 \;\; ; \;\; h(k) = 1.16(0.5)^{k-1}\cos[(k-1)1.05 - 0.53] \;\; , \;\; k \geq 1$$

2.3 High Order

Consider the digital signal y which is given by the solution of the causal nth order linear time invariant difference equation

$$y(k+n)=\sum_{p=0}^{n-1} a_p y(k+p)+\sum_{m=0}^{n} b_m u(k+m) \quad ; \quad k \geq k_0-n \tag{2.87}$$

where $\{a_p, b_m\}$ for all integers p, m are real constants with $a_0 \neq 0$, k_0 is a fixed integer, and k is a variable integer. The signal y is uniquely determined by the n initial conditions $\{y(k_0-n), y(k_0-n+1), \ldots, y(k_0-1)\}$ and the input signal u.

Following arguments similar to those in section 2.2, we conclude that (2.87) is *causal* since the output $y(k+n)$ at time $k+n$ depends only on previous values of the output $\{y(m); m<k+n\}$, and previous and current values $\{u(m); m \leq k+n\}$ of the input. The *time invariance* property of (2.87) is a consequence of the fact that all the coefficients $\{a_p, b_m\}$ are constant, and so do not depend on time k. *Linearity* is a consequence of the fact that $y(k+n)$ is a linear function of $\{y(k+p); 0 \leq p \leq n-1\}$ and $\{u(k+m); 0 \leq m \leq n\}$.

Equation (2.87) is also of order n if $a_0 \neq 0$ since in this case, given the input signal u, precisely the n initial conditions $\{y(k_0-n), y(k_0-n+1), \ldots, y(k_0-1)\}$ are required in order to determine all values of $\{y(k) \; ; \; k \geq k_0\}$. However if $a_0=0$ but $a_1 \neq 0$, then only the $n-1$ initial conditions $\{y(k_0-n+1), \ldots, y(k_0-2)\}$ are required, in which case, (2.87) is of order $n-1$. Similarly, if $\{a_p=0; 0 \leq p \leq r<n-1\}$ and $a_{r+1} \neq 0$, then (2.87) is of order $n-r-1$.

2.3.1 Zero Input Response

The zero input response y_i is given by

$$y_i(k+n)=\sum_{p=0}^{n-1} a_p y_i(k+p) \quad ; \quad k \geq k_0-n \tag{2.88}$$

subject to the n initial conditions $\{y_i(k_0-n+p)=y(k_0-n+p); \; 0 \leq p \leq n-1\}$.

Following the approach in section 2.2.1, we assume y_i is of the form

$$y_i(k)=\beta \lambda^{k-k_0+n} \; ; \; k \geq k_0-n \tag{2.89}$$

Then after substituting into (2.88), we conclude that (2.89) is a solution if

$$\beta(\lambda^n-\sum_{p=0}^{n-1} a_p \lambda^p) \lambda^{k-k_0+n}=0$$

That is, if $y_i \neq 0$ in (2.89) is to be a solution, then $\beta \neq 0$ and λ must satisfy the nth order algebraic equation

$$\lambda^n - \sum_{p=0}^{n-1} a_p \lambda^p = 0 \tag{2.90}$$

Equation (2.90) is referred to as the *characteristic equation* of the difference equation (2.87).

The characteristic equation has n roots (called the *characteristic roots*), and since all the coefficients $\{a_p, b_q\}$ are assumed to be *real*, all characteristic roots occur in *complex conjugate* pairs. That is, if $\lambda_1 = \varrho + j\omega$ is a complex root of (2.90), then so too is $\overline{\lambda}_1 = \varrho - j\omega$.

Distinct Characteristic Roots

Suppose *all* roots $\{\lambda_i; 1 \leq i \leq n\}$ of the characteristic equation (2.90) are *distinct* (and possible complex). Then we claim that the solution of (2.88) is given by

$$y_i(k) = \sum_{i=1}^{n} \beta_i \lambda_i^{k-k_0+n} \quad ; \quad k \geq k_0 - n \tag{2.91}$$

for some uniquely defined set of coefficients $\{\beta_i; 1 \leq i \leq n\}$ which are determined by the initial conditions $\{y_i(k_0 - n), y_i(k_0 - n + 1), \ldots, y_i(k_0 - 1)\}$. After substitution into (2.91), it follows that the coefficients $\{\beta_i\}$ are given by the solution of the algebraic equations

$$\boldsymbol{F}\boldsymbol{\beta} = \boldsymbol{z} \tag{2.92}$$

where the n-square matrix $\boldsymbol{F}$ and the n-vectors $\{\boldsymbol{\beta}, \boldsymbol{z}\}$ are given by

$$\boldsymbol{F} = \begin{bmatrix} 1 & 1 & \cdots & 1 \\ \lambda_1 & \lambda_2 & \cdots & \lambda_n \\ \lambda_1^2 & \lambda_2^2 & \cdots & \lambda_n^2 \\ \cdot & \cdot & \cdots & \cdot \\ \cdot & \cdot & \cdots & \cdot \\ \lambda_1^{n-1} & \lambda_2^{n-1} & \cdots & \lambda_n^{n-1} \end{bmatrix} \quad ; \quad \boldsymbol{z} = \begin{bmatrix} y_i(k_0 - n) \\ y_i(k_0 - n + 1) \\ y_i(k_0 - n + 2) \\ \cdot \\ \cdot \\ y_i(k_0 - 1) \end{bmatrix} \tag{2.93}$$

$$\boldsymbol{\beta}^T = [\beta_1\ \beta_2\ \beta_3 \cdot\cdot\ \beta_n]$$

Now any matrix $\boldsymbol{F}$ of the form (2.93) where $\lambda_k \neq \lambda_m$ for all $\{k, m\}$ is known as a *Vandermonde matrix*. In particular, it can be shown that: $det(\boldsymbol{F}) = \lambda_2 - \lambda_1$ when $n = 2$, and

$$det(\boldsymbol{F}) = (-1)^n (\lambda_n - \lambda_1) \prod_{i=1}^{n-1} (\lambda_{i+1} - \lambda_i) \quad ; \quad n \geq 3$$

and so for *distinct* characteristic roots λ_i, $\boldsymbol{F}$ in (2.93) is invertible. We therefore conclude that (2.93) has a unique solution $\boldsymbol{\beta}$.

Note that (2.92) can be solved using standard linear algebra packages (such as MATLAB) even when some (or all) of the λ_i are complex. Furthermore $\lambda_2 = \overline{\lambda}_1$ implies $\beta_2 = \overline{\beta}_1$, so that when

$$\lambda_2 = \varrho exp\{j\theta\} \ ; \ \beta_2 = rexp\{j\phi\}$$

we have

$$\beta_2 \lambda_2^{k-k_0+n} + \overline{\beta}_2 \overline{\lambda}_2^{k-k_0+n} = 2r\varrho^{k-k_0+n} \cos[(k-k_0+n)\theta + \phi]$$

That is, the expression for the zero input response y_i in (2.91) when some or all of the characteristic roots λ_i are complex can also be written in *real* form in terms of cosine functions.

Repeated Characteristic Roots

If some roots $\{\lambda_i\}$ are *not* distinct, then (2.92) does *not* have a solution $\boldsymbol{\beta}$ for all vectors $\boldsymbol{z}$ of initial conditions which means that (2.91) cannot now be a solution.

To see what happens when roots are *repeated*, suppose (say) $\lambda_1 = \lambda_2 \neq 0$ is a once repeated root, but no other roots $\{\lambda_3, \lambda_4, ..., \lambda_n\}$ are repeated. Then we claim that the zero input solution y_i has the general form: for $k \geq k_0 - n$

$$y_i(k) = \beta_1 \lambda_1^{k-k_0+n} + (k-k_0+n)\beta_2 \lambda_1^{k-k_0+n} + \sum_{i=3}^{n} \beta_i \lambda_i^{k-k_0+n} \qquad (2.94)$$

In order to establish this result, we need to show that for any set of initial conditions $\{y_i(k_0-n), y_i(k_0-n+1), ..., y_i(k_0-1)$, there exists a unique set of coefficients $\{\beta_i : 1 \leq i \leq n\}$ in (2.94). Now from (2.94), we have the system of algebraic equations

$$\boldsymbol{F}_1 \boldsymbol{\beta} = \boldsymbol{z} \qquad (2.95)$$

where the n-vectors $\{\boldsymbol{\beta}, \boldsymbol{z}\}$ are defined in (2.93), and the n-square matrix $\boldsymbol{F}_1$ is given by

$$\boldsymbol{F}_1 = \begin{bmatrix} 1 & 0 & 1 & \cdot\cdot & 1 \\ \lambda_1 & \lambda_1 & \lambda_3 & \cdot\cdot & \lambda_n \\ \lambda_1^2 & 2\lambda_1^2 & \lambda_3^2 & \cdot\cdot & \lambda_n^2 \\ \cdot & \cdot & \cdot & \cdot\cdot & \cdot \\ \cdot & \cdot & \cdot & \cdot\cdot & \cdot \\ \cdot & \cdot & \cdot & \cdot\cdot & \cdot \\ \lambda_1^{n-1} & (n-1)\lambda_1^{n-1} & \lambda_3^{n-1} & \cdot\cdot & \lambda_n^{n-1} \end{bmatrix} \qquad (2.96)$$

It also can be shown that $\boldsymbol{F}_1$ is invertible if and only if $\lambda_k \neq \lambda_m$ for all $\{k, m\}$.

More generally, if $\lambda_1 = \lambda_2 = \ldots = \lambda_m$ is a mth repeated root, but no other roots $\{\lambda_{m+1}, \lambda_{m+2}, ..., \lambda_n\}$ are repeated, then it can be shown that the zero input solution y_i has the general form: for $k \geq k_0 - n$

$$y_i(k) = \sum_{i=1}^{m} \beta_i (k - k_0 + n)^{i-1} \lambda_1^{k-k_0+n} + \sum_{i=m+1}^{n} \beta_i \lambda_i^{k-k_0+n} \tag{2.97}$$

Once again the correctness of this form for the solution can be confirmed by showing that the existence of this form is equivalent to being able to solve a system of algebraic equations $\boldsymbol{F}_m \boldsymbol{\beta} = \boldsymbol{z}$ for some invertible matrix $\boldsymbol{F}_m$.

Similarly, if there are two distinct roots that are repeated; that is, if

$$\lambda_1 = \lambda_2 = \ldots = \lambda_m \;\; ; \;\; \lambda_{m+1} = \lambda_{m+2} = \ldots = \lambda_{m+p} \;\; (\lambda_m \neq \lambda_{m+1})$$

and

$$\lambda_{m+p+1} \neq \lambda_{m+p+2} \neq \ldots \neq \lambda_n$$

then for $k \geq k_0 - n$

$$y_i(k) = \sum_{i=1}^{m} \beta_i (k - k_0 + n)^{i-1} \lambda_1^{k-k_0+n} + \sum_{i=1}^{p} \beta_i (k - k_0 + n)^{i-1} \lambda_{m+1}^{k-k_0+n} + \sum_{i=m+p+1}^{n} \beta_i \lambda_i^{k-k_0+n} \tag{2.98}$$

Unit Impulse Response

The unit impulse response h of (2.87) is given by the solution of (2.87) subject to the n initial conditions

$$h(-n) = h(-n+1) = \ldots = h(-2) = h(-1) = 0$$

when the input signal u is the unit impulse given in (2.74). In particular, after substitution in (2.87), it follows that h is given by:

(i) $h(k) = 0$ for $k < 0$; $h(0) = b_n$
(ii) $\{h(k); 1 \leq k \leq n\}$ by the solution of the system of linear algebraic equations

$$\begin{aligned} h(1) &= a_{n-1} h(0) + b_{n-1} \\ h(2) &= a_{n-1} h(1) + a_{n-2} h(0) + b_{n-2} \\ &\;\cdot \\ &\;\cdot \\ h(n) &= a_{n-1} h(n-1) + a_{n-2} h(n-2) + \ldots + a_0 h(0) + b_0 \end{aligned}$$

or equivalently by the solution of the algebraic system of equations:

$$\boldsymbol{G}\boldsymbol{x} = \boldsymbol{d} \tag{2.99}$$

where the n-square matrix $\boldsymbol{G}$ and the n-vectors $\{\boldsymbol{x}, \boldsymbol{d}\}$ are given by

$$\boldsymbol{G} = \begin{bmatrix} 1 & 0 & 0 \cdot & \cdot & 0 \\ -a_{n-1} & 1 & 0 \cdot & \cdot & 0 \\ -a_{n-2} & -a_{n-1} & 1 \cdot & \cdot & 0 \\ \cdot & \cdot & \cdot\ \cdot & \cdot & \cdot \\ \cdot & \cdot & \cdot\ \cdot & \cdot & \cdot \\ \cdot & \cdot & \cdot\ \cdot & \cdot & \cdot \\ -a_1 & -a_2 & \cdot\ \cdot & -a_{n-1} & 1 \end{bmatrix} \tag{2.100}$$

$$\boldsymbol{x} = \begin{bmatrix} h(1) \\ h(2) \\ h(3) \\ \cdot \\ \cdot \\ \cdot \\ h(n) \end{bmatrix} \;;\; \boldsymbol{d} = \begin{bmatrix} b_{n-1} + a_{n-1}b_n \\ b_{n-2} + a_{n-2}b_n \\ b_{n-3} + a_{n-3}b_n \\ \cdot \\ \cdot \\ \cdot \\ b_0 + a_0 b_n \end{bmatrix}$$

and (iii) $\{h(k); k \geq n+1\}$ is given by the initial condition response

$$h(k+n) = \sum_{p=0}^{n-1} a_p h(k+p) \;\;;\;\; k \geq 1 \tag{2.101}$$

subject to the initial conditions $\{h(1), \ldots, h(n)\}$ given in (ii) above.

Since $det(\boldsymbol{G}) = 1$, (2.99) always has a unique solution $\boldsymbol{x}$ for all coefficients $\{a_p, b_m\}$ which then gives a unique solution to (2.101) as earlier described. For example, when the characteristic equation (2.90) has n *distinct roots*, the impulse response h is of the form

$$h(k) = \sum_{i=1}^{n} \beta_i \lambda_i^{k-1} \;\;;\;\; k \geq 1 \tag{2.102}$$

In particular, from (2.92) and (2.99) with $k_0 = n+1$ and $y_i(p) = h(p)$, $\boldsymbol{x} = \boldsymbol{z}$ and so $\boldsymbol{\beta}^T = [\beta_1 \; \beta_2 \; \ldots \; \beta_n]$ is given by

$$\boldsymbol{\beta} = \boldsymbol{F}^{-1}\boldsymbol{G}^{-1}\boldsymbol{d}$$

where $\boldsymbol{F}$ is given by (2.93), and $\{\boldsymbol{G}, \boldsymbol{d}\}$ by (2.100).

Example 2.3.1. Consider the third order difference equation

$$\begin{aligned} y(k+3) = &-0.1y(k+2) + 0.07y(k+1) - 0.65y(k) \\ &+2u(k) - u(k+1) - 0.3u(k+2) + 0.5u(k+3) \end{aligned}$$

Then substituting $\{y(-3) = y(-2) = y(-1) = 0\}$, and using the definition of the unit impulse signal (2.74), it follows that the unit impulse response $\{h(n); n \leq 3\}$ is given by

$$\begin{aligned}
h(k) &= 0 \; ; \quad k < 0 \\
h(0) &= 0.5 \\
h(1) &= -0.1h(0) - 0.3 \\
h(2) &= -0.1h(1) + 0.07h(0) - 1 \\
h(3) &= -0.1h(2) + 0.07h(1) - 0.65h(0) + 2
\end{aligned}$$

or equivalently, $h(0) = 0.5$ with

$$\begin{bmatrix} 1 & 0 & 0 \\ 0.1 & 1 & 0 \\ -0.07 & 0.1 & 1 \end{bmatrix} \begin{bmatrix} h(1) \\ h(2) \\ h(3) \end{bmatrix} = \begin{bmatrix} -0.3 - 0.05 \\ -1.0 + 0.035 \\ 2.0 - 0.325 \end{bmatrix}$$

Solution of this set of linear algebraic equations gives $\{h(1) = -0.35;\ h(2) = -0.93;\ h(3) = 1.74\}$. The unit impulse response $\{h(n); n \geq 1\}$ is given by the zero input response

$$h(k+3) = -0.1h(k+2) + 0.07h(k+1) - 0.65h(k) \; ; \quad k \geq 1$$

for which the corresponding characteristic equation is

$$\lambda^3 = -0.1\lambda^2 + 0.07\lambda - 0.65$$

with corresponding characteristic roots $\{\lambda_1, \lambda_2, \lambda_3\}$

$$\lambda_1 = -0.5 \; ; \; \lambda_2 = 0.2 + j0.3 = \bar{\lambda}_3$$

Hence since the characteristic roots are all distinct, the unit impulse response h is given by

$$h(k) = \sum_{i=1}^{3} \beta_i \lambda_i^{k-1} \; ; \quad k \geq 1 \tag{2.103}$$

where after substituting for $\{h(1), h(2), h(3)\}$, we have

$$\begin{bmatrix} 1 & 1 & 1 \\ \lambda_1 & \lambda_2 & \bar{\lambda}_2 \\ \lambda_1^2 & \lambda_2^2 & (\bar{\lambda}_2)^2 \end{bmatrix} \begin{bmatrix} \beta_1 \\ \beta_2 \\ \beta_3 \end{bmatrix} = \begin{bmatrix} h(1) \\ h(2) \\ h(3) \end{bmatrix} = \begin{bmatrix} -0.35 \\ -0.93 \\ 1.74 \end{bmatrix}$$

Solution of this algebraic systems of equations using (say) MATLAB gives $\{\beta_1 = 3.48;\ \beta_2 = \bar{\beta}_3 = -1.91 - j2.56\}$. Now

$$r exp\{j\phi\} = -1.91 - j2.56 \quad ; \quad \varrho exp\{j\theta\} = 0.2 + j0.3$$

implies $\{r = 3.19; \phi = 4.07 \; rads\}$ and $\{\varrho = 0.36, \theta = 0.98 \; rads\}$. Hence h in (2.103) can be expressed in the real form

$$h(n) = 3.48(-0.5)^{n-1} + 6.38(0.36)^{n-1} \cos[(n-1)0.98 + 4.07] \; ; \quad n \geq 1$$

2.3.2 Stability Test

The zero input response y_i of the difference equation (2.87) is given by (2.88). Furthermore, given that the characteristic equation (2.90) has the n characteristic roots $\{\lambda_k\}$, all terms in y_i are of the form

$$\lambda_m^{k-k_0+n} \sum_{i=1}^{r} \beta_i (k - k_0 + n)^{i-1}$$

where the characteristic root λ_m is repeated r times.

We say that the difference equation (2.87) is asymptotically stable if: for all initial conditions $\{y_i(k_0 - n), y_i(k_0 - n + 1), \ldots, y_i(k_0 - 1)\}$, the zero input response y_i satisfies $y_i(k) \to 0$ as $k \to \infty$.

Now for all r

$$\lim_{k \to \infty} \lambda_m^{k-k_0+n} \sum_{i=1}^{r} \beta_i (k - k_0 + n)^{i-1} = 0$$

if and only if $|\lambda_m| < 1$ for all m. We therefore have the following result.

Theorem 2.3.1. *The zero input response (2.88) is asymptotically stable if and only if all roots λ_m of the characteristic equation (2.90) satisfy*

$$|\lambda_m| < 1$$

The result in Theorem 2.3.1 is unsatisfactory in the sense that one must first find all the n roots $\{\lambda_m\}$ of the characteristic equation (2.90). A computationally simpler test known as the *Marden stability test* which is based only on the *coefficients* $\{a_p\}$ of the characteristic equation is also available. [See M. Marden, *The geometry of the zeros of a plynomial in a complex variable*, Amer. Math. Soc., New York, 1949.]

Theorem 2.3.2. *Consider a polynomial*

$$\psi(\lambda) = \lambda^n - \sum_{p=0}^{n-1} a_p \lambda^p$$

and define the n parameters $\{\gamma_m\}$ for $m = n, n-1, n-2, \ldots, 1$ by

$$\gamma_m = \frac{a_0^m}{a_m^m}$$

where $a_p^n = a_p$ for $0 \le p \le n-1$, and the coefficients $\{a_{n-j}^{n-p-1}\}$ for $0 \le p \le n-2$ and $p+1 \le j \le n$ are defined by:

$$a_{n-j}^{n-p-1} = a_{n-p}^{n-p}(a_{n-j+1}^{n-p}) - a_0^{n-p}(a_{j-p-1}^{n-p}) \quad ; \quad 1 \le j \le n-p$$

Then $\psi(\lambda) = 0$ has all roots λ_i such that $|\lambda_i| < 1$ if and only if $|\gamma_m| < 1$ for $m = n, n-1, ..., 1$.

Furthermore, if $|\gamma_m| \neq 1$ for all m, then no roots of $\psi(\lambda) = 0$ occur on the unit circle $|\lambda| = 1$, and p of the n roots are inside the unit circle where p equals the number of products

$$(1-\gamma_1^2)(1-\gamma_2^2)...(1-\gamma_k^2)$$

for $k = 1, 2, ..., n$ which are positive.

The parameters $\{\gamma_m\}$ can be conveniently found by means of Table 2.1.

Table 2.1. Marden stability table

a_n^n	a_{n-1}^n	.	.	.	a_1^n	a_0^n	$\gamma_n \triangleq \frac{a_0^n}{a_n^n}$
a_0^n	a_1^n	.	.	.	a_{n-1}^n	a_n^n	
a_{n-1}^{n-1}	a_{n-2}^{n-1}	.	.	.	a_0^{n-1}		$\gamma_{n-1} \triangleq \frac{a_0^{n-1}}{a_{n-1}^{n-1}}$
a_0^{n-1}	a_1^{n-1}	.	.	.	a_{n-1}^{n-1}		
a_{n-2}^{n-2}	a_{n-3}^{n-2}	.	.	a_0^{n-2}			$\gamma_{n-2} \triangleq \frac{a_0^{n-2}}{a_{n-2}^{n-2}}$
a_0^{n-2}	a_1^{n-2}	.	.	a_{n-2}^{n-2}			
.	.						
.	.						
a_1^1	a_0^1						$\gamma_1 \triangleq \frac{a_0^1}{a_1^1}$
a_0^1	a_1^1						

Example 2.3.2. (a) Consider the second order polynomial

$$\psi_2(\lambda) = \lambda^2 + a\lambda + b$$

and Table 2.2 where

$$\gamma_1 = \frac{b}{1} \quad ; \quad \gamma_2 = \frac{a(1-b)}{1-b^2} = \frac{a}{1+b}$$

Necessary and sufficient conditions for $\psi(\lambda) = 0$ to have all roots λ_i such that $|\lambda_i| < 1$ are therefore

Table 2.2. Marden stability table for second order equation

1	a	b	γ_1
b	a	1	
$1-b^2$	$a(1-b)$		γ_2
$a(1-b)$	$1-b^2$		

$$|b| < 1 \;\; ; \;\; |a| < 1 + b$$

(Note that since $|b| < 1$, then $|1 + b| = 1 + b$.)

(b) Consider the third order polynomial

$$\psi(\lambda) = \lambda^3 + a\lambda^2 + b\lambda + c$$

and Table 2.3 where

$$a_1^1 \triangleq (1-c^2)^2 - (b-ac)^2 \;\; ; \;\; a_0^1 \triangleq (a-bc)(1-c^2-b+ac)$$

$$\gamma_1 = \frac{c}{1} \;\; ; \;\; \gamma_2 = \frac{b-ac}{1-c^2} \;\; ; \;\; \gamma_3 = \frac{a_0^1}{a_1^1}$$

The equation $\psi(\lambda) = 0$ has all roots λ_i such that $|\lambda_i| < 1$ if and only if $\{|\gamma_k| < 1;\ k = 1, 2, 3\}$; that is, if and only if

$$|c| < 1 \;\; ; \;\; |b-ac| < 1-c^2 \;\; ; \;\; |a-bc| < |1-c^2+b-ac|$$

Table 2.3. Marden stability table for third order equation

1	a	b	c	γ_1
c	b	a	1	
$1-c^2$	$a-bc$	$b-ac$		γ_2
$b-ac$	$a-bc$	$1-c^2$		
a_1^1	a_0^1			γ_3
a_0^1	a_1^1			

2.4 Linear Convolution

In this section, we first show by means of the time invariance and linearity properties that the forced response y_f in (2.87) due to *any* input signal u can be expressed in terms of a *linear convolution* of the unit impulse response h and the input signal u. We later show how the linear convolution formulation can be used to develop a general expression for the steady state response of an asymptotically stable difference equation to a sinusoidal input.

2.4.1 Forced Response

To begin, consider the *delayed* unit impulse signal δ_m defined by

$$\delta_m(n) \triangleq \begin{cases} 1 \; ; n = m \\ 0 \; ; n \neq m \end{cases} \tag{2.104}$$

(Note that the unit impulse signal u in (2.74) is given by $u = \delta_0$.)

By definition, the forced response y_f due to the unit impulse input $u = \delta_0$ beginning at the zero initial conditions

$$y(-n) = y(-n+1) = \; \ldots \; y(-1) = 0$$

is given by $y_f = h$ where $h(k) = 0$ for $k < 0$.

By the *time invariance property* of the difference equation (2.87), the forced response h_m due to the input signal δ_m for $m \geq 0$ starting from the zero initial conditions

$$y(-n) = y(-n+1) = \; \ldots \; = y(-1) = 0$$

is given by

$$h_m(n) = \begin{cases} 0 & ; n \leq m \\ h(n-m) & ; n \geq m \end{cases} \tag{2.105}$$

(Note that $h_0 = h$.)

Now observe from (2.104) that any input signal u with $u(k) = 0$ for $k < 0$ can be written as a summation of delayed unit impulse signals δ_m in the form

$$u(k) = \sum_{m=0}^{k} \delta_m(k) u(m) \tag{2.106}$$

It then follows from the *linearity property* of the difference equation (2.87) and (2.105) that the forced response y_f due to the input signal u in (2.106) is given by

$$y_f(k) = \sum_{m=0}^{k} h_m(k) u(m) = \sum_{m=0}^{k} h(k-m) u(m)$$

We make the following definition.

The linear convolution $h * u$ of two signals h and u with the property $h(k) = u(k) = 0$ for $k < 0$ is defined by

$$(h * u)(k) \triangleq \sum_{m=0}^{k} h(k-m)u(m) \tag{2.107}$$

With a change of variable $p = k - m$, the linear convolution $h * u$ in (2.107) can also be written in any of the equivalent form:

$$(h * u)(k) = \sum_{p=0}^{k} u(k-p)h(p) = (u * h)(k) \tag{2.108}$$

We summarize these results as follows:

Theorem 2.4.1. *The forced response y_f of the linear time invariant difference equation (2.87) with unit impulse response h when $u(m) = 0$ for $m < 0$ is given by*

$$y_f = h * u = u * h$$

or equivalently

$$y_f(k) = \sum_{m=0}^{k} h(k-m)u(m) = \sum_{p=0}^{k} u(k-p)h(p) \tag{2.109}$$

Observe that since $h(k) = 0$ for $k < 0$, the linear convolution requires the evaluation of $k + 1$ products for each value of k. Specifically

$$\begin{aligned}
\text{For } k = 1: \quad & y_f(1) = h(0)u(1) + h(1)u(0) \\
\text{For } k = 2: \quad & y_f(2) = h(0)u(2) + h(1)u(1) + h(2)u(0) \\
\text{For } k = 3: \quad & y_f(3) = h(0)u(3) + h(1)u(2) + h(2)u(1) + h(3)u(0)
\end{aligned}$$

and more generally

$$y_f(k) = h(0)u(k) + h(1)u(k-1) + \ \ldots \ + h(k)u(0) \tag{2.110}$$

The calculation of $\{y_f(k);\ 1 \le k \le N\}$ therefore generally requires a total of S_N multiplications where

$$S_N = \sum_{k=1}^{N} (k+1) = 0.5N(N+3)$$

We will see later in section 4.2 that a more numerically efficient way is available which only requires the order of $Nlog_2(N)$ rather than the order of N^2 multiplications.

Finite Impulse Response

The difference equation (2.87) is said to have a *finite impulse response* when the unit impulse response h only has a *finite* number of nonzero values. It follows from the discussion in section 2.3.1 that (2.87) has a *finite impulse response* if and only if $a_p = 0$ for $0 \leq p \leq n-1$; that is, the difference equation is of the form

$$y(k+n) = \sum_{m=0}^{n} b_m u(k+m)$$

In this case, we then have that the unit impulse response h is given by

$$h(k) = \begin{cases} b_{n-k} \; ; \; 0 \leq k \leq n \\ 0 \quad\; ; \; k \geq n+1 \end{cases} \tag{2.111}$$

In terms of (2.108), this implies that the forced response y_f is given by

$$y_f(k) = \begin{cases} \sum_{p=0}^{k} h(p)u(k-p) \; ; \; 0 \leq k \leq n \\ \\ \sum_{p=0}^{n} h(p)u(k-p) \; ; \; k \geq n+1 \end{cases} \tag{2.112}$$

Unit Step Response

Consider now the forced response in (2.87) when the input signal is a unit step applied at $k = 0$; that is

$$u(k) = \begin{cases} 1 \; ; \; k \geq 0 \\ 0 \; ; \; k < 0 \end{cases}$$

The resulting forced response, known as the *unit step response* s, is given from (2.108) directly in terms of the impulse response h by

$$s(k) = \sum_{m=0}^{k} h(k-m) = \sum_{p=0}^{k} h(p) \tag{2.113}$$

Example 2.4.1. Consider a second order (i.e. $n = 2$) difference equation (2.41) whose characteristic roots $\{\lambda_1, \lambda_2\}$ of the characteristic equation $\lambda^2 - a_1\lambda - a_0 = 0$ are both real and unequal. Then the unit impulse response h is given by (2.77) and (2.78). Then from (2.113) the unit step response s is given by

$$s(k) = \sum_{p=0}^{k} h(p) = b_2 + \sum_{p=1}^{k} [\beta_1 \lambda_1^{p-1} + \beta_2 \lambda_2^{p-1}] \; ; \; k \geq 1$$

This summation may be simplified using the series formula

$$\sum_{p=1}^{k} \lambda^{p-1} = \begin{cases} k & ; \lambda = 1 \\ (1-\lambda^k)(1-\lambda)^{-1} & ; \lambda \neq 1 \end{cases}$$

When the second order difference equation (2.41) is *asymptotically stable* (i.e. $|\lambda_1| < 1$, $|\lambda_2| < 1$), we have that $s(k) \to \overline{s}$ as $k \to \infty$ where

$$\overline{s} \triangleq b_2 + \frac{\beta_1}{1-\lambda_1} + \frac{\beta_2}{1-\lambda_2} \tag{2.114}$$

More generally, if the difference equation (2.87) is asymptotically stable, and the input signal u is a unit step, then the response y asymptotically approaches some finite constant $\overline{s}$. Then if in (2.87), we let $y(p) = \overline{s}$ and $u(p) = 1$ for all p, we get

$$\overline{s} = (\sum_{p=0}^{n-1} a_p)\overline{s} + \sum_{m=0}^{n} b_m$$

That is

$$\overline{s} = \frac{\sum_{m=0}^{n} b_m}{1 - \sum_{p=0}^{n-1} a_p}$$

where asymptotic stability guarantees that $\overline{s}$ is finite, and so the denominator in the expression for $\overline{s}$ is nonzero. Hence for the asymptotically stable second order equation (2.87)

$$\overline{s} = \frac{b_0 + b_1 + b_2}{1 - a_1 - a_0} \tag{2.115}$$

Both expressions in (2.114) and (2.115) are correct for distinct characteristic roots $\{\lambda_1, \lambda_2\}$. However (2.115) is more easily obtained, and moreover, is correct for *all* charateristic roots. Hence if only the steady state reponse of an asymptotically stable difference equation to a unit step input is required, it is then *not* necessary to find the characteristic roots.

2.4.2 Steady State Sinusoidal Response

Consider now the forced response y_f in (2.87) when the input signal u is an exponential signal of the form

$$u(m) = \begin{cases} z^m & ; m \geq 0 \\ 0 & ; m < 0 \end{cases}$$

Then from (2.108), $y_f = h * u$ is given by

$$y_f(k) = \sum_{p=0}^{k} h(p) z^{k-p} = z^k \sum_{p=0}^{k} h(p) z^{-p} \tag{2.116}$$

When the input u is a sinusoidal signal of the form

$$u(m) = \begin{cases} \sin \omega_0 m = Im\{exp\{j\omega_0 m\}\} & ; m \geq 0 \\ 0 & ; m < 0 \end{cases} \tag{2.117}$$

the forced response y_f is given by the imaginary part of the expression in (2.116) with $z = exp\{j\omega_0\}$; that is

$$y_f(k) = Im\left\{ exp\{jk\omega_0\} \sum_{p=0}^{k} h(p) z - p \right\} \quad ; \quad z = exp\{j\omega_0\}$$

Now suppose the difference equation (2.87) is asymptotically stable so that $|\lambda_i| < 1$ for all characteristic roots λ_i of the characteristic equation (2.90). Then since h is a summation of powers of λ_i (as in (2.102) when the λ_i are distinct), it follows that: $|\lambda_i| < 1$ for all i implies

$$\sum_{p=0}^{\infty} |h(p)| < \infty \tag{2.118}$$

which in turn implies

$$\sum_{p=0}^{\infty} |h(p) z^{-p}| < \infty$$

for all z such that $|z| \leq 1$. Now define

$$H(z) \triangleq \sum_{p=0}^{\infty} h(p) z^{-p} \tag{2.119}$$

Then for an asymptotically stable difference equation (2.87), the steady state sinusoidal response y_{ss} is given by

$$y_{ss}(k) = \varrho \sin(k\omega_0 - \phi)$$

where

$$\varrho exp\{-j\phi\} = H(z)|_{z=exp\{j\omega_0\}}$$

However rather than compute $H(exp\{j\omega_0\})$ as in (2.119) by first computing the unit impulse response h, a more convenient expression is available which we now develop.

To begin, consider the asymptotically stable second order difference equation

$$y(k+2) = a_1 y(k+1) + a_0 y(k) + b_0 u(k) + b_1 u(k+1) + b_2 u(k+2)$$

for which

$$h(0) = b_2 \quad ; \quad h(1) = a_1 h(0) + b_1 \quad ; \quad h(2) = a_1 h(1) + a_0 h(0) + b_0$$

and

$$h(k+2) = a_1 h(k+1) + a_0 h(k) \quad ; \quad k \geq 1$$

Then for $k \geq 1$

$$\sum_{k=1}^{\infty} h(k+2) z^{-k} = a_1 \sum_{k=1}^{\infty} h(k+1) z^{-k} + a_0 \sum_{k=1}^{\infty} h(k) z^{-k}$$

which after a change of variable gives

$$z^2 \sum_{p=3}^{\infty} h(p) z^{-p} = a_1 z \sum_{p=2}^{\infty} h(p) z^{-p} + a_0 \sum_{p=1}^{\infty} h(p) z^{-p}$$

Then from (2.119)

$$\begin{aligned} z^2\{H(z) - h(2)z^{-2} - h(1)z^{-1} - h(0)\} = a_1 z\{H(z) - h(1)z^{-1} - h(0)\} \\ + a_0\{H(z) - h(0)\} \end{aligned}$$

which gives

$$\begin{aligned} (z^2 - a_1 z - a_0) H(z) &= h(0) z^2 + [h(1) - a_1 h(0)] z \\ &\quad + [h(2) - a_1 h(1) - a_0 h(0)] \\ &= b_2 z^2 + b_1 z + b_0 \end{aligned}$$

That is

$$H(z) = \frac{b_2 z^2 + b_1 z + b_0}{z^2 - a_1 z - a_0}$$

Extending this analysis to the nth order difference equation gives the following result.

Theorem 2.4.2. *Consider an asymptotically stable nth order difference equation (2.87). Then the steady state response y_{ss} due to the sinusoidal input signal*

$$u(k) = \sin k\omega_0$$

is given by

$$y_{ss}(k) = \varrho_0 \sin(k\omega_0 - \phi_0) \tag{2.120}$$

where

$$\varrho_0 exp\{-j\phi_0\} = H(z)|_{z=exp\{j\omega_0\}} \tag{2.121}$$

with

$$H(z) = \frac{b_n z^n + b_{n-1} z^{n-1} + \ \ldots \ + b_1 z + b_0}{z^n - a_{n-1} z^{n-1} - \ \ldots \ - a_1 z - a_0}$$

The function $H(z)$ is called the frequency response function.

Example 2.4.2. Consider once again the second order difference equation in example 2.2.7; that is

$$y(k+2) = -0.1y(k+1) + 0.3y(k) + u(k+1) - 2u(k) + 4u(k+2) \quad (2.122)$$

The aim now is to find (i) the forced response due to exponential input u where

$$u(m) = \begin{cases} (-0.4)^m & ; m \geq 0 \\ 0 & ; m < 0 \end{cases} \quad (2.123)$$

and (ii) the forced response due to a sinusoidal input u where

$$u(m) = \begin{cases} \sin 0.2m & ; m \geq 0 \\ 0 & ; m < 0 \end{cases} \quad (2.124)$$

In order to obtain these results, first recall that in example 2.2.7, we showed that the unit impulse response h of (2.122) is given by

$$h(0) = 4 \ ; \ h(k) = -0.45(0.5)^{k-1} + 1.05(-0.6)^{k-1} \ , \ k \geq 1 \quad (2.125)$$

(i) Now that the unit impulse response h is available, the forced response y_f (i.e. the response due to an input u with $u(k) = 0$ for $k < 0$ beginning at zero initial conditions) is given by

$$y_f(k) = \sum_{m=0}^{k} h(k-m)u(m) \quad ; \quad k \geq 0$$

Hence using (2.125) with the input given by (2.123), we have: $y_f(0) = 4$ and for $k \geq 1$

$$\begin{aligned} y_f(k) &= 4(-0.4)^k + \sum_{m=0}^{k-1} [-0.45(0.5)^{k-m-1} + 1.05(-0.6)^{k-m-1}](-0.4)^m \\ &= 4(-0.4)^k - 0.45(0.5)^{k-1} \sum_{m=0}^{k-1} \left(\frac{-0.4}{0.5}\right)^m \\ &\quad +1.05(-0.6)^{k-1} \sum_{m=0}^{k-1} \left(\frac{-0.4}{-0.6}\right)^m \\ &= 4(-0.4)^k - 0.45(0.5)^{k-1} \left\{\frac{1-(-0.8)^k}{1+0.8}\right\} \\ &\quad +1.05(-0.6)^{k-1} \left\{\frac{1-(0.67)^k}{1-0.67}\right\} \end{aligned}$$

That is, for $k \geq 1$

$$\begin{aligned} y_f(k) &= 4(-0.4)^k + \frac{-0.45}{(0.5)(1.8)}\{(0.5)^k - (-0.4)^k\} \\ &\quad + \frac{1.05}{(-0.6)(0.33)}\{(-0.6)^k - (-0.4)^k\} \\ &= 4(-0.4)^k - 0.50(0.5)^k - 5.30(-0.6)^k + 5.80(-0.4)^k \end{aligned}$$

Note that it is only possible to have terms of the form $(0.5)^k$, $(-0.6)^k$ and $(-0.4)^k$ in the forced response. The first two terms arise because $\lambda_1 = 0.5$ and $\lambda_2 = -0.6$ are characteristic roots, and the last term arises due to form of the input signal.

(ii) The forced response y_f due to the sinusoidal input (2.124) is given by

$$y_f(k) = exp\{j0.2\} \sum_{p=0}^{k} h(p) z^{-p} \quad ; \quad z = exp\{j0.2\}$$

where

$$s(k) \triangleq \sum_{p=0}^{k} h(p) z^{-p}$$

is given by

$$\begin{aligned} s(k) &= 4 + \sum_{p=1}^{k} [-0.45(0.5)^{p-1} + 1.05(-0.6)^{p-1}] z^p \\ &= 4 - 0.45 z^{-1} \sum_{p=1}^{k} (0.5 z^{-1})^{p-1} + 1.05 z^{-1} \sum_{p=1}^{k} (-0.6 z^{-1})^{p-1} \\ &= 4 - 0.45 z^{-1} \left(\frac{1 - (0.5z^{-1})^k}{1 - 0.5z^{-1}} \right) + 1.05 z^{-1} \left(\frac{1 - (-0.6z^{-1})^k}{1 + 0.6z^{-1}} \right) \\ &= 4 - \frac{0.45}{z - 0.5} + \frac{1.05}{z + 0.6} + z^{-k} \left(\frac{0.45(0.5)^k}{z - 0.5} - \frac{1.05(-0.6)^k}{z + 0.6} \right) \end{aligned}$$

However if only the *steady state* sinusoidal response y_{ss} is required, then

$$y_{ss}(k) = \varrho \sin(0.2k - \phi)$$

where

$$\varrho exp\{-j\phi\} = H(exp\{j0.2\}) \quad ; \quad H(z) = \frac{4z^2 + z - 2}{z^2 + 0.1z - 0.3}$$

Example 2.4.3. Consider the third order difference equation from example 2.3.1

$$y(k+3) = -0.1y(k+2) + 0.07y(k+1) - 0.65y(k) \\ +2u(k) - u(k+1) - 0.3u(k+2) + 0.5u(k+3) \quad (2.126)$$

The aim is to find (i) the steady state response due the unit step response, and (ii) the steady response due to the sinusoidal input u in example 2.4.2.

We first need to establish if the difference equation (2.126) is asymptotically stable. In order to do this, observe that the characteristic equation is given by

$$\lambda^3 + 0.1\lambda^2 - 0.07\lambda + 0.65 = 0$$

The Marden stability table is therefore as given in Table 2.4 where $\{x_1 = 0.578,\ x_2 = 0.146,\ x_3 = -0.135,\ z_1 = 0.315,\ z_2 = 0.104\}$ which gives

$$\gamma_1 = 0.65 \ ; \quad \gamma_2 = -0.234 \ ; \quad \gamma_3 = 0.329$$

Table 2.4. Marden stability table for third order equation (2.126)

1	0.1	−0.07	0.65	γ_1
0.65	−0.07	0.1	1	
$x1$	$x2$	$x3$		γ_2
$x3$	$x2$	$x1$		
$z1$	$z2$			γ_3
$z2$	$z1$			

Since $|\gamma_k| < 1$ for $k = 1, 2, 3$, asymptotic stability is established. Note that the characteristic roots were determined in example 2.3.1 as $\{\lambda_1 = -0.5, \lambda_2 = 0.2 + j0.3 = \overline{\lambda}_3\}$. However it is not necessary to find these roots in order to determine asymptotic stability. The steady state reponse y_{ss} to a unit step is therefore given by $y_{ss}(k) = \overline{s}$ where (to 2 dec. places)

$$\overline{s} = \frac{2 - 1 - 0.3 + 0.5}{1 + 0.1 - 0.07 + 0.65} = 0.71$$

and the steady state response y_{ss} due to the sinusoidal input $u(k) = \sin 0.2k$ is given by

$$y_{ss}(k) = \varrho \sin(0.2k - \phi) \ ; \quad \varrho exp\{-j\phi\} = H(exp\{j0.2\})$$

where

$$H(z) = \frac{0.5z^3 - 0.3z^2 - z + 2}{z^3 + 0.1z^2 - 0.07z + 0.65}$$

Inverse Filtering Problem

Suppose a transmitted signal u results in a distorted signal y, and a measurement z is made according to the following model

$$y(k+n) = \sum_{p=0}^{n-1} a_p y(k+p) + \sum_{p=0}^{m} b_p u(k+p) \quad ; \quad b_m \neq 0 \tag{2.127}$$

$$z(k) = y(k) + w(k)$$

where the signal w represents a measurement error. That is, the frequency response function $H(z)$ between the transmitted signal u and the noise free signal y is given by

$$H(z) = \frac{\sum_{p=0}^{m} b_p z^p}{z^n - \sum_{p=0}^{n-1} a_p z^p}$$

The aim is to process the signal z in order to recover the transmitted signal u. To this end, consider the signal processing algorithm (which is the generalization of (2.40)) as described by

$$r(k+n) = \frac{1}{b_m} \left\{ \sum_{p=0}^{m-1} b_p r(k+p) - z(k+n) + \sum_{p=0}^{n} a_p z(k+p) \right\} \tag{2.128}$$

and define the estimation error signal $\varepsilon \triangleq r - u$. Then

$$\sum_{p=0}^{m} b_p \varepsilon(k+p) = w(k+n) - \sum_{p=0}^{n-1} a_p w(k+p)] \tag{2.129}$$

Consequently, when the error signal is constant, it follows that $|\varepsilon(k)| \to 0$ as $k \to \infty$ provided all roots ϱ_k of the algebraic equation

$$\sum_{p=0}^{m} b_p z^p = 0 \tag{2.130}$$

satisfy $|\varrho_k| < 1$. Under this condition the signal u is ultimately recovered by "inverse filtering" the signal z.

Recall that the system function $H(z)$ is asymptotically stable if all the roots λ_k of the algebraic equation

$$z^n - \sum_{p=0}^{n-1} a_p z^p = 0 \tag{2.131}$$

satisfy $|\lambda_k| < 1$. Hence the ability of being able to recover the signal u by "inverse filtering" is equivalent to the condition that the "inverse frequency response function" $H^{-1}(z)$ is asymptotically stable.

2.5 State Space Representation

In section 2.2, we showed how a second order difference equation results from a system of two coupled first order difference equations. In this section, we generalize this result by showing how the system

$$\begin{aligned} x_j(k+1) &= \sum_{m=1}^{n} f_{jm} x_m(k) + g_j u(k) \quad ; \quad 1 \leq j \leq n \\ y(k) &= \sum_{m=1}^{n} h_m x_m(k) + du(k) \end{aligned} \tag{2.132}$$

of n coupled first order difference equations is equivalent to an nth order difference equation (2.87). The system of equivalent first order difference equations (2.132) is then referred to as a *state space representation* of (2.87). The n-vector signal $\boldsymbol{x}^T = [x_1 \; x_2 \; \ldots \; x_n]$ is called the *state* signal, and the scalar signals $\{x_j\}$ as the *state components*.

In particular, we shall derive expressions for the coefficients $\{a_p, b_q\}$ in (2.87) in terms of the coefficients $\{f_{jm}, g_j, h_m\}$ of the coupled first order difference equations. The state space representation provides a standard format by which programs such as *MATLAB* can be used to numerically determine solutions of the nth order difference equations (2.87).

2.5.1 Second Order

Consider the following system of two linear first order time invariant first order difference equations described by

$$\begin{aligned} x_1(k+1) &= f_{11} x_1(k) + f_{12} x_2(k) + g_1 u(k) \\ x_2(k+1) &= f_{21} x_1(k) + f_{22} x_2(k) + g_2 u(k) \\ y(k) &= h_1 x_1(k) + h_2 x_2(k) + du(k) \end{aligned} \tag{2.133}$$

which is more general than (2.42) in that (2.133) also includes an output signal y. The first equation in (2.133) is a first order difference equation in the signal x_1 with input signal $v_1 = f_{12}x_2 + g_1 u$, while the second equation is a first order difference equation in the signal x_2 with input signal $v_2 = f_{21}x_1 + g_2 u$. The last equation is an algebraic equation which expresses the output y in terms of the state components $\{x_1, x_2\}$ and the input signal u.

This system of equations can be expressed more succinctly using linear algebra in the form

$$\begin{aligned} \boldsymbol{x}(k+1) &= \boldsymbol{F}\boldsymbol{x}(k) + \boldsymbol{g}u(k) \\ y(k) &= \boldsymbol{h}^T \boldsymbol{x}(k) + du(k) \end{aligned} \tag{2.134}$$

where the 2-vector state $\boldsymbol{x}$, the 2-square matrix $\boldsymbol{F}$, and the 2-vectors $\boldsymbol{g}$ and $\boldsymbol{h}$ are from (2.133) defined by

$$\boldsymbol{x}(k) \triangleq \begin{bmatrix} x_1(k) \\ x_2(k) \end{bmatrix} \; ; \; \boldsymbol{F} \triangleq \begin{bmatrix} f_{11} & f_{12} \\ f_{21} & f_{22} \end{bmatrix} \; ; \; \boldsymbol{g} \triangleq \begin{bmatrix} g_1 \\ g_2 \end{bmatrix} \; ; \; \boldsymbol{h} \triangleq \begin{bmatrix} h_1 \\ h_2 \end{bmatrix} \tag{2.135}$$

We have the following result.

Theorem 2.5.1. *The first order vector system of two first order linear time invariant difference equation (2.134), (2.135) has the same input-output description as the second order linear time invariant difference equation*

$$\begin{aligned} y(k+2) = {} & a_1 y(k+1) + a_0 y(k) + b_0 u(k) \\ & + b_1 u(k+1) + b_2 u(k+2) \end{aligned} \tag{2.136}$$

if and only if $\{\boldsymbol{F},\ \boldsymbol{g},\ \boldsymbol{h}\}$ *in (2.135) and* $\{a_1,\ a_0,\ b_2,\ b_1,\ b_0\}$ *are related according to*

$$b_2 = d \; ; \; \begin{bmatrix} 1 & -a_1 \\ 0 & 1 \end{bmatrix} \begin{bmatrix} \boldsymbol{h}^T \boldsymbol{F} \boldsymbol{g} \\ \boldsymbol{h}^T \boldsymbol{g} \end{bmatrix} = \begin{bmatrix} b_0 + a_0 d \\ b_1 + a_1 d \end{bmatrix} \tag{2.137}$$

with

$$det(\lambda \boldsymbol{I}_2 - \boldsymbol{F}) = \lambda^2 - a_1 \lambda - a_0 \tag{2.138}$$

The initial conditions $\{y(k_0-1), y(k_0)\}$ *of (2.136) are related to the initial state* $\boldsymbol{x}(k_0-1)$ *of (2.134) by*

$$\begin{bmatrix} y(k_0) \\ y(k_0-1) \end{bmatrix} = \begin{bmatrix} \boldsymbol{h}^T \boldsymbol{F} \\ \boldsymbol{h}^T \end{bmatrix} \boldsymbol{x}(k_0-1) + \begin{bmatrix} d & \boldsymbol{h}^T \boldsymbol{g} \\ 0 & d \end{bmatrix} \begin{bmatrix} u(k_0) \\ u(k_0-1) \end{bmatrix} \tag{2.139}$$

In order to prove this result, we make use of the *Caley-Hamilton theorem* which for a 2-square matrix states that: if

$$det(\lambda \boldsymbol{I_2} - \boldsymbol{F}) = \lambda^2 - \delta_1 \lambda - \delta_0 \tag{2.140}$$

then the matrix $\boldsymbol{F}$ satisfies the matrix equation

$$\boldsymbol{F}^2 - \delta_1 \boldsymbol{F} - \delta_0 \boldsymbol{I_2} = \boldsymbol{0_2} \tag{2.141}$$

For example, if

$$\boldsymbol{F} = \begin{bmatrix} 1 & 2 \\ -1 & 3 \end{bmatrix}$$

then

$$det(\lambda \boldsymbol{I_2} - \boldsymbol{F}) = det \begin{bmatrix} \lambda - 1 & -2 \\ 1 & \lambda - 3 \end{bmatrix} = \lambda^2 - 4\lambda + 5$$

Also

$$\boldsymbol{F}^2 = \begin{bmatrix} -1 & 8 \\ -4 & 7 \end{bmatrix} \quad ; \quad 4\boldsymbol{F} - 5\boldsymbol{I}_2 = \begin{bmatrix} -1 & 8 \\ -4 & 7 \end{bmatrix}$$

and so $\boldsymbol{F}^2 - 4\boldsymbol{F} + 5\boldsymbol{I_2} = \boldsymbol{0_2}$.

Now from (2.134)

$$\begin{aligned} y(k+1) &= \boldsymbol{h}^T \boldsymbol{x}(k+1) + du(k+1) \\ &= \boldsymbol{h}^T \boldsymbol{F}\boldsymbol{x}(k) + \boldsymbol{h}^T \boldsymbol{g}u(k) + du(k+1) \end{aligned}$$

and

$$\boldsymbol{x}(k+2) = \boldsymbol{F}\boldsymbol{x}(k+1) + \boldsymbol{g}u(k+1) = \boldsymbol{F}^2\boldsymbol{x}(k) + \boldsymbol{F}\boldsymbol{g}u(k) + \boldsymbol{g}u(k+1)$$

so that

$$\begin{aligned} y(k+2) &= \boldsymbol{h}^T \boldsymbol{x}(k+2) + du(k+2) \\ &= \boldsymbol{h}^T \boldsymbol{F}^2 \boldsymbol{x}(k) + \boldsymbol{h}^T \boldsymbol{F}\boldsymbol{g}u(k) + \boldsymbol{h}^T \boldsymbol{g}u(k+1) + du(k+2) \end{aligned}$$

Then substituting for $\boldsymbol{F}^2$ using (2.141), we have

$$\begin{aligned} y(k+2) = {} & (\delta_1 \boldsymbol{h}^T \boldsymbol{F} + \delta_0 \boldsymbol{h}^T)\boldsymbol{x}(k) \\ & + \boldsymbol{h}^T \boldsymbol{F}\boldsymbol{g}u(k) + \boldsymbol{h}^T \boldsymbol{g}u(k+1) + du(k+2) \end{aligned}$$

Then making the substitutions

$$\begin{aligned} \boldsymbol{h}^T \boldsymbol{F}\boldsymbol{x}(k) &= y(k+1) - \boldsymbol{h}^T \boldsymbol{g}u(k) - du(k+1) \\ \boldsymbol{h}^T \boldsymbol{x}(k) &= y(k) - du(k) \end{aligned}$$

we have

$$\begin{aligned} y(k+2) = {} & \delta_1[y(k+1) - \boldsymbol{h}^T \boldsymbol{g}u(k) - du(k+1)] + \delta_0[y(k) - du(k)] \\ & + \boldsymbol{h}^T \boldsymbol{F}\boldsymbol{g}u(k) + \boldsymbol{h}^T \boldsymbol{g}u(k+1) + du(k+2) \end{aligned} \tag{2.142}$$

After comparing coefficients in (2.136) and (2.142), we get (2.137). The relationship in (2.139) between the initial state $\boldsymbol{x}(k_0 - 1)$ of (2.134) and the initial conditions of (2.136) follows directly from (2.134). We have therefore proven the result.

Example 2.5.1. Find the second order difference equation corresponding to the state space equations

$$\begin{aligned} x_1(k+1) &= x_1(k) + T[\alpha x_1(k) + \beta x_2(k) + \gamma u(k)] \;\; ; \quad k \geq 2 \qquad (2.143) \\ x_2(k+1) &= x_2(k) + T x_1(k) \\ y(k) &= x_2(k) \end{aligned}$$

In (2.52), we showed that $y(k)$ in (2.143) is an approximation at time $t = kT$ of the solution $x(t)$ of the second order differential equation

$$\ddot{x}(t) = \alpha \dot{x}(t) + \beta x(t) + \gamma v(t) \;\; ; \quad t \geq 0 \qquad (2.144)$$

given the initial conditions $\{x(0), \dot{x}(0)\}$ based on the backward Euler numerical approximation scheme.)

In (2.134), corresponding to (2.143), we have

$$\boldsymbol{F} = \begin{bmatrix} 1+\alpha T & \beta T \\ T & 1 \end{bmatrix} \; ; \; \boldsymbol{g} = \begin{bmatrix} \gamma T \\ 0 \end{bmatrix} \; ; \; \boldsymbol{h} = \begin{bmatrix} 0 \\ 1 \end{bmatrix} \; ; \; d = 0$$

so that

$$\boldsymbol{h}^T \boldsymbol{g} = 0 \; ; \; \boldsymbol{h}^T \boldsymbol{F} \boldsymbol{g} = \gamma T^2$$

and

$$det(\lambda \boldsymbol{I_2} - \boldsymbol{F}) = \lambda^2 - (2 + \alpha T)\lambda + 1 + \alpha T - \beta T^2 \qquad (2.145)$$

The corresponding second order difference equation in y is therefore given from Theorem 2.5.1 by

$$\begin{aligned} y(k+2) &= a_1 y(k+1) + a_0 y(k) \\ &\quad + b_0 u(k) + b_1 u(k+1) + b_2 u(k+2) \qquad (2.146) \end{aligned}$$

where

$$a_1 = 2 + \alpha T \; ; \; a_0 = \beta T^2 - \alpha T - 1 \; ; \; b_2 = 0 \; ; \; b_1 = 0 \; ; \; b_0 = \gamma T^2$$

which was earlier derived in (2.53). The corresponding initial conditions are related by

$$\begin{bmatrix} y(1) \\ y(0) \end{bmatrix} = \begin{bmatrix} T & 1 \\ 0 & 1 \end{bmatrix} \begin{bmatrix} x_1(0) \\ x_2(0) \end{bmatrix}$$

which corresponds to

$$y(0) = x_2(0) \;\; ; \;\; y(1) = T x_1(0) + x_2(0)$$

We now make some observations about the characteristic roots $\{\lambda_1, \lambda_2\}$.

(i) Since in (2.145)

$$a_1^2 + 4a_0 = (2 + \alpha T)^2 - 4(1 + \alpha T - \beta T^2) = (\alpha^2 + 4\beta)T^2$$

it follows that: independent of the step size $T(> 0)$, the roots $\{\lambda_1, \lambda_2\}$ of the characteristic equation $det[\lambda \boldsymbol{I_2} - \boldsymbol{F}] = 0$ are both *real* if and only if $4\beta \geq -\alpha^2$.

(ii) Asymptotic stability of the second order difference equation (2.146) requires $|\lambda_1| < 1, |\lambda_2| < 1$, and by the Marden stability test (Theorem 2.3.2), we have asymptotic stability if and only if

$$|a_0| < 1 \;\; ; \;\; |a_1| < 1 - a_0$$

that is, if and only if

$$|\beta T^2 - \alpha T - 1| < 1 \;\; ; \;\; |2 + \alpha T| < 2 - \beta T^2 + \alpha T \tag{2.147}$$

Now the solution of the *differential equation* (2.144) when the input signal v is zero can be shown to be of the form

$$x(t) = c_1 exp\{\gamma_1 t\} + c_2 exp\{\gamma_2 t\}$$

where $\{c_1, c_2\}$ are constants determined by the initial conditions $\{x(0), \dot{x}(0)\}$, and $\{\gamma_1, \gamma_2\}$ are the roots of the algebraic equation

$$s^2 - \alpha s - \beta = 0$$

Hence when the input v is zero, $x(t) \rightarrow 0$ for all initial conditions $\{x(0), \dot{x}(0)\}$ (i.e. (2.144) is asymptotically stable) if and only if both roots $\{\gamma_1, \gamma_2\}$ have *negative real part*; that is, if and only if

$$\alpha < 0 \;\; ; \;\; \beta < 0 \tag{2.148}$$

Therefore

$$1 + \alpha T - \beta T^2 = 1 - |\alpha| T + |\beta| T^2 < 1$$

and

$$2 + \alpha T = 2 - |\alpha| T < 2 - |\alpha| T + |\beta| T^2$$

for $T > 0$ sufficiently small. That is, asymptotic stability of the difference equation is guaranteed for

$$0 < T < \frac{|\alpha|}{|\beta|} \triangleq T_{max}$$

However T may need to be much smaller than T_{max} in order to provide a reasonably accurate approximation the solution $x(kT)$ of the differential equation.

Example 2.5.2. In Chapter 1, we considered a marketing problem involving two shops which sell all the bicycles in a particular town. The fraction of customers $x_1(k)$ that shop #1 has at the end of the kth month with respect to the fraction of customers $x_2(k) = 1 - x_1(k)$ that shop #2 has at the end of the kth month was shown to be given by

$$\begin{aligned} \boldsymbol{x}(k+1) &= \boldsymbol{F}\boldsymbol{x}(k) \quad ; \quad k \geq 0 \\ y(k) &= \boldsymbol{h}^T\boldsymbol{x}(k) \end{aligned}$$

where

$$\boldsymbol{F} = \begin{bmatrix} a_{11} & 1 - a_{22} \\ 1 - a_{11} & a_{22} \end{bmatrix} \quad ; \quad \boldsymbol{x}(0) = \begin{bmatrix} x_{10} \\ 1 - x_{10} \end{bmatrix}$$

with

$$0 < a_{11} < 1 \ ; \ 0 < a_{22} < 1 \ ; \ 0 < x_{10} < 1$$

In Theorem 2.5.1, with $y = x_1$, this corresponds to $\{\boldsymbol{h}^T = [1 \ \ 0],\ \boldsymbol{g} = \boldsymbol{0_2}, d = 0\}$, and hence y satisfies the second order difference equation

$$y(k+2) = a_1 y(k+1) + a_0 y(k) \tag{2.149}$$

where

$$\begin{aligned} det(\lambda \boldsymbol{I_2} - \boldsymbol{F}) &= \lambda^2 - a_1\lambda - a_0 \\ &= (\lambda - a_{11})(\lambda - a_{22}) - (1 - a_{11})(1 - a_{22}) \end{aligned}$$

That is

$$a_1 = a_{11} + a_{22} \ ; \ a_0 = 1 - a_{11} - a_{22}$$

Also from (2.139), the initial conditions in (2.149) are given by

$$\begin{bmatrix} y(1) \\ y(0) \end{bmatrix} = \begin{bmatrix} a_{11} & 1 - a_{22} \\ 1 & 0 \end{bmatrix} \begin{bmatrix} x_{10} \\ 1 - x_{10} \end{bmatrix} \tag{2.150}$$

Since the roots $\{\lambda_1, \lambda_2\}$ of the characteristic equation

$$\lambda^2 - (a_{11} + a_{22})\lambda + a_{11} + a_{22} - 1 = 0$$

are given by

$$\lambda_1 = 1 \ ; \ \lambda_2 = a_{11} + a_{22} - 1$$

the solution $y = x_1$ of (2.149) is given by

$$y(k) = \beta_1 + \beta_2(a_{11} + a_{22} - 1)^k \ ; \ k \geq 0 \tag{2.151}$$

where from (2.150), $\{\beta_1, \beta_2\}$ are given from

$$y(0) = x_{10} = \beta_1 + \beta_2$$
$$y(1) = 1 - a_{22} + (1 + a_{11} - a_{22})x_{10} = \beta_1 + \beta_2(a_{11} + a_{22} - 1)$$

That is

$$\beta_1 = \frac{1 - a_{22} + 2(1 - a_{22})x_{10}}{2 - a_{11} - a_{22}} \; ; \; \beta_2 = x_{10} - \beta_1 \tag{2.152}$$

Recall that when this problem was discussed in Chapter 1, two questions were posed: Firstly, do $x_1(k)$ and $x_2(k)$ converge to some limiting values as $k \to \infty$, and secondly, if such limits exist, do they depend on the initial values $x_1(0)$ and $x_2(0)$? In order to address these questions, observe that the conditions on the a_{ij} imply that $0 < a_{11} + a_{22} < 2$. Therefore from (2.151), we have that $(a_{11} + a_{22} - 1)^k$ approaches zero in the limit as $k \to \infty$; that is, in the limit as $k \to \infty$

$$y(k) = x_1(k) \to \beta_1 \; ; \; x_2(k) \to 1 - \beta_1$$

where β_1 is given by (2.152).

In terms of the marketing problem, shop #1 is interested in whether or not it will eventually secure more that 50% of the market share (i.e $\beta > 0.5$). It is evident that this condition depends on both fractions $\{a_{11}, a_{22}\}$ that each shop retains of its own customers each month, and the *initial* fraction $x_{10} = x_1(0)$ of customers of shop #1. For example, if $\{a_{11} = 0.35;\ a_{22} = 0.55\}$ which implies $\{a_{12} = 0.65, a_{21} = 0.45\}$, then

$$\beta_1 = \frac{0.45 + 0.9x_{10}}{1.1}$$

and so the condition $\beta_1 > 0.5$ is satisfied when $x_{10} > 0.12$. On the other hand, if $\{a_{11} = 0.55;\ a_{22} = 0.35\}$ which implies $\{a_{12} = 0.45, a_{21} = 0.65\}$, then

$$\beta_1 = \frac{0.65 + 1.30x_{10}}{1.1}$$

and so the condition $\beta_1 > 0.5$ is satisfied for all $x_{10}(> 0)$.

Notice also that the market share "rises and falls" when $a_{11} + a_{22} - 1$ is *negative* since $(a_{11} + a_{22} - 1)^k$ is then negative for odd values of k, and positive for even values of k.

System Matrix and Input/Output Vectors

The matrix $\boldsymbol{F}$ is called the *system matrix*, and the vectors $\boldsymbol{g}$ and $\boldsymbol{h}$ are called the *system input* and *system output* vectors respectively of the state space representation (2.134). Since $d = b_2$ in (2.137), there are 8 remaining components of $\{\boldsymbol{F}, \boldsymbol{g}, \boldsymbol{h}\}$ in (2.135) which must be specified to match the 4 remaining coefficients $\{a_0, a_1, b_0, b_1\}$. Consequently, many choices for $\{\boldsymbol{F}, \boldsymbol{g}, \boldsymbol{h}\}$ are possible, and the available freedom can be summarized as follows:

1. Condition (2.138) imposes a constraint on the characteristic polynomial of the system matrix $\boldsymbol{F}$.
2. From (2.139), the invertibility of the 2-square matrix $\boldsymbol{M}_2$ where

$$\boldsymbol{M}_2 \triangleq \begin{bmatrix} \boldsymbol{h}^T \boldsymbol{F} \\ \boldsymbol{h}^T \end{bmatrix} \tag{2.153}$$

   is a necessary condition in order to be able to determine the initial state $\boldsymbol{x}(k_0 - 1)$ for *all* initial conditions $\{y(k_0), y(k_0 - 1)\}$. That is, if $\boldsymbol{M}_2$ is invertible, then

$$\boldsymbol{x}(k_0 - 1) = \boldsymbol{M}_2^{-1} \left\{ \begin{bmatrix} y(k_0) \\ y(k_0 - 1) \end{bmatrix} - \begin{bmatrix} d & \boldsymbol{h}^T \boldsymbol{g} \\ 0 & d \end{bmatrix} \begin{bmatrix} u(k_0) \\ u(k_0 - 1) \end{bmatrix} \right\} \tag{2.154}$$

3. Since

$$\begin{bmatrix} \boldsymbol{h}^T \boldsymbol{F} \boldsymbol{g} \\ \boldsymbol{h}^T \boldsymbol{g} \end{bmatrix} = \begin{bmatrix} \boldsymbol{h}^T \boldsymbol{F} \\ \boldsymbol{h}^T \end{bmatrix} \boldsymbol{g} = \boldsymbol{M}_2 \boldsymbol{g}$$

   the last remaining condition (2.137) which is imposed on the choice of $\{\boldsymbol{F}, \boldsymbol{g}, \boldsymbol{h}\}$ can be arranged into the form

$$\boldsymbol{D}_2 \boldsymbol{M}_2 \boldsymbol{g} = \boldsymbol{\gamma}_2$$

   where

$$\boldsymbol{D}_2 \triangleq \begin{bmatrix} 1 & -a_1 \\ 0 & 1 \end{bmatrix} \quad ; \quad \boldsymbol{\gamma}_2 \triangleq \begin{bmatrix} b_0 + a_0 d \\ b_1 + a_1 d \end{bmatrix} \tag{2.155}$$

We therefore have the following result.

Theorem 2.5.2. *Suppose the system matrix $\boldsymbol{F}$ has the characteristic polynomial (2.138), and the output vector $\boldsymbol{h}$ is chosen such that the 2-square matrix $\boldsymbol{M}_2$ defined by (2.153) is invertible.*

Then $\{\boldsymbol{F}, \boldsymbol{g}, \boldsymbol{h}, d\}$ where

$$d = b_2 \;;\; \boldsymbol{g} = \boldsymbol{M}_2^{-1} \boldsymbol{D}_2^{-1} \boldsymbol{\gamma}_2 \tag{2.156}$$

and $\{\boldsymbol{D}_2, \boldsymbol{\gamma}_2\}$ are defined by (2.155) is a state space representation of the difference equation (2.136) with initial state

$$\boldsymbol{x}(k_0 - 1) = \boldsymbol{M}_2^{-1} \left\{ \begin{bmatrix} y(k_0) \\ y(k_0 - 1) \end{bmatrix} - \begin{bmatrix} d & \boldsymbol{h}^T \boldsymbol{g} \\ 0 & d \end{bmatrix} \begin{bmatrix} u(k_0) \\ u(k_0 - 1) \end{bmatrix} \right\} \tag{2.157}$$

Phase Variable Form

We now define one representation that always works. Specifically, let $\{\boldsymbol{F} = \boldsymbol{F}_0 \ ; \ \boldsymbol{h} = \boldsymbol{h}_0\}$ where

$$\boldsymbol{F}_0 = \begin{bmatrix} 0 & a_0 \\ 1 & a_1 \end{bmatrix} \ ; \ \boldsymbol{h}_0 = \begin{bmatrix} 0 \\ 1 \end{bmatrix} \tag{2.158}$$

then

$$det(\lambda \boldsymbol{I_2} - \boldsymbol{F}_0) = \lambda(\lambda - a_1) - a_0 = \lambda^2 - a_1\lambda - a_0$$

as required, and from (2.153) - (2.156)

$$\boldsymbol{M}_2 = \begin{bmatrix} 1 & a_1 \\ 0 & 1 \end{bmatrix} \ ; \ det(\boldsymbol{M}_2) = 1$$

and

$$\boldsymbol{M}_2^{-1}\boldsymbol{D}_2^{-1} = \begin{bmatrix} 1 & a_1 \\ 0 & 1 \end{bmatrix}^{-1} \begin{bmatrix} 1 & -a_1 \\ 0 & 1 \end{bmatrix}^{-1} = \begin{bmatrix} 1 & -a_1 \\ 0 & 1 \end{bmatrix} \begin{bmatrix} 1 & a_1 \\ 0 & 1 \end{bmatrix} = \boldsymbol{I_2}$$

and so

$$\boldsymbol{g} = \boldsymbol{g}_0 = \begin{bmatrix} b_0 + a_0 d \\ b_1 + a_1 d \end{bmatrix} \tag{2.159}$$

The initial state $\boldsymbol{x}(k_0 - 1)$ is given from $\{y(k_0 - 1), y(k_0)\}$ by (2.157); that is

$$\begin{aligned} \boldsymbol{x}(k_0 - 1) &= \begin{bmatrix} 1 & a_1 \\ 0 & 1 \end{bmatrix}^{-1} \left\{ \begin{bmatrix} y(k_0) \\ y(k_0 - 1) \end{bmatrix} - \begin{bmatrix} d & b_1 + a_1 d \\ 0 & d \end{bmatrix} \begin{bmatrix} u(k_0) \\ u(k_0 - 1) \end{bmatrix} \right\} \\ &= \begin{bmatrix} y(k_0) - a_1 y(k_0 - 1) - b_1 u(k_0 - 1) - d u(k_0) \\ y(k_0 - 1) - d u(k_0 - 1) \end{bmatrix} \end{aligned} \tag{2.160}$$

The representation $\{\boldsymbol{F}_0, \boldsymbol{g}_0, \boldsymbol{h}_0, d\}$ in (2.158) and (2.159) is referred to as *phase variable form*. Alternative phase variable forms are given in exercise 2.26.

Example 2.5.3. Consider the second order difference equation

$$\begin{aligned} y(k+2) = {} & 0.4y(k+1) + 0.05y(k) \\ & -0.32u(k) - 0.80u(k+1) + 4u(k+2) \end{aligned} \tag{2.161}$$

for $k \geq 3$ where $\{y(3) = -3, y(4) = 2\}$, and $\{u(3) = 5, u(4) = 1\}$.

Then from Theorem 2.5.1 with $d = 4$, this equation has the equivalent state space description (2.134) provided $\{\boldsymbol{F}, \ \boldsymbol{g}, \ \boldsymbol{h}\}$ are chosen such that

$$h^T g = -0.64 \quad ; \quad h^T F g = -0.376$$

with

$$det(\lambda I_2 - F) = \lambda^2 - 0.4\lambda - 0.05 = (\lambda - 0.5)(\lambda + 0.1) \tag{2.162}$$

One choice for $\{F, g, h\}$ is given by the phase variable form

$$F_0 = \begin{bmatrix} 0 & 0.05 \\ 1 & 0.40 \end{bmatrix} \; ; \; g_0 = \begin{bmatrix} -0.12 \\ -0.64 \end{bmatrix} \; ; \; h_0 = \begin{bmatrix} 0 \\ 1 \end{bmatrix}$$

In this case, since $\{u(3) = 5, u(4) = 1\}$, the initial state $x(3)$ is given from $y(3)$ and $y(4)$ by

$$x(3) = \begin{bmatrix} 1 & 0.4 \\ 0 & 1 \end{bmatrix}^{-1} \left\{ \begin{bmatrix} 2 \\ -3 \end{bmatrix} - \begin{bmatrix} 4 & -0.64 \\ 0 & 4 \end{bmatrix} \begin{bmatrix} 1 \\ 5 \end{bmatrix} \right\} = \begin{bmatrix} 8 \\ -17 \end{bmatrix}$$

Jordan Form

We now consider another special state space representation of (2.136) which has three forms depending on the type of roots of the characteristic equation.

(a) When F has *unequal real eigenvalues* $\{\lambda_1, \lambda_2\}$ (or equivalently, the characteristic equation of (2.136) has unequal real roots); that is

$$det(\lambda I_2 - F) = (\lambda - \lambda_1)(\lambda - \lambda_2) = \lambda^2 - a_1\lambda - a_0$$

then the representation

$$F = \begin{bmatrix} \lambda_1 & 0 \\ 0 & \lambda_2 \end{bmatrix} \; ; \; h = \begin{bmatrix} 1 \\ 1 \end{bmatrix} \tag{2.163}$$

gives the required characteristic polynomial. The form of F in (2.163) for distinct eigenvalues $\{\lambda_1, \lambda_2\}$ is known as the *Jordan form*. Note however that the system output vector h may be any vector provided both components are non-zero.

Then in (2.153)

$$M_2 = \begin{bmatrix} \lambda_1 & \lambda_2 \\ 1 & 1 \end{bmatrix} \quad ; \quad det(M_2) = \lambda_1 - \lambda_2 \neq 0$$

Also in (2.156)

$$M_2^{-1} D_2^{-1} = \frac{1}{\lambda_1 - \lambda_2} \begin{bmatrix} 1 & -\lambda_2 \\ -1 & \lambda_1 \end{bmatrix} \begin{bmatrix} 1 & a_1 \\ 0 & 1 \end{bmatrix} \quad ; \quad a_1 = \lambda_1 + \lambda_2$$

from which

$$g = M_2^{-1} D_2^{-1} \begin{bmatrix} b_0 + a_0 d \\ b_1 + a_1 d \end{bmatrix}$$

(b) When $\boldsymbol{F}$ has a *repeated real eigenvalue* λ_0; that is

$$det(\lambda \boldsymbol{I_2} - \boldsymbol{F}) = (\lambda - \lambda_0)^2 = \lambda^2 - a_1\lambda - a_0$$

then the representation

$$\boldsymbol{F} = \begin{bmatrix} \lambda_0 & 1 \\ 0 & \lambda_0 \end{bmatrix} \; ; \; \boldsymbol{h} = \begin{bmatrix} 1 \\ 1 \end{bmatrix} \tag{2.164}$$

gives the required characteristic polynomial, and in (2.153)

$$\boldsymbol{M}_2 = \begin{bmatrix} \lambda_0 & \lambda_0 + 1 \\ 1 & 1 \end{bmatrix} \; ; \; det(\boldsymbol{M}_2) = -1 \neq 0$$

(Note that even though the choice

$$\boldsymbol{F} = \begin{bmatrix} \lambda_0 & 0 \\ 0 & \lambda_0 \end{bmatrix}$$

also gives the required characteristic polynomial, the corresponding matrix $\boldsymbol{M}_2$ as given by

$$\boldsymbol{M}_2 = \begin{bmatrix} \lambda_0 & \lambda_0 \\ 1 & 1 \end{bmatrix} \; ; \; det(\boldsymbol{M}_2) = 0$$

is *not* invertible. The choice $\boldsymbol{F}$ in (2.164) for a repeated real eigenvalue λ_0 is known as the *Jordan form* of a 2-square matrix. Continuing, we have

$$\boldsymbol{g} = \begin{bmatrix} -1 & \lambda_0 + 1 \\ 1 & -\lambda_0 \end{bmatrix} \begin{bmatrix} 1 & a_1 \\ 0 & 1 \end{bmatrix} \begin{bmatrix} b_0 + a_0 d \\ b_1 + a_1 d \end{bmatrix} \; ; \; a_1 = 2\lambda_0, \; a_0 = \lambda_0^2$$

(c) Finally, when $\boldsymbol{F}$ has *complex eigenvalues*; that is

$$det(\lambda \boldsymbol{I_2} - \boldsymbol{F}) = (\lambda - \varrho)^2 + \omega^2 = (\lambda - \lambda_1)(\lambda - \overline{\lambda}_1) = \lambda^2 - a_1\lambda - a_0$$

where $\lambda_1 = \varrho + j\omega$ for $\omega > 0$, then the representation

$$\boldsymbol{F} = \begin{bmatrix} \varrho & \omega \\ -\omega & \varrho \end{bmatrix} \; ; \; \boldsymbol{h} = \begin{bmatrix} 1 \\ 1 \end{bmatrix} \tag{2.165}$$

gives the required characteristic polynomial. For complex eigenvalues $\varrho \pm j\omega$, $\boldsymbol{F}$ in (2.165) is in the *Jordan form*. Then from (2.153)

$$\boldsymbol{M}_2 = \begin{bmatrix} \varrho - \omega & \varrho + \omega \\ 1 & 1 \end{bmatrix} \; ; \; det(\boldsymbol{M}_2) = -2\omega \neq 0$$

and from (2.156)

$$\boldsymbol{g} = \frac{1}{2\omega} \begin{bmatrix} -1 & \varrho - \omega \\ 1 & -\varrho + \omega \end{bmatrix} \begin{bmatrix} 1 & a_1 \\ 0 & 1 \end{bmatrix} \begin{bmatrix} b_0 + a_0 d \\ b_1 + a_1 d \end{bmatrix}$$

$$a_1 = 2\varrho \; ; \; a_0 = \varrho^2 + \omega^2$$

The three representations (2.163), (2.164) and (2.165) are referred to as *Jordan forms.* In each case, the initial state $\boldsymbol{x}(k_0 - 1)$ corresponding to the initial conditions $\{y(k_0 - 1), y(k_0)\}$ is determined from (2.157).

Example 2.5.4. (i) Another possible choice $\{\boldsymbol{F}, \boldsymbol{g}, \boldsymbol{h}\}$ for a state space representation of the difference equation (2.161) in example 2.5.3 is given by

$$\boldsymbol{F} = \begin{bmatrix} 0.5 & 0 \\ 0 & -0.1 \end{bmatrix} \quad ; \quad \boldsymbol{h} = \begin{bmatrix} 1 \\ 1 \end{bmatrix} \quad ; \quad d = 4$$

The components $\{g_1, g_2\}$ of $\boldsymbol{g}$ corresponding to this choice of $\boldsymbol{F}$ therefore must satisfy the equations:

$$\begin{aligned} \boldsymbol{h}^T \boldsymbol{g} &= g_1 + g_2 = -0.64 \\ \boldsymbol{h}^T \boldsymbol{F} \boldsymbol{g} &= 0.5 g_1 - 0.1 g_2 = -0.376 \end{aligned} \tag{2.166}$$

That is,

$$\begin{bmatrix} 1 & 1 \\ 0.5 & -0.1 \end{bmatrix} \begin{bmatrix} g_1 \\ g_2 \end{bmatrix} = \begin{bmatrix} -0.64 \\ -0.376 \end{bmatrix}$$

which gives $\{g_1 = 0.52 \,, g_2 = -1.16\}$.

Since $\{u(3) = 5, u(4) = 1\}$, it follows from (2.139) that the initial state $\boldsymbol{x}(3)$ in (2.161) is given from $\{y(3), y(4)\}$ by

$$\boldsymbol{x}(3) = \begin{bmatrix} 0.26 & 0.116 \\ 0.52 & -1.16 \end{bmatrix}^{-1} \left\{ \begin{bmatrix} 2 \\ -3 \end{bmatrix} - \begin{bmatrix} 4 & -0.64 \\ 0 & 4 \end{bmatrix} \begin{bmatrix} 1 \\ 5 \end{bmatrix} \right\} = \begin{bmatrix} -1.60 \\ 13.94 \end{bmatrix}$$

(ii) Consider the difference equation

$$y(k+2) = y(k+1) + 0.29 y(k) + u(k+1) \quad ; \quad k \geq 3$$

subject to the initial conditions $\{y(3) = -3, y(4) = -2\}$ with $u(k) = 0$ for $k < 3$. This equation corresponds in (2.136) to

$$a_1 = 1 \; ; \; a_0 = 0.29 \; ; \; b_2 = 0 \; ; \; b_1 = 1 \; ; \; b_0 = 0 \; ; \; d = 0$$

with $k_0 = 5$. The characteristic equation $\lambda^2 = \lambda + 0.29$ has complex roots $\lambda_1 = 0.5 + j0.2 = \overline{\lambda_2}$.

A Jordan form of the state space representation is therefore given by

$$\boldsymbol{F} = \begin{bmatrix} 0.5 & 0.2 \\ -0.2 & 0.5 \end{bmatrix} \quad ; \quad \boldsymbol{h} = \begin{bmatrix} 1 \\ 1 \end{bmatrix}$$

where $\boldsymbol{g}^T = [g_1 \;\; g_2]$ is given by:

$$\begin{aligned} \boldsymbol{h}^T \boldsymbol{g} &= g_1 + g_2 = b_1 + a_1 d = 1 \\ \boldsymbol{h}^T \boldsymbol{F} \boldsymbol{g} &= 0.7 g_1 + 0.3 g_2 = b_0 + a_0 d + a_1(b_1 + a_1 d) = 1 \end{aligned}$$

That is,

$$\begin{bmatrix} 1 & 1 \\ 0.3 & 0.7 \end{bmatrix} \begin{bmatrix} g_1 \\ g_2 \end{bmatrix} = \begin{bmatrix} 1 \\ 1 \end{bmatrix}$$

which gives $\{g_1 = -0.75, g_2 = 1.75\}$.

From (2.139), the initial state $\boldsymbol{x}(4)$ in (2.161) is given from $\{y(3), y(4)\}$ by

$$\boldsymbol{x}(3) = \begin{bmatrix} 1.025 & -0.025 \\ 1.75 & -0.75 \end{bmatrix}^{-1} \begin{bmatrix} 2 \\ -3 \end{bmatrix} = \begin{bmatrix} 2.17 \\ 9.07 \end{bmatrix}$$

2.5.2 High Order

More generally, the system of n first order difference equations (2.132) can also be represented as a first order *vector* system

$$\begin{aligned} \boldsymbol{x}(k+1) &= \boldsymbol{F}\boldsymbol{x}(k) + \boldsymbol{g}u(k) \\ y(k) &= \boldsymbol{h}^T \boldsymbol{x}(k) + du(k) \end{aligned} \tag{2.167}$$

where now $\boldsymbol{x}(k) = [x_j(k)]$, $\boldsymbol{g} = [g_j]$ and $\boldsymbol{h} = [c_j]$ are all n-vectors, and $\boldsymbol{F} = [f_{jm}]$ is an n-square matrix. The n-vector $\boldsymbol{x}$ is called the *state* of the system, and $\boldsymbol{x}_j$ for $1 \leq j \leq n$ are called the *state components.*

The output signal $y(k)$ at time k in the state space representation (2.167) can be calculated if the values of the state $\boldsymbol{x}(k)$ and the input $u(k)$ are known. Alternatively, we have

$$\begin{aligned} y(k_0 - n) &= \boldsymbol{h}^T \boldsymbol{x}(k_0 - n) + du(k_0 - n) \\ \boldsymbol{x}(k_0 - n + 1) &= \boldsymbol{F}\boldsymbol{x}(k_0 - n) + \boldsymbol{g}u(k_0 - n) \\ y(k_0 - n + 1) &= \boldsymbol{h}^T \boldsymbol{x}(k_0 - n + 1) + du(k_0 - n + 1) \\ &= \boldsymbol{h}^T \boldsymbol{F}\boldsymbol{x}(k_0 - n) + \boldsymbol{h}^T \boldsymbol{g}u(k_0 - n) + du(k_0 - n + 1) \\ \boldsymbol{x}(k_0 - n + 2) &= \boldsymbol{F}\boldsymbol{x}(k_0 - n + 1) + \boldsymbol{g}u(k_0 - n + 1) \\ &= \boldsymbol{F}^2 \boldsymbol{x}(k_0 - n) + \boldsymbol{F}\boldsymbol{g}u(k_0 - n) + \boldsymbol{g}u(k_0 - n + 1) \\ y(k_0 - n + 2) &= \boldsymbol{h}^T \boldsymbol{x}(k_0 - n + 2) + du(k_0 - n + 2) \\ &= \boldsymbol{h}^T \boldsymbol{F}^2 \boldsymbol{x}(k_0 - n) + \boldsymbol{h}^T \boldsymbol{F}\boldsymbol{g}u(k_0 - n) \\ &\quad + \boldsymbol{h}^T \boldsymbol{g}u(k_0 - n + 1) + du(k_0 - n + 2) \end{aligned}$$

Continuing this process, we obtain the following result.

Theorem 2.5.3. *The complete response y in (2.167) is given in terms of the initial state $\boldsymbol{x}(k_0 - n)$ and the input signal u by*

$$\begin{aligned} \boldsymbol{x}(k) &= \boldsymbol{F}^{k-k_0+n}\boldsymbol{x}(k_0 - n) + \sum_{m=k_0-n}^{k-1} \boldsymbol{F}^{k-1-m}\boldsymbol{g}u(m) \\ y(k) &= \boldsymbol{h}^T\boldsymbol{F}^{k-k_0+n}\boldsymbol{x}(k_0 - n) + \sum_{m=k_0-n}^{k} h(k-m)u(m) \end{aligned} \tag{2.168}$$

where the signal h is given by

$$h(m) = \begin{cases} 0 & ;\ m < 0 \\ d & ;\ m = 0 \\ \boldsymbol{h}^T\boldsymbol{F}^{m-1}\boldsymbol{g} & ;\ m \geq 1 \end{cases} \tag{2.169}$$

Recall that the *unit impulse response* is the forced response y_f beginning at zero initial conditions due to a unit impulse input u defined by

$$u(m) \triangleq \begin{cases} 1 \,;\ m = 0 \\ 0 \,;\ m \neq 0 \end{cases}$$

Therefore from (2.168), the signal h in (2.169) is in fact the *unit impulse response* of (2.167).

By definition, the *complete response* $y = y_i + y_f$ in (2.168) is the sum of the *zero input* response y_i and the *forced response* y_f where

$$\begin{aligned} y_i(k) &\triangleq \boldsymbol{h}^T\boldsymbol{F}^{k-k_0+n}\boldsymbol{x}(k_0 - n) \\ y_f(k) &\triangleq \sum_{m=k_0-n}^{k} h(k-m)u(m) \end{aligned} \tag{2.170}$$

and where the unit impulse response h is defined by (2.169). The result in Theorem 2.5.1 can be extended as follows.

Theorem 2.5.4. *The first order system of n first order linear time invariant difference equations*

$$\begin{aligned} \boldsymbol{x}(k+1) &= \boldsymbol{F}\boldsymbol{x}(k) + \boldsymbol{g}u(k) \\ y(k) &= \boldsymbol{h}^T\boldsymbol{x}(k) + du(k) \end{aligned} \tag{2.171}$$

has the same input-output description as the nth order linear time invariant difference equation

$$y(k+n) = \sum_{p=0}^{n-1} a_p y(k+p) + \sum_{m=0}^{n} b_m u(k+m) \quad ; \quad k \geq k_0 - n$$

if and only if the matrices $\{\boldsymbol{F},\ \boldsymbol{g},\ \boldsymbol{h}\}$ *and the coefficients* $\{a_p, 0 \leq p \leq n-1\}$ *and* $\{b_q, 0 \leq q \leq n\}$ *are related according to*

$$b_n = d \tag{2.172}$$

$$\begin{bmatrix} 1 & -a_{n-1} & \cdot & \cdot & -a_1 \\ 0 & 1 & -a_{n-1} & \cdot & -a_2 \\ \cdot & \cdot & \cdot & \cdot & \cdot \\ \cdot & \cdot & \cdot & \cdot & \cdot \\ \cdot & \cdot & \cdot & \cdot & \cdot \\ 0 & 0 & \cdot & \cdot & 1 \end{bmatrix} \begin{bmatrix} \boldsymbol{h}^T \boldsymbol{F}^{n-1} \boldsymbol{g} \\ \boldsymbol{h}^T \boldsymbol{F}^{n-2} \boldsymbol{g} \\ \cdot \\ \cdot \\ \cdot \\ \boldsymbol{h}^T \boldsymbol{g} \end{bmatrix} = \begin{bmatrix} b_0 + a_0 d \\ b_1 + a_1 d \\ \cdot \\ \cdot \\ \cdot \\ b_{n-1} + a_{n-1} d \end{bmatrix}$$

with

$$det(\lambda \boldsymbol{I}_{\mathbf{n}} - \boldsymbol{F}) = \lambda^n - \sum_{p=0}^{n-1} a_p \lambda^p \tag{2.173}$$

The initial conditions $\{y(k_0 - n + 1), y(k_0 - n + 2), \ \ldots \ y(k_0)\}$ *of (2.87) are related to the initial state* $\boldsymbol{x}(k_0 - n + 1)$ *of (2.167) by*

$$\boldsymbol{M}_n \boldsymbol{x}(k_0 - n + 1) = \begin{bmatrix} y(k_0) \\ y(k_0 - 1) \\ \cdot \\ \cdot \\ y(k_0 - n + 2) \\ y(k_0 - n + 1) \end{bmatrix} - \boldsymbol{H} \begin{bmatrix} u(k_0) \\ u(k_0 - 1) \\ \cdot \\ \cdot \\ u(k_0 - n + 2) \\ u(k_0 - n + 1) \end{bmatrix} \tag{2.174}$$

where for $n \geq 2$*, the* n*-square matrices* $\boldsymbol{M}_n$ *and* $\boldsymbol{H}$ *are given by*

$$\boldsymbol{M}_n \triangleq \begin{bmatrix} \boldsymbol{h}^T \boldsymbol{F}^{n-1} \\ \boldsymbol{h}^T \boldsymbol{F}^{n-2} \\ \cdot \\ \cdot \\ \boldsymbol{h}^T \boldsymbol{F} \\ \boldsymbol{h}^T \end{bmatrix} ; \ \boldsymbol{H} \triangleq \begin{bmatrix} d & \boldsymbol{h}^T \boldsymbol{g} & \boldsymbol{h}^T \boldsymbol{F} \boldsymbol{g} & \cdot & \cdot & \boldsymbol{h}^T \boldsymbol{F}^{n-2} \boldsymbol{g} \\ 0 & d & \boldsymbol{h}^T \boldsymbol{g} & \cdot & \cdot & \boldsymbol{h}^T \boldsymbol{F}^{n-3} \boldsymbol{g} \\ \cdot & \cdot & \cdot & \cdot & \cdot & \cdot \\ \cdot & \cdot & \cdot & \cdot & \cdot & \cdot \\ \cdot & \cdot & \cdot & \cdot & \cdot & \cdot \\ 0 & 0 & \cdot & \cdot & 0 & d \end{bmatrix} \tag{2.175}$$

System Matrix and Input/Output Vectors

Following on from the earlier work for second order systems, we can summarize the available freedom in the choice for the *system matrix* $\boldsymbol{F}$, the *input vector* $\boldsymbol{g}$, and the *output vector* $\boldsymbol{h}$ of the state space representation (2.171) as follows:

1. Condition (2.173) imposes a constraint on the characteristic polynomial of the system matrix $\boldsymbol{F}$.
2. From (2.174), the invertibility of the n-square matrix $\boldsymbol{M}_n$ is a necessary condition in order to be able to determine the initial state $\boldsymbol{x}(k_0 - 1)$ for *all* initial conditions $\{y(k_0), y(k_0 - 1), \ldots, y(k_0 - n + 1)\}$.
3. Since

$$\begin{bmatrix} \boldsymbol{h}^T\boldsymbol{F}^{n-1}\boldsymbol{g} \\ \boldsymbol{h}^T\boldsymbol{F}^{n-2}\boldsymbol{g} \\ \cdot \\ \cdot \\ \boldsymbol{h}^T\boldsymbol{F}\boldsymbol{g} \\ \boldsymbol{h}^T\boldsymbol{g} \end{bmatrix} = \begin{bmatrix} \boldsymbol{h}^T\boldsymbol{F}^{n-1} \\ \boldsymbol{h}^T\boldsymbol{F}^{n-2} \\ \cdot \\ \cdot \\ \boldsymbol{h}^T\boldsymbol{F} \\ \boldsymbol{h}^T \end{bmatrix} \boldsymbol{g} = \boldsymbol{M}_n\boldsymbol{g}$$

the last remaining condition (2.172) which is imposed on the choice of $\{\boldsymbol{F}, \boldsymbol{g}, \boldsymbol{h}\}$ can be arranged into the form

$$\boldsymbol{D}_n\boldsymbol{M}_n\boldsymbol{g} = \boldsymbol{\gamma}_n$$

where

$$\boldsymbol{D}_n \triangleq \begin{bmatrix} 1 & -a_{n-1} & \cdot & \cdot\cdot & -a_1 \\ 0 & 1 & -a_{n-1} & \cdot\cdot & -a_2 \\ \cdot & \cdot & \cdot & \cdot\cdot & \cdot \\ \cdot & \cdot & \cdot & \cdot\cdot & \cdot \\ \cdot & \cdot & \cdot & \cdot\cdot & -a_{n-1} \\ 0 & 0 & \cdot & \cdot\cdot & 1 \end{bmatrix} ; \ \boldsymbol{\gamma}_n \triangleq \begin{bmatrix} b_0 + a_0 d \\ b_1 + a_1 d \\ \cdot \\ \cdot \\ \cdot \\ b_{n-1} + a_{n-1} d \end{bmatrix} \tag{2.176}$$

We therefore have the following result.

Theorem 2.5.5. *Suppose the system matrix $\boldsymbol{F}$ has the characteristic polynomial (2.173), and the output vector $\boldsymbol{h}$ is chosen such that the n-square matrix $\boldsymbol{M}_n$ defined by (2.175) is invertible.*

Then $\{\boldsymbol{F}, \boldsymbol{g}, \boldsymbol{h}, d\}$ where

$$d = b_n \ ; \ \boldsymbol{g} = \boldsymbol{M}_n^{-1}\boldsymbol{D}_n^{-1}\boldsymbol{\gamma}_n \tag{2.177}$$

and $\{\boldsymbol{D}_n, \boldsymbol{\gamma}_n\}$ are defined by (2.176) is a state space representation of the difference equation (2.87) with initial state

$$\boldsymbol{x}(k_0 - 1) = \boldsymbol{M}_n^{-1} \left\{ \begin{bmatrix} y(k_0) \\ y(k_0 - 1) \\ \cdot \\ \cdot \\ y(k_0 - n + 2) \\ y(k_0 - n + 1) \end{bmatrix} - \boldsymbol{H} \begin{bmatrix} u(k_0) \\ u(k_0 - 1) \\ \cdot \\ \cdot \\ u(k_0 - n + 2) \\ u(k_0 - n + 1) \end{bmatrix} \right\} \tag{2.178}$$

where the n-square matrix $\boldsymbol{H}$ is defined by (2.175).

Phase Variable Form

The state space structure $\{\boldsymbol{F}, \boldsymbol{h}\}$ where

$$\boldsymbol{F} = \begin{bmatrix} 0 & 0 & \cdot & \cdot & a_0 \\ 1 & 0 & \cdot & \cdot & a_1 \\ 0 & 1 & \cdot & \cdot & \cdot \\ \cdot & \cdot & \cdot & \cdot & \cdot \\ 0 & 0 & \cdot & 1 & a_{n-1} \end{bmatrix} \quad ; \quad \boldsymbol{h} = \begin{bmatrix} 0 \\ 0 \\ \cdot \\ \cdot \\ 1 \end{bmatrix} \tag{2.179}$$

always satisfies conditions (2.172), (2.173), and is a direct generalization of (2.158).

The initial state $\boldsymbol{x}(k_0 - n + 1)$ is given from $\{y(k_0), y(k_0 - 1), \ldots y(k_0 - n + 1)\}$ and $\{u(k_0), u(k_0 - 1), \ldots u(k_0 - n + 1)\}$ by (2.174), (2.175) where the corresponding n-square matrix $\boldsymbol{M}_n$ is of the form

$$\boldsymbol{M}_n = \begin{bmatrix} 1 & a_{n-1} & \cdot & \cdot & * & * \\ \cdot & 1 & a_{n-1} & \cdot & \cdot & \cdot \\ \cdot & \cdot & \cdot & \cdot & \cdot & \cdot \\ \cdot & \cdot & \cdot & \cdot & \cdot & \cdot \\ 0 & 0 & \cdot & 1 & a_{n-1} & * \\ 0 & 0 & \cdot & \cdot & 1 & a_{n-1} \\ 0 & 0 & \cdot & \cdot & 0 & 1 \end{bmatrix} \quad ; \quad det(\boldsymbol{M}_n) = 1 \tag{2.180}$$

for some components $*$, and so $\boldsymbol{M}_n$ is always invertible. It also follows that

$$\boldsymbol{M}_n^{-1} \boldsymbol{D}_n^{-1} = \boldsymbol{I}_{\mathbf{n}}$$

and so from (2.177)

$$\boldsymbol{g} = \boldsymbol{g}_0 = \boldsymbol{\gamma}_n = \begin{bmatrix} b_0 + a_0 d \\ b_1 + a_1 d \\ \cdot \\ \cdot \\ b_{n-1} + a_{n-1} d \end{bmatrix} \tag{2.181}$$

The representation in (2.179), (2.181) is referred to as the *phase variable form.*

Example 2.5.5. From Theorem 2.5.4, the third order difference equation

$$\begin{aligned} y(k+3) = {} & a_2 y(k+2) + a_1 y(k+1) + a_0 y(k) \\ & + b_0 u(k) + b_1 u(k+1) + b_2 u(k+2) + b_3 u(k+3) \end{aligned} \tag{2.182}$$

has the equivalent state space description (2.167) for $n = 3$ where

$$b_3 = h(0)$$

$$\begin{bmatrix} b_0 + a_0 d \\ b_1 + a_1 d \\ b_2 + a_2 d \end{bmatrix} = \begin{bmatrix} 1 & -a_2 & -a_1 \\ 0 & 1 & -a_2 \\ 0 & 0 & 1 \end{bmatrix} \begin{bmatrix} h(3) \\ h(2) \\ h(1) \end{bmatrix} \tag{2.183}$$

with $h(k) = \boldsymbol{h}^T \boldsymbol{F}^{k-1} \boldsymbol{g}$ for $k \geq 1$, and

$$det(\lambda \boldsymbol{I_3} - \boldsymbol{F}) = \lambda^3 - a_2\lambda^2 - a_1\lambda - a_0$$

The initial conditions $\{y(k_0 - 3), y(k_0 - 2), y(k_0 - 1)\}$ of (2.182) are related to the initial state $\boldsymbol{x}(k_0 - 3)$ of (2.167) by

$$\begin{bmatrix} \boldsymbol{h}^T \boldsymbol{F}^2 \\ \boldsymbol{h}^T \boldsymbol{F} \\ \boldsymbol{h}^T \end{bmatrix} \boldsymbol{x}(k_0 - 2) = \begin{bmatrix} y(k_0) \\ y(k_0 - 1) \\ y(k_0 - 2) \end{bmatrix} - \begin{bmatrix} h(0) & h(1) & h(2) \\ 0 & h(0) & h(1) \\ 0 & 0 & h(0) \end{bmatrix} \begin{bmatrix} u(k_0 - 2) \\ u(k_0 - 1) \\ u(k_0) \end{bmatrix}$$

In particular, one possible choice for $\{\boldsymbol{F}, \boldsymbol{g}, \boldsymbol{h}\}$ is the *phase variable canonical form* given by

$$\boldsymbol{F}_0 = \begin{bmatrix} 0 & 0 & a_0 \\ 1 & 0 & a_1 \\ 0 & 1 & a_2 \end{bmatrix} \; ; \; \boldsymbol{g}_0 = \begin{bmatrix} b_0 + a_0 d \\ b_1 + a_1 d \\ b_2 + a_2 d \end{bmatrix} \; ; \; \boldsymbol{h}_0 = \begin{bmatrix} 0 \\ 0 \\ 1 \end{bmatrix} \; ; \; d = b_3$$

for which the corresponding values $\{h(k) \; ; \; 0 \leq k \leq 3\}$ of the unit impulse response h are given by

$$\begin{aligned} h(0) &= b_3 \; (= d) \\ h(1) &= \boldsymbol{h}^T \boldsymbol{g} = b_2 + a_2 d \\ h(2) &= \boldsymbol{h}^T \boldsymbol{F} \boldsymbol{g} = b_1 + a_1 d + a_2(b_2 + a_2 d) \\ h(3) &= \boldsymbol{h}^T \boldsymbol{F}^2 \boldsymbol{g} = b_0 + a_0 d + a_1(b_2 + a_2 d) + a_2[b_1 + a_1 d + a_2(b_2 + a_2 d)] \end{aligned}$$

Jordan Form

Consider a $n = 4$th order difference equation in Theorem (2.5.4) where

$$det(\lambda \boldsymbol{I}_4 - \boldsymbol{F}) = \prod_{k=1}^{4} (\lambda - \lambda_k)$$

For the moment, assume all $\{\lambda_k\}$ are real and distinct. Then a state space structure $\{\boldsymbol{F}, \boldsymbol{h}\}$ where $\boldsymbol{F}$ is in the Jordan form

$$\boldsymbol{F} = \begin{bmatrix} \lambda_1 & 0 & 0 & 0 \\ 0 & \lambda_2 & 0 & 0 \\ 0 & 0 & \lambda_3 & 0 \\ 0 & 0 & 0 & \lambda_4 \end{bmatrix} \; ; \; \boldsymbol{h} = \begin{bmatrix} 1 \\ 1 \\ 1 \\ 1 \end{bmatrix} \tag{2.184}$$

always satisfies conditions (2.172), (2.173), and is a direct generalization of (2.163). When $\{\lambda_1 = \lambda_2 = \lambda_3 \neq \lambda_4\}$, then

$$\boldsymbol{F} = \begin{bmatrix} \lambda_1 & 1 & 0 & 0 \\ 0 & \lambda_1 & 1 & 0 \\ 0 & 0 & \lambda_1 & 0 \\ 0 & 0 & 0 & \lambda_4 \end{bmatrix} \quad ; \quad \boldsymbol{h} = \begin{bmatrix} 1 \\ 1 \\ 1 \\ 1 \end{bmatrix} \tag{2.185}$$

always satisfies conditions (2.172), (2.173). When $\{\lambda_1 = \varrho + j\omega = \overline{\lambda_2},\ \lambda_3 \neq \lambda_4\}$, then

$$\boldsymbol{F} = \begin{bmatrix} \varrho & \omega & 0 & 0 \\ -\omega & \varrho & 0 & 0 \\ 0 & 0 & \lambda_3 & 0 \\ 0 & 0 & 0 & \lambda_4 \end{bmatrix} \quad ; \quad \boldsymbol{h} = \begin{bmatrix} 1 \\ 1 \\ 1 \\ 1 \end{bmatrix} \tag{2.186}$$

always satisfies conditions (2.172), (2.173). Finally, when $\lambda_1 = \varrho + j\omega = \overline{\lambda_2},\ \lambda_3 = \lambda_1, \lambda_4 = \lambda_2\}$, then

$$\boldsymbol{F} = \begin{bmatrix} \varrho & \omega & 1 & 0 \\ -\omega & \varrho & 0 & 1 \\ 0 & 0 & \varrho & \omega \\ 0 & 0 & -\omega & \varrho \end{bmatrix} \quad ; \quad \boldsymbol{h} = \begin{bmatrix} 1 \\ 1 \\ 1 \\ 1 \end{bmatrix} \tag{2.187}$$

always satisfies conditions (2.172), (2.173). All of the above 4-square matrices are examples of (real) Jordan forms whose structure depends on whether or not eigenvalues are real or complex, distinct or repeated. Extensions to even higher order difference equations follow in a similar way.

2.5.3 Steady State Sinusoidal Response

We have shown in Theorem 2.4.1 that the forced response $y_f = h * u$ is given by the linear convolution of the unit impulse response h and the input signal u. Furthermore in Theorem 2.4.2, we concluded that the steady state response y_{ss} of an asymptotically stable difference equation (2.87) as the result of a sinusoidal input signal $u(k) = \sin k\omega_0$ is given by

$$y_{ss}(k) = \varrho_0 \sin(k\omega_0 - \phi_0) \ ; \ \varrho_0 exp\{-j\phi_0\} = H(exp\{j\omega_0\})$$

where

$$H(z) = \sum_{p=0}^{\infty} h(p) z^{-p} = \frac{\sum_{m=0}^{n} b_m z^m}{z^n - \sum_{p=0}^{n-1} a_p z^p} \tag{2.188}$$

For the state space representation (2.134) of (2.87), the unit impulse response h is given by (2.169). Therefore

$$H(z) = \sum_{p=0}^{\infty} h(p) z^{-p} = d + z^{-1} \sum_{p=1}^{\infty} \boldsymbol{h}^T \boldsymbol{F}^{p-1} \boldsymbol{g} z^{-(p-1)}$$

That is

$$H(z) = d + z^{-1} \sum_{m=0}^{\infty} \boldsymbol{h}^T \boldsymbol{F}^m \boldsymbol{g} z^{-m} \tag{2.189}$$

Now

$$\begin{aligned}(z\boldsymbol{I_n} - \boldsymbol{F})^{-1} &= z^{-1}(\boldsymbol{I_n} - z^{-1}\boldsymbol{F})^{-1} \\ &= z^{-1}[\boldsymbol{I_n} + z^{-1}\boldsymbol{F} + (z^{-1}\boldsymbol{F})^2 + \ \dots \ + (z^{-1}\boldsymbol{F})^n + \ \dots]\end{aligned}$$

and so from (2.189)

$$H(z) = d + \boldsymbol{h}^T (z\boldsymbol{I_n} - \boldsymbol{F})^{-1} \boldsymbol{g} \tag{2.190}$$

Formally, the series for $(\boldsymbol{I_n} - z^{-1}\boldsymbol{F})^{-1}$ can be shown to be *convergent* for all z which guarantee that all eigenvalues of $z^{-1}\boldsymbol{F}$ have magnitude less than unity. In particular, when $z = exp\{j\omega\}$ (so that $|z| = 1$), the series is convergent provided all eigenvalues of $\boldsymbol{F}$ have magnitude less than unity. Moreover, if the system is asymptotically stable, then all eigenvalues of $\boldsymbol{F}$ have less than unity magnitude. We therefore have the following result.

Theorem 2.5.6. *The frequency response function of an asymptotically stable linear time invariant system (2.167) is given by*

$$H(z) = d + \boldsymbol{h}^T (z\boldsymbol{I_n} - \boldsymbol{F})^{-1} \boldsymbol{g} \tag{2.191}$$

Hence when the input signal u is given by $u(k) = \sin k\omega_0$, the steady state component y_{ss} of the output y is given by

$$y_{ss}(k) = \varrho_0 \sin(k\omega_0 - \phi_0) \ ; \ \varrho_0 exp\{-j\phi_0\} = H(exp\{j\omega_0\})$$

The expression for H in (2.191) is often more convenient to use than the expression in (2.188) which involves the ratio of two polynomials in z. This is because (2.191) can be evaluated directly using programs (such as MATLAB) which are based on linear algebra computations using complex arithmetic.

Summary

This chapter began with a detailed treatment of first order linear time invariant difference equations. The concepts and significance of *order*, *time invariance* and *linearity* were developed. Although it was not evident at the time, it later turned out in section 2.5 that the solution of a first order difference equation equation leads naturally to the solution of a first order *system* of difference equations. The study of second order difference equations was also considered, and again it was shown that the *complete response* can be decomposed as the sum of a *zero input response* and the *forced response.* It was also shown that the complete response can also be decomposed into a *transient response* and a *steady state response.* The *transient response* is that part of the complete response that ultimately goes to zero as time increases. The remaining part of the complete response is then the *steady state response.*

The *zero input solution* is the response to a particular set of initial conditions when the input signal is zero. The number of initial conditions that are needed to uniquely define the zero input solution is equal to the order of the difference equation. The *forced response* is the response to a particular input signal beginning at zero initial conditions. Particular forced responses were given special names: the *unit impulse response* is the forced response when the input is a unit impulse at time $k = 0$ and the *unit step response* is the forced response when the input is a unit step at time $k = 0$.

The key property that is needed for the solution of the zero input response of a difference equation is the solution of an algebraic equation known as the *characteristic equation.* A second order difference equation has an assosciated second order characteristic equation. The roots of the characteristic equation are referred to as the *characteristic roots*, and the form of the solution of a zero input response of a second order difference equation was developed when both roots are complex, when both roots are real and unequal, and when both roots are real and equal.

Once first and second order equations were covered in some detail, it was then appropriate to consider to the study of a general nth order linear time invariant difference equation. In particular, it was shown that the form of the zero input response is determined by the n characteristic roots of the nth order characteristic equation. A difference equation was said to be *asymptotically stable* if the zero input response decayed to zero for all initial conditions. It was shown that asymptotic stability is equivalent to all characteristic roots of the characteristic equation having magnitude strictly less than unity. In addition, a result was given that enabled a test for asymptotic stability (known as the *Marden stability test*) to be performed in terms of the coefficients of the characteristic equation.

Furthermore, it was shown that the forced response $y_f = h * u$ is given by the *linear convolution* of the input signal u and the unit impulse response h. This is a very important result which is fundamental to many signal processing developments. An algebraic expression was also developed which gave the magnitude response and phase response as functions of frequency of the steady state output response of an asymptotically stable difference equation when the input signal is sinusoidal. This characteristic of linear time invariant difference equations is the basis for an understanding of the design of digital filters in Chapter 3.

We showed that any nth order linear time invariant difference equation can be non-uniquely expressed as a system of n first order linear time invariant difference equations known as the *state space representation*. The state space representation provides a unified matrix algebra framework with which to numerically calculate the solutions of difference equations. Later in Chapter 5, we shall see that particular internal realizations of digital filters correspond to particular state space representations, and even though all such realizations are ideally equivalent, they are not equivalent when one examines the numerical errors that result because of finite precision arithmetic.

Exercises

2.1 (i) Show that (2.9) is a Backward Euler approximation of (2.5).
(ii) Show that (2.11) is a Forward Euler approximation of (2.5).
(iii) Graphically illustrate the integral approximation

$$\int_{kT}^{(k+1)T} z(\sigma)d\sigma \approx T[\alpha z(kT+T) + (1-\alpha)z(kT)]$$

for $\alpha = 0, 0.25, 0.5, 0.75, 1$.

(iv) Based on the above integral approximation for $0 \leq \alpha \leq 1$, find a difference equation whose solution approximates the solution of (2.5). Conclude that (i) and (ii) are special cases of this result.

2.2 What are the orders of the following difference equations in which u is the input signal and y is the output signal:

$$\begin{aligned}
&\text{(i)} \quad y(k+1) = y(k-1) + u(k-1) + u(k+1) \\
&\text{(ii)} \quad y(k+1) = y(k) + u(k-2) \\
&\text{(iii)} \quad y(k+1) = y(k-1) + u(k-2) \\
&\text{(iv)} \quad y(k+1) = y(k+2) + u(k-1) - y(k-1)
\end{aligned}$$

2.3 An attendant checks out customers from a queue at a supermarket. Suppose $u(k)$ is the time at which the kth customer arrives at the checkout, and $y(k)$ is the time at which the customer has been checked out. If each customer is served in T minutes, derive a difference equation for y in terms of u.
(i) Is the difference equation linear or nonlinear ?
(ii) Is the difference equation time invariant or time varying ?

2.4 Consider the difference equation with input u and output y:

$$y(k+1) - 0.5y(k) = u(k) + 0.6u(k+1)$$

(i) Find the unit impulse response
(ii) Find the complete response for $k \geq 0$ when $y(-1) = 2$, and

$$u(k) = \begin{cases} 1 \; ; \; k = 1 \\ 0 \; ; \; \text{otherwise} \end{cases}$$

(iii) Find the complete response for $k \geq 0$ when $y(-1) = -3$ and $u(k) = \sin(0.5k - 0.7)$.

2.5 Consider the difference equation

$$y(k+1) = y^2(k) + u(k) \quad ; \quad k \geq 0$$

Suppose $y(0) = -1$.

(i) Find $\{y(k); 0 \leq k \leq 4\}$ when: (a) $u(k) = 1$ for $k \geq 0$, and (b) $u(k) = 2$ for $k \geq 0$.

(ii) By comparing the two response signals in (a) and (b), show that the difference equation is nonlinear.

2.6 Consider the difference equation

$$y(k+1) = 2^k y(k) + u(k)$$

(i) Find $\{y(k);\ 0 \leq k \leq 3\}$ when $y(0) = 0$ and $u(k) = 1$ for $k \geq 0$.

(ii) Find $\{y(k);\ 2 \leq k \leq 5\}$ when $y(2) = 0$ and $u(k) = 1$ for $k \geq 2$.

(iii) By comparing the two resulting response signals in (i) and (ii), show that the system is time varying.

2.7 Find the unit impulse responses of the difference equations:

$$\text{(i)} \quad y(k+1) = ky(k) + u(k) \quad ; \quad \text{(ii)} \quad y(k+1) = k^2 y(k) + u(k)$$

2.8 (i) Show that the difference equation

$$y(k+1) = 0.5y(k) + y^{-1}(k)u(k)$$

is nonlinear.

(ii) Suppose $\{u(k) = 4,\ y(0) = 1\}$. Show that $y(k) \to 2$ as $k \to \infty$.

(iii) In order to solve for the roots of the equation $f(y) = 0$, one can use Newton's method which is based on the approximation

$$\frac{df(y(k))}{dy} \simeq \frac{f(y(k+1)) - f(y(k))}{y(k+1) - y(k)}$$

Show that when $f(y) = y^2 - \alpha^2$, Newton's method results in the difference equation in (i). What is the corresponding signal u ?

(iii) What is the difference equation which results from Newton's method when iteratively solving the equation $y^3 = \beta$. ?

2.9 A savings account with interest accrued quarterly can be modelled by the first order difference equation

$$y(k+1) = (1 + 0.25r)y(k) + u(k+1)$$

where $u(k)$ is the amount deposited at the beginning of the kth quarter, $y(k)$ is the balance of the account at the end of the kth quarter, and r is the annual interest rate (as a decimal).

(i) Suppose $y(0) = \$1000$ and $r = 6\%$. Compute $y(k)$ for $1 \leq k \leq 5$ when $u(k) = \$500$ for $1 \leq k \leq 5$.

(ii) Find the integer N such that $y(N) \geq \$10,000$.

2.10 Suppose a loan is obtained from a bank to purchase a car worth \$5000. Assume the monthly repayments made at the beginning of each month are \$100, and that the annual interest rate charged on the loan is r (as a decimal).

(i) Derive a difference equation for the value $y(k)$ of the debt at the end of the kth month

(ii) How many months will it take to pay off the loan if $r = 12\%$? How much will the last payment be ?

2.11 The capacitor voltage response v_c of a series first order RC circuit as the result of an applied input voltage u is given by the solution of the first order differential equation

$$RC\dot{v}_c(t) + v_c(t) = u(t)$$

(i) Find a difference equation whose solution y provides an approximation $y(k)$ of $v_c(kT)$ where T is fixed and k is an integer based on the Euler approximation of an integral.

(ii) Repeat (i) based on the *trapezoidal* approximation of an integral

(iii) Suppose $v_c(0) = 0$ and $u(t) = 1$ for $t \geq 0$. Shows that the exact solution $v_c(t)$ is given by

$$v_c(t) = 1 - exp\{-\frac{t}{RC}\}$$

(iv) Compare the accuracy of the solutions in (i) and (ii) with the exact solution in (iii) when: (a) $T = 0.1RC$ and (b) $T = RC$.

2.12 Find the complete response y of the following first order difference equations:

$$\begin{aligned} y(k+1) + 0.2y(k) &= 1 \; ; \; k \geq 1 \; , \; y(1) = 2.8 \\ 2y(k+1) + 0.6y(k) &= k^2 \; ; \; k \geq 0 \; , \; y(0) = -1.5 \\ y(k+1) - 0.1y(k) &= (0.1)^k \; ; \; k \geq 0, \; y(0) = 0 \end{aligned}$$

In each case:

(a) Plot $y(k)$ for $1 \leq k \leq 7$.

(b) Decompose $y = y_i + y_f$ where y_i is the zero input response, and y_f is the forced response.

(c) Decompose $y = y_t + y_{ss}$ where y_t is the transient response, and y_{ss} is the steady state response.

2.13 (i) Find the forced response of the difference equation:

$$y(k+1) - 0.5y(k) = 2u(k) \; ; \; k \geq 0$$

when $u(k) = \sin 0.5k$.

(ii) Find the steady state response of the difference equation:

$$2y(k+1) + y(k) = u(k) \; ; \; k \geq 1$$

when $u(k) = \sin k$.

(iii) Find the steady state response of the difference equation in (i) when

$$u(k) = 3\cos(k - 0.25\pi)$$

2.14 Find a general expression for $y(k)$ when

$$y(k+2) = y(k+1) + y(k) \quad ; \quad k \geq 0$$

with the initial conditions $\{y(0) = 0, y(1) = 1\}$. (The solution y defines the Fibonacci numbers.)

2.15 The roots of the algebraic equation $\lambda^2 - a_1\lambda - a_0 = 0$ are given by

$$\lambda_{1,2} = 0.5(a_1 \pm \sqrt{a_1^2 + 4a_0})$$

Show that $\{|\lambda_1| < 1, |\lambda_2| < 1\}$ if and only if $\{|a_0| < 1 \;\; ; \;\; |a_1| < 1 - a_0\}$.

2.15 (i) Find the unit impulse response h of the first order time varying difference equation

$$y(k+1) = (k+1)y(k) + u(k+1)$$

and express h in the form

$$h(k) = (k!)^2 p(k)$$

(ii) Find the unit impulse response $\{h(k);\ 0 \leq k \leq 4\}$ of the second order time varying difference equation

$$y(k+2) = 2\alpha y(k+1) + 2(k+1)y(k) + u(k+2)$$

[The solution $y(k)$ is known as a Hermite polynomial in α.]

2.16 Express the solution of the difference equation

$$y(k+2) = 4y(k+1) - 3y(k)$$

in terms of the initial conditions $\{y(0), y(1)\}$.

2.17 Consider the second order difference equation

$$y(k+2) = 2\cos\theta \; y(k+1) - y(k)$$

(i) Show that the solution y is given by

$$y(k) = \frac{\sin(k+1)\theta}{\sin\theta}$$

when $\{y(-2) = -1,\ y(-1) = 0\}$.

(ii) Show that the solution y is given by

$$y(k) = \cos k\theta$$

when $\{y(-2) = \cos 2\theta,\ y(-1) = \cos\theta\}$.

2.18 Find the unit impulse response of the following second order difference equations:

$$y(k+2) = y(k+1) - 0.25y(k) + u(k+1)$$
$$y(k+2) = -y(k+1) - 0.5y(k) + 2u(k+2) - 3u(k+1)$$

2.19 Consider the difference equations

$$v(k+1) = i(k+1)Z_{k+1} + v(k+2)$$
$$i(k) = v(k+1)Y_{k+1} + i(k+1)$$

for $k = n-2, n-3, \ldots, 2, 1$ where

$$v(n) = Ri(n) \;\; ; \;\; v(1) = 1$$

(i) Suppose $n = 3$. Find $v(3)$ in terms of $\{Z_m, Y_m\}$
(ii) What are the conditions on $\{Z_m, Y_m\}$ such that

$$|v(3)| < |v(2)| < 1$$

2.20 Suppose a linear time invariant difference equation has the unit step response s as given by

$$s(k) = (0.5)^k \sin 0.5\pi k \;\; ; \;\; k \geq 0$$

(i) Compute the unit impulse response h
(ii) Compute the forced response y when the input signal u is given by

$$u(k) = \begin{cases} 2^k \; ; \; 1 \leq k \leq 4 \\ 0 \; ; \; k \geq 5 \end{cases}$$

2.21 Consider the second order difference equations:

$$y(k+2) = 0.7y(k+1) - 0.1y(k) + u(k)$$
$$y(k+2) = 0.4y(k+1) + 0.32y(k) + 2u(k+1) + u(k)$$
$$y(k+2) = -0.1y(k+1) + 0.3y(k) + u(k) - u(k+1)$$

In each case:
(i) Find the *zero input response* beginning with the initial conditions $\{y(1) = -2,\; y(2) = 1\}$.
(ii) Find the unit impulse response
(iii) Find the steady state sinusoidal response when $u(k) = 2\cos(\pi/8)k$.

2.22 (i) Using the Marden stability test, determine whether or not the zero input responses of the following difference equations decay to zero for all initial conditions:

$$\begin{aligned} y(k+2) - 0.6y(k+1) - 0.7y(k) &= u(k) + 4u(k-1) \\ 2y(k+2) - 0.8y(k+1) + 0.6y(k) &= u(k) \\ y(k+1) - 0.9y(k) - 0.1y(k-2) &= u(k) - 2u(k-1) \end{aligned}$$

(ii) Suppose $u(k) = \alpha y(k)$ For some constant α. For each of the difference equations in (i), find conditions (if they exist) on the constant parameter α which guarantee asymptotic stability.

2.23 Suppose the unit impulse response of a third order difference equation is given by

$$h(k) = \begin{cases} -0.5(0.6)^k + (0.2)^k - 2(-0.1)^k & ;\ k \geq 0 \\ 0 & ;\ k < 0 \end{cases}$$

(i) Find the forced response for $k \geq 0$ when the input $u(k) = 0$ for $k < 0$, and $\{u(0) = 2;\ u(1) = -2\}$ and $u(k) = 0,\ k \geq 2$.

(ii) Find the unit step response

2.24 Consider the following third order difference equations:

$$\begin{aligned} y(k+3) &= 0.03y(k+1) + 0.002y(k) - u(k+2) + 2u(k+1) \\ y(k+3) &= 0.3y(k+2) + 0.34y(k+1) - 0.12y(k) - u(k+3) \\ y(k+3) &= 0.12y(k+2) - 0.16y(k+1) + 0.064y(k) - u(k+2) + u(k) \\ y(k+3) &= -0.8y(k+2) - 0.3y(k+1) + 0.1y(k) + u(k+1) - 4u(k) \end{aligned}$$

In each case, find:

(i) The unit impulse response

(ii) The steady state response due to the input $u(k) = 0.1\sin(0.5\pi k - 0.6)$. [Note that

$$\begin{aligned} \lambda^3 - 0.03\lambda - 0.002 &= (\lambda + 0.1)^2(\lambda - 0.2) \\ \lambda^3 - 0.3\lambda^2 - 0.34\lambda + 0.12 &= (\lambda - 0.5)(\lambda + 0.6)(\lambda - 0.4) \\ \lambda^3 - 0.12\lambda^2 + 0.16\lambda - 0.064 &= (\lambda - 0.4)^3 \\ \lambda^3 + 0.8\lambda^2 + 0.3\lambda - 0.1 &= (\lambda^2 + \lambda + 0.5)(\lambda - 0.2)\ ] \end{aligned}$$

2.25 Consider the second order difference equations as presented in Exercise 2.21. In each case:

(i) Find the *phase variable* state space representation $\{\boldsymbol{F}, \boldsymbol{g}, \boldsymbol{h}, d\}$ corresponding to the initial conditions $\{y(1) = -2,\ y(2) = 1\}$.

(ii) Repeat part (i) when $\{y(3) = -2,\ y(4) = 1\}$.

(iii) Find a *Jordan form* state space representation $\{\boldsymbol{F}, \boldsymbol{g}, \boldsymbol{h}, d\}$ when $\boldsymbol{g}^T = [1\ \ 1]$ and $\{y(1) = -2,\ y(2) = 1\}$.

(iv) Using the results of either part (i) or part (ii), calculate the frequency function $H(z)$ between the output y and the input u.

2.26 (i) Show that (alternative phase variable) state space representations $\{\boldsymbol{F}, \boldsymbol{g}, \boldsymbol{h}, d\}$ of the second order difference equation

$$y(k+2) = a_1 y(k+1) + a_0 y(k) + b_0 u(k) + b_1 u(k+1) + b_2 u(k+2)$$

are given by $d = b_2$, and

$$(a)\quad \boldsymbol{F} = \begin{bmatrix} a_1 & a_0 \\ 1 & 0 \end{bmatrix} \ ;\ \boldsymbol{g} = \begin{bmatrix} 1 \\ 0 \end{bmatrix} \ ;\ \boldsymbol{h} = \begin{bmatrix} b_0 + a_0 d \\ b_1 + a_1 d \end{bmatrix}$$

$$(b)\quad \boldsymbol{F} = \begin{bmatrix} 0 & 1 \\ a_0 & a_1 \end{bmatrix} \ ;\ \boldsymbol{g} = \begin{bmatrix} 0 \\ 1 \end{bmatrix} \ ;\ \boldsymbol{h} = \begin{bmatrix} b_1 + a_1 d \\ b_0 + a_0 d \end{bmatrix}$$

$$(c)\quad \boldsymbol{F} = \begin{bmatrix} a_0 & 0 \\ a_1 & 1 \end{bmatrix} \ ;\ \boldsymbol{g} = \begin{bmatrix} b_1 + a_1 d \\ b_0 + a_0 d \end{bmatrix} \ ;\ \boldsymbol{h} = \begin{bmatrix} 1 \\ 0 \end{bmatrix}$$

(ii) For each representation in (i), find the corresponding relationship between the initial conditions $\{y(-1), y(0)\}$ and the initial state $\boldsymbol{x}(0)$.

(iii) Extend the results in (i)-(ii) to an nth order digital filter.

2.27 Consider the second order differential equation

$$\ddot{z}(t) = -2\dot{z}(t) - 3z(t) + v(t) \ ;\ t \geq 0$$

(i) Using the trapezoidal approximation of an integral, obtain a system of two first order difference equations in the form

$$\boldsymbol{x}(k+1) = \boldsymbol{F}\boldsymbol{x}(k) + \boldsymbol{g}u(k) \ ;\ u(k) \stackrel{\Delta}{=} v(kT)$$

where

$$\boldsymbol{x}^T(k) = [x_1(k)\ \ x_2(k)] \ ;\ x_1(k) \simeq z(kT)\ ,\ x_2(k) \simeq \dot{z}(kT)$$

(ii) Based on the result in (i), obtain a second order difference equation between $y = x_1$ and u.

(iii) What is the corresponding frequency function $H_1(z)$ between $y = x_1$ and u.

2.28 Consider the third order difference equations as presented in exercise 2.24. In each case:

(i) Find the *phase variable* state space representation $\{\boldsymbol{F}, \boldsymbol{g}, \boldsymbol{h}, d\}$ corresponding to the initial conditions $\{y(1) = -2,\ y(2) = 1,\ y(3) = 3\}$.

(ii) Find one *Jordan form* $\{\boldsymbol{F}, \boldsymbol{g}, \boldsymbol{h}, d\}$ when $\boldsymbol{g}^T = [1\ \ 1\ \ 1]$ and $\{y(1) = -2,\ y(2) = 1,\ y(3) = 3\}$.

(iii) Using the results of either part (i) or part (iii), calculate the frequency function $H(z)$ between the output y and the input u.

2.29 Consider the state space equations

$$\boldsymbol{x}(k+1) = \begin{bmatrix} 1+\alpha T & \beta T \\ T & 1 \end{bmatrix} \boldsymbol{x}(k) + \begin{bmatrix} \gamma T \\ 0 \end{bmatrix} u(k) \ \ ; \ \ \boldsymbol{x}(k) = \begin{bmatrix} x_1(k) \\ x_2(k) \end{bmatrix}$$

Find the difference between y and u when $y = x_1 + Tx_2$.

2.30 The state space equations

$$\boldsymbol{x}(k+1) = \boldsymbol{A}\boldsymbol{x}(k) \ ; \ \ \boldsymbol{A} = \begin{bmatrix} a_{11} & a_{12} & a_{13} \\ a_{21} & a_{22} & a_{23} \\ a_{31} & a_{32} & a_{33} \end{bmatrix} \ \boldsymbol{x}(k) = \begin{bmatrix} x_1(k) \\ x_2(k) \\ x_3(k) \end{bmatrix}$$

corresponds to a model for a marketing problem involving three shops as described in Chapter 1. Suppose

$$a_{11} = a_{22} = a_{33} = 0.5 \ \ ; \ \ x_1(0) = 1/3,\ x_2(0) = 1/3,\ x_3(0) = 1/3$$

(i) Find the eigenvalues of $\boldsymbol{A}$ in terms of $\{a_{21}, a_{12}, a_{23}\}$.

(ii) Find the steady state values $\{\overline{x}_1, \overline{x}_2, \overline{x}_3\}$ of $\{x_1(k), x_2(k), x_3(k)\}$ in terms of $\{a_{12}, a_{23}, a_{31}\}$.

(iii) What conditions will guarantee that $\overline{x}_1 > 0.33$, or equivalently, that shop #1 will eventually capture more than 1/3 of the market share ?

2.31 Consider the first order system

$$y(k+1) = 0.2y(k) + u(k) \ \ ; \ \ k \geq 0$$

Show that the forced response due to $u(k) = k^2$ is given by

$$\begin{aligned} y^{(1)}(k) &= k\left(\frac{1-0.2^k}{0.8}\right) - \frac{1-0.2^k + 0.8k(0.2)^{k-1}}{0.64} - k(0.2)^{k-1} \\ &= \frac{25}{16} + \frac{5k}{4} + (0.2)^k\left(\frac{25}{16} - \frac{25}{2}k\right) \end{aligned}$$

2.32 Consider the forced response y_f in (2.13) when $k_0 = 0$ and the input signal u is given by

$$u(m) = m^p \quad ; \quad m \geq 0$$

for some integer $p \geq 0$.

(i) Show that the forced response $y_f = y^{(p)}$ is given by

$$y^{(p)}(k) = b \sum_{m=0}^{k-1} a^{k-m-1} m^p = ba^{k-1} \sum_{m=0}^{k-1} a^{-m} m^p \tag{2.192}$$

or in expanded form as

$$y^{(p)}(k) = ba^{k-1}\{1 + a^{-1}a^{-2}2^p + a^{-3}3^p + \ldots\}$$

(ii) by means of the ratio Test, show that the series for $y^{(p)}(k)$ is convergent for all $|a| < 1$.

3. Digital Filters

A linear time invariant digital filter is represented by an nth order linear time invariant difference equation

$$y(k+n) = \sum_{p=0}^{n-1} a_p y(k+p) + \sum_{q=0}^{n} b_q u(k+q) \tag{3.1}$$

As we have shown in Chapter 2, the steady state sinusoidal response y_{ss} of an asymptotically stable difference equation (3.1) as the result of the sinusoidal input

$$u(k) = \sin k\omega_0$$

is given by

$$y_{ss}(k) = A(\omega_0)\sin[k\omega_0 - \phi(\omega_0)] \tag{3.2}$$

where

$$A(\omega_0)exp\{-j\phi(\omega_0)\} = H(exp\{j\omega_0\}) \tag{3.3}$$

$$H(z) = \frac{\sum_{k=0}^{n} b_k z^k}{z^n - \sum_{k=0}^{n-1} a_k z^k}$$

In this Chapter, we show how a digital filter can be designed so that the magnitude function $A(\omega_0)$ and phase function $\phi(\omega_0)$ satisfy particular properties.

3.1 Overview

The aim in digital filter design is to select the filter coefficients $\{a_p, b_q\}$ in (3.1) so that some particular filter performance specifications are satisfied. If a filter is specified based on some knowledge of both the *frequency components* that are present in the input signal u, and the *frequency components* that are to be *retained* in the output signal y, then the specification of the digital filter (3.1) may be made in terms of its *frequency domain description*.

In some cases, the frequency response function $H(z)$ of the digital filter is specified in terms of an approximation to some desired frequency response function $H_d(z)$ where the desired amplitude response function A_d and the desired phase response function ϕ_d are given by

$$A_d(\omega_0)exp\{-j\phi(\omega_0)\} = H_d(exp\{j\omega_0\}) \tag{3.4}$$

For example:

- H_d corresponds to a filter (3.1) of order n_d, and the aim is to find another filter H in (3.1) of *lower* order $n < n_d$.
- H_d defines some "ideal" characteristic that is not realizable by *any* filter of the form (3.1)
- H_d is given by

  $$H_d(exp\{j\omega\}) = \mathcal{H}(j\Omega) \; ; \; \Omega = \frac{\omega}{T}$$

  where $\mathcal{H}(s)$ is the frequency response function of a given *analog* filter.

Two commonly used *performance measures* that are often used to measure how well H *approximates* H_d are given by J_1 or J_2 where

$$J_1 \triangleq \sum_{k=1}^{M} \gamma_d(\omega_k)|H_d(exp\{j\omega_k\}) - H(exp\{j\omega_k\})|^2 \tag{3.5}$$

and

$$J_2 \triangleq \max_{1\leq k\leq M} \gamma_d(\omega_k)\,|H_d(exp\{j\omega_k\}) - H(exp\{j\omega_k\})| \tag{3.6}$$

for some given positive *weighting function* γ_d (ie $\gamma_d(\omega) > 0$ for all ω) and some given set of digital frequencies $\{\omega_k;\ k = 1, 2, 3, \ \ldots\ , M\}$.

A filter design which minimizes the weighted sum of squares (3.5) is referred to as a *weighted least squares filter design*, while a design which minimizes the weighted maximum deviation (3.6) over a given set of frequencies is referred to as a *weighted minimax filter design*. When the weighting function $\gamma_d(\omega_k)$ is constant for all ω_k, we simply refer to the filter designs as *least squares* and *minimax* respectively.

As we have shown, the unit impulse response h of the digital filter (3.1) is given by:

1. $h(k) = 0$ for $k < 0$
2. The filter coefficients $\{a_p, b_q\}$ for $0 \leq k \leq n$. That is, from (3.1), with $\{y(k) = 0;\ -n \leq k \leq -1\}$ and $\{u(0) = 1;\ u(k) = 0,\ k \neq 0\}$, we have

$$\begin{aligned}
h(0) &= b_n \\
h(1) &= a_{n-1}h(0) + b_{n-1} \\
h(2) &= a_{n-1}h(1) + a_{n-2}h(0) + b_{n-2} \\
&\ldots \\
&\ldots \\
h(n) &= a_{n-1}h(n-1) + a_{n-2}h(n-2) + \ \ldots\ + a_1h(1) + a_0h(0) + b_0
\end{aligned} \tag{3.7}$$

and

3. $h(k)$ for $k \geq n+1$ by the zero input response

$$h(k+n) = \sum_{p=0}^{n-1} a_p h(k+p) \; ; \; k \geq 1 \tag{3.8}$$

Recursive Filter

If $a_p \neq 0$ for some integer $0 \leq p \leq n-1$, then the output $y(k+n)$ at time $k+n$ requires knowledge of the output $y(k+p)$ for some *previous time* $k+p$. In this case, the digital filter described by (3.1) is said to be *recursive*. For a recursive digital filter, it follows from (3.8) that the unit impulse response h contains a term of the form $\{\lambda^\ell\}$ for some root $\lambda \neq 0$ of the characteristic equation

$$\lambda^n = \sum_{p=0}^{n-1} a_p \lambda^p \tag{3.9}$$

Hence for a recursive digital filter, there exists no (finite) integer M such that the unit impulse response satisfies the property:

$$h(m) = 0 \text{ for all } m \geq M \tag{3.10}$$

A *recursive* digital filter is therefore also called an *infinite impulse response* (or IIR) digital filter.

Nonrecursive Filter

If $a_p = 0$ for *all* integers $0 \leq p \leq n-1$, then (3.1) becomes

$$y(k+n) = \sum_{q=0}^{n} b_q u(k+q) \tag{3.11}$$

In this case, since $y(k+n)$ does *not* require knowledge of $y(\ell)$ for any $\ell < k+n$, and so the filter described by (3.11) is said to be *nonrecursive*.

It follows from (3.8) for the nonrecursive filter (3.11) that

$$h(m) = 0 \text{ for all } m \geq n+1$$

which means that the unit impulse response h has only a *finite* number of nonzero values. A *nonrecursive* digital filter is therefore also called a *finite impulse response* (or FIR) filter.

Periodicity of Frequency Response Function

The possible digital frequency response functions H that can be achieved using a linear time invariant digital filter (3.1) is restricted to those functions which can be described by (3.3). Since

$$exp\{j\omega\} = exp\{j(\omega + 2\pi n)\}$$

for all ω and all integers n, any digital frequency response function H that can be realized by a digital filter is necessarily *periodic* in ω of period 2π rads; that is

$$H(exp\{j\omega\}) = H(exp\{j(\omega + 2\pi)\}) \tag{3.12}$$

for all frequencies ω measured in *radians* (rads). Equivalently, if the frequency f is measured in *cycles* where $\omega = 2\pi f$, then

$$H(exp\{j2\pi f\}) = H(exp\{j2\pi(f + 1)\}) \tag{3.13}$$

and so H is periodic in f of period 1 cycle.

This means that both the corresponding amplitude response $A(\omega)$ and the phase response $\phi(\omega)$ are periodic in ω of period 2π rads, or equivalently, periodic in f of period 1 cycle.

Magnitude, Amplitude and Phase Functions

Given the complex number $H(exp\{j\omega_0\})$ in (3.3), the solution for $\{A(\omega_0), \phi(\omega_0)\}$ is *non unique* even though the signal y_{ss} in (3.2) is *unique*. For example, the complex number $H(exp\{j\omega_0\}) = -\sqrt{2} + j\sqrt{2}$ can be written as

$$H(exp\{j\omega_0\}) = 2exp\{-j\frac{5\pi}{4}\} = 2exp\{-j\frac{-3\pi}{4}\} = -2exp\{-j\frac{\pi}{4}\}$$

That is, different solutions for $\{A(\omega_0), \phi(\omega_0)\}$ are then:

$$\{A(\omega_0) = 2, \phi(\omega_0) = 5\pi/4\}, \; \{A(\omega_0) = 2, \phi(\omega_0) = -3\pi/4\}$$
$$\{A(\omega_0) = -2, \phi(\omega_0) = \pi/4\}$$

even though

$$y_{ss}(k) = 2\sin(k\omega_0 - \frac{5\pi}{4}) = 2\sin(k\omega_0 + \frac{3\pi}{4}) = -2\sin(k\omega_0 - \frac{\pi}{4})$$

is unique. We therefore need to be more precise in our definitions about what we mean by 'amplitude' and 'phase'.

We shall now show how the frequency response function H can be decomposed into two unique ways

$$H(exp\{j\omega\}) = M(\omega)exp\{-j\delta(\omega)\} = A(\omega)exp\{-j\phi(\omega)\} \tag{3.14}$$

In the first case, $M(\omega)$ is the unique *magnitude function* which is defined to be nonnegative for all frequencies, and $\delta(\omega)$ is the unique *phase function* which is (normally) a *discontinuous function* of the frequency ω over the frequency range $-\pi < \omega \leq \pi$. In the second case, $A(\omega)$ is the unique *amplitude function* and $\phi(\omega)$ is the unique *phase function* which is always a *continuous function* of the frequency ω. Since the function ϕ is continuous, the amplitude $A(\omega)$ will generally have positive and negative values depending on the value of the frequency ω.

In order to derive the magnitude function, we write the complex number $H(exp\{j\omega\})$ in the cartesian form

$$H(exp\{j\omega\}) = R(exp\{j\omega\}) + jI(exp\{j\omega\})$$

where R and I are both real numbers, then

$$\begin{aligned} M(\omega) &\triangleq \sqrt{H(exp\{j\omega\})\ \overline{H(exp\{j\omega\})}} \\ &= \sqrt{R^2(exp\{j\omega\}) + I^2(exp\{j\omega\})} \geq 0 \end{aligned} \tag{3.15}$$

defines the *magnitude* of $H(exp\{j\omega\})$ where $\sqrt{\ }$ is the *positive* square root. Hence, by definition, the *magnitude function* is a *nonnegative function* of ω. Given the magnitude function, the *(discontinuous) phase function* δ is then *uniquely* specified by

$$H(exp\{j\omega\}) = M(\omega)exp\{-j\delta(\omega)\}\ ;\ -\pi < \delta(\omega) \leq \pi \tag{3.16}$$

The magnitude function has particular properties. Firstly, since $H(z)$ is a ratio of polynomials of z in (3.3)

$$\overline{H(exp\{j\omega\})} = H(exp\{-j\omega\})$$

and so

$$M^2(\omega) = H(exp\{j\omega\}).\overline{H(exp\{j\omega\})} = H(exp\{j\omega\}).\ H(exp\{-j\omega\})$$

Also since $exp\{j\pi\} = -1$, we have

$$\begin{aligned} M^2(\pi - \omega) &= H(exp\{j(\pi - \omega)\}).\ H(exp\{-j(\pi - \omega)\}) \\ &= H(-exp\{-j\omega\}).\ H(-exp\{j\omega\}) \end{aligned}$$

and

$$\begin{aligned} M^2(\pi + \omega) &= H(exp\{j(\pi + \omega)\}).H(exp\{-j(\pi + \omega)\}) \\ &= H(-exp\{j\omega\}).\ H(-exp\{-j\omega\}) \end{aligned}$$

We therefore conclude that $M(\pi - \omega) = M(\pi + \omega)$ is a *symmetric* function about the frequency $\omega_1 = \pi$ rads. However since $M(\omega)$ is periodic in ω of period 2π, this result implies that the *magnitude function* $M(\omega)$ for $\omega \geq 0$ is *uniquely* determined over the range $0 \leq \omega < \pi$.

In the second case, when the phase functions is restricted to be a *continuous function* of ω, the *amplitude function* in (3.14) can either be positive or negative, and $\phi(\omega)$ is not bounded. We summarise these results as follows.

Theorem 3.1.1. *Suppose the filter frequency response function $H(exp\{j\omega\})$ is expressed in two ways as follows:*

$$H(exp\{j\omega\}) = A(\omega)exp\{-j\phi(\omega)\} = M(\omega)exp\{-j\delta(\omega)\}$$

where $M(\omega) \geq 0$ for all ω, and ϕ is a continuous function of ω with $-\infty < \phi(\omega) < \infty$. Then

1.

$$A(\omega) = A(\omega + 2\pi) \ ; \ \phi(\omega) = \phi(\omega + 2\pi)$$
$$M(\omega) = M(\omega + 2\pi) \ ; \ \delta(\omega) = \delta(\omega + 2\pi)$$

2. $M(\pi - \omega) = M(\pi + \omega) \ ; \ |A(\pi - \omega)| = |A(\pi + \omega)|$
is symmetric about the frequency $\omega_1 = \pi$ radians, and so $M(\omega)$ and $A(\omega)$ are both uniquely specified over the range $0 \leq \omega < \pi$.

3. *If $n\pi < \phi(\omega) \leq (n+1)\pi$ for some integer n, then*

$$M(\omega) = (-1)^n A(\omega) \text{ when } \delta(\omega) = \phi(\omega) - n\pi$$

4. $\phi(\pi - \omega) = -\phi(\pi + \omega) \ ; \ \delta(\pi - \omega) = -\delta(\pi + \omega)$
are both skew-symmetric about the frequency $\omega_1 = \pi$ radians, and so $\delta(\omega)$ and $\phi(\omega)$ are uniquely specified over the range $0 \leq \omega < \pi$.

Example 3.1.1. (a) Consider the digital filter defined by the frequency response function

$$H(z) = \frac{1}{z - a}$$

Then

$$H(exp\{j\omega\}) = \frac{1}{\cos\omega - a + j\sin\omega}$$

$$H(exp\{-j\omega\}) = \overline{H(exp\{j\omega\})} = \frac{1}{\cos\omega - a - j\sin\omega}$$

and so

$$M(\omega) \triangleq \sqrt{H(exp\{-j\omega\})H(exp\{j\omega\})} = \frac{1}{\sqrt{1 + a^2 - 2a\cos\omega}} \geq 0$$

Also

$$\phi(\omega) = -\tan^{-1}\left\{\frac{\sin\omega}{\cos\omega - a}\right\}$$

so that

$$\phi(\pi-\omega) = -\tan^{-1}\left\{\frac{\sin(\pi-\omega)}{\cos(\pi-\omega) - a}\right\} = -\tan^{-1}\left\{\frac{\sin\omega}{-\cos\omega - a}\right\}$$

and

$$\phi(\pi+\omega) = -\tan^{-1}\left\{\frac{\sin(\pi+\omega)}{\cos(\pi+\omega) - a}\right\} = -\tan^{-1}\left\{\frac{-\sin\omega}{-\cos\omega - a}\right\}$$

which confirms $\phi(\pi-\omega) = -\phi(\pi+\omega)$. The magnitude, amplitude and phase functions for $a = 0.5$ are illustrated in Figure 3.1. For this first order filter, the magnitude and amplitude functions are the same, and the phase functions $\delta(\omega)$ and $\phi(\omega)$ are the same.

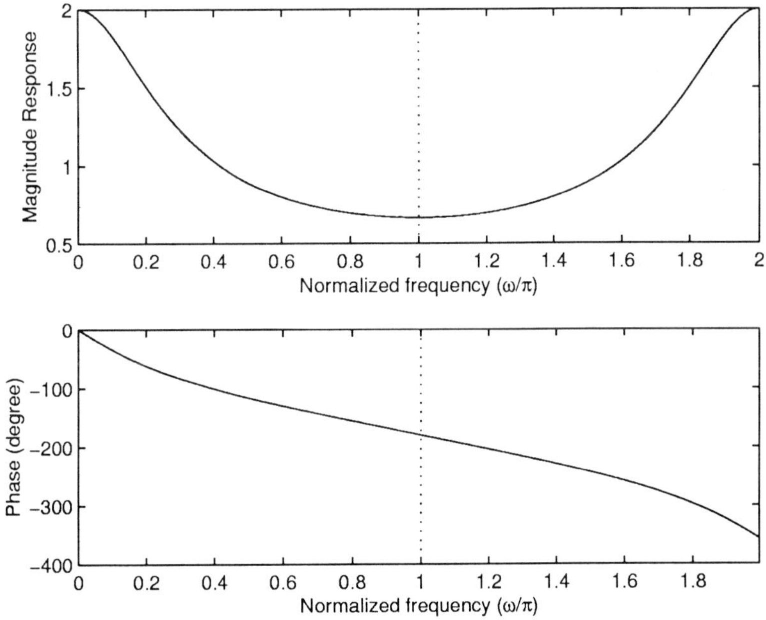

Fig. 3.1. Frequency response characteristics of first order IIR filter

(b) Now consider the FIR digital filter defined by the frequency response function

$$H(z) = \frac{z^2 + z + 1}{z^2} = 1 + z^{-1} + z^{-2}$$

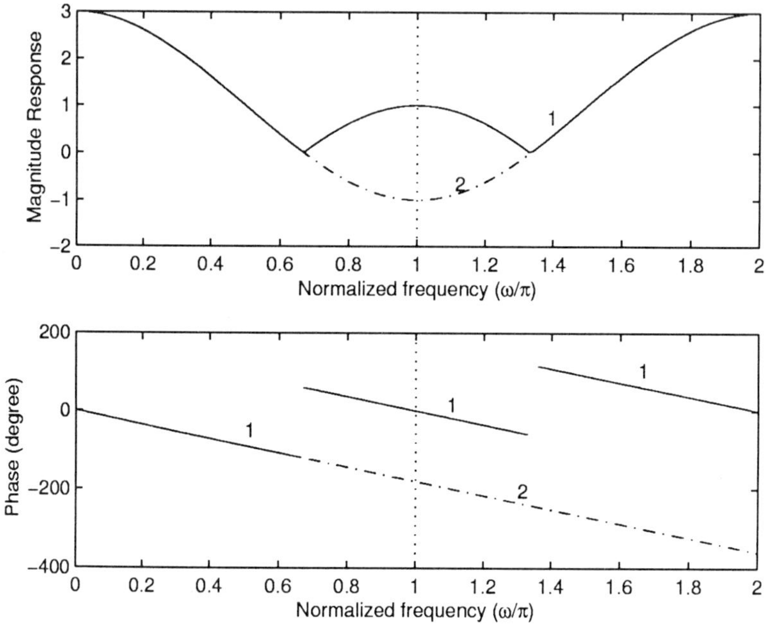

Fig. 3.2. Frequency response characteristics of FIR filter for 1: $M(\omega)$ and $\delta(\omega)$, and 2: $A(\omega)$ and $\phi(\omega)$

The plots of the magnitude $M(\omega)$, amplitude $A(\omega)$ and phase functions are illustrated in Figure 3.2. In this case, the magnitude and amplitude functions are *not* the same, and the (discontinuous) phase function $\delta(\omega)$ and the (continuous) phase function $\phi(\omega)$ are also *not* the same.

All-Pass Filter

A digital filter whose magnitude response is constant; that is

$$M(\omega) = \gamma \text{ for all } \omega \tag{3.17}$$

is called an *all-pass filter* of gain γ. The term "all-pass" is used because condition (3.17) implies that all sinusoidal signals of all frequencies are "passed through" the filter without amplitude attenuation.

Linear Phase Filter

A digital filter is said to have a *linear phase response* if the (continuous) phase response ϕ is a linear function of frequency ω; that is, if

$$\phi(\omega) = \beta_0 \omega \tag{3.18}$$

for some constant β_0. The function

$$\beta(\omega) \triangleq \frac{d\phi(\omega)}{d\omega}$$

is called the *group delay* of the digital filter, and so that a filter having a linear phase response is also said to have a *flat* or *constant group delay.*

Low Pass Filter

Frequency response characteristics can be described in terms of a *tolerance band* on the magnitude function. In particular, a *low pass digital filter* H can be specified by the condition on the magnitude function $M(\omega)$ that: for some $\omega_b \leq \omega_s$

$$\begin{aligned} 1-\varepsilon_1 \leq M(\omega) \leq 1+\varepsilon_1 \ &; \ 0 \leq \omega \leq \omega_b \\ M(\omega) \leq \varepsilon_2 \ &; \ \omega_b < \omega_s \leq \omega < \pi \end{aligned} \tag{3.19}$$

In this case, as illustrated in Figure 3.3, the *passband* is defined by the frequency band $0 \leq \omega \leq \omega_b$, the *transition band* is defined by the frequency band $\omega_b \leq \omega \leq \omega_s$, and the *stopband* by the frequency band $\omega_s \leq \omega \leq \pi$.

Besides the *width* $\omega_s - \omega_b$ of the transition band, there are normally no other specifications on the magnitude function in the transition band other than the requirement that there be a "smooth" transition from the passband to the stopband. However as illustrated in Figure 3.3, *additional constraints* of the form

$$\beta_{01} + \beta_1\omega \leq |\phi(\omega)| \leq \beta_{02} + \beta_1\omega \ ; \ 0 \leq \omega \leq \omega_b \tag{3.20}$$

for some constants $\{\beta_1, \beta_{01}, \beta_{02}\}$ with $\beta_{01} < \beta_{02}$, can also be imposed on the *phase response* ϕ in the *passband* so as to achieve an approximately linear phase response, or equivalently, an approximately flat group delay response.

An *ideal low pass* digital filter which *cannot* be realized by any filter (3.1) is defined as a filter which has *zero transition band width*, and *zero magnitude tolerance.* These ideal filter requirements corresponds in (3.19) to $\omega_p = \omega_s$ and $\varepsilon_1 = \varepsilon_2 = 0$; that is, for an ideal low pass filter, the magnitude response is given by

$$M_d(\omega) = \begin{cases} 1 \ ; \ 0 \leq \omega \leq \omega_b \\ 0 \ ; \ \omega_b = \omega_s < \omega \leq \pi \end{cases} \tag{3.21}$$

Sometimes, an *ideal low pass filter* is also specified to have a linear phase response in the pass-band; that is, $\phi(\omega) = \phi_1\omega$ for all $0 \leq \omega \leq \omega_b$. If such an ideal filter existed, then the steady state output response due to the input signal $u(k) = \sin k\omega_0$ would be given by

$$y_{ss}(k) = \begin{cases} \sin(\omega_0(k-\phi_1)) & ; \ 0 \leq \omega \leq \omega_b \\ 0 & ; \ \omega_s \leq \omega \leq \pi \end{cases}$$

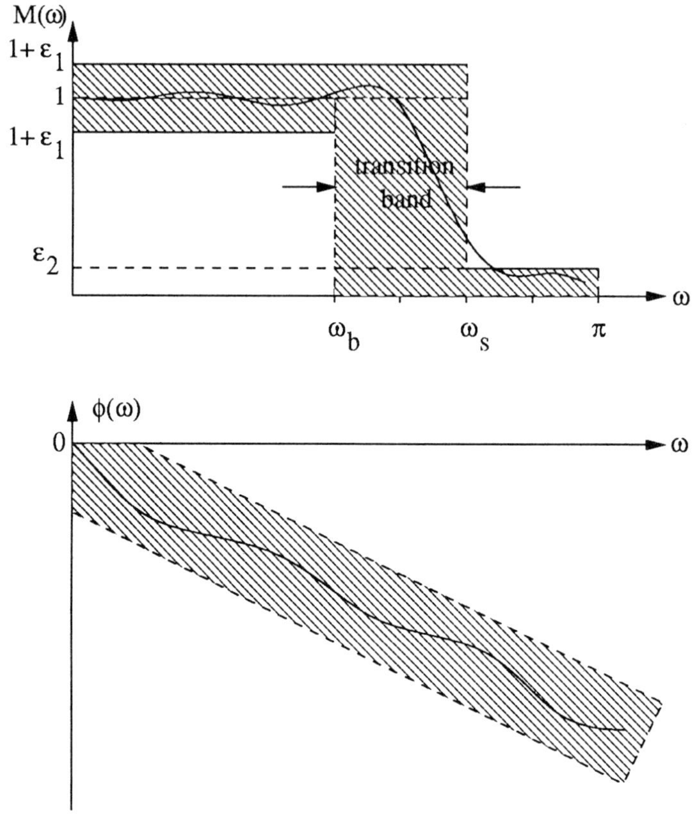

Fig. 3.3. Frequency response characteristics of low pass filter

That is, sinusoidal inputs having a frequency ω_0 *below* the cut-off frequency ω_b are passed through the ideal low pass filter with a *fixed time delay* ϕ_1 (and hence no phase distortion) and *no change in magnitude*, whereas sinusoidal inputs of a frequency ω_0 *above* the cut-off frequency $\omega_s = \omega_b$ are completely stopped, or rejected, by the filter.

High Pass Filter

The magnitude response $M(\omega)$ of a *high pass digital filter* with transition bandwidth $\omega_2 - \omega_1$ is defined for some $\omega_1 < \omega_2$ by

$$\begin{aligned} M(\omega) &\le \varepsilon_2 \ ; \ 0 \le \omega \le \omega_1 \\ 1 - \varepsilon_1 \le M(\omega) &\le 1 + \varepsilon_1 \ ; \ \omega_1 < \omega_2 \le \omega < \pi \end{aligned} \tag{3.22}$$

for some sufficiently small $\{\varepsilon_1 \ge 0,\ \varepsilon_2 \ge 0\}$. In the *ideal* case (which is not realizable)

$$\varepsilon_1 = \varepsilon_2 = 0 \; ; \; \omega_1 = \omega_2 \tag{3.23}$$

The magnitude characteristics are illustrated in Figure 3.4(a).

Bandpass Filter

A *bandpass digital filter* H which has two transition bands $\omega_2 - \omega_1$ and $\omega_4 - \omega_3$ is defined by

$$\begin{aligned} M(\omega) &\leq \varepsilon_2 \; ; \; 0 \leq \omega \leq \omega_1 < \omega_2, \; \omega_3 < \omega_4 \leq \omega < \pi \\ 1 - \varepsilon_1 \leq M(\omega) &\leq 1 + \varepsilon_1 \; ; \; \omega_2 \leq \omega \leq \omega_3 \end{aligned} \tag{3.24}$$

for some sufficiently small $\{\varepsilon_1 \geq 0, \; \varepsilon_2 \geq 0\}$. In the *ideal* case (which cannot be realizable)

$$\varepsilon_1 = \varepsilon_2 = 0 \; ; \; \omega_1 = \omega_2 \, , \; \omega_3 = \omega_4 \tag{3.25}$$

The magnitude characteristics are illustrated in Figure 3.4(b).

Bandstop Filter

An *bandstop digital filter* H which has two transition bands $\omega_2 - \omega_1$ and $\omega_4 - \omega_3$ is defined by

$$\begin{aligned} 1 - \varepsilon_1 \leq M(\omega) &\leq 1 + \varepsilon_1 \; ; \; 0 \leq \omega \leq \omega_1 < \omega_2, \; \omega_3 < \omega_4 \leq \omega \leq \pi \\ M(\omega) &\leq \varepsilon_2 \; ; \; \omega_2 \leq \omega \leq \omega_3 \end{aligned} \tag{3.26}$$

In the ideal case, (3.25) is satisfied. The magnitude characteristics are illustrated in Figure 3.4(c). More generally, one can specify a filter characteristic to have *several* passbands and *several* stopbands.

Notch Filter

A filter H where

$$M(\omega_0) = 0$$

for some $0 < \omega_0 < \pi$ is said to have a "notch" at the frequency $\omega = \omega_0$. A bandstop filter in which the frequency range $\omega_1 \leq \omega \leq \omega_4$ in (3.26) is small or narrow is called a *notch filter*.

Finite Wordlength Implementation

Once a digital filter (3.1) is designed to satisfy some particular frequency response specifications, consideration must then be given to the fact that the difference equation must be implemented using *finite precision arithmetic*

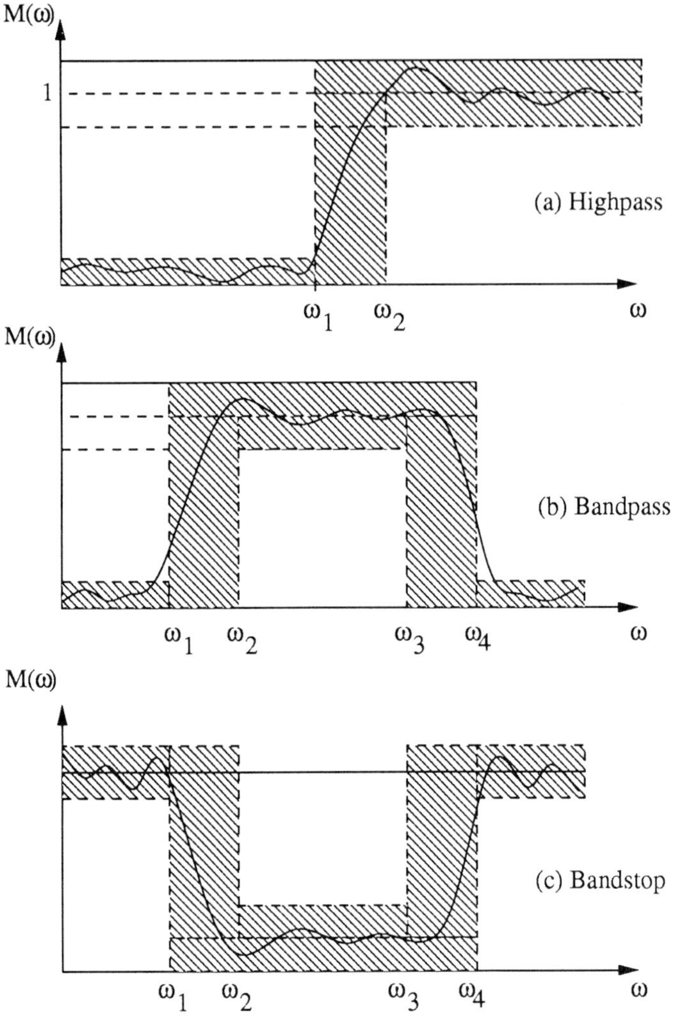

Fig. 3.4. Magnitude characteristics of highpass, bandpass & bandstop filters

either on some general purpose microprocessor, or on some special purpose digital signal processing (i.e. DSP) device. When implementing a digital filter, there are many issues that must be addressed including the type of arithmetic (floating point or fixed point), numerical scaling and filter structure.

For example, we showed that the difference equation (3.1) has an infinite number of state space representations which *ideally* (i.e. assuming *infinite* precision arithmetic) are all equivalent. However once a particular *finite wordlength* is assigned to both the filter states and filter coefficients, these state space realizations are *no longer* equivalent. For example, consider possible state space realizations of a fourth order IIR digital filter with characteristic polynomial

$$det(\lambda \boldsymbol{I}_4 - \boldsymbol{F}) = (\lambda - \beta_1)^4 \ ; \ \beta_1 = 0.99$$

Since

$$det(\lambda \boldsymbol{I}_4 - \boldsymbol{F}) = \lambda^4 - a_3\lambda^3 - a_2\lambda^2 - a_1\lambda - a_0$$

where

$$a_3 = 3.96 \ ; \ a_2 = -5.8806 \ ; \ a_1 = 3.8811960 \ ; \ a_0 = -0.96059601$$

Then as described in Chapter 2, two possible choices for the system matrix $\boldsymbol{F}$ are then the phase variable form $\boldsymbol{F}_1$ (see (2.179)) and the Jordan form $\boldsymbol{F}_2$ (see (2.185)) where

$$\boldsymbol{F}_1 = \begin{bmatrix} 0 & 0 & 0 & a_0 \\ 1 & 0 & 0 & a_1 \\ 0 & 1 & 0 & a_2 \\ 0 & 0 & 1 & a_3 \end{bmatrix} \ ; \ \boldsymbol{F}_2 = \begin{bmatrix} \beta_1 & 1 & 0 & 0 \\ 0 & \beta_1 & 1 & 0 \\ 0 & 0 & \beta_1 & 1 \\ 0 & 0 & 0 & \beta_1 \end{bmatrix}$$

When the digital filter is implemented with finite wordlength, all of the coefficients $\{f_{ij}\}$ of the matrices $\{\boldsymbol{F}_1, \boldsymbol{F}_2\}$ must be quantized to the available wordlength. For example, if all coefficients are quantized to a fixed point representation in which the fraction is represented by $B = 12$ bits, then (to 10 dec. places)

$$\hat{a_3} = 3.9599609375 \ ; \ \hat{a_2} = -5.8806152344 \ ; \ \hat{a_1} = 3.8811035156$$
$$\hat{a_0} = -0.9606933594 \ ; \ \hat{\beta_1} = 0.9899902344$$

Then if one recomputes the eigenvalues $\{\hat{\lambda}_k\}$ of the corresponding matrices $\{\hat{\boldsymbol{F}}_1, \hat{\boldsymbol{F}}_2\}$ which are obtained by replacing a_k by $\hat{a}_k$ and β_m by $\hat{\beta}_m$, we find that the eigenvalues of the matrix $\hat{\boldsymbol{F}}_1$ are given by

$$\hat{\lambda}_1 = \overline{\hat{\lambda}_2} = 0.9019646884 + j0.0844961392$$
$$\hat{\lambda}_3 = \overline{\hat{\lambda}_4} = 1.0780157803 + j0.0921338448$$

and the eigenvalues of the matrix $\hat{\boldsymbol{F}}_2$ are given by

$$\hat{\lambda}_1 = \hat{\lambda}_2 = \hat{\lambda}_3 = \hat{\lambda}_4 = 0.9899902344$$

Since all eigenvalues of the matrix $\hat{\boldsymbol{F}}_2$ satisfy $|\lambda_k| < 1$, we still have an asymptotically stable digital filter. However two eigenvalues of the matrix $\hat{\boldsymbol{F}}_1$ satisfy $|\lambda_k| > 1$ which then gives an *unstable* digital filter.

Issues which arise as a result of a finite wordlength implementation of a digital filter are considered in Chapter 5.

3.2 Design of FIR Filters

As previously defined, a *nonrecursive* or *finite impulse response* (FIR) filter is described by (3.1) with $a_k = 0$ for all k. The frequency response function of an FIR filter is then given from (3.3) by

$$H(z) = \frac{\sum_{p=0}^{n} b_p z^p}{z^n} = b_n + b_{n-1}z^{-1} \ \dots \ + b_1 z^{-n+1} + b_0 z^{-n} \tag{3.27}$$

Since the number of filter coefficients is $n+1$, we have that when n is *even*, the number of coefficients is *odd*, and when n is *odd*, the number of coefficients is *even*.

Since $a_p = 0$ for all p, all the roots λ_k of the characteristic equation (3.9) are given by $\lambda_k = 0$, and so an FIR filter (3.27) is always *asymptotically stable.* Although this condition imposes a constraint on the class of possible filters, there are important advantages which follow from this choice. In particular, a numerical routine which optimizes some performance measure over the filter coefficients $\{b_q\}$ is more likely to converge since unlike the case of IIR filters, there is no need to verify a stability condition at each step of the iteration.

3.2.1 Linear Phase

From (3.27), we have

$$\begin{aligned} H(exp\{j\omega\}) &= b_n + \sum_{k=1}^{n} b_{n-k} exp\{-jk\omega\} \\ &= b_n + \sum_{k=1}^{n} b_{n-k}\cos k\omega - j\sum_{k=1}^{n} b_{n-k}\sin k\omega \end{aligned} \tag{3.28}$$

and so the magnitude response is given by

$$M^2(\omega) = (b_n + \sum_{k=1}^{n} b_{n-k}\cos k\omega)^2 + (\sum_{k=1}^{n} b_{n-k}\sin k\omega)^2$$

In terms of trying to achieve a constant magnitude response, it follows that if $M(\omega) = 1$ over *any* frequency range $\omega_1 < \omega < \omega_2$, then $b_n = 1$ and $b_k = 0$ for all $0 \le k \le n-1$, which gives the trivial solution $H(z) = 1$. That is, the magnitude response of any (nontrivial) FIR filter can *never* be constant over any frequency range. In particular, unlike the case of IIR filters, an *all-pass* filter *cannot* be achieved using a FIR filter.

From (3.28), the phase response ϕ is given by

$$\phi(\omega) = \tan^{-1}\left(\frac{\sum_{k=1}^{n} b_{n-k}\sin k\omega}{b_0 + \sum_{k=1}^{n} b_{n-k}\cos k\omega}\right) \tag{3.29}$$

In this form, it is not obvious as to whether or not the coefficients $\{b_q\}$ can be chosen such that ϕ can be made *linear* over any range $-\pi \leq \omega_1 \leq \omega \leq \omega_2 \leq \pi$.

We therefore consider the problem of achieving a linear phase response from another perspective. To begin, consider a *four coefficient* (i.e. $n = 3$) FIR filter (3.27) so that H is given by

$$H(exp\{j\omega\}) = b_3 + b_2 exp\{-j\omega\} + b_1 exp\{-j2\omega\} + b_0 exp\{-j3\omega\}$$

which can be arranged into the form

$$\begin{aligned} H(exp\{j\omega\}) = exp\{-\frac{j3\omega}{2}\} \Big[& b_3 exp\{\frac{j3\omega}{2}\} + b_2 exp\{\frac{j\omega}{2}\} \\ & + b_1 exp\{-\frac{j\omega}{2}\} + b_0 exp\{-\frac{j3\omega}{2}\} \Big] \end{aligned} \tag{3.30}$$

We now make the following two observations:

1. If $\{b_3 = b_0 \ ; \ b_2 = b_1\}$, then in (3.30)

$$H(exp\{j\omega\}) = exp\{-\frac{j3\omega}{2}\} \left[2b_0 \cos\left(\frac{3\omega}{2}\right) + 2b_1 \cos\left(\frac{\omega}{2}\right) \right] \tag{3.31}$$

That is,

$$\begin{aligned} H(exp\{j\omega\}) &= H_2(exp\{j\omega\}) \stackrel{\Delta}{=} A_2(\omega) exp\{-j(\phi_0 + \phi_1\omega)\} \\ A_2(\omega) &= 2b_0 \cos\left(\frac{3\omega}{2}\right) + 2b_1 \cos\left(\frac{\omega}{2}\right) \ ; \ \phi_0 = 0 \ , \ \phi_1 = \frac{3}{2} \end{aligned}$$

describes a *linear phase* FIR filter

2. If $\{b_0 = -b_3 \ ; \ b_1 = -b_2\}$, then from (3.30)

$$H(exp\{j\omega\}) = -j exp\{-\frac{j3\omega}{2}\}[2b_0 \sin\left(\frac{3\omega}{2}\right) + 2b_1 \sin\left(\frac{\omega}{2}\right)] \tag{3.32}$$

That is,

$$\begin{aligned} H(exp\{j\omega\}) &= H_4(exp\{j\omega\}) \stackrel{\Delta}{=} A_4(\omega) exp\{-j(\phi_0 + \phi_1\omega)\} \\ A_4(\omega) &= 2b_0 \sin\left(\frac{3\omega}{2}\right) + 2b_1 \sin\left(\frac{\omega}{2}\right) \ ; \ \phi_0 = \frac{\pi}{2} \ , \ \phi_1 = \frac{3}{2} \end{aligned}$$

describes a *linear phase* FIR filter

Now consider a *five coefficient* (i.e. $n = 4$) FIR filter (3.27) so that H is given by

$$H(exp\{j\omega\}) = b_4 + b_3 exp\{-j\omega\} + b_2 exp\{-j2\omega\}$$
$$+ b_1 exp\{-j3\omega\} + b_0 exp\{-j4\omega\}$$

which can be arranged into the form

$$H(exp\{j\omega\}) = exp\{-j2\omega\}\,[b_4 exp\{j2\omega\} + b_3 exp\{j\omega\} + b_2 + b_1 exp\{-j\omega\} + b_0 exp\{-j2\omega\}] \tag{3.33}$$

It then follows from (3.33) that:

1. If $\{b_4 = b_0 \ ; \ b_3 = b_1\}$, then

$$H(exp\{j\omega\}) = exp\{-j2\omega\}\,[b_2 + 2b_0 \cos 2\omega + 2b_1 \cos \omega] \tag{3.34}$$

   That is,

$$H(exp\{j\omega\}) = H_1(exp\{j\omega\}) \triangleq A_1(\omega) exp\{-j(\phi_0 + \phi_1\omega)\}$$
$$A_1(\omega) = b_2 + 2b_0 \cos 2\omega + 2b_1 \cos \omega \ ; \ \phi_0 = 0 \ , \ \phi_1 = 2$$

   describes a *linear phase* FIR filter
2. If $\{b_4 = -b_0 \ ; \ b_3 = -b_1\}$, then from (3.33)

$$H(exp\{j\omega\}) = -j exp\{-j2\omega\}[-jb_2 + 2b_0 \sin 2\omega + 2b_1 \sin \omega] \tag{3.35}$$

   Then for $b_2 = 0$,

$$H(exp\{j\omega\}) = H_3(exp\{j\omega\}) \triangleq A_3(\omega) exp\{-j(\phi_0 + \phi_1\omega)\}$$
$$A_3(\omega) = 2b_0 \sin 2\omega + 2b_1 \sin \omega \ ; \ \phi_0 = \frac{\pi}{2} \ , \ \phi_1 = 2$$

   describes a *linear phase* FIR filter

More generally, we have the following result.

Theorem 3.2.1. *There are four types of linear phase FIR filters:*

$$H_p(z) = b_n + b_{n-1}z^{-1} \ \ldots \ + b_1 z^{-n+1} + b_0 z^{-n}$$

having $n+1$ coefficients corresponding to $p = 1, 2, 3, 4$ defined as follows:

1. *Type 1. If for n even*

$$b_m = b_{n-m} \ ; \ m = 0, 1, 2, \ \ldots \ , \frac{n}{2}$$

   then

$$H_1(exp\{j\omega\}) = A_1(\omega) exp\{-j\phi_1(\omega)\}$$

where

$$A_1(\omega) = b_{0.5n} + 2\sum_{m=0}^{0.5n-1} b_m \cos(\frac{n}{2} - m)\omega \ ; \ \phi_1(\omega) = \frac{n}{2}\omega$$

The amplitude function A_1 has the properties:

$$A_1(\omega) = A_1(-\omega) = A_1(\omega + 2\pi) \ ; \ A_1(\pi + \omega) = A_1(\pi - \omega)$$

2. *Type 2. If for n odd*

$$b_m = b_{n-m} \ ; \ m = 0, 1, 2, \ \ldots \ , \frac{n-1}{2}$$

then

$$H_2(exp\{j\omega\}) = A_2(\omega)exp\{-j\phi_2(\omega)\}$$

where

$$A_2(\omega) = 2\sum_{m=0}^{0.5(n-1)} b_m \cos(\frac{n}{2} - m)\omega \ ; \ \phi_2(\omega) = \frac{n}{2}\omega$$

The amplitude function A_2 has the properties:

$$A_2(\omega) = A_2(-\omega) = A_2(\omega + 4\pi) \ ; \ A_2(\pi + \omega) = -A_2(\pi - \omega)$$

3. *Type 3. If for n even*

$$b_{0.5n} = 0 \ ; \ b_m = -b_{n-m} \ ; \ m = 0, 1, 2, \ \ldots \ , \frac{n}{2}$$

then

$$H_3(exp\{j\omega\}) = A_3(\omega)exp\{-j\phi_3(\omega)\}$$

where

$$A_3(\omega) = 2\sum_{m=0}^{0.5n-1} b_m \sin(\frac{n}{2} - m)\omega \ ; \ \phi_3(\omega) = \frac{\pi}{2} + \frac{n}{2}\omega$$

The amplitude function $A_3(\omega)$ has the properties:

$$A_3(\omega) = -A_3(-\omega) = A_3(\omega + 2\pi) \ ; \ A_3(\pi + \omega) = -A_3(\pi - \omega)$$

4. *Type 4. If for n odd*

$$b_m = -b_{n-m} \; ; \; m = 0, 1, 2, \; \ldots \; , \frac{n-1}{2}$$

then

$$H_4(exp\{j\omega\}) = A_4(\omega)exp\{-j\phi_4(\omega)\}$$

where

$$A_4(\omega) = 2 \sum_{m=0}^{0.5(n-1)} b_m \sin(\frac{n}{2} - m)\omega \; ; \; \phi_4(\omega) = \frac{\pi}{2} + \frac{n}{2}\omega$$

The amplitude function A_4 *has the properties:*

$$A_4(\omega) = -A_4(-\omega) = A_4(\omega + 4\pi) \; ; \; A_4(\pi + \omega) = A_4(\pi - \omega)$$

The frequency response functions $\{H_1, H_2, H_3, H_4\}$ defined in (3.34), (3.31), (3.35) and (3.32) respectively are examples of *Type 1, Type 2, Type 3* and *Type 4 linear phase FIR filters.* Since from Theorem 3.2.1, we have $A_3(0) = A_4(0) = 0$, both Type 3 and Type 4 linear phase filters are unsuitable for *low pass* design, and so either a Type 1 or Type 2 FIR filter must be used. A Type 3 or Type 4 FIR filter is used for a *high pass* design.

Example 3.2.1. Consider the following four FIR filters:

$$\begin{aligned}
H_1(z) &= 2 + 3z^{-1} + 4z^{-2} + 3z^{-3} + 2z^{-4} \quad (n = 4) \\
H_2(z) &= 2 + 4z^{-1} + 4z^{-2} + 2z^{-3} \quad (n = 3) \\
H_3(z) &= 2 + 3z^{-1} - 3z^{-3} - 2z^{-4} \quad (n = 4) \\
H_4(z) &= 2 + 4z^{-1} - 4z^{-2} - 2z^{-3} \quad (n = 3)
\end{aligned}$$

Then by Theorem 3.2.1, the filter $H_m(z)$ for $m = 1, 2, 3, 4$ is of Type m. The corresponding unit impulse responses $\{h_m\}$ are given by

$$\begin{aligned}
h_1 &= \{2, 3, 4, 3, 2\} \text{ with } h_1(k) = 0, \; k \geq 5 \\
h_1 &= \{2, 4, 4, 2\} \text{ with } h_2(k) = 0, \; k \geq 4 \\
h_1 &= \{2, 3, -3, -2\} \text{ with } h_3(k) = 0, \; k \geq 4 \\
h_1 &= \{2, 4, -4, -2\} \text{ with } h_4(k) = 0, \; k \geq 4
\end{aligned}$$

The magnitude $M(\omega)$, amplitude $A(\omega)$, discontinuous phase $\delta(\omega)$ and continuous phase characteristics $\phi(\omega)$ are illustrated in Figures 3.5, 3.6, 3.7 and Figure 3.8.

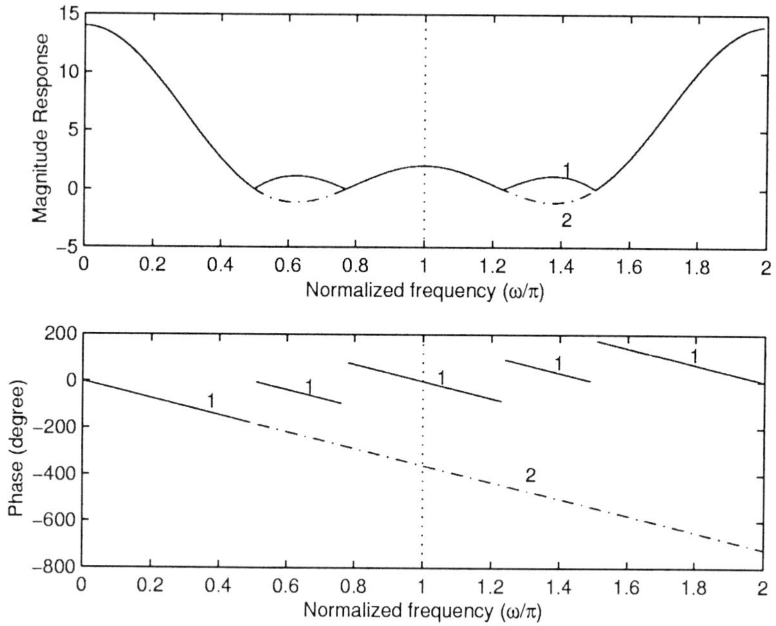

Fig. 3.5. Characteristics of Type 1 linear phase FIR filter for 1: $M(\omega)$ and $\delta(\omega)$, and 2: $A(\omega)$ and $\phi(\omega)$

3.2.2 Windowing

The unit impulse response h of an IIR filter $H(z)$ is such that there is no (finite) integer M such that $h(k) = 0$ for all $k > M$. Consequently, given the frequency response function

$$H(exp\{j\omega\}) = \sum_{k=0}^{\infty} h(k)exp\{-jk\omega\} \tag{3.36}$$

there is no finite set $\{h_N(k) : 0 \leq k \leq N < \infty\}$ of coefficients such that: for all ω

$$H_N(exp\{j\omega\}) = H(exp\{j\omega\})$$

where

$$H_N(exp\{j\omega\}) = \sum_{k=0}^{N} h_N(k)exp\{-jk\omega\} \tag{3.37}$$

However one important approach to the design of a FIR filter is to find such a set of coefficients such that in some well defined sense, $H_N(exp\{j\omega\})$ is a good approximation of $H(exp\{j\omega\})$.

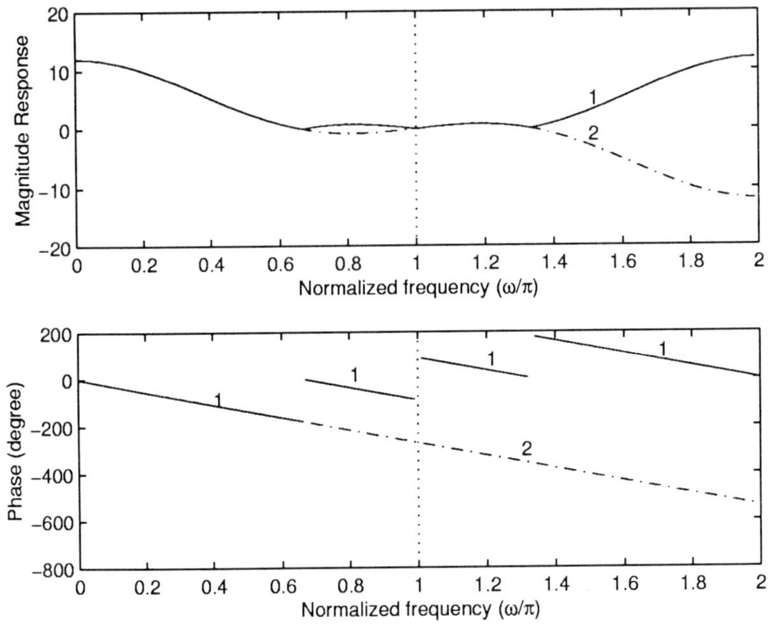

Fig. 3.6. Characteristics of Type 2 linear phase FIR filter for 1: $M(\omega)$ and $\delta(\omega)$, and 2: $A(\omega)$ and $\phi(\omega)$

This problem may be stated another way: Given the infinite set of coefficients $\{h(k);\ k \geq 0\}$ in (3.36), find a *window function* w such that $H_N(exp\{j\omega\})$ provides a good approximation to $H(exp\{j\omega\})$ when

$$h_N(k) = h(k)w(k)\ ;\ k \geq 0 \tag{3.38}$$

where

$$w(k) = 0\ ;\ k \geq N+1 \tag{3.39}$$

Gibbs Phenomenon

Perhaps the most obvious candidate is the *rectangular window function* w_R as defined by

$$w_R(k) \triangleq \begin{cases} 1\ ;\ 0 \leq k \leq N \\ 0\ ;\ k \geq N+1 \end{cases} \tag{3.40}$$

Truncation of a given IIR impulse response function h as implied by the use of a rectangular window function while providing an approximation to $H(exp\{j\omega\}$ often leads to a problem known as the *Gibbs phenomenon.* This problem manifests itself when the given magnitude function $|H(exp\{j\omega\}|$

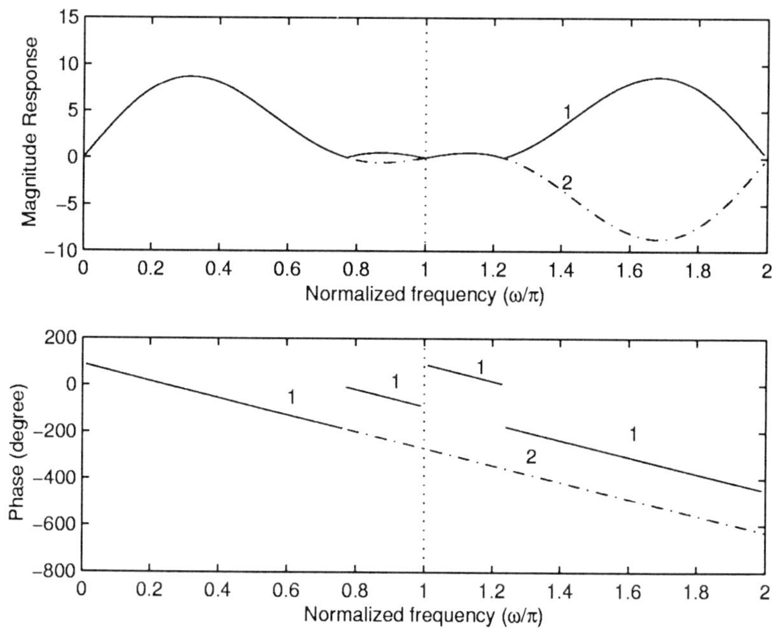

Fig. 3.7. Characteristics of Type 3 linear phase FIR filter for 1: $M(\omega)$ and $\delta(\omega)$, and 2: $A(\omega)$ and $\phi(\omega)$

which is to be approximated has sudden changes in values which occur when there is a sharp transition from a pass band to a stop band (or vice versa). The effect of the so called Gibbs phenomenon is to cause "ripples" in the approximating magnitude function $|H_N(exp\{j\omega\}|$ near the boundary of this transition.

For example, it can be shown that the analog signal u as defined by

$$u(t) \triangleq \begin{cases} 1 & ; \ 0 < t < \pi \\ -1 & ; \ \pi < t < 2\pi \end{cases}$$

can be approximated by the signal u_N as defined by

$$u_N(t) \triangleq \frac{4}{\pi} \sum_{m=0}^{N-1} \frac{\sin(2m+1)t}{2m+1}$$

in the sense that

$$\lim_{N \to \infty} |u(t) - u_N(t)| = 0 \ ; \ 0 < t < \pi, \ \pi < t < 2\pi$$

However, it can also be shown that

$$\lim_{N \to \infty} u(t_{N,1}) \simeq 1.18 \quad \text{when} \quad t_{N,1} = \frac{1}{2N+1}, \frac{2N\pi}{2N+1}$$

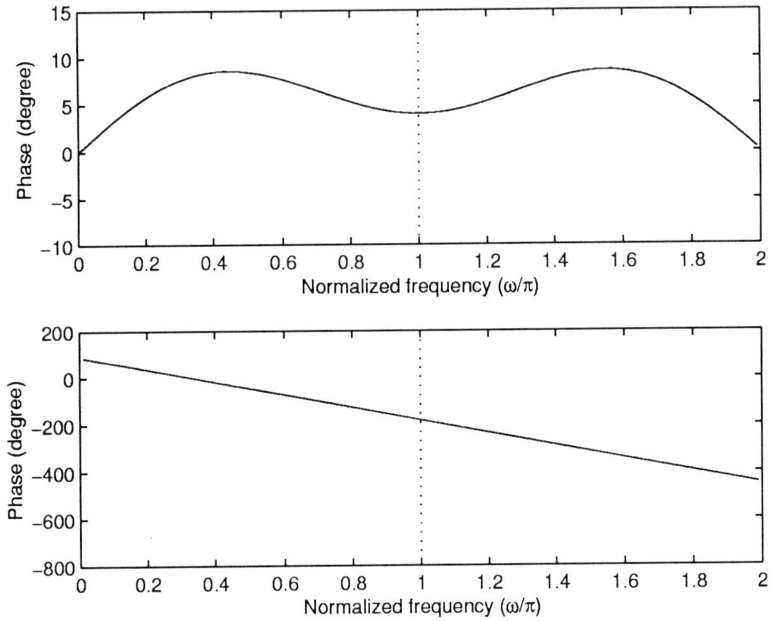

Fig. 3.8. Characteristics of Type 4 linear phase FIR filter for 1: $M(\omega)$ and $\delta(\omega)$, and 2: $A(\omega)$ and $\phi(\omega)$

and

$$\lim_{N\to\infty} u(t_{N,2}) \simeq -1.18 \quad \text{when} \quad t_{N,2} = \frac{(2N+2)\pi}{2N+1}, \frac{4N\pi}{2N+1}$$

Graphs of the signals u_N for $N = 2, 4, 8$ are illustrated in Figure 3.9. The key to reducing the Gibbs phenomenon is to provide a window function which has a "smooth" rather than a "sudden" transition.

One possibility is to use a *triangular window function* $w_{T,1}$ as defined by

$$w_{T,1}(k) \triangleq \begin{cases} 1 & ; \; 1 \leq k \leq N-1 \\ 0.5 & ; \; k = 0, N, \; k \geq N+1 \end{cases} \tag{3.41}$$

Another possible *triangular window function* $w_{T,2}$ is defined by

$$w_{T,2}(k) \triangleq \begin{cases} 1 & ; \; 2 \leq k \leq N-2 \\ 0.5 & ; \; k = 1, N-1 \\ 0 & ; \; k = 0, N \end{cases} \tag{3.42}$$

Many different window functions w have been examined in terms of their ability to generate good approximations of $H(exp\{j\omega\})$ while at the same time avoiding the rippling effects associated with the Gibbs phenomenon. The most useful function in this regard is the so called generalized Hamming Window $w_{H,\alpha}$ as defined by

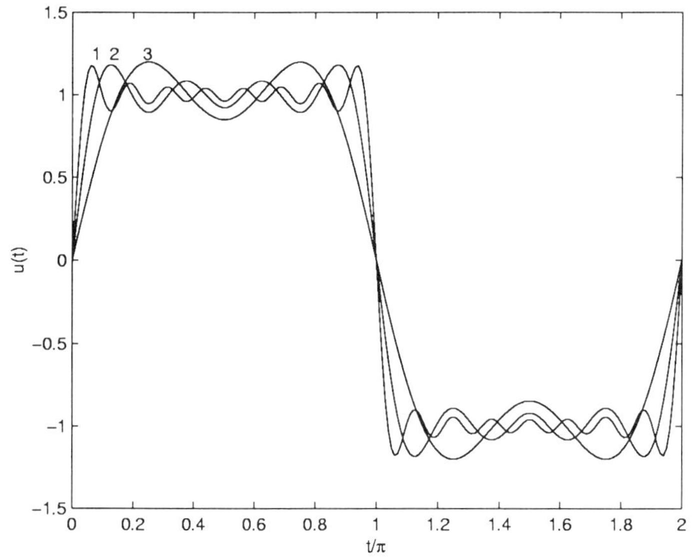

Fig. 3.9. Illustration of Gibbs phenomenon

$$w_{H,\alpha}(k) \triangleq \begin{cases} \alpha - (1-\alpha)\cos\{2\pi k(N-1)^{-1}\} & ;\ 0 \le k \le N-1 \\ 0 & ;\ k \ge N \end{cases} \tag{3.43}$$

for some $0 < \alpha \le 1$. The *rectangular window* $w_R = w_{H,1}$ corresponds to $\alpha = 1$, while the *Hanning window* $w_h = w_{H,0.5}$ is given by $\alpha = 0.5$, and the *Hamming window* $w_H = w_{H,0.54}$ is given by $\alpha = 0.54$. Since $w_h(0) = w_h(N-1) = 0$, the Hann window has only $N-2$ nonzero terms while the Hamming window has N nonzero terms. When $N = 15$, the corresponding Hanning and Hamming window functions are illustrated in Figure 3.10. Other window functions (which are all available in MATLAB) are the Bartlett triangular window, the Blackman window and the Kaiser window.

Often approximations to frequency response characteristics of digital filters also need to also take into account that the required approximating FIR filter must also have *linear phase*. For example, when n is *even*, we have from Theorem 3.2.1 that a linear phase Type 1 FIR filter $H(z)$ is given by

$$H(z) = \sum_{p=0}^{n} b_{n-p} z^{-p} \ ;\ b_m = b_{n-m}\ ,\ m = 0, 1, \ldots, 0.5n$$

where

$$|H(exp\{j\omega\})| = A(\omega) exp\{-j0.5n\omega\} \tag{3.44}$$

$$A(\omega) = b_{0.5n} + 2\sum_{m=0}^{0.5n-1} b_m \cos(0.5n - m)\omega$$

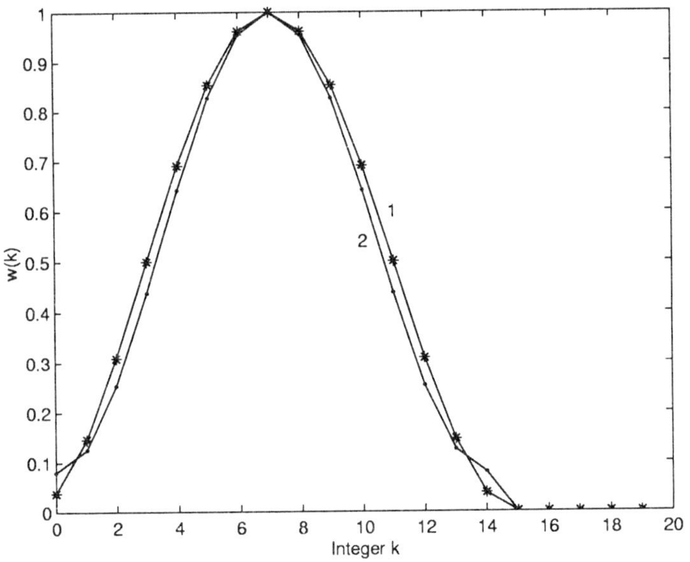

Fig. 3.10. Hanning (=1) and Hamming (=2) window functions

That is, the $0.5n$ coefficients $\{b_m; 0 \leq m \leq 0.5n - 1\}$ determine the $0.5n + 1$ coefficients $\{b_m; 0.5n \leq m \leq n\}$. For example, a window function w can be used to approximate $A(\omega)$ in (3.44) by the function $\hat{A}(\omega)$ where

$$\hat{A}(\omega) = b_{50}w(50) + 2\sum_{m=0}^{49} b_m w(m)\cos(25 - m)\omega$$

3.3 Design of IIR Filters

In this section, we present methods for the design of a digital IIR filter beginning with an analog filter. We therefore start this section with an overview of the design of some classical analog low pass filters. Analog filters are described by differential equations whereas digital filters are described by difference equations. Consequently, any method which results in a difference equation whose solution approximates sampled values of the solution of a differential equation provides a method for transforming an analog filter to a digital filter. The resulting transformation methods are most efficiently and conveniently expressed in algebraic form in terms of transformations between the system matrix, the input vector and the output vector of an analog state space representation and the system matrix, the input vector and the output vector of the corresponding discrete state space representation.

As previously defined, a *recursive* or *infinite impulse response* (IIR) filter is described by (3.1) with $a_k \neq 0$ for some k. This condition means that all the characteristic roots λ_p of the characteristic equation (3.9) do not necessarily have magnitude less than unity, and so apriori, the filter (3.1) is not necessarily *asymptotically stable.*

The asymptotic stability requirement (namely, $|\lambda_p| < 1$ for *all* p) can be expressed in terms of *nonlinear inequality constraints* on the coefficients $\{a_p\}$ via the *Marden Stability Test.* However these nonlinear inequalities often cause a problem with numerical routines which attempt to design an IIR filter which optimizes some filter performance measure J expressed in terms of the filter coefficients $\{a_p\}$. For example, if at some iterative step in the algorithm, the filter turns out to be *unstable* (i.e. some $|\lambda_p| \geq 1$), then one needs to somehow "project" the current parameter values to new values corresponding to an *asymptotically stable* filter without (substantially) increasing the current value of the performance measure J. This can prove to be a difficult problem to overcome, and as a consequence, any numerical optimization of IIR filters can be very difficult to achieve. Numerical iteration schemes cannot be guaranteed to always find an optimal solution.

3.3.1 All Pass Filter

Consider the asymptotically stable first order IIR filter

$$y(k+1) = a_0 y(k) + b_0 u(k) + b_1 u(k+1) \ ; \ 0 < |a_0| < 1 \tag{3.45}$$

which in (3.1) corresponds to $n = 1$. The corresponding digital frequency function H is given by

$$H(z) = \frac{b_0 + b_1 z}{z - a_0} \ ; \ 0 < |a_0| < 1 \tag{3.46}$$

and so the magnitude response is given by

$$\begin{aligned} M^2(\omega) &\triangleq H(exp\{j\omega\})H(exp\{-j\omega\}) \\ &= \frac{b_0 + b_1\cos\omega + jb_1\sin\omega}{\cos\omega - a_0 + j\sin\omega} \cdot \frac{b_0 + b_1\cos\omega - jb_1\sin\omega}{\cos\omega - a_0 - j\sin\omega} \\ &= \frac{b_0^2 + b_1^2 + 2b_0b_1\cos\omega}{1 - 2a_0\cos\omega + a_0^2} \end{aligned}$$

First Order: In particular, if

$$b_0 b_1 = -a_0 \ ; \ b_0^2 + b_1^2 = 1 + a_0^2 \tag{3.47}$$

then

$$M(\omega) = 1 \text{ for all } \omega$$

in which case (3.45) is a (unity gain) *all-pass filter.*

Since $a_0 \neq 0$ for a recursive first order filter, the condition (3.47) cannot be satisfied unless $b_1 b_0 \neq 0$. After substituting for $b_1 = -a_0 b_0^{-1}$, it follows that all solutions of (3.47) are then given by one of the following four possibilities:

$$\{b_0 = 1, b_1 = -a_0\} \;\; ; \;\; \{b_0 = -1, b_1 = a_0\}$$
$$\{b_0 = a_0, b_1 = -1\} \;\; ; \;\; \{b_0 = -a_0, b_1 = 1\}$$

The case $\{b_0 = a_0, b_1 = -1\}$ implies $H(z) = -1$, while the case $\{b_0 = -a_0, b_1 = 1\}$ implies $H(z) = 1$. Both these cases are trivial since then either $y(k) = -u(k)$ or $y(k) = u(k)$ for all k which means that in both these cases, there is in fact no filtering at all.

The first two possibilities however provide *nontrivial all-pass filters.* Specifically, from (3.46), we then have

$$H(z) = H_0(z) \triangleq \frac{1 - a_0 z}{z - a_0} \;\; ; \;\; \{b_0 = 1, b_1 = -a_0\} \tag{3.48}$$

and $H(z) = -H_0(z)$ for $\{b_0 = -1, b_1 = a_0\}$. Now

$$H_0(exp\{j\omega\}) = \frac{1 - a_0 \cos\omega - j a_0 \sin\omega}{\cos\omega - a_0 + j\sin\omega}$$

and after multiplying top and bottom by $\cos\omega - a_0 - j\sin\omega$, it follows that the corresponding phase response $\phi_0(\omega)$ of $H_0(z)$ in (3.48) is given by

$$\phi_0(\omega) = \tan^{-1}\left\{\frac{(1 - a_0^2)\sin\omega}{(1 + a_0^2)\cos\omega - 2a_0}\right\} \tag{3.49}$$

This phase function is illustrated in Figure 3.11 for a_0 = -0.9, -0.5, -0.1, 0.1, 0.5, 0.9.

Second Order: Now consider the second order difference equation

$$y(k+2) = a_1 y(k+1) + a_0 y(k) + b_0 v(k) + b_1 v(k+1) + b_2 v(k+2)$$

which corresponds in (3.1) to $n = 2$. The corresponding frequency response $H(z)$ is given by

$$H(z) = \frac{b_2 z^2 + b_1 z + b_0}{z^2 - a_1 z - a_0} \tag{3.50}$$

Now, without any loss of generality, let

$$z^2 - a_1 z - a_0 = (z - \lambda_1)(z - \lambda_2)$$
$$b_2 z^2 + b_1 z + b_0 = \gamma_0(\gamma_1 z - 1)(\gamma_2 z - 1)$$

where $\{\lambda_1, \lambda_2, \gamma_1, \gamma_2\}$ may be complex. Then

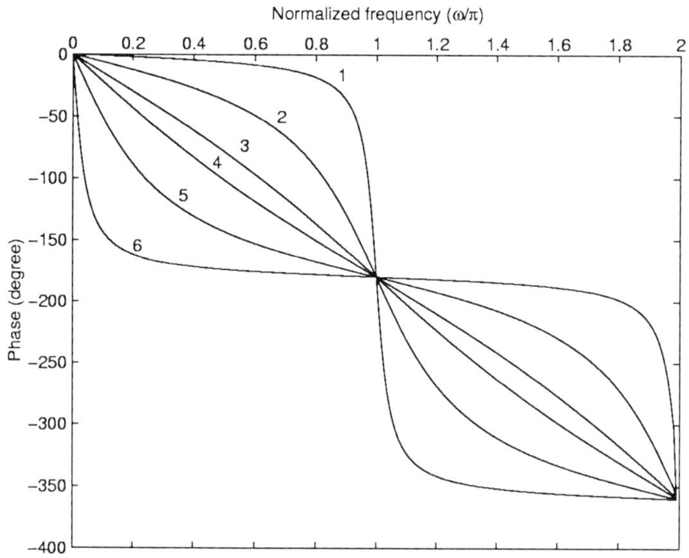

Fig. 3.11. Phase response characteristics for a first order all-pass IIR filter for 1: $a_0 = -0.9$ 2: $a_0 = -0.5$ 3: $a_0 = -0.1$ 4: $a_0 = 0.1$ 5: $a_0 = 0.5$ and 6: $a_0 = 0.9$

$$H(z) = \gamma_0 \frac{\gamma_1 z - 1}{z - \lambda_1} \cdot \frac{\gamma_2 z - 1}{z - \lambda_2}$$

and so from (3.48), $|H(exp\{j\omega\})| = \gamma_0$ for all ω when $\{\gamma_1 = \lambda_1, \gamma_2 = \lambda_2\}$, in which case $H(z)$ describes an all-pass filter of gain γ_0. It can be shown that if $H(z)$ in (3.50) defines an *asymptotically stable all-pass* IIR filter, then there are no other possible choices for $\{\gamma_1, \gamma_2\}$ (other than $\{\gamma_1 = \lambda_2, \gamma_2 = \lambda_1\}$). If λ_1 is complex, then in order that the coefficients of H in (3.50) are *real*, it is necessary that $\lambda_2 = \overline{\lambda}_1$.

A unity gain second order all-pass filter $H_0(z)$ is therefore defined by

$$H_0(z) = \frac{(\lambda_1 z - 1)(\lambda_2 z - 1)}{(z - \lambda_1)(z - \lambda_2)} = \frac{b_2 z^2 + b_1 z + b_0}{z^2 - a_1 z - a_0}$$

where

$$b_2 = \lambda_1 \lambda_2 = -a_0 \ ; \ b_1 = -(\lambda_1 + \lambda_2) = -a_1 \ ; \ b_0 = 1$$

That is

$$H_0(z) = \frac{b_2 z^2 + b_1 z + b_0}{z^2 + b_1 z + b_2}$$

From (3.48), (3.49), the corresponding all-pass phase response when $\{\lambda_1, \lambda_2\}$ are *both real* is given by

$$\phi(\omega) = \tan^{-1}\left\{\frac{(1 - \lambda_1^2)\sin\omega}{(1 + \lambda_1^2)\cos\omega - 2\lambda_1}\right\} + \tan^{-1}\left\{\frac{(1 - \lambda_2^2)\sin\omega}{(1 + \lambda_2^2)\cos\omega - 2\lambda_2}\right\}$$

When the second order all-pass digital filter (3.50) has *complex characteristic roots* $\{\lambda_1 = \varrho_1 + j\omega_1, \bar{\lambda}_1\}$, then

$$H(z) = \frac{b_2 z^2 + b_1 z + 1}{z^2 + b_1 z + b_2} \ ; \ b_1 = -2\varrho_1 \ ; \ b_2 = \varrho_1^2 + \omega_1^2$$

Then after substituting for $z = exp\{j\omega\}$, it follows that the corresponding all-pass phase response is given by

$$\phi(\omega) = \tan^{-1}\{\alpha_1\} - \tan^{-1}\{\alpha_2\}$$

where

$$\alpha_1 = \frac{(\varrho_1^2 + \omega_1^2)\sin 2\omega - 2\varrho_1 \sin\omega}{(\varrho_1^2 + \omega_1^2)\cos 2\omega - 2\varrho_1 \cos\omega + 1}$$

$$\alpha_2 = \frac{\sin 2\omega - 2\varrho_1 \sin\omega}{\cos 2\omega - 2\varrho_1 \cos\omega + \varrho_1^2 + \omega_1^2}$$

More generally, we have the following result.

Theorem 3.3.1. *(a) The complete class of asymptotically stable unity gain all-pass digital filters H of order n are defined by*

$$H(z) = \pm \prod_{k=1}^{n} \left(\frac{\lambda_k z - 1}{z - \lambda_k} \right) \tag{3.51}$$

with $|\lambda_k| < 1$ for all k such that: if λ_k is complex for some k, then $\lambda_m = \overline{\lambda_k}$ for some $1 \leq m \leq n$.

(b) The frequency response function $H(z)$ of an all-pass filter can also be expressed in the form

$$H(z) = \pm \frac{\sum_{p=0}^{n} b_{n-p} z^p}{\sum_{p=0}^{n} b_p z^p} \tag{3.52}$$

(c) If the all-pass filter

$$H(z) = \pm \prod_{k=1}^{p} \frac{\lambda_k z - 1}{z - \lambda_k} \prod_{k=1}^{q} \frac{(\varrho_k^2 + \omega_k^2) z^2 - 2\varrho_k z + 1}{z^2 - 2\varrho_k z + \varrho_k^2 + \omega_k^2}$$

of order $n = p + 2q$ has p real characteristic roots $\{\lambda_1, \lambda_2, \ldots \lambda_p\}$ and q pairs of complex characteristic roots $\{\varrho_1 \pm j\omega_1, \varrho_2 \pm j\omega_2, \ldots \varrho_q \pm j\omega_q\}$, then the phase response is given by

$$\phi(\omega) = \sum_{k=1}^{p} \tan^{-1}(\beta_k) + \sum_{k=1}^{q} \tan^{-1}(\gamma_{1k}) - \sum_{k=1}^{q} \tan^{-1}(\gamma_{2k})$$

where

$$\beta_k = \frac{(1-\lambda_k^2)\sin\omega}{(1+\lambda_k^2)\cos\omega - 2\lambda_k} \;;\; \gamma_{1k} = \frac{(\varrho_k^2+\omega_k^2)\sin 2\omega - 2\varrho_k \sin\omega}{(\varrho_k^2+\omega_k^2)\cos 2\omega - 2\varrho_k\cos\omega + 1}$$

$$\gamma_{2k} = \frac{\sin 2\omega - 2\varrho_k\sin\omega}{\cos 2\omega - 2\varrho_k\cos\omega + \varrho_k^2 + \omega_k^2}$$

Even though it is possible to design an all-pass digital IIR filter, it is *not* possible to design a filter which has a constant magnitude response over some frequency interval *less* than π. That is, any low-pass filter characteristic (3.21) for $\omega_p < \pi$ must *necessarily* have some "ripple" in the passband.

Phase Response

For an IIR filter, it is *not* possible to have a *linear phase* characteristic over *any* frequency range. In fact, the phase response of an IIR filter can be *highly nonlinear* which causes severe phase distortion. Since an IIR *all-pass filter* has a constant gain for all frequencies ω, one approach which is used to overcome phase distortion in an IIR filter is to *cascade* the original IIR filter $H_1(z)$ with a unity gain all-pass filter $H_0(z)$ to give a new filter with frequency response function $H_2(z)$ given by

$$H_2(z) = H_1(z).\ H_0(z)$$

Since then

$$|H_2(exp\{j\omega\})| = |H_1(exp\{j\omega\})|.\ |H_0(exp\{j\omega\})| = |H_1(exp\{j\omega\})|$$

the filters H_1 and H_2 have the *same magnitude response* function. However the phase response characteristic ϕ_2 of H_2 will be the *sum* of the phase characteristic ϕ_1 of H_1 and the phase characteristic ϕ_0 of the all-pass filter H_0; that is

$$\phi_2(\omega) = \phi_1(\omega) + \phi_0(\omega)$$

A method for reducing phase distortion in the original filter H_1 is then to adjust $\phi_0(\omega)$ using a numerical procedure so as to make $\phi_2(\omega)$ approximately linear over some desired frequency range $\omega_1 \le \omega \le \omega_2$.

3.3.2 Analog Filters

A linear time invariant *differential equation*

$$z^{(n)}(t) = \alpha_{n-1}z^{(n-1)}(t) + \alpha_{n-2}z^{(n-2)}(t) + \ \ldots\ + \alpha_1\dot{z}(t) + \alpha_0 z(t) + \beta_0 v(t) + \beta_1\dot{v}(t)\ \ldots\ + \beta_{n-1}v^{(n-1)} + \beta_n v^{(n)}(t) \qquad (3.53)$$

where $z^{(m)}$ denotes the mth derivative of the analog signal z represents a *linear time invariant analog filter*. It can be shown that the steady state analog response z_{ss} of an asymptotically stable differential equation (3.53) resulting from the analog input signal v where

$$v(t) = \sin \Omega_0 t$$

is given by

$$z_{ss}(t) = \mathcal{A}(\Omega_0) \sin[\Omega_0 t - \Phi(\Omega_0)] \tag{3.54}$$

where the *analog magnitude response* $\mathcal{A}(\Omega)$ and the *analog phase response* $\Phi(\Omega)$ are given by

$$\mathcal{A}(\Omega_0) exp\{-j\Phi(\Omega_0)\} = \mathcal{H}(j\Omega_0) \; ; \; \mathcal{H}(s) = \frac{\sum_{k=0}^{n} \beta_k s^k}{s^n - \sum_{k=0}^{n-1} \alpha_k s^k} \tag{3.55}$$

The *analog group delay* $d_a(\Omega)$ is also defined by

$$d_a(\Omega) \triangleq \frac{d\Phi(\Omega)}{d\Omega}$$

An analog filter (3.53) is *asymptotically stable* if the response due to the input signal v when $v(t) = 0$ for all t decays to zero as $t \to \infty$ for all initial conditions $\{z(0), \dot{z}(0), \ldots, z^{(n-1)}(0)\}$. It can be shown that asymptotic stability of an analog filter is equivalent to the condition that all roots λ_k of the algebraic equation

$$s^n - \alpha_{n-1}s^{n-1} - \alpha_{n-2}s^{n-2} - \; \ldots \; - \alpha_1 s - \alpha_0 = 0$$

satisfy the condition that $Re\{\lambda_k\} < 0$ for all k.

Note that we use the symbol Ω for *analog frequency* (in rads/sec), and the symbol ω for *discrete frequency* (in rads). If the analog signal $v(t) = \sin \Omega_0 t$ of frequency Ω_0 rads/sec is sampled at the uniform time instants $t_k = kT$ (k an integer) for some constant sampling period T (in secs), then the digital signal u defined by

$$u(k) \triangleq v(t = kT) = \sin k\omega_0 \; ; \; \omega_0 = \Omega_0 T$$

has digital frequency ω_0 ($= \Omega_0 T$) rads.

We now show how a first order digital filter can be derived from a first order analog filter. Consider the analog filter as described by the first order differential equation

$$\dot{z}(t) = -\alpha z(t) + \alpha v(t) \; ; \; \alpha > 0 \tag{3.56}$$

and suppose the input signal v is given by

$$v(t) = \sin \Omega_0 t = Im\{exp\{j\Omega_0 t\}\}$$

Then by substituting into (3.56), it can be verified that the solution z is given by

$$z(t) = \beta_0 exp\{-\alpha t\} + \mathcal{A}(\Omega_0)\sin[\Omega_0 t - \Phi(\Omega_0)] \tag{3.57}$$

where

$$\mathcal{A}(\Omega) = \frac{\alpha}{\sqrt{\alpha^2 + \Omega^2}} \;;\; \Phi(\Omega\} = -\tan^{-1}\left\{\frac{\Omega}{\alpha}\right\} \tag{3.58}$$

and

$$\beta_0 = z(0) - \mathcal{A}(\Omega_0)\sin[-\Phi(\Omega_0)]$$

The *amplitude response* $\mathcal{A}(\Omega)$ and the *phase response* $\Phi(\Omega)$ can also be expressed in the form

$$\mathcal{A}(\Omega) exp\{-j\Phi(\Omega\} = \mathcal{H}(j\Omega) \tag{3.59}$$

where the *continuous frequency response* $\mathcal{H}$ is defined by

$$\mathcal{H}(s) = \frac{\alpha}{s + \alpha} \tag{3.60}$$

Since $\alpha > 0$, it follows from (3.57) that $\beta_0 exp\{-\alpha t\} \to 0$ as $t \to \infty$, and so the *steady state component* z_{ss} of z is given by

$$z_{ss}(t) = \mathcal{A}(\Omega_0)\sin[\Omega_0 t - \Phi(\Omega_0)] \tag{3.61}$$

Suppose now the aim is to replace the first order *analog filter* (3.56) by a first order *digital filter*

$$y(k+1) = ay(k) + bu(k) \;;\; |a| < 1, b > 0 \tag{3.62}$$

such that when the digital input signal u to the *digital filter* is a sampled version of the analog input signal v to the *analog filter*; that is, when

$$u(k) = v(t = kT) = \sin k\omega_0 \;;\; \omega_0 \triangleq \Omega_0 T \tag{3.63}$$

then the steady state value $y_{ss}(k)$ of y in (3.62) at time k *approximates* the steady state solution $z_{ss}(t = kT)$ in (3.61).

Now since $|a| < 1$, we have shown that: for the first order digital filter (3.62)

$$y_{ss}(k) = A(\omega_0)\sin[k\omega_0 - \phi(\omega_0)] \tag{3.64}$$

where

$$A(\omega) exp\{-j\phi(\omega)\} = H(exp\{j\omega\}) \tag{3.65}$$

and the *digital frequency response* function H is given by

$$H(z) = \frac{b}{z-a} \tag{3.66}$$

That is, in (3.64)

$$A(\omega) = \frac{b}{\sqrt{1+a^2-2a\cos\omega}} \; ; \; \phi(\omega) = -\tan^{-1}\left\{\frac{\sin\omega}{\cos\omega - a}\right\} \tag{3.67}$$

If we now compare (3.58), (3.61) with (3.64), (3.67), then we see that

$$y_{ss}(k) \approx z_{ss}(t = kT)$$

for all k provided there exists a "good" magnitude and/or a "good" phase approximation; that is, if

$$A(\Omega_0 T) \approx \mathcal{A}(\Omega_0) \; ; \; \phi(\Omega_0 T) \approx \Phi(\Omega_0) \tag{3.68}$$

The relative importance of having a either "good" magnitude or a "good" phase approximation (or both) depends on the particular filtering application. For example, if the dc gain is to be preserved (i.e. $A(0) = \mathcal{A}(0)$), then since $\mathcal{A}(0) = 1$, this constraint implies that $|b| = |1-a|$. If in addition, the sampling period T is sufficiently small so that

$$\cos \Omega_0 T \approx 1 - 0.5\Omega_0^2 T^2$$

then from (3.67), we have that

$$A(\Omega_0 T) \simeq \frac{b}{\sqrt{1+a^2-2a+a\Omega_0^2T^2}} = \frac{b}{T\sqrt{a}\sqrt{(1-a)^2/(aT^2)+\Omega_0^2}}$$

Therefore from (3.58), we have that $A(\Omega_0 T) \simeq \mathcal{A}(\Omega_0)$ if the digital filter parameters $\{a, b\}$ are selected in terms of the analog filter parameter $\{\alpha > 0\}$ in (3.56) - (3.58) and the (sufficiently small) sampling period $T > 0$ by

$$\alpha = \frac{1-a}{aT} \; ; \; \frac{b}{T\sqrt{a}} = \alpha$$

That is,

$$a = \frac{1}{1+\alpha T} \; ; \; b = \frac{1-a}{\sqrt{a}}$$

Since $|a| < 1$ (actually $0 < a < 1$) for all $\{\alpha > 0, T > 0\}$, the approximating digital filter is asymptotically stable.

Ideal Low Pass Filter

The *passband* of an ideal analog low pass filter is defined to be the range of frequencies $0 < \Omega < \Omega_p$ in which the analog magnitude response is constant. In the *stopband* $\Omega > \Omega_p$, $|\mathcal{H}(j\Omega)| = 0$ for $\Omega > \Omega_p$.

These ideal filter characteristics *cannot* be realized in practice. In addition, the *transition* between the passband and the stopband in a practical filter *cannot* be instantaneous, and the magnitude response cannot be constant in the passband. Since for practical analog filters, the passband frequency Ω_p is not clearly defined, it is common to define the *passband frequency* Ω_p of an analog low pass filter by the frequency Ω_p for which

$$|\mathcal{H}(j\Omega)| \leq \frac{1}{\sqrt{2}}|\mathcal{H}(0)| \; ; \; \Omega \geq \Omega_p \tag{3.69}$$

The *magnitude loss function* $\mathcal{M}_L$ (in dB) is also defined by

$$\begin{aligned} \mathcal{M}_L(\Omega) &\triangleq 10\log_{10}\left(\frac{|\mathcal{H}(0)|}{|\mathcal{H}(j\Omega)|}\right)^2 \\ &= 20\log_{10}|\mathcal{H}(0)| - 20\log_{10}|\mathcal{H}(j\Omega)| \end{aligned} \tag{3.70}$$

Equivalently, the analog filter $\mathcal{H}(s)$ is said to have an *attenuation* of $\mathcal{M}_L(\Omega)$ dB. Consequently, from (3.69), $\mathcal{M}_L(\Omega)$ at $\Omega = \Omega_p$ is given by

$$\mathcal{M}_L(\Omega_p) = 10\log_{10}2 \simeq 3\text{dB}$$

The frequency Ω_p which defines the passband is therefore also commonly referred to as the 3dB point.

Low Pass Butterworth Filter

An nth order *analog low pass Butterworth filter* $\mathcal{H}(s)$ having unity dc gain is defined by

$$|\mathcal{H}(j\Omega)|^2 = \frac{1}{1 + \varepsilon_p^2 \left(\Omega/\Omega_p\right)^{2n}} \tag{3.71}$$

for some constants $\{\varepsilon_p, \Omega_p\}$. These filters are characterized by the property that the magnitude characteristic is *maximally flat* of degree $2n - 1$ at the origin, or in other words, the first $2n - 1$ derivatives of $|\mathcal{H}(j\Omega)|^2$ are zero at $\Omega = 0$. As the filter order n increases, the magnitude characteristic of a Butterworth filter becomes 'flatter' over the passband and 'sharper' near the frequency Ω_p as illustrated in Figure 3.12 for $\Omega_p = 0.75$ rads/sec. The phase delay also increases as the order of the filter increases.

Since

$$|\mathcal{H}(j\Omega)|^2 = \mathcal{H}(j\Omega)\mathcal{H}(-j\Omega)$$

it can be shown that $\mathcal{H}(s)$ in (3.71) is given by

$$\mathcal{H}(s) = \mathcal{H}_1(\frac{s}{R}) \; ; \; R \triangleq \varepsilon_p^{-1/n}\Omega_p \tag{3.72}$$

where

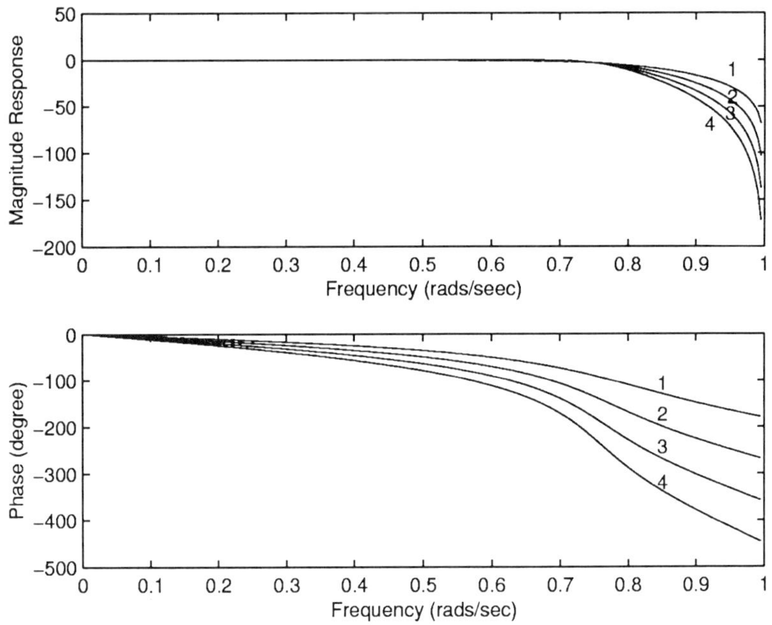

Fig. 3.12. Characteristics of nth analog Butterworth filter for 1: $n = 2$, 2: $n = 3$, 3: $n = 4$, and 4: $n = 5$

$$\mathcal{H}_1(s) = \begin{cases} \prod_{k=1}^{n/2} (s^2 + 2\sin\theta_k s + 1)^{-1} & ; \quad n \text{ even} \\ (s+1)^{-1} \prod_{k=1}^{(n-1)/2} (s^2 + 2\sin\theta_k s + 1)^{-1} & ; \quad n \text{ odd} \end{cases} \tag{3.73}$$

with

$$\theta_k \triangleq \left(\frac{2k-1}{2n} \right) \pi$$

It then follows that

$$|\mathcal{H}(j\Omega)|^2 \begin{cases} \geq (1 + \varepsilon_p^2)^{-1} \; ; \; \Omega \leq \Omega_p \\ \leq (1 + \varepsilon_s^2)^{-1} \; ; \; \Omega \geq \Omega_s \end{cases} \tag{3.74}$$

for *any* given $\{\Omega_p, \Omega_s, \varepsilon_p, \varepsilon_s\}$ provided the filter order n is such that

$$n > \frac{\log_{10}\left(\frac{\varepsilon_s}{\varepsilon_p}\right)}{\log_{10}\left(\frac{\Omega_s}{\Omega_p}\right)} \tag{3.75}$$

Example 3.3.1. A unit dc gain maximally flat (i.e. Butterworth) analog filter is to be designed to meet the following specifications:

(i) Passband frequency $\Omega_p = 7.5k$ rad/sec with maximum attenuation at Ω_p of 1 dB, and (ii) stopband frequency $\Omega_s = 10.5k$ rad/sec with minimum attenuation at Ω_s of 10 dB.

From (3.74),

$$10\log_{10}1 - 10\log_{10}\left(\frac{1}{1+\varepsilon_p^2}\right) = 1$$

implies $\varepsilon_p = 0.5$, and

$$10\log_{10}1 - 10\log_{10}\left(\frac{1}{1+\varepsilon_s^2}\right) = 10$$

implies $\varepsilon_s = 3.0$ and on substituting into (3.75), $n > 5.32$. That is, a sixth order Butterworth filter is necessary, and from (3.72) $R = 2^6 \times 10.5 \times 10^3 = 6.72 \times 10^5$. The required filter $\mathcal{H}(s)$ can be expressed in the form

$$\mathcal{H}(s) = \mathcal{H}_6(\frac{s}{R})$$

where

$$\mathcal{H}_6(s) = \frac{1}{s^6 + q_5 s^5 + q_4 s^4 + q_3 s^3 + q_2 s^2 + q_1 s + 1} \; ; \; |\mathcal{H}_6(j1)| = \frac{1}{\sqrt{2}}$$

The coefficients q_k (to 5 dec. places) are given in Table 3.1 where

$$q_k = q_{n-k} \; ; \; 0 < k < [0.5n], \; [0.5n] \stackrel{\Delta}{=} \begin{cases} 0.5n & ; n \text{ even} \\ 0.5(n-1) & ; n \text{ odd} \end{cases}$$

Table 3.1. Normalized Butterworth filter coefficients

n	q_{n-1}	q_{n-2}	q_{n-3}	q_{n-4}
1	1			
2	1.41421			
3	2.00000	2.00000		
4	2.61313	3.41421		
5	3.23606	5.23607		
6	3.86370	7.46410	9.14162	
7	5.12583	13.1370	21.84615	25.68836

Low Pass Chebyshev Filter

A *Chebyshev analog low pass filter* has the property that the *peak magnitude error* (with respect to the perfectly flat magnitude response of an ideal filter) is *minimized* over a prescribed frequency band. Because the magnitude response oscillates an equal distance above and below the nominal ideal flat response, the Chebyshev filter is said to be *equiripple*. The filter is then further classified as being either *Type 1* (when the equiripple frequency band is in the *passband*), or *Type 2* (when the equiripple frequency is in the *stopband*).

A Type 1 Chebyshev filter $\mathcal{H}_1(s)$ having unity dc gain is defined by

$$|\mathcal{H}_1(j\Omega)|^2 = \frac{1}{1+\varepsilon_p^2 C_n^2\left(\Omega/\Omega_p\right)} \tag{3.76}$$

where $C_n(x)$ is the nth order Chebyshev polynomial defined recursively by

$$\begin{aligned} C_0(x) &= 1\ ;\ C_1(x) = x \\ C_{n+1}(x) &= 2xC_n(x) - C_{n-1}(x)\ ;\ n > 1 \end{aligned} \tag{3.77}$$

The parameter ε_p controls the magnitude of the passband ripple, and Ω_p the passband edge. It can be shown that the frequency response function of the filter is of the form

$$\mathcal{H}_1(s) = \frac{1}{\prod_{k=1}^{m}(s+\sigma_k)^2+\Omega_k^2}$$

where $\{\sigma_k, \Omega_k\}$ are given by

$$\frac{\sigma_k}{\Omega_p} = -\sin\left[\frac{(2k+1)\pi}{2n}\right]\sinh\xi\ ;\quad \frac{\Omega_k}{\Omega_p} = \cos\left[\frac{(2k+1)\pi}{2n}\right]\cosh\xi$$

In addition, it follows that the *magnitude constraints* (3.76) are satisfied provided the order n of the Type 1 Chebyshev filter satisfies

$$n > \frac{\cosh^{-1}\left(\varepsilon_s\varepsilon_p^{-1}\right)}{\cosh^{-1}\left(\Omega_s\Omega_p^{-1}\right)} \tag{3.78}$$

For the *same* specifications $\{\varepsilon_p, \varepsilon_s, \Omega_p, \Omega_s\}$, the expression (3.78) gives a *smaller* bound on the filter order n than the corresponding expression for the Butterworth filter as given by (3.75).

The normalized coefficients of the denominator polynomial of $\mathcal{H}_1(s)$ for levels of passband ripple between 0.1 dB and 3 dB are available. The magnitude and phase characteristics of Chebyshev filters of orders 2,3,4 and 5 corresponding to 0.5 dB ripple are presented in Figure 3.13 for $\Omega_p = 0.75$ rads/sec.

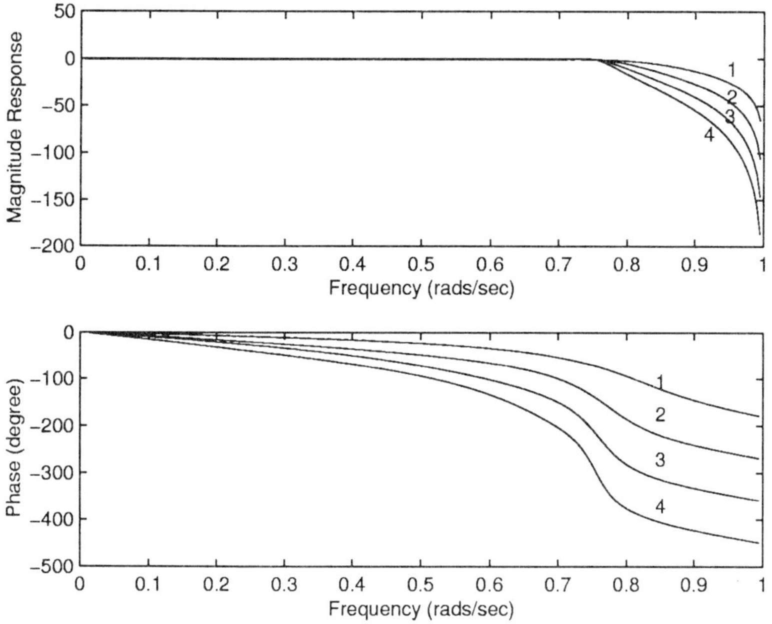

Fig. 3.13. Characteristics of nth order analog Chebyshev filter for 1: $n = 2$, 2: $n = 3$, 3: $n = 4$, and 4: $n = 5$

Example 3.3.2. Design a Type 1 Chebyshev filter (i.e. equiripple in the passband) to meet the following specifications:

(i) passband frequency $\Omega_p = 7.5k$rad/sec with maximum attenuation at Ω_p of 1 dB, and (ii) stopband frequency $\Omega_s = 10.5k$rad/sec with minimum attenuation at Ω_s of 10 dB.

Suppose as in example 3.3.1, $\varepsilon_p = 0.5$ and $\varepsilon_s = 3.0$. Then on substituting into (3.78), we get $n > 2.18$ so that a 3rd order filter is necessary. Note that for the same values of $\{\varepsilon_p, \varepsilon_s\}$ a sixth order Butterworth filter was required. From (3.78), the Chebyshev filter is given by $\xi = 0.481$ with

$$\frac{\sigma_0}{\Omega_p} = -0.250 \; ; \quad \frac{\Omega_0}{\Omega_p} = 0.968$$

$$\frac{\sigma_1}{\Omega_p} = -0.500 \; ; \quad \Omega_1 = 0 \; ; \quad \sigma_2 = \sigma_0 \; ; \quad \Omega_2 = -\Omega_0$$

Thus the normalized filter transfer function $\mathcal{H}_1(s/\Omega_p)$ is given by

$$\mathcal{H}_1\left(\frac{s}{\Omega_p}\right) = \frac{1}{(s+0.5)(s+0.5s+1)}$$

Low Pass Elliptic Filter

The characteristic of an *analog lowpass elliptic filter* is such that the *peak magnitude error* is *minimized* over prescribed frequency bands in *both* the passband and stopband. It can also be shown for a given filter order n and ripple error magnitudes ε_p and ε_s that an elliptic filter achieves the *fastest transition* (i.e. minimizes $\Omega_s - \Omega_p$) between the passband and stopband.

An elliptic filter is defined by

$$|\mathcal{H}(j\Omega)|^2 = \frac{1}{1 + \varepsilon_p^2 R_n^2(\Omega/\Omega_c)} \tag{3.79}$$

where $R_n(x)$ is a rational function with the property $R_n(x^{-1}) = R_n^{-1}(x)$ of the form

$$R_n(x) = x \prod_{k=1}^{N} \frac{x_k^2 - x^2}{1 - x_k^2 x^2} \; ; \; N = 0.5(n-1)$$

when n is odd, and

$$R_n(x) = \prod_{k=1}^{N} \frac{x_k^2 - x^2}{1 - x_k^2 x^2} \; ; \; N = 0.5n$$

when n is even for some $0 < x_k < 1$. The full details may be found in references cited in the Preface. As with Butterworth and Chebychev filters, the design of an elliptic filter may be achieved using MATLAB without explicit knowledge of such details. However a proper understanding of analog filter design is required when these standard filter characteristics don't satisfy the design specifications.

For appropriate $\{\Omega_k\}$

$$|\mathcal{H}(j\Omega)| = \begin{cases} 1 \text{ at } \Omega/\Omega_c = \Omega_k \\ \\ 0 \text{ at } \Omega/\Omega_c = \Omega_k^{-1} \end{cases}$$

Also for n odd, the zero frequency gain is unity. It further follows from the form of $R_n(x)$ that $\mathcal{H}(j\Omega)$ oscillates between 1 and $(1 + \varepsilon_p^2)^{-1}$ for $0 < \Omega < \Omega_p < \Omega_c$, and between 0 and $(1 + \varepsilon_s^2)^{-1}$ for $\Omega > \Omega_s > \Omega_c$ where

$$\Omega_c = \sqrt{\Omega_s \Omega_p}$$

It can also be shown that the *transition band* is given by

$$(\Omega_s - \Omega_p) = \Omega_c \left[\frac{\Omega_c}{\Omega_p} - \frac{\Omega_p}{\Omega_c} \right] \tag{3.80}$$

For this filter, no simple formula exists which describes the relationship between the ripple parameters $\{\varepsilon_p, \varepsilon_s\}$ and the required filter order n. The frequency response characteristics are illustrated in Figure 3.14 for $\Omega_p = 0.75$ rads/sec, and a maximum ripple in passband of 0.5dB.

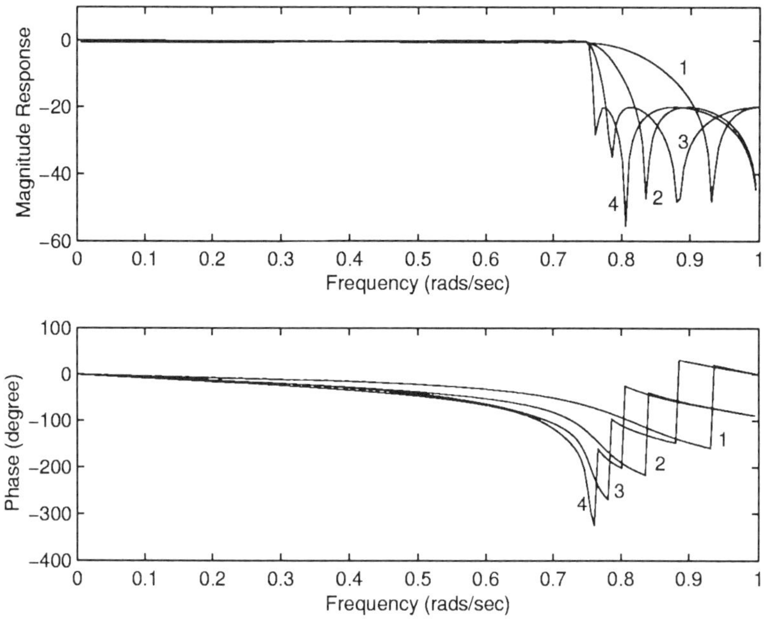

Fig. 3.14. Characteristics of nth order analog elliptic filter for 1: $n = 2$, 2: $n = 3$, 3: $n = 4$, and 4: $n = 5$

Example 3.3.3. Design a third order elliptic filter to meet the following specifications:

(i) passband frequency $\Omega_c = 7.5k\text{rad/sec}$ with maximum attenuation at Ω_p of 1 dB, and (ii) stopband frequency $\Omega_s = 10.5k\text{rad/sec}$ with minimum attenuation at Ω_s of 10 dB

The passband edge Ω_p should be as near as possible to Ω_c. Once again, $\varepsilon_p = 0.5$ and $\varepsilon_s = 3.0$. Then $n = 3$, and

$$\frac{\Omega_p}{\Omega_c} = 0.961$$

Overall, the *maximally flat* amplitude response of the Butterworth filter is better than the Chebyshev amplitude characteristics. However, the transition and stopband magnitude characteristics of the Chebyshev filter are superior. The Butterworth filter, however, generally has a flatter group delay, and as the order of the Chebyshev filter increases, the group delay characteristics of this filter become corresponding more distorted. If both the phase and amplitude responses are important, then the Butterworth filter may be preferred. If the *transistion band width* is critical, then elliptic filters are superior. For a given filter order and allowable passband and stopband deviations, the elliptic filters minimize the transition bandwidth.

Low Pass to Low Pass Transformation

For low pass filters, the process of *frequency scaling* has been used to transform a low pass filter with frequency response $\mathcal{H}_1(s)$ having a bandwidth of 1 rad/sec to another low pass filter $\mathcal{H}_{lp}(s)$ having a bandwidth of Ω_c rad/sec using the frequency transformation:

$$s \Rightarrow \frac{s}{\Omega_c} \tag{3.81}$$

so that

$$\mathcal{H}_{lp}(s) \triangleq \mathcal{H}_1(\frac{s}{\Omega_c}) \tag{3.82}$$

For example, the second order lowpass Butterworth filter

$$\mathcal{H}_1(s) = \frac{1}{s^2 + \sqrt{2}s + 1} \; ; \; |\mathcal{H}_1(j1)| = \frac{1}{\sqrt{2}} \tag{3.83}$$

has a bandwidth of 1 rad/sec. Then

$$\mathcal{H}_{lp}(s) = \mathcal{H}_1(\frac{s}{\Omega_c}) = \frac{\Omega_c^2}{s^2 + \sqrt{2}\Omega_c s + \Omega_c^2} \; ; \; |\mathcal{H}_{lp}(j\Omega_c)| = \frac{1}{\sqrt{2}} \tag{3.84}$$

is a lowpass filter which has a bandwidth of Ω_c rads/sec.

Low Pass to Highpass Transformation

A *analog highpass filter* has a passband from some frequency Ω_p to infinity with a stop band from zero to Ω_s for some $0 < \Omega_s < \Omega_p$. An analog low pass filter with frequency response $\mathcal{H}_1(s)$ may be transformed to *highpass filter* $\mathcal{H}_{hp}(s)$ having a cut-off frequency of Ω_c rad/sec using the *frequency transformation*:

$$s \Rightarrow \frac{\Omega_c}{s} \tag{3.85}$$

that is;

$$\mathcal{H}_{hp}(s) \triangleq \mathcal{H}_1(\frac{\Omega_c}{s}) \tag{3.86}$$

For example, given the second order analog *lowpass Butterworth filter* (3.83) having a bandwidth of 1 rad/sec, we have a second order analog *highpass Butterworth filter* given by

$$\mathcal{H}_{hp}(s) = \frac{1}{(\Omega_c s^{-1})^2 + \sqrt{2}\,(\Omega_c s^{-1}) + 1} = \frac{s^2}{s^2 + \sqrt{2}\Omega_c s + \Omega_c^2} \tag{3.87}$$

with

$$|\mathcal{H}_{hp}(j\Omega_c)| = \frac{1}{\sqrt{2}}$$

The lowpass frequency response $\mathcal{H}_{lp}(s)$ and the transformed highpass frequency response $\mathcal{H}_{hp}(s)$ when $\Omega_c = 1$ rad/sec are illustrated in Figure 3.15 and Figure 3.16 respectively.

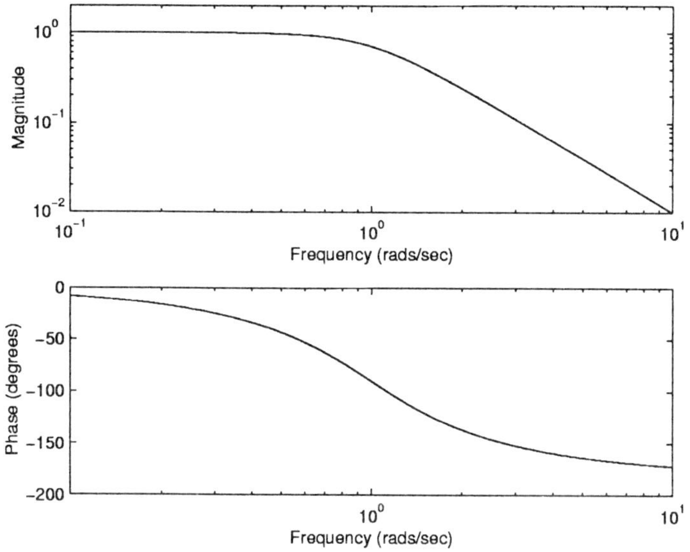

Fig. 3.15. Characteristics of analog lowpass filter

Low Pass to Bandpass Transformation

A *analog bandpass filter* has a passband from a frequency Ω_{p1} to another frequency Ω_{p2}, and stopbands from zero to Ω_{s1} and Ω_{s2} to infinity for some $0 < \Omega_{s1} < \Omega_{p1} < \Omega_{p2} < \Omega_{s2}$. A low pass pass filter with frequency response $\mathcal{H}_1(s)$ may be transformed to *bandpass filter* $\mathcal{H}_{bp}(s)$ using the frequency transformation:

$$s \Rightarrow \frac{s^2 + \Omega_0^2}{Ws} \tag{3.88}$$

that is;

$$\mathcal{H}_{bp}(s) = \mathcal{H}_1(\frac{s^2 + \Omega_0^2}{Ws}) \tag{3.89}$$

Hence if $\mathcal{H}_1(s)$ is of order n, then $\mathcal{H}_{bp}(s)$ is of order $2n$.

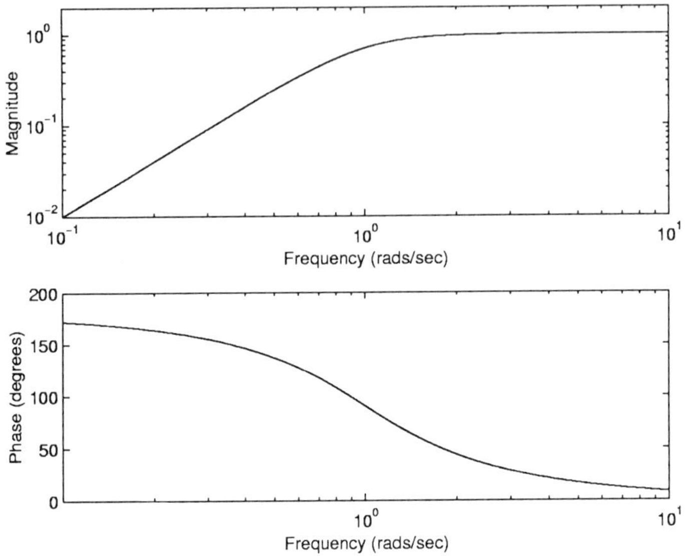

Fig. 3.16. Characteristics of analog highpass filter

For example, given the analog second order lowpass Butterworth filter in (3.83) which has a bandwidth of 1 rad/sec, a fourth order bandpass Butterworth filter is given by

$$\begin{aligned}\mathcal{H}_{bp}(s) &= \frac{1}{\left(\frac{s^2+\Omega_0^2}{Ws}\right)^2 + \sqrt{2}\left(\frac{s^2+\Omega_0^2}{Ws}\right) + 1} \\ &= \frac{W^2s^2}{s^4 + \sqrt{2}Ws^3 + (2\Omega_0^2 + W^2)s^2 + \sqrt{2}W\Omega_0^2 s + \Omega_0^4}\end{aligned} \tag{3.90}$$

The bandpass frequency response $\mathcal{H}_{bp}(s)$ which results from transforming the lowpass characteristic illustrated in Figure 3.15 is illustrated in Figure 3.17 for $\Omega_0 = 1$ and $W = 0.1, 1, 10$.

Low Pass to Bandstop Transformation

A *analog bandstop filter* has passbands from zero to Ω_{p1} and Ω_{p2} to infinity, and a stopband from Ω_{s1} to Ω_{s2} for some $0 < \Omega_{p1} < \Omega_{s1} < \Omega_{s2} < \Omega_{p2}$. A low pass pass filter with frequency response $\mathcal{H}_1(s)$ may be transformed to *bandstop filter* $\mathcal{H}_{bs}(s)$ using the frequency transformation:

$$s \Rightarrow \frac{Ws}{s^2 + \Omega_0^2} \tag{3.91}$$

with

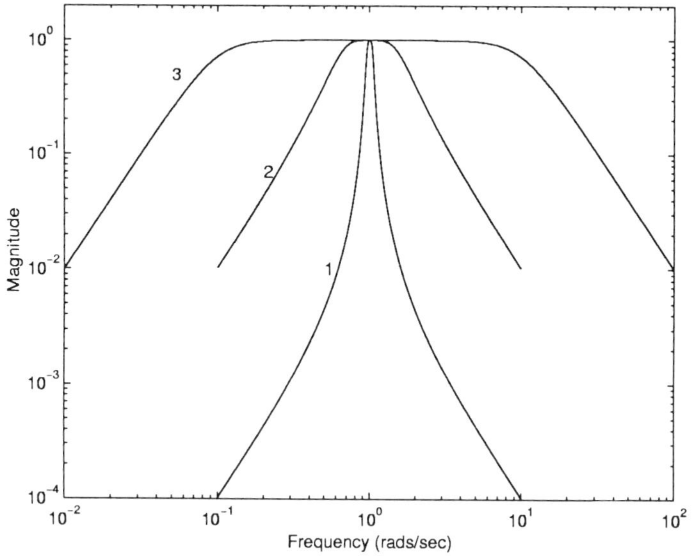

Fig. 3.17. Magnitude characteristics of analog bandpass filter for 1: $W = 0.1$, 2: $W = 1$, and 3: $W = 10$

$$\mathcal{H}_{bs}(s) = \mathcal{H}_1(\frac{Ws}{s^2 + \Omega_0^2}) \tag{3.92}$$

Hence if $\mathcal{H}_1(s)$ has order n, then $\mathcal{H}_{bs}(s)$ has order $2n$.

For example, given the second order (lowpass) Butterworth filter in (3.83) which has a bandwidth of 1 rad/sec, we have

$$\begin{aligned}\mathcal{H}_{bs}(s) &= \frac{1}{\left(\frac{Ws}{s^2+\Omega_0^2}\right)^2 + \sqrt{2}\left(\frac{Ws}{s^2+\Omega_0^2}\right) + 1} \\ &= \frac{(s^2 + \Omega_0^2)^2}{s^4 + \sqrt{2}Ws^3 + (2\Omega_0^2 + W^2)s^2 + \sqrt{2}W\Omega_0^2 s + \Omega_0^4}\end{aligned} \tag{3.93}$$

The bandstop frequency response $\mathcal{H}_{bs}(s)$ which results from transforming the lowpass characteristic illustrated in Figure 3.15 is illustrated in Figure 3.18 for $\Omega_0 = 1$ and $W = 0.1, 1, 10$.

3.3.3 Analog State Space Representation

As we shall see, an nth order analog filter (3.53) can also be represented in the form of a system of n first order differential equations

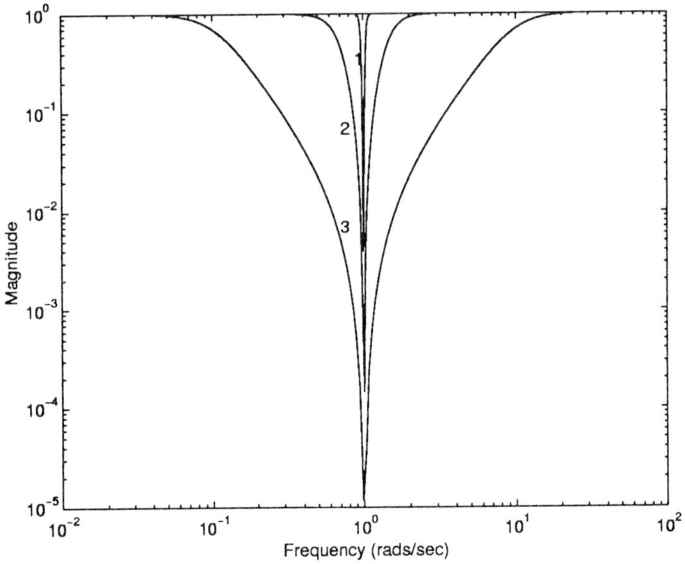

Fig. 3.18. Magnitude characteristics of analog bandstop filter for 1: $W = 0.1$, 2: $W = 1$, and 3: $W = 10$

$$\dot{x}_p(t) = \sum_{k=1}^{n} a_{pk} x_k(t) + b_p v(t) \ ; \ p = 1, 2, \ \ldots \ , n \tag{3.94}$$
$$z(t) = \sum_{k=1}^{n} c_k x_k(t) + dv(t)$$

which is called an *analog state space representation.*

One advantage of the state space representation compared to (3.53) is that the state space representation (3.94) does not require the determination of any derivatives of v. As we shall see, a state space representation also provides a convenient means for obtaining a numerical approximation of the solution of a differential equation (3.53). Many of the algebraic derivations and results which are useful for obtaining an analog state space representation of an nth order differential equation are similar to those which were derived in Chapter 2 for representing an nth order *difference equation* by means of a *discrete state space representation.* We begin with a second order example.

Second Order Representation

Consider the following *system of two linear time invariant first order differential equations*

$$\begin{aligned} \dot{x}_1(t) &= a_{11}x_1(t) + a_{12}x_2(t) + b_1v(t) \\ \dot{x}_2(t) &= a_{21}x_1(t) + a_{22}x_2(t) + b_2v(t) \\ z(t) &= c_1x_2(t) + c_2x_2(t) + dv(t) \end{aligned} \tag{3.95}$$

which defines a second order analog filter.

The first equation in (3.95) is a first order differential equation in the signal x_1 with input signal $a_{12}x_2 + b_1v$, while the second equation is a first order differential equation in the signal x_2 with input signal $a_{21}x_1 + b_2v$. The internal signals $\{x_1, x_2\}$ are the *state components* (or states) of the analog filter, while $\{x_1(t), x_2(t)\}$ are the states at time t. The vector $\boldsymbol{x}$ where $\boldsymbol{x}^T = [x_1 \ \ x_2]$ is the *state vector* (or state) of the system. The output signal y is a linear combination of the states and the input signal.

The system of differential equations (3.95) having input v and output y can be expressed more succinctly using linear algebra in the form

$$\begin{aligned} \dot{\boldsymbol{x}}(t) &= \boldsymbol{A}\boldsymbol{x}(t) + \boldsymbol{b}v(t) \\ z(t) &= \boldsymbol{c}^T\boldsymbol{x}(t) + dv(t) \end{aligned} \tag{3.96}$$

where the 2-vector $\boldsymbol{x}(t)$, the 2-square matrix $\boldsymbol{A}$, and the 2-vectors $\boldsymbol{b}$ and $\boldsymbol{c}$ are from (3.95) defined by

$$\boldsymbol{x}(t) = \begin{bmatrix} x_1(t) \\ x_2(t) \end{bmatrix} \ ; \ \boldsymbol{A} = \begin{bmatrix} a_{11} & a_{12} \\ a_{21} & a_{22} \end{bmatrix} \ ; \ \boldsymbol{b} = \begin{bmatrix} b_1 \\ b_2 \end{bmatrix} \ ; \ \boldsymbol{c} = \begin{bmatrix} c_1 \\ c_2 \end{bmatrix} \tag{3.97}$$

Equation (3.96) is a *first order vector system* of two first order differential equations with initial condition $\boldsymbol{x}(t_0)$.

Now from (3.96), we have

$$\begin{aligned} \dot{z}(t) &= \boldsymbol{c}^T\dot{\boldsymbol{x}}(t) + d\dot{v}(t) = \boldsymbol{c}^T\boldsymbol{A}\boldsymbol{x}(t) + \boldsymbol{c}^T\boldsymbol{b}v(t) + d\dot{v}(t) \\ \ddot{\boldsymbol{x}}(t) &= \boldsymbol{A}\dot{\boldsymbol{x}}(t) + \boldsymbol{b}\dot{v}(t) = \boldsymbol{A}^2\boldsymbol{x}(t) + \boldsymbol{A}\boldsymbol{b}v(t) + \boldsymbol{b}\dot{v}(t) \\ \ddot{z}(t) &= \boldsymbol{c}^T\ddot{\boldsymbol{x}}(t) + d\ddot{v}(t) = \boldsymbol{c}^T\boldsymbol{A}^2\boldsymbol{x}(t) + \boldsymbol{c}^T\boldsymbol{A}\boldsymbol{b}v(t) + \boldsymbol{c}^T\boldsymbol{b}\dot{v}(t) + d\ddot{v}(t) \end{aligned} \tag{3.98}$$

We now use (3.98) to show that z in (3.96) can be expressed as the solution of a second order linear time invariant differential equation

$$\ddot{z}(t) = \alpha_1\dot{z}(t) + \alpha_0 z(t) + \beta_0 v(t) + \beta_1\dot{v}(t) + \beta_2\ddot{v}(t) \tag{3.99}$$

with initial conditions $\{z(t_0), \dot{z}(t_0)\}$.

In order to derive this result, we again (as in section 2.5) make use of the *Caley-Hamilton* theorem. Thus when $\boldsymbol{A}$ is a 2-square matrix, we have that if:

$$\det(\lambda\boldsymbol{I}_2 - \boldsymbol{A}) = \lambda^2 - \delta_1\lambda - \delta_0 = 0 \tag{3.100}$$

then

$$\boldsymbol{A}^2 - \delta_1\boldsymbol{A} - \delta_0\boldsymbol{I}_2 = \boldsymbol{0} \tag{3.101}$$

Then after substituting for $\boldsymbol{A}^2$ in (3.98) using (3.101), we have

$$\ddot{z}(t) = (\delta_1 \boldsymbol{c}^T \boldsymbol{A} + \delta_0 \boldsymbol{c}^T)\boldsymbol{x}(t) + \boldsymbol{c}^T \boldsymbol{A}\boldsymbol{b}v(t) + \boldsymbol{c}^T \boldsymbol{b}\dot{v}(t) + d\ddot{v}(t)$$

Also from (3.98) and (3.96), we have

$$\boldsymbol{c}^T \boldsymbol{A}\boldsymbol{x}(t) = \dot{z}(t) - \boldsymbol{c}^T \boldsymbol{b}v(t) - d\dot{v}(t) \quad ; \quad \boldsymbol{c}^T \boldsymbol{x}(t) = z(t) - dv(t)$$

and so

$$\begin{aligned}\ddot{z}(t) = \delta_1[\dot{z}(t) - \boldsymbol{c}^T \boldsymbol{b}v(t) - d\dot{v}(t)] + \delta_0[z(t) - dv(t)] \\ +\boldsymbol{c}^T \boldsymbol{A}\boldsymbol{b}v(t) + \boldsymbol{c}^T \boldsymbol{b}\dot{v}(t) + d\ddot{v}(t)\end{aligned} \tag{3.102}$$

After comparing coefficients in (3.99) and (3.102), we have the following result.

Theorem 3.3.2. *The first order vector system of two first order linear time invariant differential equation (3.96) has the same input-output description between the input v and the output z as the second order linear time invariant differential equation (3.99) if and only if $\{\boldsymbol{A},\ \boldsymbol{b},\ \boldsymbol{c}, d\}$ and $\{\alpha_1,\ \alpha_0,\ \beta_0,\ \beta_1\}$ are related according to*

$$d = \beta_2 \ ; \quad \begin{bmatrix} 1 & -\alpha_1 \\ 0 & 1 \end{bmatrix} \begin{bmatrix} \boldsymbol{c}^T \boldsymbol{A}\boldsymbol{b} \\ \boldsymbol{c}^T \boldsymbol{b} \end{bmatrix} = \begin{bmatrix} \beta_0 + \alpha_0 d \\ \beta_1 + \alpha_1 d \end{bmatrix} \tag{3.103}$$

with

$$\det(\lambda \boldsymbol{I}_2 - \boldsymbol{A}) = \lambda^2 - \alpha_1 \lambda - \alpha_0 \tag{3.104}$$

The initial conditions $\{z(t_0), \dot{z}(t_0)\}$ of (3.99) are related to the initial state $\boldsymbol{x}(t_0)$ of (3.96) by

$$\begin{bmatrix} \dot{z}(t_0) \\ z(t_0) \end{bmatrix} = \begin{bmatrix} \boldsymbol{c}^T \boldsymbol{A} \\ \boldsymbol{c}^T \end{bmatrix} \boldsymbol{x}(t_0) + \begin{bmatrix} d & \boldsymbol{c}^T \boldsymbol{b} \\ 0 & d \end{bmatrix} \begin{bmatrix} \dot{v}(t_0) \\ v(t_0) \end{bmatrix} \tag{3.105}$$

This result is similar to that of Theorem 2.5.1 in Chapter 2 which provides a state space representation for a second order difference equation.

Choice of System Matrix and Input/Output Vectors

Since $d = \beta_2$, there are 8 remaining components of $\{\boldsymbol{A}, \boldsymbol{b}, \boldsymbol{c}\}$ in (3.97) which must be specified to match the 4 remaining coefficients $\{\alpha_0, \alpha_1, \beta_0, \beta_1\}$ in (3.99). Consequently, many choices for the system matrix $\boldsymbol{A}$, the input vector $\boldsymbol{b}$ and the output vector $\boldsymbol{c}$ of the analog state space representation (3.96) are possible. More specifically, we the available freedom can be summarized as follows:

1. Condition (3.104) imposes a constraint on the characteristic polynomial of the system matrix $\boldsymbol{A}$.
2. From (3.105), the invertibility of the 2-square matrix $\boldsymbol{P}_c$ where

$$\boldsymbol{P}_c \triangleq \begin{bmatrix} \boldsymbol{c}^T\boldsymbol{A} \\ \boldsymbol{c}^T \end{bmatrix} \tag{3.106}$$

is a necessary condition in order to be able to determine the initial state $\boldsymbol{x}(t_0)$ for *all* initial conditions $\{z(t_0), \dot{z}(t_0)\}$. That is, if $\boldsymbol{P}_c$ is invertible, then

$$\boldsymbol{x}(t_0) = \boldsymbol{P}_c^{-1}\left\{\begin{bmatrix} \dot{z}(t_0) \\ z(t_0) \end{bmatrix} - \begin{bmatrix} d & \boldsymbol{c}^T\boldsymbol{b} \\ 0 & d \end{bmatrix}\begin{bmatrix} \dot{v}(t_0) \\ v(t_0) \end{bmatrix}\right\} \tag{3.107}$$

3. Since

$$\begin{bmatrix} \boldsymbol{c}^T\boldsymbol{A}\boldsymbol{b} \\ \boldsymbol{c}^T\boldsymbol{b} \end{bmatrix} = \begin{bmatrix} \boldsymbol{c}^T\boldsymbol{A} \\ \boldsymbol{c}^T \end{bmatrix}\boldsymbol{b}$$

the last remaining condition (3.103) which is imposed on the choice of $\{\boldsymbol{A}, \boldsymbol{b}, \boldsymbol{c}\}$ can be arranged into the form

$$\boldsymbol{D}_2\boldsymbol{P}_c\boldsymbol{b} = \boldsymbol{\gamma}_2$$

where

$$\boldsymbol{D}_2 \triangleq \begin{bmatrix} 1 & -\alpha_1 \\ 0 & 1 \end{bmatrix} \; ; \; \boldsymbol{\gamma}_2 \triangleq \begin{bmatrix} \beta_0 + \alpha_0 d \\ \beta_1 + \alpha_1 d \end{bmatrix} \tag{3.108}$$

We therefore have the following result.

Theorem 3.3.3. *Suppose the system matrix* $\boldsymbol{A}$ *has the characteristic polynomial (3.104), and the output vector* $\mathbf{c}$ *is chosen such that the 2-square matrix* $\boldsymbol{P}_c$ *defined by (3.106) is invertible.*

Then $\{\boldsymbol{A}, \boldsymbol{b}, \mathbf{c}, d\}$ *where*

$$d = \beta_2 \; ; \; \boldsymbol{b} = \boldsymbol{P}_c^{-1}\boldsymbol{D}_2^{-1}\boldsymbol{\gamma}_2 \tag{3.109}$$

with $\{\boldsymbol{D}_2, \boldsymbol{\gamma}_2\}$ *defined by (3.108) is an analog state space representation of the differential equation (3.99) with initial state*

$$\boldsymbol{x}(t_0) = \boldsymbol{P}_c^{-1}\left\{\begin{bmatrix} \dot{z}(t_0) \\ z(t_0) \end{bmatrix} - \begin{bmatrix} d & \boldsymbol{c}^T\boldsymbol{b} \\ 0 & d \end{bmatrix}\begin{bmatrix} \dot{v}(t_0) \\ v(t_0) \end{bmatrix}\right\} \tag{3.110}$$

Notice the similarity between this result and Theorem 2.5.1 in section 2.5.1 for the state space representation of a second order digital filter.

Example 3.3.4. Consider the second order analog filter

$$\ddot{z}(t) + 0.6\dot{z}(t) + 0.05z(t) = 0.08v(t) - 0.80\dot{v}(t)$$

where $\{z(4) = -3, \dot{z}(4) = 2\}$ and $\{\dot{v}(4) = v(4) = 0\}$. Then from Theorem 3.3.2 with

$$\alpha_1 = -0.6 \ ; \ \alpha_0 = -0.05 \ ; \ \beta_0 = 0.08 \ ; \ \beta_1 = -0.80 \ ; \ \beta_2 = 0$$

this analog filter has an equivalent state space description (3.96) provided $\{\boldsymbol{A},\ \boldsymbol{b},\ \boldsymbol{c}, d\}$ are chosen such that

$$d = 0 \ ; \quad \begin{bmatrix} 1 & 0.6 \\ 0 & 1 \end{bmatrix} \begin{bmatrix} \boldsymbol{c}^T\boldsymbol{A}\boldsymbol{b} \\ \boldsymbol{c}^T\boldsymbol{b} \end{bmatrix} = \begin{bmatrix} 0.08 \\ -0.80 \end{bmatrix}$$

or, $\{\boldsymbol{c}^T\boldsymbol{b} = -0.80;\ \boldsymbol{c}^T\boldsymbol{A}\boldsymbol{b} = 0.56\}$ with

$$\det(\lambda\boldsymbol{I}_2 - \boldsymbol{A}) = \lambda^2 + 0.6\lambda + 0.05 = (\lambda + 0.5)(\lambda + 0.1) \tag{3.111}$$

(i) One choice for the matrix $\boldsymbol{A}$ having the required characteristic polynomial (3.111) is given by the phase variable form (recall also (2.158) and the related discussion in Chapter 2)

$$\boldsymbol{A} = \begin{bmatrix} 0 & -0.05 \\ 1 & -0.6 \end{bmatrix} \ ; \ \boldsymbol{b} = \begin{bmatrix} b_1 \\ b_2 \end{bmatrix} \ ; \ \boldsymbol{c} = \begin{bmatrix} c_1 \\ c_2 \end{bmatrix}$$

In this case, the components $\{b_1, b_2\}$ of $\boldsymbol{b}$ and $\{c_1, c_2\}$ of $\boldsymbol{c}$ must satisfy

$$\begin{aligned} \boldsymbol{c}^T\boldsymbol{b} &= c_1b_1 + c_2b_2 = -0.80 \\ \boldsymbol{c}^T\boldsymbol{A}\boldsymbol{b} &= c_2b_1 - (0.05c_1 + 0.6c_2)b_2 = 0.56 \end{aligned}$$

One possible (nonunique) choice for $\{\boldsymbol{b}, \boldsymbol{c}\}$ is

$$\boldsymbol{c} = \begin{bmatrix} 0 \\ 1 \end{bmatrix} \ ; \ \boldsymbol{b} = \begin{bmatrix} 0.08 \\ -0.80 \end{bmatrix}$$

The initial state $\boldsymbol{x}(4)$ is then given from $\{z(4), \dot{z}(4)\}$ by

$$\begin{bmatrix} 1 & -0.6 \\ 0 & 1 \end{bmatrix} \boldsymbol{x}(4) = \begin{bmatrix} 2 \\ -3 \end{bmatrix}$$

or

$$\boldsymbol{x}(4) = \begin{bmatrix} 1 & -0.6 \\ 0 & 1 \end{bmatrix}^{-1} \begin{bmatrix} 2 \\ -3 \end{bmatrix} = \begin{bmatrix} 0.2 \\ -3 \end{bmatrix}$$

(ii) Another possible matrix $\boldsymbol{A}$ which also has the required characteristic polynomical (3.111) is given by the Jordan form (recall also (2.163) and the related discussion in Chapter 2)

$$\boldsymbol{A} = \begin{bmatrix} -0.5 & 0 \\ 0 & -0.1 \end{bmatrix} \; ; \; \boldsymbol{b} = \begin{bmatrix} b_1 \\ b_2 \end{bmatrix} \; ; \; \boldsymbol{c} = \begin{bmatrix} c_1 \\ c_2 \end{bmatrix} \tag{3.112}$$

The components $\{b_1, b_2\}$ of $\boldsymbol{b}$ and $\{c_1, c_2\}$ of $\boldsymbol{c}$ corresponding to this choice of $\boldsymbol{A}$ must also satisfy the equations (3.103); that is, now

$$\begin{aligned} \boldsymbol{c}^T\boldsymbol{b} &= c_1b_1 + c_2b_2 = -0.80 \\ \boldsymbol{c}^T\boldsymbol{A}\boldsymbol{b} &= -0.5c_1b_1 - 0.1c_2b_2 = 0.56 \end{aligned}$$

which gives

$$\begin{bmatrix} -0.5 & -0.1 \\ 1 & 1 \end{bmatrix} \begin{bmatrix} c_1b_1 \\ c_2b_2 \end{bmatrix} = \begin{bmatrix} 0.56 \\ -0.80 \end{bmatrix}$$

or $\{c_1b_1 = -1.2 \; ; \; c_2b_2 = 0.4\}$. One possible choice is

$$\boldsymbol{b} = \begin{bmatrix} 1 \\ 1 \end{bmatrix} \; ; \; \boldsymbol{c} = \begin{bmatrix} -1.2 \\ 0.4 \end{bmatrix}$$

The initial state $\boldsymbol{x}(4)$ is then given from $\{z(4), \dot{z}(4)\}$ by

$$\boldsymbol{x}(4) = \begin{bmatrix} 0.6 & -0.04 \\ -1.2 & 0.4 \end{bmatrix}^{-1} \begin{bmatrix} 2 \\ -3 \end{bmatrix} = \begin{bmatrix} 3.54 \\ 3.13 \end{bmatrix}$$

High Order Representation

More generally, the system of n first order linear time invariant differential equations

$$\begin{aligned} \dot{x}_j(t) &= \sum_{m=1}^{n} a_{jm}x_m(t) + b_jv(t) \quad ; \quad 1 \le j \le n \\ z(t) &= \sum_{m=1}^{n} c_mx_m(t) + dv(t) \end{aligned} \tag{3.113}$$

can also be expressed using linear algebra as a *first order vector system*

$$\begin{aligned} \dot{\boldsymbol{x}}(t) &= \boldsymbol{A}\boldsymbol{x}(t) + \boldsymbol{b}v(t) \\ z(t) &= \boldsymbol{c}^T\boldsymbol{x}(t) + dv(t) \end{aligned} \tag{3.114}$$

where now $\boldsymbol{x}(t) = [x_j(k)]$, $\boldsymbol{b} = [b_j]$ and $\boldsymbol{c} = [c_j]$ are all n-vectors, and $\boldsymbol{A} = [a_{jm}]$ is an n-square matrix. The result in Theorem 3.3.2 can be generalized as follows.

Theorem 3.3.4. *The first order system of n first order linear time invariant differential equations (3.114) has the same input-output description as the nth order linear time invariant differential equation*

$$z^{(n)}(t) = \sum_{p=0}^{n-1} \alpha_p z^{(p)}(t) + \sum_{q=0}^{n} \beta_q v^{(q)}(t) \tag{3.115}$$

if and only if $\{\boldsymbol{A},\ \boldsymbol{b},\ \mathbf{c},\ d\}$ *and the coefficients* $\{\alpha_p, 0 \leq p \leq n-1\ ;\ \beta_q, 0 \leq q \leq n\}$ *are related by*

$$d = \beta_n \tag{3.116}$$

$$\begin{bmatrix} 1 & -\alpha_{n-1} & \cdot & \cdot & -\alpha_1 \\ 0 & 1 & -\alpha_{n-1} & \cdot & -\alpha_2 \\ \cdot & \cdot & \cdot & \cdot & \cdot \\ \cdot & \cdot & \cdot & \cdot & \cdot \\ \cdot & \cdot & \cdot & \cdot & \cdot \\ 0 & 0 & \cdot & \cdot & 1 \end{bmatrix} \begin{bmatrix} \mathbf{c}^T \boldsymbol{A}^{n-1} \boldsymbol{b} \\ \mathbf{c}^T \boldsymbol{A}^{n-2} \boldsymbol{b} \\ \cdot \\ \cdot \\ \cdot \\ \mathbf{c}^T \boldsymbol{b} \end{bmatrix} = \begin{bmatrix} \beta_0 + \alpha_0 d \\ \beta_1 + \alpha_1 d \\ \cdot \\ \cdot \\ \cdot \\ \beta_{n-1} + \alpha_{n-1} d \end{bmatrix}$$

with

$$det(\lambda \boldsymbol{I}_{\mathbf{n}} - \boldsymbol{A}) = \lambda^n - \sum_{p=0}^{n-1} \alpha_p \lambda^p \tag{3.117}$$

The initial conditions $\{z(t_0), \dot{z}(t_0), \ \ldots\ z^{(n-1)}(t_0)\}$ *of (3.115) are related to the initial state* $\boldsymbol{x}(t_0)$ *of (3.114) by*

$$\begin{bmatrix} z^{(n-1)}(t_0) \\ z^{(n-2)}(t_0) \\ \cdot \\ \cdot \\ \dot{z}(t_0) \\ z(t_0) \end{bmatrix} = \begin{bmatrix} \mathbf{c}^T \boldsymbol{A}^{n-1} \\ \mathbf{c}^T \boldsymbol{A}^{n-2} \\ \cdot \\ \cdot \\ \mathbf{c}^T \boldsymbol{A} \\ \mathbf{c}^T \end{bmatrix} \boldsymbol{x}(t_0) + \tag{3.118}$$

$$\begin{bmatrix} d & \mathbf{c}^T\boldsymbol{b} & \mathbf{c}^T\boldsymbol{A}\boldsymbol{b} & \cdot & \mathbf{c}^T\boldsymbol{A}^{n-3}\boldsymbol{b} & \mathbf{c}^T\boldsymbol{A}^{n-2}\boldsymbol{b} \\ 0 & d & \mathbf{c}^T\boldsymbol{b} & \mathbf{c}^T\boldsymbol{A}\boldsymbol{b} & \cdot & \mathbf{c}^T\boldsymbol{A}^{n-3}\boldsymbol{b} \\ \cdot & \cdot & \cdot & \cdot & \cdot & \cdot \\ \cdot & \cdot & \cdot & \cdot & \cdot & \cdot \\ \cdot & \cdot & \cdot & \cdot & d & \mathbf{c}^T\boldsymbol{b} \\ 0 & 0 & \cdot & 0 & \cdot & d \end{bmatrix} \begin{bmatrix} v^{(n-1)}(t_0) \\ v^{(n-2)}(t_0) \\ \cdot \\ \cdot \\ \dot{v}(t_0) \\ v(t_0) \end{bmatrix}$$

Notice the similarity between this result and the result in Theorem 2.5.3 in section 2.5.2 for the state space representation of an nth order digital filter.

Choice of System Matrix and Input/Output Vectors

We summarize the available freedom in the choice for the system matrix $\boldsymbol{A}$, the input vector $\boldsymbol{b}$, and the output vector $\boldsymbol{c}$ as follows:

1. Condition (3.117) imposes a constraint on the characteristic polynomial of the system matrix $\boldsymbol{A}$.
2. From (3.118), the invertibility of the n-square matrix $\boldsymbol{P}_c$ where

$$\boldsymbol{P}_c \triangleq \begin{bmatrix} \boldsymbol{c}^T\boldsymbol{A}^{n-1} \\ \boldsymbol{c}^T\boldsymbol{A}^{n-2} \\ \cdot \\ \cdot \\ \boldsymbol{c}^T\boldsymbol{A} \\ \boldsymbol{c}^T \end{bmatrix} \tag{3.119}$$

is a necessary condition in order to be able to determine the initial state $\boldsymbol{x}(t_0)$ for *all* initial conditions $\{z(t_0), \dot{z}(t_0), \ldots, y^{(n-1)}(t_0)\}$.
3. Since

$$\begin{bmatrix} \boldsymbol{c}^T\boldsymbol{A}^{n-1}\boldsymbol{b} \\ \boldsymbol{c}^T\boldsymbol{A}^{n-2}\boldsymbol{b} \\ \cdot \\ \cdot \\ \boldsymbol{c}^T\boldsymbol{A}\boldsymbol{b} \\ \boldsymbol{c}^T\boldsymbol{b} \end{bmatrix} = \begin{bmatrix} \boldsymbol{c}^T\boldsymbol{A}^{n-1} \\ \boldsymbol{c}^T\boldsymbol{A}^{n-2} \\ \cdot \\ \cdot \\ \boldsymbol{c}^T\boldsymbol{A} \\ \boldsymbol{c}^T \end{bmatrix} \boldsymbol{b} = \boldsymbol{P}_c\boldsymbol{b}$$

the last remaining condition (3.116) which is imposed on the choice of $\{\boldsymbol{A}, \boldsymbol{b}, \boldsymbol{c}\}$ can be arranged into the form

$$\boldsymbol{D}_n\boldsymbol{P}_c\boldsymbol{b} = \gamma_n$$

where

$$\boldsymbol{D}_n \triangleq \begin{bmatrix} 1 & -\alpha_{n-1} & \cdot & \cdot\cdot & -\alpha_1 \\ 0 & 1 & -\alpha_{n-1} & \cdot\cdot & -\alpha_2 \\ \cdot & \cdot & \cdot & \cdot\cdot & \cdot \\ \cdot & \cdot & \cdot & \cdot\cdot & \cdot \\ \cdot & \cdot & \cdot & \cdot\cdot & -\alpha_{n-1} \\ 0 & 0 & \cdot & \cdot\cdot & 1 \end{bmatrix} ; \; \gamma_n \triangleq \begin{bmatrix} \beta_0 + \alpha_0 d \\ \beta_1 + \alpha_1 d \\ \cdot \\ \cdot \\ \cdot \\ \beta_{n-1} + \alpha_{n-1} d \end{bmatrix} \tag{3.120}$$

Note that $det(\boldsymbol{D}_n) = 1$, and hence $\boldsymbol{D}_n^{-1}$ exists for all values of the coefficients $\{\alpha_1, \alpha_2, \ldots, \alpha_{n-1}\}$. We therefore have the following result.

Theorem 3.3.5. *Suppose the system matrix $\boldsymbol{A}$ has the characteristic polynomial (3.117), and the output vector $\boldsymbol{c}$ is chosen such that the n-square matrix $\boldsymbol{P}_c$ defined by (3.119) is invertible. Then $\{\boldsymbol{A}, \boldsymbol{b}, \boldsymbol{c}, d\}$ where*

$$d = \beta_n \; ; \; \boldsymbol{b} = \boldsymbol{P}_c^{-1}\boldsymbol{D}_n^{-1}\gamma_n \tag{3.121}$$

with $\{D_n, \gamma_n\}$ defined by (3.120) is a state space representation of the difference equation (3.115) with initial state

$$
\boldsymbol{x}(t_0) = \boldsymbol{P}_c^{-1} \left\{ \begin{bmatrix} z^{(n-1)}(t_0) \\ z^{(n-2)}(t_0) \\ \cdot \\ \cdot \\ \cdot \\ \cdot \\ \dot{z}(t_0) \\ z(t_0) \end{bmatrix} - \right. \tag{3.122}
$$

$$
\left. \begin{bmatrix} d & \boldsymbol{c}^T\boldsymbol{b} & \boldsymbol{c}^T\boldsymbol{A}\boldsymbol{b} & \cdot & \boldsymbol{c}^T\boldsymbol{A}^{n-3}\boldsymbol{b} & \boldsymbol{c}^T\boldsymbol{A}^{n-2}\boldsymbol{b} \\ 0 & d & \boldsymbol{c}^T\boldsymbol{b} & \boldsymbol{c}^T\boldsymbol{A}\boldsymbol{b} & \cdot & \boldsymbol{c}^T\boldsymbol{A}^{n-3}\boldsymbol{b} \\ \cdot & \cdot & \cdot & \cdot & \cdot & \cdot \\ \cdot & \cdot & \cdot & \cdot & \cdot & \cdot \\ \cdot & \cdot & \cdot & \cdot & \cdot & \cdot \\ \cdot & \cdot & \cdot & \cdot & \cdot & \cdot \\ \cdot & \cdot & \cdot & \cdot & d & \boldsymbol{c}^T\boldsymbol{b} \\ 0 & 0 & \cdot & \cdot & 0 & d \end{bmatrix} \begin{bmatrix} v^{(n-1)}(t_0) \\ v^{(n-2)}(t_0) \\ \cdot \\ \cdot \\ \cdot \\ \dot{v}(t_0) \\ v(t_0) \end{bmatrix} \right\}
$$

Notice the similarity between this result and the result in Theorem 2.5.5 in section 2.5.2 for the state space representation of an nth order digital filter. As in the case for difference equations, an analog state space representation can always be defined with $\boldsymbol{A}$ in either phase variable or Jordan form.

3.3.4 Analog to Digital Transformation

In this section, we develop some algebraic methods for designing a digital IIR filter via a transformation of an analog filter. The transformations are based on different ways to numerically approximate the solution of a differential equation. Specifically, given an analog filter as represented by an nth order differential equation

$$
\begin{aligned}
z^{(n)}(t) = \alpha_{n-1} z^{(n-1)}(t) + \; \dots \; + \alpha_1 \dot{z}(t) + \alpha_0 z(t) \\
+ \beta_0 v(t) + \beta_1 \dot{v}(t) + \; \dots \; + \beta_n v^{(n)}(t)
\end{aligned} \tag{3.123}
$$

the approach to be taken in this section is as follows:

1. First, obtain an analog state space representation

$$
\begin{aligned}
\dot{\boldsymbol{x}}(t) &= \boldsymbol{A}\boldsymbol{x}(t) + \boldsymbol{b}v(t) \; ; \; \boldsymbol{x}(t) \in \mathcal{R}^n \\
z(t) &= \boldsymbol{c}^T \boldsymbol{x}(t) + dv(t)
\end{aligned} \tag{3.124}
$$

   of the analog filter (3.123)

2. Write the system of differential equations (3.124) in the integral form

$$\boldsymbol{x}(t_{k+1}) = \boldsymbol{x}(t_k) + \int_{t_k}^{t_{k+1}} \{\boldsymbol{A}\boldsymbol{x}(\sigma) + \boldsymbol{b}v(\sigma)\}d\sigma \tag{3.125}$$
$$z(t_k) = \boldsymbol{c}^T\boldsymbol{x}(t_k) + dv(t_k)$$

and then use a numerical scheme for approximating the integral with $t_k = kT$. This approach results in a discrete state space representation

$$\boldsymbol{x}_e(k+1) = \boldsymbol{F}_e\boldsymbol{x}_e(k) + \boldsymbol{g}_e u(k) \ ; \ \boldsymbol{x}_e(k) \in \mathcal{R}^n \tag{3.126}$$
$$y_e(k) = \boldsymbol{h}_e^T\boldsymbol{x}_e(k) + d_e u(k)$$

where the solution y_e provides an approximation $y_e(k) \simeq z(t_k)$ of the differential equation (3.123).

3. The digital filter having discrete frequency function $H_e(z)$ which results from the given analog filter having analog frequency function $\mathcal{H}(s)$ where

$$\mathcal{H}(s) = d + \boldsymbol{c}^T(s\boldsymbol{I} - \boldsymbol{A})^{-1}\boldsymbol{b} \tag{3.127}$$

is then given by

$$H_e(z) = d_e + \boldsymbol{h}_e^T(z\boldsymbol{I} - \boldsymbol{F}_e)^{-1}\boldsymbol{g}_e \tag{3.128}$$

Backward Euler Approximation of Integral

The *backward Euler* approximation of an integral of a vector signal $\boldsymbol{f}$ is given by

$$\int_{t_k}^{t_{k+1}} \boldsymbol{f}(\sigma)d\sigma \approx \boldsymbol{f}(t_k)(t_{k+1} - t_k) \tag{3.129}$$

In particular, using this approximation in (3.125) with $t_k = kT$ (so that $t_{k+1} - t_k = T$ is constant for all k) with $u(k) \stackrel{\Delta}{=} v(t = kT)$, we conclude that $\boldsymbol{x}_{e1}(k) \simeq \boldsymbol{x}(t_k)$ is given by

$$\boldsymbol{x}_{e1}(k+1) = \boldsymbol{x}_{e1}(k) + T[\boldsymbol{A}\boldsymbol{x}_{e1}(k) + \boldsymbol{g}u(k)]$$

We have the following result.

Theorem 3.3.6. *When $t_k = kT$, an approximation $y_{e1}(k)$ of $z(t = kT)$ in (3.124) based on using the backward Euler approximation of the integral in (3.125) with $u(k) \stackrel{\Delta}{=} v(kT)$ is given by*

$$\boldsymbol{x}_{e1}(k+1) = \boldsymbol{F}_{e1}\boldsymbol{x}_{e1}(k) + \boldsymbol{g}_{e1}u(k) \tag{3.130}$$
$$y_{e1}(k) = \boldsymbol{c}^T\boldsymbol{x}_{e1}(k) + d_{e1}u(k)$$

where

$$F_{e1} \triangleq I_n + AT \ ; \ g_{e1} \triangleq Tb \ ; \ d_{e1} \triangleq d \tag{3.131}$$

Forward Euler Approximation of Integral

The *forward Euler* approximation of an integral of a vector signal f is given by

$$\int_{t_k}^{t_{k+1}} f(\sigma)d\sigma \approx f(t_{k+1})(t_{k+1} - t_k) \tag{3.132}$$

When compared to the *backward* Euler scheme (3.129), it can be seen that the approximation uses the value of f at $t = t_{k+1}$ rather than f at $t = t_k$.

In particular, when $t = kT$ (so that $t_{k+1} - t_k = T$ is constant for all k), we have that an approximation $\tilde{x}_{e2}(k)$ of $x(t_k)$ and $y_{e2}(k)$ of $z(kT)$ in (3.125) with $u(k) = v(kT)$ is given by

$$\tilde{x}_{e2}(k+1) = \tilde{x}_{e2}(k) + T[A_{e2}\tilde{x}_{e2}(k+1) + gu(k+1)]$$

That is

$$\begin{aligned} \tilde{x}_{e2}(k+1) &= F_{e2}\tilde{x}_{e2}(k) + gu(k+1) \\ y_{e2}(k) &= c^T\tilde{x}_{e2}(k) + du(k) \end{aligned} \tag{3.133}$$

$$F_{e2} \triangleq (I_n - TA)^{-1} \ ; \ g \triangleq T(I_n - TA)^{-1}b$$

provided the matrix $I_n - AT$ is invertible.

Now the required inverse exists provided $I_n - AT$ has no zero eigenvalues; equivalently, there exists no vector $v \neq 0$ such that

$$(I_n - TA)v = 0$$

This condition for invertibility is therefore equivalent to the condition that the matrix A has no eigenvalue at T^{-1}.

This vector difference equation for $\tilde{x}_{e2}$ is *not* in the standard state space form of (3.126) since the right hand side of (3.133) involves $u(k+1)$ rather than $u(k)$. In order to obtain a *standard* form, define

$$x_{e2}(k) \triangleq \tilde{x}_{e2}(k) - gu(k)$$

Then from (3.133), we have

$$\begin{aligned} x_{e2}(k+1) &= \tilde{x}_{e2}(k+1) - gu(k+1) \\ &= F_{e2}\tilde{x}_{e2}(k) + gu(k+1) - gu(k+1) = F_{e2}[x_{e2}(k) + gu(k)] \end{aligned}$$

We then have the following result.

Theorem 3.3.7. *When* $t_k = kT$, *an approximation* $y_{e2}(k)$ *of* $z(t = kT)$ *in (3.123) based on using the forward Euler approximation of the integral in (3.125) with* $u(k) \triangleq v(kT)$ *is given by*

$$\begin{aligned} \boldsymbol{x}_{e2}(k+1) &= \boldsymbol{F}_{e2}\boldsymbol{x}_{e2}(k) + \boldsymbol{g}_{e2}u(k) \\ y_{e2}(k) &= \boldsymbol{c}^T\boldsymbol{x}_{e2}(k) + d_{e2}u(k) \end{aligned} \tag{3.134}$$

where

$$\begin{aligned} \boldsymbol{F}_{e2} &\triangleq (\boldsymbol{I}_n - T\boldsymbol{A})^{-1} \ ; \ \boldsymbol{g}_{e2} \triangleq T(\boldsymbol{I}_n - T\boldsymbol{A})^{-2}\boldsymbol{b} \\ d_{e2} &\triangleq d + T\boldsymbol{c}^T(\boldsymbol{I}_n - T\boldsymbol{A})^{-1}\boldsymbol{b} \end{aligned} \tag{3.135}$$

provided $\boldsymbol{A}$ *has no eigenvalue at* T^{-1}.

One noticeable difference between the *backward Euler* approximation (3.130) and the *forward Euler* approximation (3.134) concerns the direct feedthrough coefficient. That is, in the backward Euler scheme, the direct feedthrough coefficient $d_{e1} = 0$ when $d\ (= \beta_n) = 0$ in the differential equation (3.123). On the other hand, in the forward Euler scheme, the direct feedthrough coefficient $d_{e2} \neq 0$ when $d = 0$.

Trapezoidal Approximation of Integral

The trapezoidal approximation of an integral of a vector signal $\boldsymbol{f}$ is given by

$$\int_{t_k}^{t_{k+1}} \boldsymbol{f}(\sigma)d\sigma \approx 0.5[\boldsymbol{f}(t_{k+1}) + \boldsymbol{f}(t_k)](t_{k+1} - t_k) \tag{3.136}$$

Then using the integral form (3.125), an approximation $\tilde{\boldsymbol{x}}_{e3}(k)$ of $\boldsymbol{x}(t_k)$ $y_{e3}(k)$ of $z(t_k)$ when $t_k = kT$ (which implies $t_{k+1} - t_k = T$ is constant) with $u(k) \triangleq v(kT)$ is given by

$$\tilde{\boldsymbol{x}}_{e3}(k+1) = \tilde{\boldsymbol{x}}_{e3}(k) + 0.5T[\boldsymbol{A}\boldsymbol{x}_{e3}(k) + \boldsymbol{g}u(k) + \boldsymbol{A}\boldsymbol{x}_{e3}(k+1) + \boldsymbol{g}u(k+1)]$$

That is

$$\begin{aligned} \tilde{\boldsymbol{x}}_{e3}(k+1) &= \boldsymbol{F}_{e3}\tilde{\boldsymbol{x}}_{e3}(k) + \boldsymbol{g}_1u(k) + \boldsymbol{g}_1u(k+1) \\ y_{e3}(k) &= \boldsymbol{c}^T\tilde{\boldsymbol{x}}_{e3}(k) + du(k) \end{aligned}$$

where

$$\boldsymbol{F}_{e3} \triangleq (\boldsymbol{I}_n - 0.5T\boldsymbol{A})^{-1}(\boldsymbol{I}_n + 0.5T\boldsymbol{A}) \ ; \ \boldsymbol{g}_1 \triangleq 0.5T(\boldsymbol{I}_n - 0.5T\boldsymbol{A})^{-1}\boldsymbol{b}$$

provided $(\boldsymbol{I}_{\mathrm{n}} - 0.5T\boldsymbol{A})^{-1}$ exists.

In order to reduce this equation to the standard state space form (3.126), define

$$\boldsymbol{x}_{e3}(k) \triangleq \tilde{\boldsymbol{x}}_{e3}(k) - \boldsymbol{g}_1 u(k)$$

which implies

$$\begin{aligned}
\boldsymbol{x}_{e3}(k+1) &= \tilde{\boldsymbol{x}}_{e3}(k+1) - \boldsymbol{g}_1 u(k+1) \\
&= \boldsymbol{F}_{e3}\tilde{\boldsymbol{x}}_{e3}(k) + \boldsymbol{g}_1 u(k) + \boldsymbol{g}_1 u(k+1) - \boldsymbol{g}_1 u(k+1) \\
&= \boldsymbol{F}_{e3}\boldsymbol{x}_{e3}(k) + (\boldsymbol{F}_{e3}\boldsymbol{g}_1 + \boldsymbol{g}_1)u(k) \\
y_{e3}(k) &= \boldsymbol{c}^T\boldsymbol{x}_{e3} + (\boldsymbol{c}^T\boldsymbol{g}_1)u(k) + du(k)
\end{aligned}$$

Now with $\gamma = 0.5T$, we have

$$\begin{aligned}
\boldsymbol{F}_{e3}\boldsymbol{g}_1 + \boldsymbol{g}_1 &= \gamma(\boldsymbol{I}_{\mathrm{n}} - \gamma\boldsymbol{A})^{-1}\left\{(\boldsymbol{I}_{\mathrm{n}} + \gamma\boldsymbol{A})(\boldsymbol{I}_{\mathrm{n}} - \gamma\boldsymbol{A})^{-1} + \boldsymbol{I}\right\}\boldsymbol{b} \\
&= \gamma(\boldsymbol{I}_{\mathrm{n}} - \gamma\boldsymbol{A})^{-1}\left\{(\boldsymbol{I}_{\mathrm{n}} + \gamma\boldsymbol{A}) + (\boldsymbol{I}_{\mathrm{n}} - \gamma\boldsymbol{A})\right\}(\boldsymbol{I}_{\mathrm{n}} - \gamma\boldsymbol{A})^{-1}\boldsymbol{b} \\
&= 2\gamma(\boldsymbol{I}_{\mathrm{n}} - \gamma\boldsymbol{A})^{-2}\boldsymbol{b}
\end{aligned}$$

We then have the following result.

Theorem 3.3.8. *When* $t_k = kT$, *an approximation* $y_{e3}(k)$ *of* $z(t = kT)$ *based on using the trapezoidal approximation of the integral in (3.125) with* $u(k) \triangleq v(kT)$ *is given by*

$$\begin{aligned}
\boldsymbol{x}_{e3}(k+1) &= \boldsymbol{F}_{e3}\boldsymbol{x}_{e3}(k) + \boldsymbol{g}_{e3}u(k) \\
y_{e3}(k) &= \boldsymbol{c}^T\boldsymbol{x}_{e3}(k) + d_{e3}u(k)
\end{aligned} \tag{3.137}$$

where with $\gamma = 0.5T$

$$\begin{aligned}
\boldsymbol{F}_{e3} &\triangleq (\boldsymbol{I}_{\mathrm{n}} - \gamma\boldsymbol{A})^{-1}(\boldsymbol{I}_{\mathrm{n}} + \gamma\boldsymbol{A}) \\
\boldsymbol{g}_{e3} &\triangleq 2\gamma(\boldsymbol{I}_{\mathrm{n}} - \gamma\boldsymbol{A})^{-2}\boldsymbol{b} \\
d_{e3} &\triangleq d + \gamma\boldsymbol{c}^T(\boldsymbol{I}_{\mathrm{n}} - \gamma\boldsymbol{A})^{-1}\boldsymbol{b}
\end{aligned} \tag{3.138}$$

provided $\boldsymbol{A}$ *has no eigenvalue at* $\gamma^{-1} = 2T^{-1}$.

After some algebraic manipulations, it can be shown (see exercise 3.36) that another state space representation $\{\tilde{\boldsymbol{F}}_{e3}, \tilde{\boldsymbol{g}}_{e3}, \tilde{\boldsymbol{h}}_{e3}, d_{e3}\}$ is given by

$$\begin{aligned}
&\tilde{\boldsymbol{F}}_{e3} = (\boldsymbol{I}_{\mathrm{n}} + \gamma\boldsymbol{A})(\boldsymbol{I}_{\mathrm{n}} - \gamma\boldsymbol{A})^{-1} \; ; \; \tilde{\boldsymbol{g}}_{e3} = \sqrt{2\gamma}(\boldsymbol{I}_{\mathrm{n}} - \gamma\boldsymbol{A})^{-1}\boldsymbol{b} \\
&\boldsymbol{h}_{e3}^T = \sqrt{2\gamma}\boldsymbol{c}^T(\boldsymbol{I}_{\mathrm{n}} - \gamma\boldsymbol{A})^{-1} \; ; \; d_{e3} = d + \gamma\boldsymbol{c}^T(\boldsymbol{I}_{\mathrm{n}} - \gamma\boldsymbol{A})^{-1}\boldsymbol{b}
\end{aligned} \tag{3.139}$$

Example 3.3.5. Consider the first order analog filter

$$\dot{z}(t) = -2z(t) + 3v(t) \ ; \ t \geq 0$$

with $z(0) = -0.55$ where the aim is to obtain different numerical approximations for $z(t = kT)$. We first write the differential equation in the form

$$\dot{x}(t) = -2x(t) + 3v(t) \ ; \ z(t) = x(t)$$

The approximations $y_{em}(k)$ for $m = 1, 2, 3$ with $u(k) \triangleq v(kT)$ are given as follows:

Backward Euler: (See Theorem 3.3.6.)

$$x_{e1}(k+1) = (1 - 2T)x_{e1}(k) + 3Tu(k) \ ; \ y_{e1}(k) = x_{e1}$$

Forward Euler: (See Theorem 3.3.7.)

$$\begin{aligned} x_{e2}(k+1) &= \left(\frac{1}{1+2T}\right) x_{e2}(k) + \left(\frac{3T}{1+2T}\right) u(k) \\ y_{e2}(k) &= x_{e2}(k) + \left(\frac{3T}{1+2T}\right) u(k) \end{aligned}$$

Trapezoidal: (See Theorem 3.3.8.)

$$\begin{aligned} x_{e3}(k+1) &= \left(\frac{1-T}{1+T}\right) x_{e3}(k) + \frac{3T}{(1+T)^2} u(k) \\ y_{e3}(k) &= x_{e3}(k) + \left(\frac{1.5T}{1+T}\right) u(k) \end{aligned}$$

Numerical solutions with $z(0) = 0$ for $T = 0.1, 0.25$ and $v(t) = 1$ are illustrated in Figure 3.19.

Example 3.3.6. Consider the second order analog filter

$$\ddot{z}(t) + 0.6\dot{z}(t) + 0.05z(t) = 0.08v(t) - 0.80\dot{v}(t) \ ; \ t \geq 0 \tag{3.140}$$

where $\{z(0) = 0, \dot{z}(0) = 0, v(0) = \dot{v}(0) = 0\}$. Then like in example 3.3.4, part (ii) a state space representation is given by (3.96) where

$$\boldsymbol{A} = \begin{bmatrix} -0.5 & 0 \\ 0 & -0.1 \end{bmatrix} \ ; \ \boldsymbol{b} = \begin{bmatrix} 1 \\ 1 \end{bmatrix} \ ; \ \boldsymbol{c} = \begin{bmatrix} -1.2 \\ 0.4 \end{bmatrix} \ ; \ \boldsymbol{x}(0) = \begin{bmatrix} 0 \\ 0 \end{bmatrix} \tag{3.141}$$

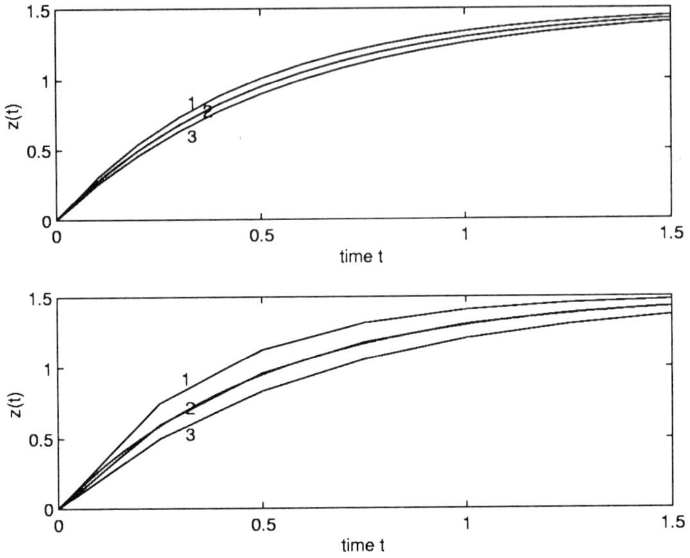

Fig. 3.19. Approximations of first order analog filter for 1: Backward Euler, 2: Forward Euler, and 3: Trapezoidal

Approximate solutions y_e of the analog signal z at the sampling instants $t_k = kT$ where $u(k) = v(kT)$ are then given by

$$\begin{aligned} \boldsymbol{x}_e(k+1) &= \boldsymbol{F}_e \boldsymbol{x}_e(k) + \boldsymbol{g}_e u(k) \\ y_e(k) &= \boldsymbol{c}^T \boldsymbol{x}_e(k) + d_e u(k) \end{aligned} \tag{3.142}$$

where $\{\boldsymbol{F}_e, \boldsymbol{g}_e, \boldsymbol{c}, d_e\}$ are given as follows:

Backward Euler Scheme: (See Theorem 3.3.6.)

$$\boldsymbol{F}_{e1} = \begin{bmatrix} 1-0.5T & 0 \\ 0 & 1-0.1T \end{bmatrix} \; ; \; \boldsymbol{g}_{e1} = T \begin{bmatrix} 1 \\ 1 \end{bmatrix}$$

$$\boldsymbol{c}_{e1} = \begin{bmatrix} -1.2 \\ 0.4 \end{bmatrix} \; ; \; d_{e1} = 0$$

Forward Euler Scheme: (See Theorem 3.3.7.)

$$\boldsymbol{F}_{e2} = \begin{bmatrix} (1+0.5T)^{-1} & 0 \\ 0 & (1+0.1T)^{-1} \end{bmatrix} \; ; \; \boldsymbol{g}_{e2} = T \begin{bmatrix} (1+0.5T)^{-2} \\ (1+0.1T)^{-2} \end{bmatrix}$$

$$\boldsymbol{c}_{e2} = \begin{bmatrix} -1.2 \\ 0.4 \end{bmatrix} \; ; \; d_{e2} = 0.4T(1+0.1T)^{-1} - 1.2T(1+0.5T)^{-1}$$

Trapezoidal Scheme: (See Theorem 3.3.8.)

$$\boldsymbol{F}_{e3} = \begin{bmatrix} (1-0.25T)(1+0.25T)^{-1} & 0 \\ 0 & (1-0.05T)(1+0.05T)^{-1} \end{bmatrix}$$

$$\boldsymbol{c}_{e3} = \begin{bmatrix} -1.2 \\ 0.4 \end{bmatrix} \; ; \; \boldsymbol{g}_{e3} = T \begin{bmatrix} (1+0.25T)^{-2} \\ (1+0.05T)^{-2} \end{bmatrix}$$

$$d_{e3} = 0.2T(1+0.05T)^{-1} - 0.6T(1+0.25T)^{-1}$$

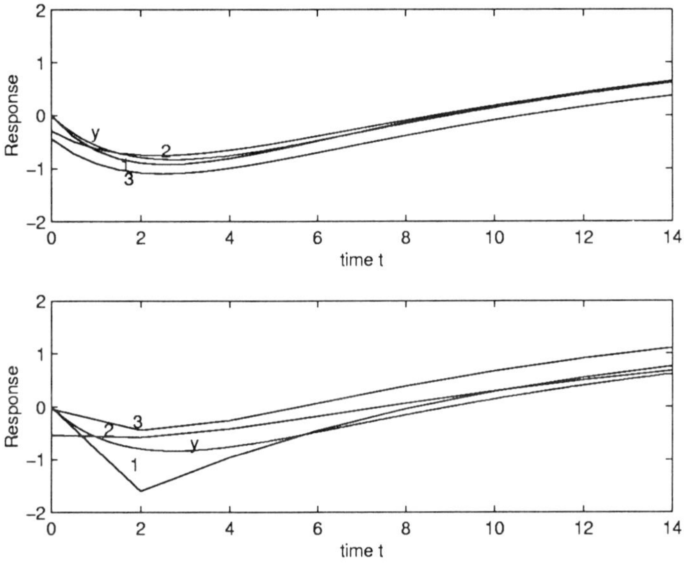

Fig. 3.20. Approximate solutions of second order analog filter for 1: Backward Euler, 2: Forward Euler, and 3: Trapezoidal.

These responses as well as the exact response $y(kT)$ for $T = 0.1, 0.25$ for zero initial conditions, and $\{v(t) = 1; t \geq 0\}$ are illustrated in Figure 3.20.

Stability Considerations

The analog filter $\mathcal{H}(s)$ in (3.127) is asymptotically stable if all roots $\{\lambda_k(\boldsymbol{A})\}$ of the characteristic equation

$$det(\lambda \boldsymbol{I}_n - \boldsymbol{A}) = \lambda^n - \sum_{p=0}^{n-1} \alpha_p \lambda^p = 0 \qquad (3.143)$$

have negative real part. This condition is equivalent to the condition that all eigenvalues $\lambda_k(\boldsymbol{A})$ of the continuous system matrix $\boldsymbol{A}$ have a negative real part.

Given that it is necessary that the resulting digital filter $H_e(z)$ in (3.128) be asymptotically stable; that is, all roots $\{\lambda_k(\boldsymbol{F})\}$ of the characteristic equation

$$det(\lambda \boldsymbol{I}_n - \boldsymbol{F}) = 0$$

have less than unity magnitude, a *necessary condition* is that the sampling period $T(> 0)$ be selected such that:

$$Re\{\lambda_k(\boldsymbol{A})\} < 0 \quad \text{for all} \quad k \tag{3.144}$$

implies

$$|\lambda_k(\boldsymbol{F}_e)| < 1 \quad \text{for all} \quad k \tag{3.145}$$

We now test each of the analog to digital transformation schemes with respect to this property.

Backward Euler Integration: From Theorem 3.3.6, $\boldsymbol{F}_{e1} = \boldsymbol{I}_{\mathbf{n}} + T\boldsymbol{A}$, and so

$$\boldsymbol{A}\boldsymbol{v} = \lambda_A \boldsymbol{v} \;;\; \lambda_A = \lambda_k(\boldsymbol{A}) \tag{3.146}$$

implies

$$(\boldsymbol{I}_{\mathbf{n}} + T\boldsymbol{A})\boldsymbol{v} = (1 + \lambda_A T)\boldsymbol{v}$$

Then if $\lambda_A = \alpha + j\beta$ (α, β real), we have

$$|1 + \lambda_A T| = \sqrt{(1 + \alpha T)^2 + \beta^2 T^2}$$

Hence if $Re(\lambda_A) = \alpha < 0$, it follows that $|1 + \lambda_A T| < 1$ for sufficiently small T. The maximum allowable value T_{max} of the sampling period is given by the value of T such that

$$max_k |1 + \lambda_k(\boldsymbol{A})T| < 1$$

Forward Euler Integration: From Theorem 3.3.7, $\boldsymbol{F}_{e2} = (\boldsymbol{I}_{\mathbf{n}} - T\boldsymbol{A})^{-1}$, and so (3.146) implies

$$(\boldsymbol{I}_{\mathbf{n}} - T\boldsymbol{A})^{-1}\boldsymbol{v} = \frac{1}{1 - \lambda_A T}\boldsymbol{v}$$

Then $\lambda_A = \alpha + j\beta$ (α, β real) implies

$$\left|\frac{1}{1 - \lambda_A T}\right| = \frac{1}{\sqrt{(1 - \alpha T)^2 + \beta^2 T^2}}$$

Hence if $Re(\lambda_A) = \alpha < 0$, it follows that $|(1 - \lambda_A T)^{-1}| < 1$ for sufficiently small T. The maximum allowable value T_{max} of the sampling period is given by the value of T such that

$$max_k \left| \frac{1}{1-\lambda_k(\boldsymbol{A})T} \right| < 1$$

Trapezoidal Integration: From Theorem 3.3.8, $\boldsymbol{F}_{e3} = (\boldsymbol{I}_{\mathbf{n}} - 0.5T\boldsymbol{A})^{-1}(\boldsymbol{I}_{\mathbf{n}} + 0.5T\boldsymbol{A})$. Now (3.146) implies

$$(\boldsymbol{I}_n + 0.5T\boldsymbol{A})\boldsymbol{v} = (1 + 0.5T\lambda_A)\boldsymbol{v}$$

so that

$$\begin{aligned}(\boldsymbol{I}_n - 0.5T\boldsymbol{A})^{-1}(\boldsymbol{I}_n + 0.5T\boldsymbol{A})\boldsymbol{v} &= (1 + 0.5T\lambda_A)(\boldsymbol{I}_n - 0.5T\boldsymbol{A})^{-1}\boldsymbol{v} \\ &= \frac{1 + 0.5T\lambda_A}{1 - 0.5T\lambda_A}\boldsymbol{v}\end{aligned}$$

Hence $\lambda_A = \alpha + j\beta$ (α, β real) implies

$$\left| \frac{1 + 0.5T\lambda_A}{1 - 0.5T\lambda_A} \right| = \sqrt{\frac{(1 + 0.5\alpha T)^2 + 0.25\beta^2T^2}{(1 - 0.5\alpha T)^2 + 0.25\beta^2T^2}}$$

and so it follows that if $Re(\lambda_A) = \alpha + j\beta$ (α, β real), then

$$\left| \frac{1 + 0.5T\lambda_A}{1 - 0.5T\lambda_A} \right| < 1 \quad \text{for all} \quad T$$

That is, the approximation of an asymptotically stable analog filter based on a trapezoidal integral approximation always results in an asymptotically stable digital filter.

Frequency Selectivity

The analog filter (3.124) has the analog frequency function $\mathcal{H}(s)$ given by

$$\mathcal{H}(s) = d + \boldsymbol{c}^T(s\boldsymbol{I} - \boldsymbol{A})^{-1}\boldsymbol{b} \tag{3.147}$$

whereas the resulting digital filter (3.126) has the discrete frequency function $H_e(z)$ given by

$$H_e(z) = d_e + \boldsymbol{c}^T(z\boldsymbol{I} - \boldsymbol{F}_e)^{-1}\boldsymbol{g}_e \tag{3.148}$$

We now examine the relationship between $\mathcal{H}(s)$ and $H_e(z)$ for each analog to digital transformation scheme.

Backward Euler Integration: In this case from Theorem 3.3.6

$$H_e(z) = d + T\boldsymbol{c}^T(z\boldsymbol{I} - \boldsymbol{I} - T\boldsymbol{A})^{-1}\boldsymbol{b}$$

Now

$$(z\boldsymbol{I} - \boldsymbol{I} - T\boldsymbol{A}) = T[T^{-1}(z-1)\boldsymbol{I} - \boldsymbol{A}]$$

and so

$$H_e(z) = d + \boldsymbol{c}^T[T^{-1}(z-1)\boldsymbol{I} - \boldsymbol{A}]^{-1}\boldsymbol{b} \tag{3.149}$$

Hence if we now compare (3.149) with (3.147), we conclude that the analog to digital transformation scheme which results from backward Euler integration corresponds to a mapping

$$\mathcal{H}(s) \Rightarrow H_e(z) \text{ in which } s \Rightarrow T^{-1}(z-1)$$

However since $T^{-1}(exp\{j\omega\} - 1)$ is not purely imaginary, there is no direct correspondence between an analog frequency Ω and a digital frequency ω.

Forward Euler Integration: In this case from Theorem 3.3.7, we have

$$H_e(z) = d + T\boldsymbol{c}^T(\boldsymbol{I} - T\boldsymbol{A})^{-1}\boldsymbol{b} + H_f(z) \tag{3.150}$$

where

$$H_f(z) = T\boldsymbol{c}^T[z\boldsymbol{I} - (\boldsymbol{I} - T\boldsymbol{A})^{-1}]^{-1}(\boldsymbol{I} - T\boldsymbol{A})^{-2}\boldsymbol{b}$$

Then (see exercise 3.19) it follows that

$$H_e(z) = d + \boldsymbol{c}^T[T^{-1}(1 - z^{-1})\boldsymbol{I} - \boldsymbol{A}]^{-1}\boldsymbol{b} \tag{3.151}$$

Hence in this case, we conclude that the analog to digital transformation scheme which results from forward Euler integration corresponds to a mapping

$$\mathcal{H}(s) \Rightarrow H_e(z) \text{ in which } s \Rightarrow T^{-1}(1 - z^{-1})$$

However since $T^{-1}(1 - exp\{-j\omega\})$ is not purely imaginary, there is once again no direct correspondence between analog and digital frequencies.

Trapezoidal Integration: In this case from Theorem 3.3.8

$$H_e(z) = d + \gamma\boldsymbol{c}^T(\boldsymbol{I} - \gamma\boldsymbol{A})^{-1}\boldsymbol{b} + H_f(z) \ ; \ \gamma = 0.5T \tag{3.152}$$

where

$$H_f(z) = \gamma\boldsymbol{c}^T[z\boldsymbol{I} - (\boldsymbol{I} - \gamma\boldsymbol{A})^{-1}(\boldsymbol{I} + \gamma\boldsymbol{A})]^{-1}(\boldsymbol{I} + \gamma\boldsymbol{A})(\boldsymbol{I} - \gamma\boldsymbol{A})^{-2}\boldsymbol{b}$$

It then follows (see the exercises) that

$$H_e(z) = d + \boldsymbol{c}^T[2T^{-1}(z-1)(z+1)^{-1}\boldsymbol{I} - \boldsymbol{A}]^{-1}\boldsymbol{b} \tag{3.153}$$

Hence in this case, we conclude that the analog to digital transformation scheme which results from trapezoidal integration corresponds to a mapping

$$\mathcal{H}(s) \Rightarrow H_e(z) \text{ in which } s \Rightarrow 2T^{-1}(z-1)(z+1)^{-1}$$

Now when $z = exp\{j\omega\}$,

$$\frac{z-1}{z+1} = \frac{(exp\{j\omega\}-1)}{(exp\{j\omega\}+1)} = j\left(\frac{\sin\omega}{1+\cos\omega}\right)$$

and so in this case there is a one-to-one correspondence between Ω over the range $0 \le \Omega < \infty$ and ω over the range $0 \le \omega < \pi$ as given by

$$\Omega = 2T^{-1}\left(\frac{\sin\omega}{1+\cos\omega}\right)$$

The relationship between the *normalized analog frequency ΩT* (in rads), and the discrete frequency ω (in rads) is illustrated in Figure 3.21. The transformation which results based on a trapezoidal approximation of an integral is also known as the *bilinear transformation.*

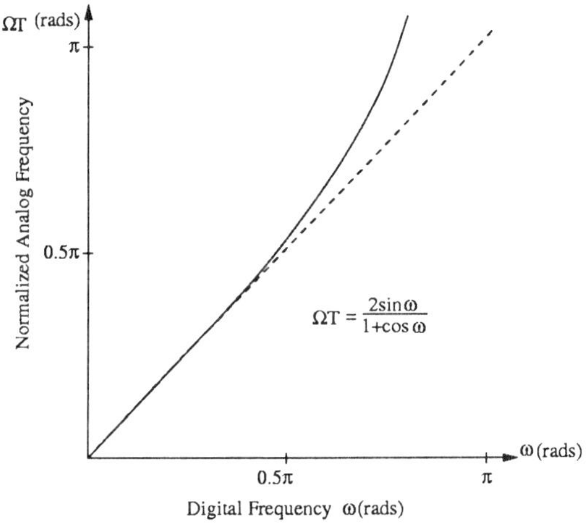

Fig. 3.21. Frequency transformation for trapezoidal integration (commonly known as the bilinear transformation)

For "sufficiently small" values of the sampling period, or equivalently, 'sufficiently large" values of the sampling frequency $f_s = T^{-1}$, it follows that $\omega \simeq \Omega T$. That is, if the analog filter $\mathcal{H}(s)$ has a cut-off frequency of Ω_c rads/sec, then for "sufficiently small' T, the cut-off frequency ω_c of the transformed digital filter $H(z)$ is given by $\omega_c = \Omega_c T$ rads.

Example 3.3.7. A second order lowpass Butterworth filter $\mathcal{H}(s) = d + c^T(s\boldsymbol{I} - \boldsymbol{A})^{-1}\boldsymbol{b}$ with a cut-off frequency of Ω_c rads/sec is given from (3.84) by

$$\mathcal{H}(s) = \frac{\Omega_c^2}{s^2 + \sqrt{2}\Omega_c s + \Omega_c^2}$$

A corresponding analog state space representation $\{\boldsymbol{A}, \boldsymbol{b}, \boldsymbol{c}, d\}$ of $\mathcal{H}(s)$ is then given by

$$\boldsymbol{A} = 0.5\sqrt{2}\Omega_c \begin{bmatrix} -1 & 1 \\ -1 & -1 \end{bmatrix} \; ; \; \boldsymbol{b} = \sqrt{2}\Omega_c \begin{bmatrix} 0 \\ 1 \end{bmatrix} \; ; \; \boldsymbol{c} = \begin{bmatrix} 1 \\ 0 \end{bmatrix} \; ; \; d = 0$$

The cut-off frequency ω_c (in rads.) of the corresponding transformed digital low pass Butterworth filter $H(z)$ obtained by the trapezoidal (or bilinear) transformation is then given by

$$\frac{\sin \omega_c}{1 + \cos \omega_c} = 0.5\Omega_c T$$

The corresponding state space representation of $H(z) = d_1 + \boldsymbol{h}^T(z\boldsymbol{I} - \boldsymbol{F})^{-1}\boldsymbol{g}$ is given by

$$\begin{aligned} \boldsymbol{F} &= (\boldsymbol{I} - \gamma\boldsymbol{A})^{-1}(\boldsymbol{I} + \gamma\boldsymbol{A}) \; ; \; \boldsymbol{h} = \boldsymbol{c} \\ \boldsymbol{g} &= 2\gamma(\boldsymbol{I} - \gamma\boldsymbol{A})^{-2}\boldsymbol{b} \; ; \; d_1 = \gamma\boldsymbol{c}^T(\boldsymbol{I} - \gamma\boldsymbol{A})^{-1}\boldsymbol{b} \end{aligned}$$

where

$$\boldsymbol{I} + \gamma\boldsymbol{A} = \begin{bmatrix} 1-\delta & \delta \\ -\delta & 1-\delta \end{bmatrix} \; ; \; \boldsymbol{I} - \gamma\boldsymbol{A} = \begin{bmatrix} 1+\delta & -\delta \\ \delta & 1+\delta \end{bmatrix}$$

$$\delta = 0.25\sqrt{2}\Omega_c T = \frac{0.5\sqrt{2}\sin\omega_c}{1 + \cos\omega_c}$$

3.3.5 Digital to Digital Transformation

Frequency transformations are also available for transforming a given digital filter $H(z)$ to another digital filter $H_Q(z)$ as defined by

$$H_Q(z) \triangleq H(Z) \; ; \; Z = Q^{-1}(z) \tag{3.154}$$

In this section, we are interested in transforming a given lowpass digital filter $H_0(z)$ with cut-off frequency ω_0 to:

- another lowpass filter with cut-off frequency ω_{c0},
- a highpass filter with cut-off frequency ω_{c0},
- a bandpass filter with bandpass from ω_{c1} to ω_{c2}, and
- a bandstop filter with stopband from ω_{c1} to ω_{c2}.

To this end, consider first an asymptotically stable first order digital filter with frequency response function $H_0(z)$ given by

$$H_0(z) = \frac{b_1 z + b_0}{z - a_0} \tag{3.155}$$

Then the transformation

$$Z = \frac{\alpha_1 z + \alpha_0}{\beta_1 z + \beta_0} = Q^{-1}(z) \tag{3.156}$$

results in the first order transformed filter $H_Q(z)$ where

$$\begin{aligned} H_Q(z) &= \frac{b_1(\alpha_1 z + \alpha_0)(\beta_1 z + \beta_0)^{-1} + b_0}{(\alpha_1 z + \alpha_0)(\beta_1 z + \beta_0)^{-1} - a_0} \\ &= \frac{(b_1\alpha_1 + b_0\beta_1)z + (\beta_1\alpha_0 + b_0\beta_0)}{(\alpha_1 - a_0\beta_1)z - (a_0\beta_0 - \alpha_0)} \end{aligned} \tag{3.157}$$

Since we require there to be a *one-to-one correspondence* between the digital frequency ω of $H(z)$ and the digital frequency ω_Q of $H_Q(z)$, it is necessary that

$$Q^{-1}(exp\{j\omega\}) = exp\{j\omega_Q\}$$

or equivalently

$$\frac{\alpha_1 exp\{j\omega\} + \alpha_0}{\beta_1 exp\{j\omega\} + \beta_0} = exp\{j\omega_Q\} \tag{3.158}$$

Low Pass to Low Pass Transformation

If $H_Q(z)$ is to be a *lowpass filter*, then $\omega = 0$ in $H(z)$ must correspond to $\omega_Q = 0$ in $H_Q(z)$; that is, from (3.158), we conclude

$$\alpha_1 + \alpha_0 = \beta_1 + \beta_0 \tag{3.159}$$

Also, since $\omega = \pi$ must correspond to $\omega_Q = \pi$, we conclude

$$-\alpha_1 + \alpha_0 = -(-\beta_1 + \beta_0) \tag{3.160}$$

Then (3.159) and (3.160) together imply $\{\beta_1 = \alpha_0,\ \beta_0 = \alpha_1\}$, and hence from (3.156) for $\beta_1 \neq 0$

$$Z = \frac{(\alpha_1\beta_1^{-1})\, z + (\alpha_0\beta_1^{-1})}{z + (\beta_0\beta_1^{-1})} = \frac{\gamma z + 1}{z + \gamma} \;;\; \gamma = \alpha_1\alpha_0^{-1} \tag{3.161}$$

Then from (3.158) with $\{\beta_1 = \alpha_0,\ \beta_0 = \alpha_1,\ \gamma = \alpha_1\alpha_0^{-1}\}$, we have

$$\frac{\gamma exp\{j\omega\} + 1}{exp\{j\omega\} + \gamma} = exp\{j\omega_Q\}$$

or equivalently

$$\gamma = \frac{exp\{j0.5(\omega + \omega_Q)\} - exp\{-j0.5(\omega + \omega_Q)\}}{exp\{j0.5(\omega - \omega_Q)\} - exp\{-j0.5(\omega - \omega_Q)\}} = \frac{\sin 0.5(\omega + \omega_Q)}{\sin 0.5(\omega - \omega_Q)}$$

Also, when Z is given by (3.161), we have from (3.157) that the transformed lowpass filter $H_Q(z)$ is asymptotically stable if and only if

$$|a_0\left(\frac{\beta_0}{\beta_1}\right) - \left(\frac{\alpha_0}{\beta_1}\right)| < |a_0 - \left(\frac{\alpha_1}{\beta_1}\right)|$$

That is, if and only if

$$|a_0\gamma - 1| < |a_0 - \gamma|$$

Since $|a_0| < 1$, this condition is satisfied for all $|\gamma| > 1$.

Low Pass to High Pass Transformation

If $H_Q(z)$ is to be a *highpass filter*, then $\omega = 0$ in (3.158) must correspond to $\omega_Q = \pi$; that is, we require

$$\alpha_1 + \alpha_0 = -(\beta_1 + \beta_0) \tag{3.162}$$

Also since $\omega = \pi$ must correspond to $\omega_Q = 0$, we require

$$-\alpha_1 + \alpha_0 = -\beta_1 + \beta_0 \tag{3.163}$$

Then (3.162) and (3.163) together imply $\{\beta_1 = -\alpha_0; \beta_0 = -\alpha_1\}$, and hence from (3.156) for $\beta_1 \neq 0$

$$Z = \frac{\left(\alpha_1\beta_1^{-1}\right)z + \left(\alpha_0\beta_1^{-1}\right)}{z + \left(\beta_0\beta_1^{-1}\right)} = \frac{\gamma z - 1}{z - \gamma} \; ; \; \gamma = -\alpha_1\alpha_0^{-1} \tag{3.164}$$

Then from (3.158) with $\{\beta_1 = -\alpha_0, \beta_0 = -\alpha_1, \gamma = -\alpha_1\alpha_0^{-1}\}$, we have

$$\frac{\gamma exp\{j\omega\} - 1}{exp\{j\omega\} - \gamma} = exp\{j\omega_Q\}$$

or equivalently

$$\gamma = \frac{exp\{j0.5(\omega + \omega_Q)\} + exp\{-j0.5(\omega + \omega_Q)\}}{exp\{j0.5(\omega - \omega_Q)\} + exp\{-j0.5(\omega - \omega_Q)\}} = \frac{\cos 0.5(\omega + \omega_Q)}{\cos 0.5(\omega - \omega_Q)}$$

Also when Z is given by (3.161), we have from (3.157) that the transformed highpass filter $H_Q(z)$ is asymptotically stable if and only if

$$|a_0\left(\frac{\beta_0}{\beta_1}\right) - \left(\frac{\alpha_0}{\beta_1}\right)| < |a_0 - \left(\frac{\alpha_1}{\beta_1}\right)|$$

That is, if and only if

$$|-a_0\gamma + 1| < |a_0 + \gamma|$$

Since $|a_0| < 1$, this condition is satisfied for all $|\gamma| > 1$. Now since any asymptotically stable filter can be written in the form

$$H(z) = \prod_k \left(\frac{b_{1k} z + b_{0k}}{z - a_k} \right) \; ; \; |a_k| < 1$$

where if a_p is complex, then $a_q = \overline{a}_p$ for some $q \neq p$, we have the following result.

Theorem 3.3.9. *Given any asymptotically stable lowpass filter $H_0(z)$ with cut-off frequency ω_0, define the transformed filter $H_Q(z)$ by*

$$H_Q(z) = H_0(Z) \; ; \; Z = Q^{-1}(z)$$

Then:

1. *The transformed filter $H_Q(z)$ is an asymptotically stable lowpass filter with cut-off frequency ω_{Q0} when*

$$Q^{-1}(z) = \frac{\gamma z + 1}{z + \gamma} \; ; \; |\gamma| > 1$$

and

$$\gamma = \frac{\sin 0.5(\omega_0 + \omega_{Q0})}{\sin 0.5(\omega_0 - \omega_{Q0})}$$

2. *The transformed filter $H_Q(z)$ is an asymptotically stable highpass filter with cut-off frequency ω_{Q0} when*

$$Q^{-1}(z) = \frac{\gamma z - 1}{z - \gamma} \; ; \; |\gamma| > 1$$

and

$$\gamma = \frac{\cos 0.5(\omega_0 + \omega_{Q0})}{\cos 0.5(\omega_0 - \omega_{Q0})}$$

Example 3.3.8. A second order digital filter $H(z)$ which results from the transformation of the analog filter

$$\mathcal{H}(s) = \frac{3}{s^2 + 3s + 3}$$

using the bilinear transformation $s \Rightarrow 2T^{-1}(z-1)(z+1)^{-1}$ is of the form

$$H(z) = \frac{d(z+1)^2}{z^2 - a_1 z - a_0} \tag{3.165}$$

Then the second order transformed lowpass digital filter $H_Q(z)$ is given by

$$H_Q(z) = H\left(\frac{\gamma z + 1}{z + \gamma} \right) = \frac{d(\gamma z + 1 + z + \gamma)^2}{(\gamma z + 1)^2 - a_1(\gamma z + 1)(z + \gamma) - a_0(z + \gamma)^2}$$

This hand manipulation can be messy especially for high order filters, and so later we develop an algebraic approach based on state space representations.

Low Pass to Bandpass and Bandstop Transformations

In general, it may be shown that if the function $Q^{-1}(z)$ is to be *frequency selective*, then $Q^{-1}(z)$ must be of the form

$$Q^{-1}(z) = \pm \prod_k \left(\frac{\gamma_k + 1}{z + \gamma_k} \right)$$

where if γ_p is complex, then $\gamma_p = \overline{\gamma}_q$ for some $p \neq q$. Asymptotic stability of the transformed filter $H_Q(z)$ then requires $|\gamma_k| > 1$ for all k. We have the following result.

Theorem 3.3.10. *Given any asymptotically stable lowpass filter $H_0(z)$ with cut-off frequency ω_0, define the transformed filter $H_Q(z)$ by*

$$H_Q(z) = H_0(Z) \ ; \ Z = Q^{-1}(z)$$

Then:

- *The transformed filter $H_Q(z)$ is an asymptotically stable bandpass filter with passband ω_{Q1} to ω_{Q2} when*

$$Q^{-1}(z) = \frac{\beta_1 - \delta_1 z - z^2}{1 + \delta_1 z - \beta_1 z^2} \ ; \ \beta_1 \triangleq \frac{1 + k_1}{1 - k_1}, \ \delta_1 \triangleq \frac{2\alpha_1 k_1}{1 - k_1}$$

and

$$\alpha_1 = \frac{\cos 0.5(\omega_{Q2} + \omega_{Q1})}{\cos 0.5(\omega_{Q2} - \omega_{Q1})} \ ; \ k_1 = \frac{\tan 0.5\omega_0}{\tan 0.5(\omega_{Q2} - \omega_{Q1})}$$

- *The transformed filter $H_Q(z)$ is an asymptotically stable bandstop filter with stopband ω_{Q1} to ω_{Q2} when*

$$Q^{-1}(z) = \frac{\beta_2 - \delta_2 z + z^2}{1 - \delta_2 z + \beta_2 z^2} \ ; \ \beta_2 = \frac{1 + k_2}{1 - k_2}, \ \delta_2 = \frac{2\alpha_2}{1 - k_2}$$

and

$$\alpha_2 = \frac{\cos 0.5(\omega_{Q2} + \omega_{Q1})}{\cos 0.5(\omega_{Q2} - \omega_{Q1})} \ ; \ k_2 = \tan 0.5(\omega_{Q2} - \omega_{Q1}) \tan 0.5\omega_0$$

Example 3.3.9. Suppose a second order digital filter $H(z)$ is given by (3.165). Then the fourth order bandpass filter $H_Q(z)$ is given by

$$H_Q(z) = H\left(\frac{\beta_1 - \delta_1 z - z^2}{1 + \delta_1 z - \beta_1 z^2} \right) = \frac{p(z)}{q(z)}$$

where

$$p(z) = d(\beta_1 - \delta_1 z - z^2 + 1 + \delta_1 z - \beta_1 z^2)^2$$
$$q(z) = (\beta_1 - \delta_1 z - z^2)^2 - a_1(\beta_1 - \delta_1 z - z^2)(1 + \delta_1 z - \beta_1 z^2)$$
$$-a_0(1 + \delta_1 z - \beta_1 z^2)^2$$

Once again, we note that there are better methods available based on state space representations.

State Space Representation

Suppose that the digital filter with frequency function $H_0(z)$ has a state space representation

$$\begin{aligned} \boldsymbol{x}(k+1) &= \boldsymbol{F}_0 \boldsymbol{x}(k) + \boldsymbol{g}_0 u(k) \; ; \; \boldsymbol{x}(k) \in \mathcal{R}^n \\ y(k) &= \boldsymbol{h}_0^T \boldsymbol{x}(k) + d_0 u(k) \end{aligned}$$

and the frequency function $Q(z)$ has a state space representation

$$\begin{aligned} \boldsymbol{q}(k+1) &= \boldsymbol{F}_1 \boldsymbol{q}(k) + \boldsymbol{g}_1 r(k) \; ; \; \boldsymbol{q}(k) \in \mathcal{R}^q \\ w(k) &= \boldsymbol{h}_1^T \boldsymbol{q}(k) + d_1 r(k) \end{aligned}$$

We now develop the corresponding state space representation for the transformed filter $H_Q(z)$ defined by (3.154).

To begin, observe that in the state space representation for $H(z)$, the frequency response function between each state component $\{x_m(k+1);\ 1 \le m \le n\}$ of $\boldsymbol{x}(m+1)$ (regarded as the *input signal*) and $x_m(k)$ (regarded as the *output signal*) is z^{-1}. Since in the transformed filter, z is replaced by $Q^{-1}(z)$, or equivalently, z^{-1} is replaced by $Q(z)$, we conclude that the state components $\boldsymbol{z}^{(m)}$ in $H_Q(z)$ for $1 \le m \le n$ are given by

$$\begin{aligned} \boldsymbol{z}^{(m)}(k+1) &= \boldsymbol{F}_1 \boldsymbol{z}^{(m)}(k) + \boldsymbol{g}_1 x_m(k+1) \\ x_m(k) &= \boldsymbol{h}_1^T \boldsymbol{z}^{(m)}(k) + d_1 x_m(k+1) \end{aligned}$$

That is, each $x_m(k)$ is the output of a system $Q(z)$ with input $x_m(k+1)$.

Now define the $q \times n$ matrix $\boldsymbol{Z}(k)$ by

$$\boldsymbol{Z}(k) \triangleq [\boldsymbol{z}^{(1)}(k) \quad \boldsymbol{z}^{(2)}(k) \quad \dots \quad \boldsymbol{z}^{(n)}(k)]$$

Then

$$\begin{aligned} \boldsymbol{Z}(k+1) &= \boldsymbol{F}_1 \boldsymbol{Z}(k) + \boldsymbol{g}_1 [\boldsymbol{F}_0 \boldsymbol{x}(k) + \boldsymbol{g}_0 u(k)]^T \\ \boldsymbol{x}(k) &= \boldsymbol{Z}^T(k) \boldsymbol{h}_1 + d_1 [\boldsymbol{F}_0 x(k) + \boldsymbol{g}_0 u(k)] \end{aligned}$$

Hence

$$x(k) = (I_n - d_1 F_0)^{-1}[Z^T(k)h_1 + d_1 g_0 u(k)]$$

and so $x(k)$ can be eliminated from the equation for $Z(k+1)$. Also, the scalar output $y(k)$ of $H_Q(z)$ is given by

$$y(k) = h_0^T(I_n - d_1 F_0)^{-1} Z^T(k)h_1 + [d_0 + d_1 h_0^T(I_n - d_1 F_0)^{-1} g_0]u(k)$$

Now the *Kronecker product* of the matrix $A = [a_{ij}] \in R^{m\times n}$ with the matrix $B = [b_{ij}] \in R^{p\times q}$ (denoted $A \otimes B$) is defined by

$$A \otimes B = \begin{bmatrix} a_{11}B & a_{12}B & \cdots & a_{1n}B \\ \vdots & & & \\ a_{m1}B & a_{m2}B & \cdots & a_{mn}B \end{bmatrix} \in R^{mp\times nq}$$

(The terms *tensor product* and *direct product* are also used.) The *Kronecker sum* of the n-square matrix A and the p-square matrix B (denoted $A \oplus B$) is defined by

$$A \oplus B = A \otimes I_p + I_n \otimes B$$

We then deduce the following result. [See C.T. Mullis and R.A. Roberts, "Synthesis of minimum roundoff noise in fixed point digital filters", *IEEE Trans. Acoust. Speech and Signal Process.*, vol. ASSP-24, pp538-550, 1976.]

Theorem 3.3.11. *Consider a state space representation of a digital filter* $H_0(z)$ *defined by*

$$\begin{aligned} x(k+1) &= F_0 x(k) + g_0 u(k) \; ; \; x(k) \in \mathcal{R}^n \\ y(k) &= h_0^T x(k) + d_0 u(k) \end{aligned}$$

and define

$$H_Q(z) = H_0(Z) \; ; \; Z = Q^{-1}(z)$$

where a state space representation of $Q(z)$ *is given by*

$$\begin{aligned} q(k+1) &= F_1 q(k) + g_1 r(k) \; ; \; q(k) \in \mathcal{R}^q \\ w(k) &= h_1^T q(k) + d_1 r(k) \end{aligned}$$

Then a state space description of the system $H_Q(z)$ *is given by*

$$\begin{aligned} z(k+1) &= F_Q z(k) + g_Q u(k) \; ; \; z(k) \in R^{nq} \\ y(k) &= h_Q^T z(k) + d_Q u(k) \end{aligned}$$

where

$$
\begin{aligned}
\boldsymbol{F}_Q &= \boldsymbol{I}_{\mathbf{n}} \otimes \boldsymbol{F}_1 + \boldsymbol{F}_0(\boldsymbol{I}_{\mathbf{n}} - d_1\boldsymbol{F}_0)^{-1} \otimes \boldsymbol{g}_1\boldsymbol{h}_1^T \\
\boldsymbol{g}_Q &= (\boldsymbol{I}_{\mathbf{n}} - d_1\boldsymbol{F}_0)^{-1}\boldsymbol{g}_0 \otimes \boldsymbol{g}_1 \\
\boldsymbol{h}_Q^T &= \boldsymbol{h}_0^T(\boldsymbol{I}_{\mathbf{n}} - d_1\boldsymbol{F}_0)^{-1} \otimes \boldsymbol{h}_1^T \\
d_Q &= d_0 + d_1\boldsymbol{h}_0^T(\boldsymbol{I}_{\mathbf{n}} - d_1\boldsymbol{F}_0)^{-1}\boldsymbol{g}_0
\end{aligned}
$$

Low Pass to Low Pass Transformation: In this case

$$Q(z) = \frac{z+\gamma}{\gamma z + 1} = \gamma^{-1} + \frac{1-\gamma^{-2}}{z+\gamma^{-1}} \; ; \; |\gamma| > 1$$

and so one state space representation $\{\boldsymbol{F}_1, \boldsymbol{g}_1, \boldsymbol{h}_1, d_1\}$ of $Q(z)$ is given by

$$\boldsymbol{F}_1 = -\gamma^{-1} \; ; \; \boldsymbol{g}_1 = 1 - \gamma^{-2} \; ; \; \boldsymbol{h}_1 = 1 \; ; \; d_1 = \gamma^{-1}$$

Then given a state space representation $\{\boldsymbol{F}_0, \boldsymbol{g}_0, \boldsymbol{h}_0, d_0\}$ of $H(z)$, a state space representation $\{\boldsymbol{F}_Q, \boldsymbol{g}_Q, \boldsymbol{h}_Q, d\}$ of $H_Q(z)$ in Theorem 3.3.11 is given by

$$
\begin{aligned}
\boldsymbol{F}_Q &= -\gamma^{-1}\boldsymbol{I}_{\mathbf{n}} + 1 - \gamma^{-2}\boldsymbol{F}_0(\boldsymbol{I}_{\mathbf{n}} - \gamma^{-1}\boldsymbol{F}_0)^{-1} \\
\boldsymbol{g}_Q &= 1 - \gamma^{-2}(\boldsymbol{I}_n - \gamma^{-1}\boldsymbol{F}_0)^{-1}\boldsymbol{g}_0 \\
\boldsymbol{h}_Q^T &= \boldsymbol{h}_0^T(\boldsymbol{I}_{\mathbf{n}} - \gamma^{-1}\boldsymbol{F}_0)^{-1} \; ; \; d_Q = d_0 + \gamma^{-1}\boldsymbol{h}_0(\boldsymbol{I}_{\mathbf{n}} - \gamma^{-1}\boldsymbol{F}_0)^{-1}\boldsymbol{g}_0
\end{aligned}
$$

Low Pass to High Pass Transformation: In this case

$$Q(z) = \frac{z-\gamma}{\gamma z - 1} = \gamma^{-1} + \frac{-1+\gamma^{-2}}{z-\gamma^{-1}}$$

and so one state space representation $\{\boldsymbol{F}_1, \boldsymbol{g}_1, \boldsymbol{h}_1, d_1\}$ of $Q(z)$ is given by

$$\boldsymbol{F}_1 = \gamma^{-1} \; ; \; \boldsymbol{g}_1 = -1 + \gamma^{-2} \; ; \; \boldsymbol{h}_1 = 1 \; ; \; d_1 = \gamma^{-1}$$

Then given a state space representation $\{\boldsymbol{F}_0, \boldsymbol{g}_0, \boldsymbol{h}_0, d_0\}$ of $H(z)$, a state space representation $\{\boldsymbol{F}_Q, \boldsymbol{g}_Q, \boldsymbol{h}_Q, d\}$ of $H_Q(z)$ in Theorem 3.3.11 is given by

$$
\begin{aligned}
\boldsymbol{F}_Q &= \gamma^{-1}\boldsymbol{I}_{\mathbf{n}} + 1 - \gamma^{-2}\boldsymbol{F}_0(\boldsymbol{I}_{\mathbf{n}} - \gamma^{-1}\boldsymbol{F}_0)^{-1} \\
\boldsymbol{g}_Q &= 1 - \gamma^{-2}(\boldsymbol{I}_n - \gamma^{-1}\boldsymbol{F}_0)^{-1}\boldsymbol{g}_0 \\
\boldsymbol{h}_Q^T &= \boldsymbol{h}_0^T(\boldsymbol{I}_{\mathbf{n}} - \gamma^{-1}\boldsymbol{F}_0)^{-1} \; ; \; d_Q = d_0 + \gamma^{-1}\boldsymbol{h}_0(\boldsymbol{I}_{\mathbf{n}} - \gamma^{-1}\boldsymbol{F}_0)^{-1}\boldsymbol{g}_0
\end{aligned}
$$

Low Pass to Bandpass Transformation: In this case

$$\begin{aligned} Q(z) &= \frac{1+\delta_1 z - \beta_1 z^2}{\beta_1 - \delta_1 z - z^2} = \frac{\beta_1 z^2 - \delta_1 z - 1}{z^2 + \delta_1 z - \beta_1} \\ &= \beta_1 + \frac{-(\delta_1 + \beta_1\delta_1)z + \beta_1^2 - 1}{z^2 + \delta_1 z - \beta_1} \end{aligned}$$

and so one state space representation $\{\boldsymbol{F}_1, \boldsymbol{g}_1, \boldsymbol{h}_1, d_1\}$ of $Q(z)$ is given by

$$\boldsymbol{F}_1 = \begin{bmatrix} 0 & \beta_1 \\ 1 & -\delta_1 \end{bmatrix} \; ; \; \boldsymbol{g}_1 = \begin{bmatrix} \beta_1^2 - 1 \\ -\delta_1(1+\beta_1) \end{bmatrix} \; ; \; \boldsymbol{h}_1 = \begin{bmatrix} 0 \\ 1 \end{bmatrix} \; ; \; d_1 = \beta_1$$

Then given a state space representation $\{\boldsymbol{F}_0, \boldsymbol{g}_0, \boldsymbol{h}_0, d_0\}$ of $H(z)$, a state space representation $\{\boldsymbol{F}_Q, \boldsymbol{g}_Q, \boldsymbol{h}_Q, d\}$ of $H_Q(z)$ in Theorem 3.3.11 is given by

$$\begin{aligned} \boldsymbol{F}_Q &= \boldsymbol{I}_{\mathbf{n}} \otimes \begin{bmatrix} 0 & \beta_1 \\ 1 & -\delta_1 \end{bmatrix} - \delta_1(1+\beta_1)\boldsymbol{F}_0(\boldsymbol{I}_{\mathbf{n}} - \beta_1 \boldsymbol{F}_0)^{-1} \otimes \begin{bmatrix} 0 & 0 \\ 1 & 1 \end{bmatrix} \\ \boldsymbol{g}_Q &= (\boldsymbol{I}_{\mathbf{n}} - \beta_1 \boldsymbol{F}_0)^{-1} \boldsymbol{g}_0 \otimes \begin{bmatrix} \beta_1^2 - 1 \\ -\delta_1(1+\beta_1) \end{bmatrix} \\ \boldsymbol{h}_Q^T &= \boldsymbol{h}_0^T (\boldsymbol{I}_{\mathbf{n}} - \beta_1 \boldsymbol{F}_0)^{-1} \otimes \begin{bmatrix} 0 & 1 \end{bmatrix} \\ d_Q &= d_0 + \beta_1 \boldsymbol{h}_0^T (\boldsymbol{I}_n - \beta_1 \boldsymbol{F}_0)^{-1} \boldsymbol{g}_0 \end{aligned}$$

If, for example, $H(z)$ is a third (i.e. $n = 3$) order filter, then

$$\boldsymbol{I}_3 \otimes \begin{bmatrix} 0 & \beta_1 \\ 1 & -\delta_1 \end{bmatrix} = \begin{bmatrix} 0 & \beta_1 & 0 & 0 & 0 & 0 \\ 1 & -\delta_1 & 0 & 0 & 0 & 0 \\ 0 & 0 & 0 & \beta_1 & 0 & 0 \\ 0 & 0 & 1 & -\delta_1 & 0 & 0 \\ 0 & 0 & 0 & 0 & 0 & \beta_1 \\ 0 & 0 & 0 & 0 & 1 & -\delta_1 \end{bmatrix}$$

$$\begin{bmatrix} p_{11} & p_{12} & p_{13} \\ p_{21} & p_{22} & p_{23} \\ p_{31} & p_{32} & p_{33} \end{bmatrix} \otimes \begin{bmatrix} 0 & 0 \\ 1 & 1 \end{bmatrix} = \begin{bmatrix} 0 & 0 & 0 & 0 & 0 & 0 \\ p_{11} & p_{11} & p_{12} & p_{12} & p_{13} & p_{13} \\ 0 & 0 & 0 & 0 & 0 & 0 \\ p_{21} & p_{21} & p_{22} & p_{22} & p_{23} & p_{23} \\ 0 & 0 & 0 & 0 & 0 & 0 \\ p_{31} & p_{31} & p_{32} & p_{32} & p_{33} & p_{33} \end{bmatrix}$$

$$\begin{bmatrix} q_1 & q_2 & q_3 \end{bmatrix} \otimes \begin{bmatrix} 0 & 1 \end{bmatrix} = \begin{bmatrix} 0 & q_1 & 0 & q_2 & 0 & q_3 \end{bmatrix}$$

Low Pass to Bandstop Transformation: In this case

$$Q(z) = \frac{1 - \delta_2 z + \beta_2 z^2}{\beta_2 - \delta_2 z + z^2} = \frac{\beta_2 z^2 - \delta_2 z + 1}{z^2 - \delta_2 z + \beta_2}$$
$$= \beta_2 + \frac{(\beta_2\delta_2 - \delta_2)z + 1 - \beta_2^2}{z^2 - \delta_2 z + \beta_2}$$

and so one state space representation $\{\boldsymbol{F}_1, \boldsymbol{g}_1, \boldsymbol{h}_1, d_1\}$ of $Q(z)$ is given by

$$\boldsymbol{F}_1 = \begin{bmatrix} 0 & -\beta_2 \\ 1 & \delta_2 \end{bmatrix} \; ; \; \boldsymbol{g}_1 = \begin{bmatrix} 1 - \beta_2^2 \\ \delta_2(\beta_2 - 1) \end{bmatrix} \; ; \; \boldsymbol{h}_1 = \begin{bmatrix} 0 \\ 1 \end{bmatrix} \; ; \; d_1 = \beta_2$$

Then given a state space representation $\{\boldsymbol{F}_0, \boldsymbol{g}_0, \boldsymbol{h}_0, d_0\}$ of $H(z)$, a state space representation $\{\boldsymbol{F}_Q, \boldsymbol{g}_Q, \boldsymbol{h}_Q, d\}$ of $H_Q(z)$ in Theorem 3.3.11 is given by

$$\boldsymbol{F}_Q = \boldsymbol{I}_\mathbf{n} \otimes \begin{bmatrix} 0 & -\beta_2 \\ 1 & \delta_2 \end{bmatrix} + \delta_2(\beta_2 - 1)\boldsymbol{F}_0(\boldsymbol{I}_\mathbf{n} - \beta_2 \boldsymbol{F}_0)^{-1} \otimes \begin{bmatrix} 0 & 0 \\ 1 & 1 \end{bmatrix}$$
$$\boldsymbol{g}_Q = (\boldsymbol{I}_\mathbf{n} - \beta_1 \boldsymbol{F}_0)^{-1} \boldsymbol{g}_0 \otimes \begin{bmatrix} 1 - \beta_2^2 \\ \delta_2(\beta_2 - 1) \end{bmatrix}$$
$$\boldsymbol{h}_Q^T = \boldsymbol{h}_0^T (\boldsymbol{I}_\mathbf{n} - \beta_2 \boldsymbol{F}_0)^{-1} \otimes \begin{bmatrix} 0 & 1 \end{bmatrix}$$
$$d_Q = d_0 + \beta_2 \boldsymbol{h}_0^T (\boldsymbol{I}_n - \beta_2 \boldsymbol{F}_0)^{-1} \boldsymbol{g}_0$$

In a particular numerical example, the state space transformations enables both a transformed bandpass and a transformed bandstop filter $H_Q(z)$ to be easily determined via MATLAB.

Summary

This Chapter began by examining the properties of the frequency response function of an nth order digital filter. Filters were classified as recursive (or IIR) and nonrecursive (or FIR). A FIR filter has all characteristic roots at zero, and so are always asymptotically stable. FIR filters can be used to achieve a linear phase characteristic. A method known as windowing was used to derive a FIR filter as an approximation of a given IIR filter. A number of windowing schemes were presented. A least squares approximation method is presented later in section 4.3.1.

IIR filters were asymptotically stable if all characteristic roots were inside the unit circle. Under such conditions, an IIR filter can be used to design an all-pass filter in which the magnitude response is constant over all frequencies. An all-pass filter is useful for correcting the phase response of a given filter. A digital IIR filter can be designed by means of a transformation of a given low pass analog filter known as an analog to digital transformation. The design methodology is based on finding a difference equation whose solution approximates sampled values of the continuous time solution of the differential equation which represents the analog filter. Three methods based on the backward Euler, the forward Euler and the trapezoidal approximation of an integral were presented. Other numerical approximation schemes are explored in the exercises. When the matching of frequency response characteristics are important then the preferred method is the trapezoidal (or bilinear) transformation. The corresponding digital filter may be derived from the analog filter by direct substitution. However another approach based on state space transformations which make use of MATLAB may be more useful in particular numerical applications.

A digital IIR filter can also be designed by means of a transformation of a given digital filter known as a digital to digital transformation. Since the requirement was to maintain a one-to-one correspondence in the frequency characteristics between the two filters, the type of transformation was restricted to a particular class of all-pass transformations. In particular, a transformation was derived to transform a given low pass digital filter of order n to another low pass filter of order n having a different cut-off frequency, a high pass filter of order n, a bandpass filter of order $2n$, and a bandstop filter of order $2n$. Once again the designated transformations could be derived either by hand manipulations, or via state space transformation methods which make use of MATLAB.

Exercises

3.1 An analog signal having a maximum frequency component of 100kHz is to be filtered so as to remove frequency components in the range 20kHz to 40kHz using a digital filter.

(i) What are the minimum and maximum sampling rates that should be used ? Justify your answer.

(ii) With a sampling rate f_s given by the average of the two sampling rates determined in (i), what is the range of digital frequencies that must be eliminated ?

(iii) What is the magnitude response of the ideal filter to achieve the results ?

3.2 Consider the function

$$s(k) = \begin{cases} (-1)^k (kT)^{-1} & ; \ k \neq 0 \\ 0 & ; \ k = 0 \end{cases}$$

(i) Compute $\{s(k); \ -5 \leq k \leq 5\}$ for $T = 1$

(ii) Define the FIR filter

$$H(z) = \sum_{n=0}^{11} h(n) z^{-n}$$

where $h(k) = s(k-5)$. Plot the magnitude frequency response and explain how the FIR filter acts like a digital differentiator.

(iii) Define the signal h_w where $h_w(k) = h(k) w_h(k)$ where w_h is the Hann window. Plot the corresponding frequency response

$$H_w(z) = \sum_{n=0}^{11} h_w(n) z^{-n}$$

and compare with the results in (ii).

(iv) Repeat (iii) using the Hamming window.

3.3 What are the conditions (if possible) on the coefficients $\{a_1, a_0, \lambda, b_1, b_0\}$ so that the digital filter defined by

$$H(z) = \frac{(0.5z^2 + b_1 z + b_0)(2z - 1)}{(z^2 - a_1 z - a_0)(z - \lambda)}$$

is: (a) a FIR filter, (b) a stable IIR filter (c) a linear phase FIR filter, and (d) a stable all-pass filter.

3.4 Consider the FIR filter

$$H(z) = b_5 + b_4 z^{-1} + b_3 z^{-2} + b_2 z^{-3} + b_1 z^{-4} + b_0 z^{-5}$$

Using first principles, derive *all* conditions on the coefficients $\{b_m\}$ such that $H(z)$ defines a linear phase filter. What are the corresponding amplitude and phase functions ?

3.5 (i) Calculate the five coefficients of a linear phase FIR filter

$$H(z) = \sum_{n=0}^{4} h_n z^{-n}$$

which approximates the amplitude response

$$A(\omega) = \begin{cases} 1 \ ; \ 0 \leq \omega < 0.25\pi \\ 0 \ ; \ 0.25\pi < \omega < \pi \end{cases}$$

(ii) Calculate the modified coefficients in (i) based on: (a) a Hanning window, and (b) a Hamming window. Plot and compare your results.

3.6 Consider the ideal low pass filter

$$H(exp\{j\omega\}) = \begin{cases} 1 \ ; \ 0 \leq \omega \leq 0.5\pi \\ 0 \ ; \ 0.5\pi < \omega \leq \pi \end{cases}$$

which can be written in the form

$$H(exp\{j\omega\}) = \sum_{k=0}^{\infty} h(k) exp\{-jk\omega\} \ ; \ h(k) = \frac{1}{2\pi} \int_{-0.5\pi}^{0.5\pi} exp\{jk\omega\} d\omega$$

(i) Show that

$$h(k) = \begin{cases} (k\pi)^{-1} \sin 0.5k\pi \ ; \ k \neq 0 \\ 0.5 \ ; \ k = 0 \end{cases}$$

(ii) Using MATLAB, investigate the use of the Rectangular, Hann, Hamming, Bartlett Triangular, and Blackmann windows for achieving FIR approximations $\hat{H}_N(exp\{j\omega\})$ of $H(exp\{j\omega\})$ where

$$\hat{H}_N(exp\{j\omega\}) = \sum_{k=0}^{N-1} h(k) w(k) exp\{-jk\omega\}$$

3.7 Find the amplitude and the (discontinuous) phase function of the digital filter with frequency function H given by

$$H(z) = \left(\frac{0.7z + 1}{z + 0.7}\right) \left(\frac{1}{z - 0.1}\right)$$

3.8 The frequency function H of a digital filter is given by

$$H(z) = \left(\frac{z + 0.1}{z + 0.2}\right) \left(\frac{1}{z - 0.1}\right)$$

Find the steady state response when the input signal u is given by

$$u(k) = \cos 0.2k$$

3.9 Consider the second order digital filter

$$y(k+2) = 2r_0 \cos\theta_0 y(k+1) + r_0^2 y(k) + u(k+2) - 2\cos\theta_0 u(k+1) + u(k)$$

(i) What are the conditions on $\{r_0, \theta_0\}$ for stability ?
(ii) Find values for $\{r_0, \theta_0\}$ so as to produce a stable *digital notch filter* at $\omega = \pi/4$ rads.

3.10 Consider a digital filter with frequency response function given by

$$H(z) = \frac{b_3 z^3 + b_2 z^2 + b_1 z + b_0}{z^3 + 0.1z^2 + 0.2z + 0.3}$$

(i) Show that the digital filter is stable
(ii) Find the coefficients $\{b_k\}$ so that $H(z)$ defines an all-pass filter.

3.11 A low pass digital filter has a unit impulse response h given by

$$h(k) = (0.9)^k - (0.8)^k; \; n \geq 0$$

Find (i) the unit step response, and (ii) the steady state response due to the input u where $u(k) = \cos 0.25\pi k$

3.12 A first order digital notch filter is given by

$$y(k+2) = -0.90y(k) + u(k+2) + u(k)$$

Find the response due to the input signal u where $u(k) = \alpha_0 + \alpha_1 \cos 0.5\pi k$.

3.13 A double bandpass filter has a frequency response $H(z)$ given by

$$H(z) = \frac{z^4 - 2z^2 + 1}{z^4 + 0.9025z^2 + 0.815}$$

Show that $z^4 + 0.9025z^2 + 0.815 = 0$ has the roots $\{0.95exp\{\pm j\pi/3\}, 0.95exp\{\pm j2\pi/3\}\}$. Plot the amplitude and phase characteristics of $H(z)$ for $0 \leq \omega < \pi$.

3.14 (i) Show that a digital filter $H(z)$ where

$$H(z) = \frac{z^2 - 1}{(z + 0.47 + j0.814)(z + 0.47 - j0.814)}$$

acts as a bandpass filter. What is the digital centre frequency ? What defines the bandpass frequency limits ?

(ii) Suppose the bandpass filter is to filter a sampled analog signal based on a sampling rate of 10kHz. What is corresponding centre frequency of the analog signal which is to be bandpass filtered ?

3.15 Two second order digital filters are defined by

$$\begin{aligned}&(a)\ \ y(k+2) = -1.8y(k+1) - 0.81y(k) + u(k+2) - 2u(k+1) + u(k)\\&(b)\ \ y(k+2) = 1.8y(k+1) - 0.81y(k) + u(k+2) + 2u(k+1) + u(k)\end{aligned}$$

In each case, find:
(i) The unit impulse response
(ii) The steady state response to the input signal u where

$$u(k) = \alpha_0 + \alpha_1 \sin 0.5\pi k + \alpha_2 \sin \pi k$$

Using the results in (ii), conclude which filter describes a low pass filter and which filter describes a high pass filter.

3.16 An analog filter $\mathcal{H}(s)$ is described by the analog state space equations

$$\begin{aligned}\dot{\boldsymbol{x}}(t) &= \begin{bmatrix} -2 & -2 \\ 1 & 0 \end{bmatrix} \boldsymbol{x}(t) + \begin{bmatrix} 10 \\ 0 \end{bmatrix} v(t)\\ z(t) &= \begin{bmatrix} 1 & -1 \end{bmatrix} \boldsymbol{x}(t)\end{aligned}$$

(i) Find the analytical expression for $\{z(t); t \geq 0\}$ when $\{v(t) = 0; t \geq 0\}$ for all $t \geq 0$ and $\boldsymbol{x}^T(0) = [2 \ \ 4]$.
(ii) What is the frequency response function $\mathcal{H}(s)$?
(iii) Find the differential equation for z for $t \geq 0$ in terms of the input v when $\boldsymbol{x}(0) = \mathbf{0}$.

3.17 Two digital filters are described by the difference equations

$$\begin{aligned}y(k+2) &= -0.81y(k) + u(k+2) - u(k)\\ y(k+2) &= 0.81y(k) + u(k+2) + u(k)\end{aligned}$$

In each case, find the steady state response for the input signal u where

$$u(k) = 8 + 10\cos 0.5\pi k - 2\cos(\pi k + 0.25\pi)$$

What conclusions can you draw from the characteristics of each of these two filters ?

3.18 (i) Plot the unit impulse response h when:

(a) $h(k) = (0.9)^k;\ k \geq 0$,
(b) $h(k) = (-0.9)^k;\ k \geq 0$, and
(c) $h(k) = (0.9)^k \cos 0.5\pi k;\ k \geq 0$.

(ii) Plot the magnitude response for each of the filters in (i).
(iii) Which filter acts as a low pass, a highpass, a bandpass or a bandstop filter ?

3.19 Suppose

$$H_e(z) = Tc^T(I - TA)^{-1}b + Tc^T[zI - (I - TA)^{-1}]^{-1}(I - TA)^{-2}b$$

Show by algebraic manipulations that

$$H_e(z) = c^T[T^{-1}(1 - z^{-1})I - A]^{-1}b$$

Hint: Begin using the fact that

$$zI - (I - TA)^{-1} = (I - TA)^{-1}[z(I - TA) - I]$$

3.20 Consider the second order analog filter

$$\ddot{z}(t) = \alpha_1 \dot{z}(t) + \alpha_0 z(t) + \beta_0 v(t) + \beta_1 \dot{v}(t)$$

and suppose

$$s^2 - \alpha_1 s - \alpha_0 = (s - \lambda_1)(s - \lambda_2)$$

(i) Show that an analog state space representation

$$\dot{x}(t) = Ax(t) + bv(t) \ ; \ z(t) = c^T x(t)$$

is given by

$$A = \begin{bmatrix} 0 & \alpha_0 \\ 1 & \alpha_1 \end{bmatrix} \ ; \ b = \begin{bmatrix} \beta_0 \\ \beta_1 \end{bmatrix} \quad c = \begin{bmatrix} 0 \\ 1 \end{bmatrix}$$

(ii) Show when $\{\dot{v}(t_0) = v(t_0) = 0\}$ that the initial state $x(t_0)$ of the analog state space representation is related to the initial conditions $\{z(t_0), \dot{z}(t_0)\}$ of the analog filter by

$$x(t_0) = \begin{bmatrix} z(t_0) \\ \dot{z}(t_0) - \alpha_1 z(t_0) \end{bmatrix}$$

This representation is referred to as an *analog phase variable form.*

(iii) Suppose $\lambda_1 \neq \lambda_2$ are real. Show that an analog state space representation of the analog filter is given by

$$A = \begin{bmatrix} \lambda_1 & 0 \\ 0 & \lambda_2 \end{bmatrix} \ ; \ c = \begin{bmatrix} 1 \\ 1 \end{bmatrix} \ ; \ b = \frac{1}{\lambda_1 - \lambda_2} \begin{bmatrix} 1 & -\lambda_2 \\ -1 & \lambda_1 \end{bmatrix} \begin{bmatrix} \beta_0 + \alpha_1 \beta_1 \\ \beta_1 \end{bmatrix}$$

(iv) Show when $\{\dot{v}(t_0) = v(t_0) = 0\}$ in part (iii) that the initial state $x(t_0)$ is given by

$$x(t_0) = \frac{1}{\lambda_1 - \lambda_2} \begin{bmatrix} 1 & -\lambda_2 \\ -1 & \lambda_1 \end{bmatrix} \begin{bmatrix} \dot{z}(t_0) \\ z(t_0) \end{bmatrix}$$

(v) Show when $\lambda_1 = \lambda_2 = \lambda_0$ that an analog state space representation is given by

$$\boldsymbol{A} = \begin{bmatrix} \lambda_0 & 1 \\ 0 & \lambda_0 \end{bmatrix} \quad ; \quad \boldsymbol{c} = \begin{bmatrix} 1 \\ 0 \end{bmatrix} \quad ; \quad \boldsymbol{b} = \begin{bmatrix} 0 & 1 \\ 1 & -\lambda_0 \end{bmatrix} \begin{bmatrix} \beta_0 + \alpha_1\beta_1 \\ \beta_1 \end{bmatrix}$$

(vi) Show when $\{\dot{v}(t_0) = v(t_0) = 0\}$ in part (v) that

$$\boldsymbol{x}(t_0) = \begin{bmatrix} z(t_0) \\ \dot{z}(t_0) - \lambda_0 z(t_0) \end{bmatrix}$$

(vii) Show when $\lambda_1 = \overline{\lambda}_2 = \rho + j\omega$ is complex, the an analog state space representation is given by

$$\boldsymbol{A} = \begin{bmatrix} \rho & \omega \\ -\omega & \rho \end{bmatrix} \quad ; \quad \boldsymbol{c} = \begin{bmatrix} 0 \\ 1 \end{bmatrix} \quad ; \quad \boldsymbol{b} = \frac{1}{\omega} \begin{bmatrix} -1 & \rho \\ 0 & \omega \end{bmatrix} \begin{bmatrix} \beta_0 + \alpha_1\beta_1 \\ \beta_1 \end{bmatrix}$$

(viii) Show in part (vii) that when $\{\dot{v}(t_0) = v(t_0) = 0\}$, then

$$\boldsymbol{x}(t_0) = \frac{1}{\omega} \begin{bmatrix} -\dot{z}(t_0) + \rho z(t_0) \\ \omega z(t_0) \end{bmatrix}$$

3.21 Show that an analog state space representation for the following second order analog filters

(a) $\ddot{z}(t) = -2\dot{z}(t) - 1.04z(t) + \dot{v}(t) \quad ; \quad t \geq 0$
(b) $\ddot{z}(t) = -\dot{z}(t) - 0.25z(t) + 3v(t) \quad ; \quad t \geq 0$

are given respectively by

$$(a) \quad \boldsymbol{A} = \begin{bmatrix} -1 & 0.2 \\ -0.2 & -1 \end{bmatrix} \quad ; \quad \boldsymbol{c} = \begin{bmatrix} 0 \\ 1 \end{bmatrix} \quad ; \quad \boldsymbol{b} = \begin{bmatrix} 5 \\ 1 \end{bmatrix}$$

and

$$(b) \quad \boldsymbol{A} = \begin{bmatrix} -0.5 & 1 \\ 0 & -0.5 \end{bmatrix} \quad ; \quad \boldsymbol{c} = \begin{bmatrix} 1 \\ 1 \end{bmatrix} \quad ; \quad \boldsymbol{b} = \begin{bmatrix} -3 \\ 3 \end{bmatrix}$$

3.22 (i) Suppose an object is travelling at *constant velocity* V has position $z(t)$ at time t is given by the signal model

$$z(t) = Vt + D$$

where D is the position measured with respect to some reference at time $t = 0$. Define

$$x_1(t) = z(t) \ ; \ x_2(t) = V$$

(i) Show that an analog state space representation of the constant velocity signal model is given by

$$\boldsymbol{A} = \begin{bmatrix} 0 & 1 \\ 0 & 0 \end{bmatrix} \ ; \ \boldsymbol{c} = \begin{bmatrix} 1 \\ 0 \end{bmatrix} \ ; \ \boldsymbol{x}(t) = \begin{bmatrix} x_1(t) \\ x_2(t) \end{bmatrix} \ ; \ \boldsymbol{x}(0) = \begin{bmatrix} D \\ V \end{bmatrix}$$

[Hence the initial position D and the constant velocity V are determined by the choice of the initial state $\boldsymbol{x}(0)$.]

(ii) Suppose now the object at position $z(t)$ at time t is travelling at *constant acceleration* A where $\{V, D\}$ are respectively the velocity and position at time $t = 0$. Show that a state space representation of this constant acceleration signal model is given by

$$\boldsymbol{A} = \begin{bmatrix} 0 & 1 & 0 \\ 0 & 0 & 1 \\ 0 & 0 & 0 \end{bmatrix} \; ; \; \boldsymbol{c} = \begin{bmatrix} 1 \\ 0 \\ 0 \end{bmatrix} \; ; \; \boldsymbol{x}(t) = \begin{bmatrix} x_1(t) \\ x_2(t) \\ x_3(t) \end{bmatrix} \; ; \; \boldsymbol{x}(0) = \begin{bmatrix} D \\ V \\ A \end{bmatrix}$$

[Hence the initial position D, the initial velocity V, and the constant acceleration A are determined by the choice of the initial state $\boldsymbol{x}(0)$.]

3.23 Consider an analog sinusoidal signal z given by

$$z(t) = A_1 \cos(\Omega_1 t + \phi_1) \; ; \; A_1 \neq 0, \; \Omega_1 > 0$$

(i) Show that z is given by the solution of the second order differential equation

$$\ddot{z}(t) + \Omega_1^2 z(t) = 0 \; ; \; t \geq 0$$

with initial conditions $\{z(0), \dot{z}(0)\}$ given by

$$z(0) = A_1 \cos\phi_1 \; ; \; \dot{z}(0) = -A_1 \Omega_1 \sin\phi_1$$

(ii) Show that a zero input analog state space representation of the signal z is given by

$$\boldsymbol{A} = \begin{bmatrix} 0 & \Omega_1 \\ -\Omega_1 & 0 \end{bmatrix} \; ; \; \boldsymbol{c} = \begin{bmatrix} 1 \\ 1 \end{bmatrix}$$

where

$$\boldsymbol{x}(0) = \frac{A_1}{2\Omega_1} \begin{bmatrix} -\cos\phi_1 - \Omega_1^2 \sin\phi_1 \\ \cos\phi_1 - \Omega_1^2 \sin\phi_1 \end{bmatrix}$$

[Hence if the frequency Ω_1 is fixed, then the amplitude A_1 and the phase ϕ_1 of the sinusoidal signal z are determined by the choice of the initial state $\boldsymbol{x}(0)$.]

3.24 Show that the third order analog filter

$$z^{(3)}(t) + 0.8\ddot{z}(t) + 1.20\dot{z}(t) + 0.416z(t) = v(t) - 0.3\dot{v}(t) + 2\ddot{v}(t) \quad ; \quad t \geq 0$$

where $\{z(0) = 0 \; ; \; \dot{z}(0) = -1 \; ; \; \ddot{z}(0) = 1.5\}$ and $\{v(0) = \dot{v}(0) = \ddot{v}(0) = 0\}$ has an analog state space representation given by

$$\boldsymbol{A} = \begin{bmatrix} 0 & 0 & -0.416 \\ 1 & 0 & -1.20 \\ 0 & 1 & -0.80 \end{bmatrix} \; ; \; \boldsymbol{b} = \begin{bmatrix} 1 \\ -0.3 \\ 2 \end{bmatrix} \; ; \; \boldsymbol{c} = \begin{bmatrix} 0 \\ 0 \\ 1 \end{bmatrix}$$

where the initial state $\boldsymbol{x}(0)$ is given by $\boldsymbol{x}^T(0) = [0.7 \;\; -1 \;\; 0]$.

3.25 Consider a mechanical system consisting of two masses with moments of inertia J_1 and J_2 connected by a spring with torque constant κ and viscous damping constant d. The equations of motion are described by

$$\begin{aligned} J_1\ddot{\theta}_1(t) + d[\dot{\theta}_1(t) - \dot{\theta}_2(t)] + \kappa[\theta_1(t) - \theta_2(t)] &= v(t) \\ J_2\ddot{\theta}_2(t) + d[\dot{\theta}_2(t) - \dot{\theta}_1(t)] + \kappa[\theta_2(t) - \theta_1(t)] &= 0 \end{aligned}$$

where v is the applied input torque. Define

$$x_1 \triangleq \theta_1 \;\; ; \;\; x_2 \triangleq \dot{\theta}_1 \;\; ; \;\; x_3 \triangleq \theta_2 \;\; ; \;\; x_4 \triangleq \dot{\theta}_2$$

(i) Show that a state space representation is given by

$$\boldsymbol{A} = \begin{bmatrix} 0 & 1 & 0 & 0 \\ -\kappa J_1^{-1} & -dJ_1^{-1} & \kappa J_1^{-1} & dJ_1^{-1} \\ 0 & 0 & 0 & 1 \\ \kappa J_2^{-1} & dJ_2^{-1} & -\kappa J_2^{-1} & -dJ_2^{-1} \end{bmatrix} \; ; \; \boldsymbol{b} = \begin{bmatrix} 0 \\ J_1^{-1} \\ 0 \\ 0 \end{bmatrix} \; ; \; \boldsymbol{c} = \begin{bmatrix} 0 \\ 0 \\ 1 \\ 0 \end{bmatrix}$$

(ii) Show that the fourth order differential equation between the output (angle) signal $y = \theta_2 = x_3$ and the input (torque) signal v is given by

$$\begin{aligned} y^{(4)}(t) &= \alpha_3 y^{(3)}(t) + \alpha_2\ddot{y}(t) + \alpha_1\dot{y}(t) + \alpha_0 y(t) \\ &\quad + \beta_0 v(t) + \beta_1\dot{v}(t) + \beta_2\ddot{v}(t) + \beta_3 v^{(3)}v(t) \end{aligned}$$

where

$$\alpha_3 = -[1 + d(J_1^{-1} + J_2^{-1})] \;\; ; \;\; \alpha_2 = -(d+\kappa)J_1^{-1} \;\; ; \;\; \alpha_1 = \alpha_0 = 0$$

$$\beta_3 = \beta_2 = 0 \;\; ; \;\; \beta_1 = dJ_1^{-1}J_2^{-1} \;\; ; \;\; \beta_0 = (\kappa + d)J_1^{-1}J_2^{-1}$$

3.26 (i) The *forward approximation* for the derivative $\dot{z}$ at the time instant $t = kT$ (k an integer and T a fixed constant) is given by

$$\dot{z}(t = kT) \simeq \frac{y(k+1) - y(k)}{T} \;\; ; \;\; y(k) \triangleq z(t = kT)$$

Based on this approximation, show that the second derivative $\ddot{z}$ at time $t = kT$ can be approximated by

$$\ddot{z}(t = kT) \simeq \frac{y(k+2) - 2y(k+1) + y(k)}{T^2}$$

(ii) Using part (i), show that the solution $z(t = kT)$ of the second order analog filter

$$\ddot{z}(t) = \alpha_1\dot{z}(t) + \alpha_0 z(t) + \beta_0 v(t) + \beta_1\dot{v}(t)$$

is given approximately by the solution $y_1(k)$ of the second order digital filter

$$y_1(k+2) = a_1 y_1(k+1) + a_0 y_1(k) + b_0 u(k) + b_1 u(k+1)$$

where

$$a_1 = \alpha_1 T + 2 \;\; ; \;\; a_0 = \alpha_0 T^2 - \alpha_1 T - 1 \;\; ; \;\; b_0 = \beta_0 T^2 - \beta_1 T \;\; ; \;\; b_1 = \beta_1 T$$

3.27 (i) The *backward approximation* of the derivative $\dot{z}$ is given by

$$\dot{z}(t = kT) \simeq \frac{y(k) - y(k-1)}{T} \; ; \; y(k) \triangleq y(t = kT)$$

Based on this approximation, show that the second derivative $\ddot{z}$ at time $t = kT$ can be approximated by

$$\ddot{z}(t = kT) \simeq \frac{y(k+1) - 2y(k) + y(k-1)}{T^2}$$

(ii) Using part (i), show that the solution $z(t = kT)$ of the second order analog filter in exercise 3.25 is approximately given by the solution $y_2(k)$ of the second order digital filter

$$y_2(k+1) = a_1 y_2(k) + a_0 y_2(k-1) + b_0 u(k-1) + b_1 u(k)$$

where

$$a_1 = \alpha_0 T^2 + \alpha_1 T + 2 \; ; \; a_0 = -(\alpha_1 T + 1)$$
$$b_0 = -\beta_1 T \; ; \; b_1 = \beta_0 T^2 + \beta_1 T$$

3.28 Suppose a state space representation of an analog filter $\mathcal{H}(s)$ is given by

$$\dot{\boldsymbol{x}}(t) = \boldsymbol{A}\boldsymbol{x}(t) + \boldsymbol{b}v(t) \; ; \; z(t) = \boldsymbol{c}^T\boldsymbol{x}(t)$$

(i) Using the forward approximation for the derivative $\dot{\boldsymbol{x}}(t)$ of $\boldsymbol{x}(t)$ at the time instants $t = kT$ given by

$$\dot{\boldsymbol{x}}(t = kT) \simeq \frac{\boldsymbol{x}(k+1) - \boldsymbol{x}(k)}{T} \; ; \; \boldsymbol{x}(k) \triangleq \boldsymbol{x}(t = kT)$$

show that the solution $z(t = kT)$ is given approximately by $y_1(k)$ and $\boldsymbol{x}(t = kT)$ by $\boldsymbol{x}_1(k)$ where

$$\boldsymbol{x}_1(k+1) = \boldsymbol{F}_1\boldsymbol{x}_1(k) + \boldsymbol{g}_1 u(k) \; ; \; y_1(k) = \boldsymbol{c}^T\boldsymbol{x}_1(k) + du(k)$$

where

$$\boldsymbol{F}_1 = \boldsymbol{I} + T\boldsymbol{A} \; ; \; \boldsymbol{g}_1 = T\boldsymbol{b}$$

3.29 (i) From the definition of the exponential function $exp\{\boldsymbol{A}t\}$, show that

$$\frac{d}{dt}(exp\{\boldsymbol{A}t\}) = \boldsymbol{A}exp\{\boldsymbol{A}t\} = exp\{\boldsymbol{A}t\}\boldsymbol{A}$$

(ii) By direct substitution, show that the solution of the analog state space equations

$$\dot{\boldsymbol{x}}(t) = \boldsymbol{A}\boldsymbol{x}(t) + \boldsymbol{b}v(t); \; z(t) = \boldsymbol{c}^T\boldsymbol{x}(t) + dv(t)$$

for $t \geq t_0$ beginning with the initial state $\boldsymbol{x}(t_0)$ at time $t = t_0$ is given by the *variations of constants formula*

$$\boldsymbol{x}(t) = exp\{\boldsymbol{A}(t - t_0)\}\boldsymbol{x}(t_0) + \int_{t_0}^{t} exp\{\boldsymbol{A}(t - \sigma)\boldsymbol{b}v(\sigma)d\sigma$$

(iii) Using (ii), show that the state $\boldsymbol{x}(t_{k+1})$ at time $t = t_{k+1}$ is given exactly in terms of the state $\boldsymbol{x}(t_k)$ at time $t = t_k$ by

$$\boldsymbol{x}(t_{k+1}) = exp\{\boldsymbol{A}(t_{k+1} - t_k)\}\boldsymbol{x}(t_k) + \int_{t_k}^{t_{k+1}} exp\{\boldsymbol{A}(t_{k+1} - \sigma)\}\boldsymbol{b}v(\sigma)d\sigma$$

$$z(t_k) = \boldsymbol{c}^T\boldsymbol{x}(t_k) + dv(t_k)$$

3.30 (i) Using the results in exercise 3.28, show that an approximation $y_2(k)$ of $z(t = kT)$ based on using the *backward Euler* approximation of the integral with $u(k) \triangleq v(kT)$ is given by

$$\boldsymbol{x}_2(k+1) = \boldsymbol{F}_2\boldsymbol{x}_2(k) + \boldsymbol{g}_2u(k) \; ; \; y_2(k) = \boldsymbol{c}^T\boldsymbol{x}_2(k) + d_2u(k)$$

where

$$\boldsymbol{F}_2 \triangleq exp\{\boldsymbol{A}T\} \; ; \; \boldsymbol{g}_2 \triangleq Texp\{\boldsymbol{A}T\}\boldsymbol{b} \; ; \; d_2 = 0 \tag{3.166}$$

(ii) Show that up to order T

$$\boldsymbol{F}_2 \simeq \boldsymbol{I} + T\boldsymbol{A} \quad ; \quad \boldsymbol{g}_2 \simeq T\boldsymbol{b}$$

3.31 (i) Using the results in exercise 3.28, show that an approximation based on using the *forward Euler* approximation of an integral results in the approximation $y_3(k)$ of $z(kT)$ as given by

$$\tilde{\boldsymbol{x}}_3(k+1) = \boldsymbol{F}_3\tilde{\boldsymbol{x}}_3(k) + \boldsymbol{g}u(k+1) \; ; \; y_3(k) = \boldsymbol{c}^T\tilde{\boldsymbol{x}}_3(k)$$

where

$$\boldsymbol{F}_3 \triangleq exp\{\boldsymbol{A}T\} \; ; \; \boldsymbol{g} \triangleq T\boldsymbol{b}$$

(ii) In order to reduce the equation in part (i) into the standard state space form, define $\boldsymbol{x}_3(k) \triangleq \tilde{\boldsymbol{x}}_3(k) - \boldsymbol{g}u(k)$, and show

$$\boldsymbol{x}_3(k+1) = \boldsymbol{F}_3\boldsymbol{x}_3(k) + \boldsymbol{g}_3u(k) \; ; \; y_3(k) = \boldsymbol{c}^T\boldsymbol{x}_3(k) + d_3u(k)$$

where

$$\boldsymbol{F}_3 \triangleq exp\{\boldsymbol{A}T\} \quad ; \quad \boldsymbol{g}_3 \triangleq Texp\{\boldsymbol{A}T\}\boldsymbol{b} \quad ; \quad d_3 \triangleq T\boldsymbol{c}^T\boldsymbol{b}$$

(iii) Show that up to order T

$$\boldsymbol{F}_3 \simeq \boldsymbol{I} + T\boldsymbol{A} \quad ; \quad \boldsymbol{g}_3 \simeq T\boldsymbol{b}$$

3.32 (i) Using the *trapezoidal approximation* of the integral in the *variations of constants formula* in exercise 3.28, show that an approximate solution $y_4(k)$ of $z(t = kT)$ is given by

$$\tilde{\boldsymbol{x}}_4(k+1) = \boldsymbol{F}_4\tilde{\boldsymbol{x}}_4(k) + \boldsymbol{g}_0 u(k) + \boldsymbol{g}_1 u(k+1) \ ; \ y_4(k) = \boldsymbol{c}^T\tilde{\boldsymbol{x}}_4(k)$$

where

$$\boldsymbol{F}_4 \triangleq exp\{\boldsymbol{A}T\} \ ; \ \boldsymbol{g}_0 \triangleq 0.5Texp\{\boldsymbol{A}T\}\boldsymbol{b} \ ; \ \boldsymbol{g}_1 \triangleq 0.5T\boldsymbol{b}$$

(ii) In order to obtain the standard state space form in (i), define $\boldsymbol{x}_4(k) \triangleq \tilde{\boldsymbol{x}}_4(k) - \boldsymbol{g}_1 u(k)$, and show

$$\boldsymbol{x}_4(k+1) = \boldsymbol{F}_4\boldsymbol{x}_4(k) + \boldsymbol{g}_4 u(k) \ ; \ y_4(k) = \boldsymbol{c}^T\boldsymbol{x}_4(k) + d_4 u(k)$$

where

$$\boldsymbol{F}_4 \triangleq exp\{\boldsymbol{A}T\} \ ; \ \boldsymbol{g}_4 \triangleq Texp\{\boldsymbol{A}T\}\boldsymbol{b} \ ; \ d_4 \triangleq 0.5T\boldsymbol{c}^T\boldsymbol{b}$$

(iii) Show that up to order T

$$\boldsymbol{F}_4 \simeq \boldsymbol{I} + T\boldsymbol{A} \quad ; \quad \boldsymbol{g}_4 \simeq T\boldsymbol{b}$$

3.33 [As we have seen, discrete approximations of the solution z of an nth order differential equation can be developed based on the numerical approximation of an integral. We now use the *variations of constants formula* as developed in exercise 3.28 to obtain a discrete approximation of z by determining an exact expression for the integral based on an approximation of the input signal v.]

One approximation $\hat{v}$ of the input signal v is obtained by approximating v by means of a piecewise constant signal of the form

$$\hat{v}(t) = \sum_k u(m)p_T(t - mT) \quad ; \quad u(m) \triangleq v(mT)$$

where the *pulse signal* p_T is defined by

$$p_T(t) \triangleq \begin{cases} 1 \ ; \ 0 \le t \le T \\ 0 \ \text{ otherwise} \end{cases}$$

(i) Show that when $t_k = kT$, an approximation $y_5(k)$ of $z(t = kT)$ with $u(k) \triangleq v(kT)$ is given by

$$\boldsymbol{x}_5(k+1) = \boldsymbol{F}_5\boldsymbol{x}_5(k) + \boldsymbol{g}_5 u(k) \ ; \ y_5(k) = \boldsymbol{c}^T\boldsymbol{x}_5(k) + du(k)$$

where

$$\boldsymbol{F}_5 \triangleq exp\{\boldsymbol{A}T\} \ ; \ \boldsymbol{g}_5 \triangleq \int_0^T exp\{\boldsymbol{A}(T-\sigma)\}\boldsymbol{b}d\sigma$$

(ii) Show that up to order T

$$\boldsymbol{F}_5 \simeq \boldsymbol{I} + T\boldsymbol{A} \quad ; \quad \boldsymbol{g}_5 \simeq T\boldsymbol{b}$$

(iii) Show that when $\boldsymbol{A}^{-1}$ exists, then

$$\boldsymbol{g}_5 = \boldsymbol{A}^{-1}(exp\{\boldsymbol{A}T\} - \boldsymbol{I}_{\mathbf{n}})\boldsymbol{b}$$

3.34 Consider the first order analog filter $\mathcal{H}(s)$ defined by

$$\dot{z}(t) = -4z(t) + 4v(t) \quad ; \quad t \geq 0$$

with $z(0) = -1$.

(i) Using the results in exercises 3.29 - 3.32, find four first order digital filters $H(z)$ which approximate $\mathcal{H}(s)$. Are they all different ?

(ii) Plot the responses of these filters and compare with the exact response $z(kT)$ beginning at the initial condition $z(0) = 0$ for $T = 0.1, 0.5$ when $\{v(t) = 5; t \geq 4\}$.

(iii) Plot the magnitude response function of these four digital filters for $T = 0.1, 0.5$, and compare with the analog magnitude response function.

3.35 Consider an analog filter $\mathcal{H}(s)$ as given by

$$\mathcal{H}(s) = \frac{4}{s + 0.1} + \frac{1}{s + 0.5}$$

(i) Based on a Jordan analog state space representation, and using the results in exercises 3.29 - 3.32, find four first order digital filters $H(z)$ which approximate $\mathcal{H}(s)$. Are they all different ?

(ii) Plot the responses of these filters and compare with the exact response $z(kT)$ beginning at the initial condition $z(0) = 0$ for $T = 0.1, 0.5$ when $v(t) = 5; t \geq 4$.

(iii) Plot the magnitude response function of these four digital filters for $T = 0.1, 0.5$, and compare with the analog magnitude response function.

3.36 Suppose

$$H(z) = d_{e3} + \boldsymbol{h}_{e3}^T(z\boldsymbol{I} - \boldsymbol{F}_{e3})^{-1}\boldsymbol{g}_{e3}$$

where $\{\boldsymbol{F}_{e3}, \boldsymbol{g}_{e3}, \boldsymbol{h}_{e3}, d_{e3}\}$ are given by (3.138). Show that

$$H(z) = d_{e3} + \tilde{\boldsymbol{h}}_{e3}^T(z\boldsymbol{I} - \tilde{\boldsymbol{F}}_{e3})^{-1}\tilde{\boldsymbol{g}}_{e3}$$

where $\{\tilde{\boldsymbol{F}}_{e3}, \tilde{\boldsymbol{g}}_{e3}, \tilde{\boldsymbol{h}}_{e3}\}$ are given by (3.139).
Hint: Consider a state space transformation $\boldsymbol{P}$ where

$$\boldsymbol{P} = \frac{1}{\sqrt{2\gamma}}(\boldsymbol{I} - \gamma\boldsymbol{A})$$

3.37 Given a function $Q^{-1}(z)$, a digital filter $H_Q(z)$ may be obtained by a transformation of a given digital filter $H(z)$ according to $H_Q(z) = H(Q^{-1}(z))$. In this exercise, we consider the digital to digital transformation as defined by

$$Q^{-1}(z) = \gamma z \; ; \; \gamma > 0$$

(i) Suppose

$$H(z) = \frac{3}{z - 0.5}$$

Find the transformed filter $H_Q(z)$.

(ii) Show that a state space representation $\{\boldsymbol{F}_\gamma, \boldsymbol{g}_\gamma, \boldsymbol{h}_\gamma, d_\gamma\}$ of $H_Q(z)$ when $H(z) = d + \boldsymbol{h}^T(z\boldsymbol{I} - \boldsymbol{F})^{-1}\boldsymbol{g}$ is given by

$$\boldsymbol{F}_\gamma = \boldsymbol{F} \; ; \; \boldsymbol{g}_\gamma = \sqrt{\gamma}\boldsymbol{g} \; ; \; \boldsymbol{h}_\gamma = \sqrt{\gamma}\boldsymbol{h} \; ; \; d_\gamma = d$$

(iii) Suppose the filter $H(z)$ has unit impulse response h. Show that the transformed filter $H_Q(z)$ has unit impulse response h_Q where $h_Q(k) = \gamma^k h(k)$. Conclude that the transformation $Q(z)$ corresponds to a *discrete time scaling* of the given filter $H(z)$.

(iv) Suppose $H(z)$ is asymptotically stable. Show that the transformed filter $H_Q(z)$ is asymptotically stability when $0 < \gamma < 1$.

(v) Given any filter $H(z)$, what are necessary and sufficient conditions on the parameter γ in terms of the properties of $H(z)$ such that the transformed filter $H_Q(z)$ is asymptotically stability ?

3.38 Show that an nth order Butterworth filter $\mathcal{H}(s)$ with cut-off frequency $\Omega_c = 1$ rad/sec for $n = 3, 4, 5, 6$ is given by

$$\mathcal{H}(s) = \frac{1}{p_n(s)}$$

where

$$\begin{aligned}
p_3(s) &= (s^2 + s + 1)(s + 1) \\
p_4(s) &= (s^2 + 0.76536s + 1)(s^2 + 1.84776s + 1) \\
p_5(s) &= (s + 1)(s^2 + 0.6180s + 1)(s^2 + 1.6180s + 1) \\
p_6(s) &= (s^2 + 0.5176s + 1)(s^2 + \sqrt{2}s + 1)(s^2 + 1.9318s + 1)
\end{aligned}$$

3.39 (i) Design a second order digital low pass Butterworth filter so that $\omega_c = 0.25\pi$ rads.

(ii) What is the corresponding stopband frequency ω_s ?

(iii) What is the level of ripple in the passband ?

(iv) Repeat (i)-(iii) for a third order digital low pass Butterworth filter and compare the results.

3.40 Suppose $\mathcal{H}(s) = \boldsymbol{c}^T(s\boldsymbol{I} - \boldsymbol{A})^{-1}\boldsymbol{b}$ is a low pass filter.

(i) Show that a state space representation of a low pass filter $\mathcal{H}_{\ell p}(s) = \mathcal{H}(\Omega_c^{-1}s)$ in (3.82) is given by

$$\mathcal{H}_{\ell p}(s) = \boldsymbol{c}^T(s\boldsymbol{I} - \boldsymbol{A}_{\ell p})^{-1}\boldsymbol{b}_{\ell p} \ ; \ \boldsymbol{A}_{\ell p} = \Omega_c \boldsymbol{A} \ , \ \boldsymbol{b}_{\ell p} = \Omega_c \boldsymbol{b}$$

(ii) Can you find a state space representation of the high pass filter $\mathcal{H}_{hp}(s) = \mathcal{H}(\Omega_c s^{-1})$ in terms of $\{\boldsymbol{A}, \boldsymbol{b}, \boldsymbol{c}\}$? If not, then why not ?

(iii) Find a state space representation for the transformed analog filter $\mathcal{H}_T(s)$ in terms of $\{\boldsymbol{A}, \boldsymbol{b}, \boldsymbol{c}\}$ where

$$\mathcal{H}_T(s) = \mathcal{H}(\mathcal{Q}^{-1}(s)) \ ; \ \mathcal{Q}(s) = \frac{s}{\Omega_c(1 + \alpha s)}$$

Relate the result to your answer in (ii).

3.41 Suppose $\mathcal{H}(s) = \boldsymbol{c}^T(s\boldsymbol{I} - \boldsymbol{A})^{-1}\boldsymbol{b}$ a low pass filter.

(i) Find a state space representation of the band pass filter $\mathcal{H}_{bp}(s)$ in (3.90) where

$$\mathcal{H}_{bp}(s) = \mathcal{H}(\mathcal{Q}^{-1}(s)) \ ; \ \mathcal{Q}(s) = \frac{Ws}{s^2 + \Omega_0^2}$$

(ii) Can you find a state space representation of the band stop filter $\mathcal{H}_{bs}(s)$ in (3.93) where

$$\mathcal{H}_{bs}(s) = \mathcal{H}(\mathcal{Q}^{-1}(s)) \ ; \ \mathcal{Q}(s) = \frac{s^2 + \Omega_0^2}{Ws}$$

If not, then why not ?

(iii) Find a state space representation of the transformed filter $\mathcal{H}_T(s)$ in terms of $\{\boldsymbol{A}, \boldsymbol{b}, \boldsymbol{c}\}$ where

$$\mathcal{H}_T(s) = \mathcal{H}(\mathcal{Q}^{-1}(s)) \ ; \ \mathcal{Q}(s) = \frac{s^2 + \Omega_0^2}{Ws(1 + \alpha)}$$

Relate the result to your answer in (ii).

3.42 [Another method for transforming an analog filter $\mathcal{H}(s)$ to a digital filter $H(z)$ which is known as the impulse invariant method.]

Suppose h_a is the unit impulse response of an analog filter $\mathcal{H}(s)$, and define the unit impulse response h of the digital filter by

$$h(k) \stackrel{\Delta}{=} h_a(t = kT)$$

for some constant sampling period T. Show that if $\mathcal{H}(s) = \boldsymbol{c}^T(s\boldsymbol{I} - \boldsymbol{A})^{-1}\boldsymbol{b}$, then

$$H(z) = \boldsymbol{c}^T(z\boldsymbol{I} - exp\{\boldsymbol{A}T\})^{-1}\boldsymbol{b}$$

3.43 (i) What is the cut-off frequency of the analog low pass filter

$$\mathcal{H}_0(s) = \frac{1}{(s+1)(s^2+\sqrt{2}s+1)}$$

(ii) Transform $\mathcal{H}_0(s)$ in (i) to a low pass filter $\mathcal{H}_1(s)$ with cut-off frequency of Ω_1 rad/sec by direct substitution for s

(iii) Based on a Jordan state space representation for $\mathcal{H}_0(s)$ in (i), find a state space representation for $\mathcal{H}_1(s)$ in (ii).

(iv) By means of the bilinear transformation, transform $\mathcal{H}_0(s)$ to a digital low pass filter $H_0(z)$ with cut-off frequency of ω_1 rad.

(vi) By means of the bilinear transformation, transform $\mathcal{H}_1(s)$ in (ii) to a digital low pass filter $H_1(z)$ with cut-off frequency of ω_1 rad. Compare your result with (iv).

(vii) Based on the state space representation for $\mathcal{H}_1(s)$ in (iii), find a state space representation for $H_1(z)$ in (vi).

3.44 (i) Design a high pass analog Butterworth filter with a cut-off frequency of 100Hz

(ii) Transform the analog filter in (i) to obtain a high pass digital Butterworth filter based on a sampling frequency of 250Hz. What is the cut-off frequency of the resulting digital filter ?

(iii) Design a high pass digital Butterworth filter with a cut-off frequency ω_c rads where

$$\omega_c = \frac{2\pi \times 100}{250}$$

Relate your answer to the result in (ii)

4. Signal Processing

In Chapter 2, we showed how a digital signal can be defined as the solution of a linear time invariant difference equation. For example, the signal v where $\{v(k) = ha^k; k \geq 0\}$ which is defined by the two parameters $\{h, a\}$ can also be described as the initial condition response of the first order difference equation $\{v(k+1) = av(k) \ ; \ v(0) = h\}$. More generally, the signal

$$v(k) = h_1\lambda_1^k + h_2\lambda_2^k + \ \ldots \ + h_n\lambda_n^k \ ; \ k \geq 0$$

which is represented by the $2n$ (possibly) complex parameters $\{h_\ell, \lambda_\ell\}$ for $1 \leq \ell \leq n$ can also be described as the initial condition response of an nth order difference equation.

In signal processing applications, such parameteric representations of signals are generally preferred rather than having to deal with of the signal values $\{v(k)\}$ themselves. For example, for the purpose of storing a signal v, less memory is required to store two parameters $\{h = 0.5, a = -0.75\}$ rather than the 1000 values $\{v(k) = 0.5(-0.75)^k; 0 \leq k \leq 999\}$. However it is not always possible (or convenient) to describe signals as the solution of difference equations. This Chapter considers other parameteric representations of signals.

4.1 Fundamental Properties

As described in Chapter 1, a scalar valued signal $v : \mathcal{T} \rightarrow \mathcal{S}$ from a mathematical point of view, is a function from the domain $\mathcal{T}$ to the range $\mathcal{S}$. If the signal is *real*, then $\mathcal{S} = \mathcal{R}$ is the set of real numbers, and if the signal is *complex*, then $\mathcal{S} = \mathcal{C}$ is the set of complex numbers. The signal v is an *analog* (or *continuous*) signal when the time set $\mathcal{T}$ is a continuous subset of $\mathcal{R}$, and a *digital* (or *discrete time*) signal when $\mathcal{T}$ is a subset of the integers $\mathcal{Z}$.

An *analog* signal is said to have *finite length* when the time set $\mathcal{T}$ is a *finite subset* of $\mathcal{R}$, and a *digital* signal is said to have *finite length* when the time set $\mathcal{T}$ is a *finite subset* of $\mathcal{Z}$. A signal that does not have finite length is said to be of infinite length. For example, if $\mathcal{T} = [0, 1] \in \mathcal{R}$, then v is a finite length analog signal, and if $\mathcal{T} = \mathcal{Z}_+$, then v is an infinite length digital signal.

Bounded Signal

One measure of the size $|v|$ of a signal v is given by the largest of all the absolute values of the signal values $v(k)$. We have the following definition.

A scalar valued signal $v : \mathcal{T} \to \mathcal{S}$ is said to be a bounded signal if there exists a (finite) real number L such that

$$|v(t)| \leq L \quad ; \quad t \in \mathcal{T} \tag{4.1}$$

If no such number L exists, then the signal v is said to be unbounded on $\mathcal{T}$.

If for a bounded signal $v : \mathcal{T} \to \mathcal{S}$, we define

$$|v| \triangleq \max_{t \in \mathcal{T}} |v(t)| \tag{4.2}$$

then it follows that $|v|$ satisfies the following three properties:

- $|v| \geq 0$ and $|v| = 0$ if and only if $v = 0$
- $|\alpha v| = |\alpha||v|$ for any scalar α
- $|v + w| \leq |v| + |w|$ for any two signals v and w.

These properties are all consistent with what we intuitively understand by the notion of "size".

Example 4.1.1. (i) The real digital signal $v(k) = k^2$ is *bounded* on any domain $\mathcal{T}$ of the form

$$\mathcal{T} \triangleq \{k \in \mathcal{Z} : 0 \leq k \leq N_1\}$$

when N_1 is *finite*. However, $v(k) = k^2$ is *unbounded* on the domain $\mathcal{T} = \mathcal{Z}$.

(ii) The digital signal $v(k) = (0.2)^k \cos 4k$ is bounded on $\mathcal{T} = \mathcal{Z}_+$ with $|v| = 1$, but is unbounded on $\mathcal{T} = \mathcal{Z}_-$.

(iii) The analog signal $f(t) = \tan t$ is bounded on the domain

$$\mathcal{T} \triangleq \{t \in \mathcal{R} : 0 \leq t \leq T_1\}$$

for any number $T_1 < 0.5\pi$, but is *unbounded* on a set which includes any number of the form $(n + 0.5)\pi$ for any integer n.

(iv) The digital signal $v(k) = f(kT) = \tan kT$ (which is obtained by sampling the analog signal $f(t) = \tan t$ at the digital time instants $t = kT$) is bounded on the domain $\mathcal{T} = \mathcal{Z}$ provided the sampling period T is selected so that $kT \neq 0.5n\pi$ for any integers k, n.

Independent Signals

Bounded signals defined on a domain $\mathcal{T}$ can be added, subtracted and multiplied pointwise. That is, if f and g are two bounded signals, and α is a (finite) scalar, then bounded signals $f+g$, $f-g$ and αf are defined by

$$\begin{aligned}(f+g)(t) &\triangleq f(t)+g(t) \quad ; \quad t \in \mathcal{T} \\ (f-g)(t) &\triangleq f(t)-g(t) \quad ; \quad t \in \mathcal{T} \\ (\alpha f)(t) &\triangleq \alpha \,.f(t) \quad ; \quad t \in \mathcal{T}\end{aligned} \tag{4.3}$$

The set of all signals which are *bounded* and defined on the same domain $\mathcal{T}$ are said to form a *linear space* $\mathcal{L}_{\mathcal{T}}$; that is, if $f,g \in \mathcal{L}_{\mathcal{T}}$, then $\alpha f + \beta g \in \mathcal{L}_{\mathcal{T}}$ for all (finite) scalars α, β. We have the following definition.

A real (or complex) bounded signal f defined on a domain $\mathcal{T}$ is said to be linearly dependent on two real (or complex) bounded signals g and h defined on $\mathcal{T}$ if: for all $t \in \mathcal{T}$

$$f(t) = \alpha g(t) + \beta h(t) \tag{4.4}$$

for some real (or complex) scalars α and β. Otherwise f is said to be linearly independent of g and h on $\mathcal{T}$.

More generally, a real (or complex) signal f_k defined on a domain $\mathcal{T}$ is said to be linearly dependent on the $p-1$ real (or complex) signals $\{f_1, f_2, \cdots, f_{k-1}, f_{k+1}, \cdots, f_p\}$ defined on $\mathcal{T}$ if there exist real (or complex) scalars $\{\alpha_1,\ \alpha_2, \cdots,\ \alpha_{k-1},\ \alpha_{k+1},\ \cdots,\ \alpha_p\}$ such that: for all $t \in \mathcal{T}$

$$f_k(t) = \alpha_1 f_1(t) + \cdots + \alpha_{k-1} f_{k-1}(t) + \alpha_{k+1} f_{k+1}(t) + \cdots + \alpha_p f_p(t) \tag{4.5}$$

Example 4.1.2. (i) Consider the two real digital signals

$$f, g : \mathcal{Z} \to \mathcal{R}$$

where

$$f(k) = \sin k \ ; \ g(k) = \cos k$$

Since no real scalar α exists such that $f(k) = \alpha g(k)$ for all $k \in \mathcal{Z}$, the real signals f and g are *independent* on $\mathcal{Z}$.

(ii) Define the digital signal $h(k) = 2\sin(k+\theta)$ for some constant θ. Now since

$$2\ \sin(k+\theta) = \alpha \ \sin k + \beta \ \cos k$$

where $\alpha = 2\cos\theta$ and $\beta = 2\sin\theta$, the signal h is *linearly dependent* on the digital signals f and g on $\mathcal{Z}$.

(iii) Consider the two real analog signals f and g as defined in Figure 4.1. Then since $f = -g$ on the interval [0, 1], and $f = -0.5g$ on the interval $[1, 4]$, we conclude that the two signals f and g are *dependent* on each of the intervals [0, 1] and [1, 4]. However f and g are *independent* on the interval $[0, 4]$, or on any (nonzero length) interval which includes the point 1.

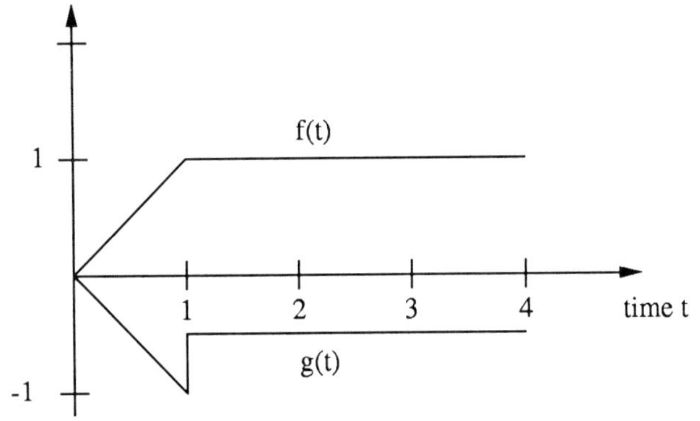

Fig. 4.1. Signals in example 4.1.2

Periodic Signal

An infinite length digital signal is one in which the domain $\mathcal{T}$ is an infinite subset of the integers $\mathcal{Z}$. A *periodic signal* is an infinite length signal which is defined by a finite number of distinct values. We have the following definition.

A complex infinite length signal $f : \mathcal{T} \to \mathcal{C}$ is said to be periodic if there exists a $T_0 \in \mathcal{T}$ so that $t + T_0 \in \mathcal{T}$ for all $t \in \mathcal{T}$ and

$$f(t) = f(t + T_0) \ ; \ t \in \mathcal{T} \tag{4.6}$$

The periodic signal f is said to have period T_0 if in (4.6): $T_0 > 0$ and there exists no number T_1 with $0 < T_1 < T_0$ such that

$$f(t) = f(t + T_1) \ ; \ t \in \mathcal{T}$$

Note that:

- If f and g are bounded periodic signals of period T_0, then $\alpha f + \beta g$ is a bounded periodic signal of period T_0 for any bounded scalars α and β.

- If f and g are bounded periodic signals of period T_0 and T_1 respectively for some rational numbers $\{T_0, T_1\}$, then $\alpha f + \beta g$ is a bounded periodic signal of period T_2 for any bounded scalars α and β where T_2 is the *lowest common multiple* of T_0 and T_1. (That is, the smallest number that is exactly divisible by both T_0 and T_1.) For example, if $T_0 = 0.3$ and $T_1 = 0.4$, then $T_2 = 1.2$.

Example 4.1.3. (i) Since

$$exp\{\frac{j2\pi(t+T_0)}{T_0}\} = exp\{\frac{j2\pi t}{T_0} + j2\pi\} = exp\{\frac{j2\pi t}{T_0}\}$$

a *complex analog* periodic signal $f : \mathcal{R} \to \mathcal{C}$ of period $T_0 \in \mathcal{R}$ is defined by

$$f(t) = exp\{\frac{j2\pi t}{T_0}\} \; ; \; t \in \mathcal{R}$$

(ii) A *real analog* periodic signal $f : \mathcal{R} \to \mathcal{R}$ of period $T_0 \in \mathcal{R}$ is defined by

$$f(t) = \cos\{\frac{2\pi t}{T_0}\} \; ; \; t \in \mathcal{R}$$

Consider the infinite length digital signal v defined by $\{v(k) = \sin k\Omega_0 T; k \in \mathcal{Z}\}$ which is obtained from the sampled values of the analog signal $\{f(t) = \sin \Omega_0 t; t \in \mathcal{R}\}$ at the digital instants $\{t = kT; k \in \mathcal{Z}\}$. Now

$$v(k) = \sin k\Omega_0 T = \sin \Omega_0 T(k + \frac{2\pi n}{\Omega_0 T})$$

for any integer n. However v is *not* periodic unless $v(k) = v(k + N_0)$ for all $k \in \mathcal{Z}$ and some *integer* N_0. That is, the digital signal v is *only* periodic if $2\pi n/(\Omega_0 T)$ is an *integer* for some integer n. For example, if $\Omega_0 T = 3\pi$, then

$$\frac{2\pi n}{\Omega_0 T} = \frac{2n}{3} = 2 \text{ when } n = 3$$

and so $v(k) = \sin 3\pi k$ is periodic (of period $N_0 = 2$). More generally, we have the following result.

Theorem 4.1.1. *Consider the infinite length digital signal*

$$v(k) = exp\{\frac{j2\pi k}{R_0}\} \; ; \; k \in \mathcal{Z} \tag{4.7}$$

Then:

- *v is periodic of period R_0 when R_0 is an integer*

- *v is periodic of period p_0 when R_0 is a rational number of the form*

$$R_0 = \frac{p_0}{q_0} \; ; \; \{p_0, q_0\} \text{ irreducible}$$

- *v is not periodic when R_0 is an irrational number*

Example 4.1.4. (a) Suppose $R_0 = \pi$, then $v(k) = exp\{j2k\}$ is not periodic.

(b) Suppose $R_0 = 3$. Then from (4.7), $v(k) = v(k+3)$ for all $k \in \mathcal{Z}$ where

$$v(0) = 1 \; ; \; v(1) = exp\{\frac{j2\pi}{3}\} \; ; \; v(2) = exp\{\frac{j4\pi}{3}\}$$

The three distinct complex numbers $\{v(k)\}$ can be represented in the complex plane as in Figure 4.2(a).

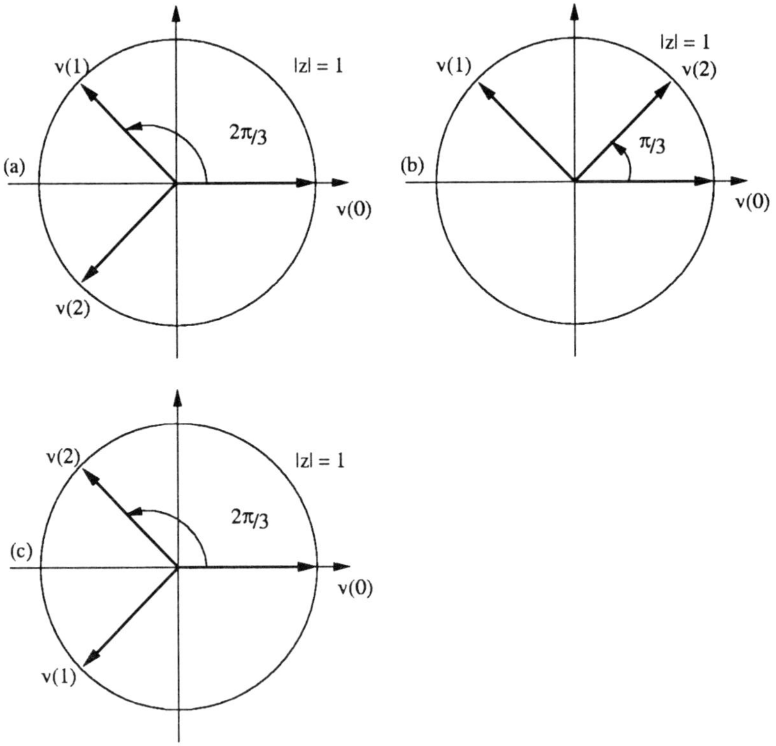

Fig. 4.2. Signals of period 3

(c) Suppose $R_0 = 3/4$. Then from (4.7), we have that $v(k) = v(k+3)$ for all $k \in \mathcal{Z}$ where

$$v(0) = 1 \; ; \; v(1) = exp\{\frac{j8\pi}{3}\} = exp\{\frac{j2\pi}{3}\}$$
$$v(2) = exp\{\frac{j16\pi}{3}\} = exp\{\frac{j\pi}{3}\}$$

The three distinct complex numbers $\{v(k)\}$ can be represented in the complex plane as in Figure 4.2(b).

(d) Suppose $R_0 = 3/2$. Then from (4.7), we have that $v(k) = v(k+3)$ for all $k \in \mathcal{Z}$ where

$$v(0) = 1 \; ; \; v(1) = exp\{\frac{j4\pi}{3}\} \; ; \; v(2) = exp\{\frac{j8\pi}{3}\} = exp\{\frac{j2\pi}{3}\}$$

The three distinct complex numbers $\{v(k)\}$ can be represented in the complex plane as in Figure 4.2(c).

(e) Consider the two complex digital periodic signals $\{f, g\}$ where

$$f(k) = exp\{\frac{j2\pi q_0 k}{p_0}\} \; ; \; g(k) = exp\{\frac{j2\pi q_1 k}{p_1}\} \; ; \; k \in \mathcal{Z}$$

in which $\{q_0, p_0, q_1, p_1\}$ are all integers with q_0/p_0 and q_1/p_1 both irreducible. Then f has period p_0, g has period p_1, and the signal $f+g$ is periodic of period N_2 where N_2 is the *lowest common multiple* of p_0 and p_1. For example, if $\{p_0 = 3, p_1 = 2\}$ (secs), then $N_2 = 6$ (secs), and if $\{p_0 = 4, p_1 = 6\}$ (secs), then $N_2 = 12$ (secs).

4.1.1 Signal Norm

Besides $|v|$, there are other measures known as *signal norms* which measure the "size" of a bounded signal. There are many possible definitions for signal norms, and the mathematical form of the definition depends on whether or not the signal is digital or analog, and finite length or infinite length. Norms of *digital* signals involve *summations*, while norms of *analog* signals involve *integrals*. In this text, we are only concerned with norms of digital signals. We have the following definition.

A norm $||v||$ of a bounded digital signal v is a nonnegative number which satisfies the following conditions:

1. $||v|| \geq 0$, *and* $||v|| = 0$ *if and only if* $v = 0$ *is the zero signal.*
2. $||v|| = |\alpha| \quad ||v||$ *where* $|\alpha|$ *denotes the absolute value of the complex number* α.
3. $||v + w|| \leq ||v|| + ||w||$ *for any two bounded signals* v *and* w.

Finite Length Signal

A digital signal $f : \mathcal{T} \to \mathcal{C}$ when the domain $\mathcal{T}$ is a *finite* digital set can be represented as a *finite dimensional vector*. For example, if

$$\mathcal{T} \triangleq \{n \in Z_+ \; ; \; 0 \leq n \leq L-1\}$$

then the *norm* $||f||$ of a *bounded finite length* signal f can then be defined by

$$||f||_p \triangleq \left\{ \sum_{n=0}^{L-1} |f(n)|^p \right\}^{\frac{1}{p}} \; ; \; p \geq 1 \tag{4.8}$$
$$||f||_\infty \triangleq \max_{0 \leq n \leq L-1} |f(n)|$$

Both these definitions are well-defined since the summations are finite, and $|f(k)|$ is assumed to be bounded for all k. Both definitions can also be shown to satisfy the three properties which are required to define a norm.

Note that $||f||_\infty$ (which is commonly referred to as the *infinity norm* and is given by $|f|$ in (4.2)), can be considered as the limiting case of the (so called) p-norm $||f||_p$ in the limit as p tends to infinity. It should also be noted that $||f||_p$ for $0 < p < 1$ is *not* a norm. For example, suppose $\{L = 2, p = 0.5\}$ and

$$f(0) = f(1) = 1 \; ; \; g(0) = 1, \; g(1) = 0$$

Then

$$||f + g||_{0.5} = (\sqrt{2} + \sqrt{1})^2 = 3 + 2\sqrt{2} \approx 5.82$$
$$||f||_{0.5} = (\sqrt{1} + \sqrt{1})^2 = 4 \; ; \; ||g||_{0.5} = (\sqrt{1} + \sqrt{0})^2 = 1$$

and so

$$||f + g||_{0.5} > ||f||_{0.5} + ||g||_{0.5}$$

which violates property 3 for a norm.

Two important examples of (4.8) (known respectively as the 1-norm and 2-norm) are given by

$$||f||_1 \triangleq \sum_{n=0}^{L-1} |f(n)| \; ; \; ||f||_2 \triangleq \sqrt{\sum_{n=0}^{L-1} |f(n)|^2}$$

Different norms of the same signal generally have different values. For example, consider the signal f where $f(k) = 1$ for all $0 \leq k \leq L-1$. Then

$$||f||_\infty = 1 \; ; \; ||f||_2 = \sqrt{L} \; ; \; ||f||_1 = L$$

and for the signal f where $f(0) = 1$ with $f(k) = 0$ for all $k \neq 0$, then

$$||f||_\infty = ||f||_2 = ||f||_1 = 1$$

More generally, it may be shown that: for any bounded signal f

$$||f||_\infty \leq ||f||_p \leq ||f||_q \; ; \; 1 \leq q \leq p < \infty$$

The choice of which norm to use depends on the particular application. For example, suppose the digital signal y where

$$y(k+1) = ay(k) + b_0 u(k) + b_1 u(k+1) \; ; \; u(k) \triangleq v(kT)$$

represents an approximation of $z(kT)$ of the continuous signal z where

$$z(kT) = z(0) + \int_0^{kT} v(\sigma) d\sigma \tag{4.9}$$

Define the error signal e where $e(k) \triangleq y(k) - z(kT)$. Then the "closeness" of the approximation can be measured by an error norm $||e||$. Different norms measure different properties of this approximation. The infinity norm

$$||e||_\infty \triangleq \max_{0 \leq n \leq N} |e(n)|$$

of the error signal e measures the largest error over the range $0 \leq k \leq N$ in (4.9). The 2-norm

$$||e||_2 = \sqrt{\sum_{n=0}^{N} e^2(n)}$$

of the error signal e measures the sum of squares of all the N errors. The *weighted* 2-norm defined by

$$||e|| \triangleq \sqrt{\sum_{n=0}^{N} \lambda_n e^2(n)} \; ; \; \lambda_n > 0 \text{ for all } n$$

adjusts the relative importance of each error according to the choice of λ_n. For example, if $\lambda_n = a^n$ for some $0 < a < 1$, then $e^2(n)$ is weighted more heavily than $e^2(m)$ for $n < m$.

Infinite Length Signal

Consider a *bounded* digital signal f defined on an *infinite* domain $\mathcal{T}$. Then since $|f(p)|$ is finite for all p, the *infinity norm* $||f||_\infty$ of the signal v where

$$||f||_\infty \triangleq \max_{n \geq 0} |f(n)| \tag{4.10}$$

is well-defined.

If the sum

$$||f||_1 \triangleq \sum_{n=0}^{\infty} |f(n)| \tag{4.11}$$

is *bounded*, then $||f||_1$ defines the *1-norm* of the infinite length digital signal f. Note, however, that the existence of $||f||_\infty$ does *not* guarantee that $||f||_1$ exists. For example, if $f(k) = 1$ for all k, then $||f||_\infty = 1$, but $\sum_{k=0}^{\infty} |f(k)|$ is *infinite*, and so $||f||_1$ does not exist.

Likewise, if

$$||f||_2 \triangleq \sqrt{\sum_{n=0}^{\infty} |f(n)|^2} \tag{4.12}$$

is *bounded*, then $||f||_2$ defines the *2-norm* of the digital signal f. But the existence of $||f||_2$ does *not* imply the existence of $||f||_1$. However since

$$\sum_{n=0}^{\infty} |f(n)|^2 \leq (\sum_{n=0}^{\infty} |f(n)|)^2$$

the converse statement *is* true; that is, if $||f||_1$ exists, then $||f||_2$ exists. Conditions under which the infinite series (4.11) and (4.12) are convergent are summarized in Theorem 1.1.1 in Chapter 1. More generally, we have the following result.

Theorem 4.1.2. *Suppose the signal norm $||f||_r$ exists for some $r \geq 1$. Then $||f||_p$ exists for all $p \geq r$.*

Example 4.1.5. (i) Consider the digital signal $f : Z_+ \rightarrow R$ where $f(0) = 0$ and

$$f(n) = n^{-1}\ ;\ \ n \geq 1$$

Then

$$\max_{n \geq 1} |f(n)| = 1$$

and so by (4.10), $||f||_\infty = 1$.

In this case, the summation $||f||_1$ in (4.11) is *infinite*, but the summation $||f||_2$ in (4.12) is *finite*; specifically

$$\sum_{n=0}^{\infty} |f(n)|^2 = \frac{\pi^2}{6}\ .$$

Hence $||f||_1$ does *not* exist, while $||f||_2 = \pi/\sqrt{6}$.

(ii) Consider the digital signal defined by

$$f(n) = a^n \; ; \quad n \in Z_+ \quad .$$

for some (possibly) complex number a where $|a| < 1$. Then $||f||_\infty = 1$ which is well-defined even when $|a| = 1$. However

$$\sum_{n=0}^{\infty} |a^n| = \frac{1}{1-|a|} = ||f||_1$$

is *only* defined for $|a| < 1$. For $|a| < 1$

$$\sqrt{\sum_{n=0}^{\infty} |a^n|^2} = \frac{1}{\sqrt{1-|a|^2}} = ||f||_2$$

is also finite.

Periodic Signal

Consider now a (bounded) periodic digital signal f of period N_0. Then unless $f(n) = 0$ for all $n \in Z_+$, the summation

$$||f||_p = \{\sum_{n=0}^{\infty} |f(n)|^p\}^{\frac{1}{p}}$$

will be *infinite* for any p. Hence *this* summation *cannot* be used to define the norm of a periodic signal.

Instead, we assert that: for a *periodic signal* f, the p-norm is defined by

$$||f||_p \triangleq \{\lim_{N\to\infty} \frac{1}{N} \sum_{n=0}^{N-1} |f(n)|^p\}^{\frac{1}{p}} \tag{4.13}$$

is indeed a norm. Note that this definition *cannot* be used to define the norm of any signal $f \neq 0$ in which the summation

$$\sum_{n=0}^{\infty} |f(n)|^p$$

is *finite*. In particular, if this summation is finite, then the limiting summation

$$||f||_p^p = \lim_{N\to\infty} \frac{1}{N} \sum_{n=0}^{N-1} |f(n)|^p$$

will be zero with $f \neq 0$ which violates property 1 for $||f||_p$ to be a p-norm.

We now simplify the expression for the p-norm $||f||_p$ in (4.13) of a periodic signal f. Specifically, since f is *periodic* of period N_0, it follows that

$$\sum_{n=n_0}^{n_0+N_0-1} |f(n)|^p = \sum_{n=0}^{N_0-1} |f(n)|^p \tag{4.14}$$

for *any* $n_0 \in \mathcal{Z}_+$. Then with $N = kN_0$ for some integer k

$$\frac{1}{N}\sum_{n=0}^{N-1} |f(n)|^p = \frac{1}{kN_0}\sum_{n=0}^{kN_0-1} |f(n)|^p = \frac{1}{kN_0}\sum_{m=0}^{k-1}\sum_{n=mN_0}^{(m+1)N_0-1} |f(n)|^p$$

and hence from (4.14) with $n_0 = mN_0$, it follows that

$$\lim_{N\to\infty}\frac{1}{N}\sum_{n=0}^{N-1} |f(n)|^p = \lim_{k\to\infty}\frac{1}{kN_0}\,k\sum_{n=0}^{N_0-1} |f(n)|^p = \frac{1}{N_0}\sum_{n=0}^{N_0-1} |f(n)|^p$$

We have the following result.

Theorem 4.1.3. *The p-norm $||f||_p$ of a bounded periodic signals f of period N_0 is defined by*

$$||f||_p \triangleq \left\{\frac{1}{N_0}\sum_{n=0}^{N_0-1} |f(n)|^p\right\}^{\frac{1}{p}} \quad ; \; p \geq 1 \tag{4.15}$$

(It is left as an exercise to confirm that each of the three norm properties are indeed satisfied.)

4.1.2 Signal Inner Product

A *signal inner product* provides a measure of the "angle" between two signals. An angle of 0^0 will imply that the two signals are *dependent*, and an angle of 90^0 will mean that two signals are perpendicular or *orthogonal*. There are many possible definitions for inner products, but as is the case with a norm, the mathematical form of the definition depends on whether or not the signals are digital or analog, and finite length or infinite length. Discrete inner products involve summations, while continuous inner products involve integrals. In this text, we are only concerned with inner products between digital signals. We have the following definition.

An inner product $\langle v,\ w\rangle$ between two digital signals v and w is a (possibly complex) scalar satisfying the following properties:

- *(Positivity)*

$$\langle v, v\rangle > 0 \text{ if and only if } v \neq 0 \qquad (4.16)$$
$$\langle v, v\rangle = 0 \text{ if and only if } v = 0$$

- *(Symmetry)*

$$\langle v, w\rangle = \overline{\langle w, v\rangle} \qquad (4.17)$$

- *(Linearly) For all (possibly) complex scalars* α *and* β

$$\langle \alpha v + \beta w, u\rangle = \alpha\langle v, u\rangle + \beta\langle w, u\rangle$$
$$\langle u, \alpha v + \beta w\rangle = \bar{\alpha}\langle u, v\rangle + \bar{\beta}\langle u, w\rangle \qquad (4.18)$$

The following result shows that an inner product between a signal v and itself *induces a norm* on that signal, and an inner product between two signals v and w defines an *angle* between these two signals.

Theorem 4.1.4. *Given an inner product* $\langle \cdot, \cdot\rangle$, *define*

$$||v|| \triangleq \sqrt{\langle v, v\rangle} \qquad (4.19)$$

Then:

- $||v||$ *is a norm*
- $$|\langle v, w\rangle| \leq ||v|| \; ||w|| \qquad (4.20)$$
  with equality if and only if $v = \alpha w$ *for some scalar* α. *Hence* θ *with* $0 \leq \theta \leq \pi$ *is uniqueley defined by*

$$\cos\theta \triangleq \frac{\langle v, w\rangle}{||v|| \; ||w||} \qquad (4.21)$$

To prove the first part of Theorem 4.1.4, first observe from (4.16) that $||v|| > 0$ for $v \neq 0$ and $||v|| = 0$ when $v = 0$ which gives property 1 for a norm. Also, from (4.18) with $\alpha = 1$ and $w = 0$, it follows that $\langle 0, u\rangle = \langle u, 0\rangle = 0$. Then from (4.18) with $w = 0$, we have $\langle u, \alpha v\rangle = \overline{\alpha}\langle u, v\rangle$. Hence with $w = 0_n$ and $u = \alpha v$ in (4.18), $\langle \alpha v, \alpha v\rangle = \alpha\langle v, \alpha v\rangle = \alpha\overline{\alpha}\langle v, v\rangle$, or from (4.19)

$$||\alpha v||^2 = |\alpha|^2 \; ||v||^2 \; .$$

which gives property 2 for a norm.

Finally, from (4.18), (4.19) and (4.20) with $\alpha = \beta = 1$ and $u = v + w$, we have that

$$\begin{aligned}\langle v+w,\ v+w\rangle &= \langle v, v+w\rangle + \langle w, v+w\rangle \\ &= \langle v,v\rangle + \langle v,w\rangle + \langle w,v\rangle + \langle w,w\rangle \\ &\leq \langle v,v\rangle + 2||v||\ ||w|| + \langle w,w\rangle\end{aligned}$$

That is, from (4.19)

$$||v+w||^2 \leq (||v|| + ||w||)^2 \ \text{ or } \ ||v+w|| \leq ||v|| + ||w||$$

which gives property 3 for a norm.

To prove the second part in Theorem 4.1.4, first assume $w = 0$ is the zero signal. Then $\langle v,\ 0\rangle = 0$ and $||v||\ ||0|| = 0$ so (4.20) is true with equality. Now assume $w \neq 0$. Then proving (4.20) is equivalent to proving

$$|\langle v,\ x\rangle| \leq ||v|| \ ; \quad x = \frac{w}{||w||} \ .$$

That is, without any loss of generality, we need only prove (4.20) is true when $||w|| = 1$. Now

$$\begin{aligned}0 &\leq ||v - \langle v,\ w\rangle w||^2 \\ &\leq \langle v - \langle v,\ w\rangle w,\ v - \langle v,\ w\rangle w\rangle \\ &= \langle v,\ v\rangle - \overline{\langle v,\ w\rangle}\langle v,\ w\rangle - \langle v,\ w\rangle\overline{\langle v,\ w\rangle} + \overline{\langle v,\ w\rangle}\langle v,\ w\rangle\langle w,\ w\rangle \\ &= \langle v,\ v\rangle - \langle v,\ w\rangle\overline{\langle v,\ w\rangle}\end{aligned}$$

since $||w|| = 1$. That is,

$$0 \leq ||v||^2 - |\langle v,\ w\rangle|^2 \ ; \quad ||w|| = 1$$

with equality if and only if $v = \langle v,\ w\rangle w$ which gives the result. Since from (4.20)

$$-1 \leq \frac{\langle v, w\rangle}{||v||\ ||w||} \leq 1$$

$\cos\theta$ (and hence θ) is well-defined.

Note that when $v = w$, we have from (4.21) that

$$\cos\theta = \frac{\langle v, v\rangle}{||v||\ ||v||} = 1$$

or the angle θ between a signal v and itself is 0^0. Equivalently, when $\theta = 0^0$, the two vectors $\{v, w\}$ are *dependent*. When

$$\langle v, v\rangle = 0$$

then $\theta = 90^0$; that is, v is perpendicular (or *orthogonal*) to w.

Finite Length Signal

When $f, g : \mathcal{T} \to \mathcal{C}$ where $\mathcal{T} = \{n \in Z_+ \;;\; 0 \leq n \leq L-1\}$, then one possible definition of an inner product between the two *finite length* signals f and g is given by

$$\langle f,\, g\rangle \triangleq \sum_{n=0}^{L-1} f(n)\overline{g(n)} \tag{4.22}$$

where $\overline{g(n)}$ is the complex congugate of $g(n)$. If $f,\ g : \mathcal{T} \to \mathcal{R}$ are *real* signals, then (4.22) reduces to

$$\langle f,\, g\rangle \triangleq \sum_{n=0}^{L-1} f(n)g(n) \ . \tag{4.23}$$

However in general, note that (4.23) is *not* an inner product when $\{f, g\}$ are *complex* signals.

Another valid definition for an inner product between two finite length signals f and g is given by

$$\langle f,\, g\rangle \triangleq \sum_{n=0}^{L-1} \lambda_n f(n)\overline{g(n)} \ ; \ \lambda_n > 0 \tag{4.24}$$

Example 4.1.6. Suppose $L = 2$ with $\{f(0) = 2, f(1) = j3\}$ and $\{g(0) = 2 + j, g(1) = 4 - j6\}$. Then from (4.22)

$$\begin{aligned}\langle f,\, g\rangle &= f(0)\overline{g(0)} + f(1)\overline{g(1)} \\ &= 2(2-j) + j3(4+j6) = -14 + j10\end{aligned}$$

Also

$$\langle g,\, f\rangle = \overline{\langle f,\, g\rangle} = -14 - j10$$

$$||f|| = \sqrt{\langle f,\, f\rangle} = \sqrt{|f(0)|^2 + |f(1)|^2} = \sqrt{4+9} = \sqrt{13}$$

and

$$||g|| = \sqrt{\langle g,\, g\rangle} = \sqrt{|g(0)|^2 + |g(1)|^2} = \sqrt{5+52} = \sqrt{57}$$

Infinite Length Signal

When $\mathcal{T}$ has an *infinite* number of elements, one can still *sometimes* define an inner product $\langle f,\, g\rangle$. For example, if $\mathcal{T} = Z_+$, then one possible definition is

$$\langle f, g\rangle \triangleq \sum_{n=0}^{\infty} f(n)\overline{g(n)} \tag{4.25}$$

However as is evident by the example when $f(n) = 1$ and $g(n) = 1$ for all $n \in Z_+$, this infinite summation in general is *not* finite. Nevertheless, in the special case when both the norms $||f||_2$ and $||g||_2$ are *finite* where

$$||f||_2 \triangleq \sqrt{\langle f, f\rangle} = \sqrt{\sum_{n=0}^{\infty} f(n)\overline{f(n)}} = \sqrt{\sum_{n=0}^{\infty} |f(n)|^2} \ .$$

we have from (4.20) that

$$|\langle f, g\rangle| \leq ||f||_2 \ ||g||_2 \tag{4.26}$$

and so $\langle f, g\rangle$ is also finite.

Also when $||f||_2$ and $||g||_2$ are finite, the *angle* θ with respect to the inner product (4.25) between two digital signals $\{f, g\}$ can then be defined by

$$\cos\theta \triangleq \frac{\langle f, g\rangle}{||f||_2 \ ||g||_2} \ . \tag{4.27}$$

More generally

$$\langle f, g\rangle = \sum_{n=0}^{\infty} \lambda_n f(n)\overline{g(n)} \ ; \ \lambda_n > 0 \tag{4.28}$$

is a well defined inner product when $||f||$ and $||g||$ are both bounded where now the norm induced by the inner product (4.28) (if it exists) is given by

$$||f|| = \sqrt{\sum_{n=0}^{\infty} \lambda_n |f(n)|^2} \ ; \ \lambda_n > 0$$

The *angle* between f and g with respect to the inner product (4.28) is then defined by

$$\cos\theta \triangleq \frac{\langle f, g\rangle}{||f|| \ ||g||} \ ; \ ||f|| = \sqrt{\langle f, f\rangle} \tag{4.29}$$

Periodic Signal

The summation

$$\langle f, g\rangle = \sum_{n=0}^{\infty} f(n)\overline{g(n)}$$

will always be *infinite* for any two non zero periodic signals $\{f, g\}$. Therefore this formula *cannot* be used to define the *inner product* between two *periodic* signals. Instead, we assert that

$$\langle f, g \rangle \stackrel{\Delta}{=} \lim_{N\to\infty} \frac{1}{N} \sum_{n=0}^{N-1} f(n)\overline{g(n)}$$

is the appropriate definition for the inner product between periodic signals. In particular, if $\{f, g\}$ are both periodic of period N_0, then

$$\langle f, g \rangle = \frac{1}{N_0} \sum_{n=0}^{N_0-1} f(n)\overline{g(n)}$$

We summarize (and extend) these results as follows.

Theorem 4.1.5. *A well-defined inner product between two period signals $\{f, g\}$ of period N_0 is given by*

$$\langle f, g \rangle \stackrel{\Delta}{=} \frac{1}{N_0} \sum_{n=0}^{N_0-1} \lambda_n f(n)\overline{g(n)} \ ; \ \lambda_n > 0 \ \text{ for all } n$$

More generally, if f is periodic of period N_0, and g is periodic of period N_1, then an inner product between the signals $\{f, g\}$ is given by

$$\langle f, g \rangle \stackrel{\Delta}{=} \frac{1}{N} \sum_{n=0}^{N-1} \lambda_n f(n)\overline{g(n)} \ ; \ \lambda_n > 0 \ \text{ for all } n$$

where N is the lowest common multiple of N_0 and N_1.

Example 4.1.7. (a) Consider the periodic signal f where $\{f(k) = f(k+3); k \in \mathcal{Z}\}$ with

$$f(0) = 1 \ ; \ f(1) = j3 \ ; \ f(2) = 2 - j$$

Then since f has period $N_0 = 3$, we have

$$||f||_2 = \sqrt{\frac{1}{3} \sum_{n=0}^{2} |f(n)|^2} = \sqrt{\frac{1}{3}(1 + 9 + 5)} = \sqrt{5}$$

(b) Consider the periodic signal g where $\{g(k) = g(k + 2); k \in \mathcal{Z}\}$ with

$$g(0) = -1 \ ; \ g(1) = 4 - j3$$

Then since g has period $N_1 = 2$, we have

$$||g||_2 = \sqrt{\frac{1}{2} \sum_{n=0}^{1} |g(n)|^2} = \sqrt{\frac{1}{2}(1 + 25)} = \sqrt{13}$$

(c) Since from (a) and (b)

$$N = gcm\{N_0, N_1\} = gcm\{2, 3\} = 6$$

an inner product between the periodic signals f in (a) and g in (b) is defined by

$$\begin{aligned}\langle f, g\rangle &= \frac{1}{6}\sum_{n=0}^{5} f(n)\overline{g(n)} \\ &= \frac{1}{6}[(1)(-1) + (j3)(4 + j3) + (2 - j)(-1) + \\ &\qquad (1)(4 + j3) + (j3)(-1) + (2 - j)(4 + j3)] \\ &= 1/2 + j5/3\end{aligned}$$

4.1.3 Orthogonal Signals

Two signals $\{f, g\}$ are said to be *orthogonal* with respect to a given inner product if

$$\langle f, g\rangle = 0$$

However two signals which are orthogonal with respect to one inner product are not necessarily orthogonal with respect to some other inner product.

Example 4.1.8. (a) Suppose $\{f, g\}$ are finite signals of length 3. Define the inner product

$$\langle f, g\rangle = f(0)\overline{g(0)} + f(1)\overline{g(1)} + f(2)\overline{g(2)}$$

Then the signal $\{f(0) = 2, f(1) = -3, f(2) = 0\}$ is orthogonal to the signal $\{g(0) = 4, g(1) = 8/3, g(2) = 0\}$. Also, $\{f(0) = 1, f(1) = -1 + j, f(2) = -1 - j2\}$ is orthogonal to $\{g(0) = 3, g(1) = 2, g(2) = 1\}$

(b) Suppose now we define the new inner product

$$\langle f, g\rangle = f(0)\overline{g(0)} + 2f(1)\overline{g(1)} + f(2)\overline{g(2)}$$

Then in terms of this inner product, the signal $\{f(0) = 2, f(1) = -3, f(2) = 0\}$ is *not* orthogonal to the signal $\{g(0) = 4, g(1) = 8/3, g(2) = 0\}$.

We have the following definition.

A set of bounded signals $S_N = \{\phi_0, \phi_1, \ldots, \phi_{N-1}\}$ which has a well-defined inner product for all $\phi_p, \phi_m \in S_N$ is said to form an orthogonal set (with respect to that inner product) if the angle θ between any two distinct signals as defined by (4.29) is 90^0; that is, if

$$\langle \phi_p, \phi_m \rangle = 0 \quad ; \ p \neq m \tag{4.30}$$

An orthogonal set $\mathcal{S}_N$ is said to be orthonormal if: $||\phi_p|| = 1$ for all $0 \leq p \leq N-1$ where

$$||\phi|| \stackrel{\Delta}{=} \sqrt{\langle \phi, \phi \rangle}$$

is the norm induced by the inner product.

Suppose it is known that a digital signal $v : \mathcal{Z} \to \mathcal{C}$ can be represented as a linear combination of the two (nonzero) orthogonal signals $\{\phi_0, \phi_1\}$; that is

$$v(k) = \beta_0 \phi_0(k) + \beta_1 \phi_1(k) \ ; \ \langle \phi_0, \phi_1 \rangle = 0$$

Then

$$\begin{aligned} ||v||^2 &= \langle \beta_0\phi_0 + \beta_1\phi_1, \beta_0\phi_0 + \beta_1\phi_1 \rangle \\ &= \beta_0 \langle \phi_0, \beta_0\phi_0 + \beta_1\phi_1 \rangle + \beta_1 \langle \phi_1, \beta_0\phi_0 + \beta_1\phi_1 \rangle \\ &= \beta_0 \overline{\langle \beta_0\phi_0 + \beta_1\phi_1, \phi_0 \rangle} + \beta_1 \overline{\langle \beta_0\phi_0 + \beta_1\phi_1, \phi_1 \rangle} \\ &= \beta_0\overline{\beta_0} \langle \phi_0, \phi_0 \rangle + \beta_0\overline{\beta_1}\ \overline{\langle \phi_1, \phi_0 \rangle} + \beta_1\overline{\beta_0}\ \overline{\langle \phi_0, \phi_1 \rangle} + \beta_1\overline{\beta_1} \langle \phi_1, \phi_1 \rangle \\ &= \beta_0\overline{\beta_0} \langle \phi_0, \phi_0 \rangle + \beta_1\overline{\beta_1} \langle \phi_1, \phi_1 \rangle \end{aligned}$$

This derivation follows from the properties of an inner product.

Also since $\langle \phi_0, \phi_1 \rangle = 0$, we have that

$$\langle v, \phi_0 \rangle = \langle \beta_0\phi_0 + \beta_1\phi_1, \phi_0 \rangle = \beta_0 \langle \phi_0, \phi_0 \rangle = \beta_0 ||\phi_0||^2$$

and similarly

$$\langle v, \phi_1 \rangle = \langle \beta_0\phi_0 + \beta_1\phi_1, \phi_1 \rangle = \beta_1 \langle \phi_1, \phi_1 \rangle = \beta_1 ||\phi_1||^2$$

Furthermore, $\beta_0 = \beta_1 = 0$ (and so v is *independent* of $\{\phi_0, \phi_1\}$) if and only if

$$\langle v, \phi_0 \rangle = \langle v, \phi_1 \rangle = 0$$

We therefore have the following result.

Theorem 4.1.6. *Suppose a digital signal $v : \mathcal{Z} \to \mathcal{C}$ can be represented as a linear combination of the N orthogonal signals $\{\phi_0, \phi_1, \ \dots \ , \phi_{N-1}\}$ with respect to an inner product $\langle \cdot, \cdot \rangle$; that is*

$$v(k) = \sum_{n=0}^{N-1} \beta_n \phi_n(k) \ ; \ k \in \mathcal{Z} \tag{4.31}$$

and $||\phi_n|| = \sqrt{\langle \phi_n, \phi_n \rangle}$. Then

$$||v|| = \sqrt{\sum_{n=0}^{N-1} |\beta_n|^2 \; ||\phi_n||^2} \tag{4.32}$$

where $||\cdot||$ *is the norm induced by the inner product, and*

$$\beta_n = \frac{\langle v, \phi_n \rangle}{||\phi_n||^2} \tag{4.33}$$

The representation of the signal v as in (4.31) by the coefficients $\{\beta_n\}$ with respect to some given orthogonal set $\{\phi_n\}$ may be thought of as a *transformation* of the signal v. This result also provides a relationship between the norm $||v||$ of the signal v as measured in the *time domain* in terms of the component values $\{v(k)\}$, and the norm of the signal measured in the *transform domain* in terms of the coefficients $\{\beta_n\}$. Note that for an *orthonormal* set $\{\phi_n\}$, $\langle \phi_n, \phi_n \rangle = 1$ for all n in which case we have

$$||v|| = \sqrt{\sum_{n=0}^{N-1} |\beta_n|^2} \; ; \; \beta_n = \langle v, \phi_n \rangle$$

Walsh Signals

A particular class of real finite length signals are known as *Walsh signals.* In continuous time, Walsh signals are finite length signals piecewise constant signals equal to either 1 or -1 over each time interval $[kT, (k+1)T)$ for some fixed time period T, and for a finite range of integers $k_0 \leq k \leq k_1$. Such orthogonal signals have practical significance since they can be easily generated using digital logic circuitry.

A Walsh signal can also be equivalently represented as a finite length digital signal w in which $w(k) = \pm 1$. An example of four Walsh analog signals and their digital equivalents are illustrated in Figure 4.3. There are 2^L Walsh signals $\{w_{p,L} : 0 \leq p \leq 2^L - 1\}$ of length 2^L which are real and orthogonal with respect to the inner product

$$\langle x, y \rangle = \sum_{k=0}^{2^L-1} x(k)\overline{y(k)}$$

For example, when $L = 1$, the two Walsh signals $\{w_{0,1}, w_{0,2}\}$ are defined by

$$w_{0,1} = \{1, 1\} \; ; \; w_{1,1} = \{1, -1\}$$

and for $L = 2$, the four Walsh signals $\{w_{m,2}; 0 \leq m \leq 3\}$ are defined by

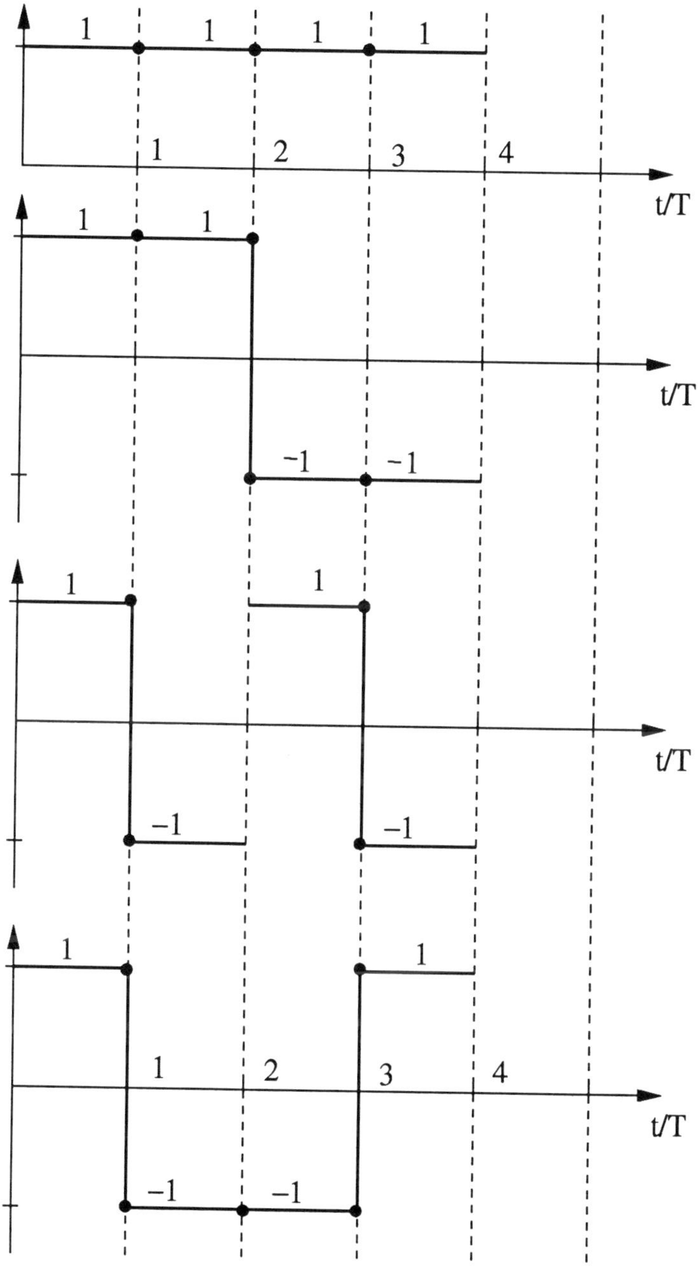

Fig. 4.3. Orthogonal Walsh signals

$$w_{0,2} = \{1,1,1,1\} \ ; \ w_{1,2} = \{1,1,-1,-1\} \qquad (4.34)$$
$$w_{2,2} = \{1,-1,1,-1\} \ ; \ w_{3,2} = \{1,-1,-1,1\}$$

More generally, the 2^{L+1} Walsh signals $\{w_{m,L+1}; 0 \leq m \leq 2^{L+1}-1\}$ can be generated from the 2^L Walsh signals $\{w_{m,L}; 0 \leq m \leq 2^L - 1\}$ as follows. For each $0 \leq p \leq 2^L - 1$, use the signal $w_{p,L}$ of length 2^L to generate two signals $\{w_{2p,L+1}, w_{2p+1,L+1}\}$ of length 2^{L+1}. The first signal $w_{2p,L+1}$ is generated by concatenating the signal $w_{p,L}$ with itself, and the second signal $w_{2p+1,L+1}$ is generated by concatenating the signal $w_{p,L}$ with $-w_{p,L}$.

For example, if $L = 2$, then from the above procedure the $2^2 = 4$ Walsh signals $\{w_{m,2}\}$ defined in (4.34) generates the $2^3 = 8$ Walsh signals $\{w_{m,3}\}$ as follows:

- $w_{0,2}$ generates $\{w_{0,3}, \ w_{1,3}\}$ where

$$w_{0,3} = \{1,1,1,1,1,1,1,1\} \ ; \ w_{1,3} = \{1,1,1,1,-1,-1,-1,-1\}$$

- $w_{1,2}$ generates $\{w_{2,3}, \ w_{3,3}\}$ where

$$w_{2,3} = \{1,1,-1,-1,1,1,-1,-1\} \ ; \ w_{3,3} = \{1,1,-1,-1,-1,-1,1,1\}$$

- $w_{2,2}$ generates $\{w_{4,3}, \ w_{5,3}\}$ where

$$w_{4,3} = \{1,-1,1,-1,1,-1,1,-1\} \ ; \ w_{5,3} = \{1,-1,1,-1,-1,1,-1,1\}$$

and finally

- $w_{3,2}$ generates $\{w_{6,3}, \ w_{7,3}\}$ where

$$w_{6,3} = \{1,-1,-1,1,1,-1,-1,1\} \ ; \ w_{7,3} = \{1,-1,-1,1,-1,1,1,-1\}$$

Least Norm Property

Another important advantage of having an orthogonal representation is that it provides a computationally simple way of *approximating* any given signal v with respect to this representation. In particular, we have the following result.

Theorem 4.1.7. *Let v be any given signal, and $\{\phi_n; 0 \leq n \leq N-1\}$ be any set of signals which is orthogonal with respect to some given inner product $\langle \cdot, \cdot \rangle$ having the same range and domain as v. Let v_N where*

$$v_N(k) = \sum_{n=0}^{N-1} \beta_n \phi_n(k) \qquad (4.35)$$

be any linear combination of the orthogonal signals $\{\phi_n\}$. Then the particular signal $\hat{v}_N$ which minimizes

$$J \triangleq ||v - v_N|| \qquad (4.36)$$

where $||x|| = \sqrt{\langle x, x \rangle}$ *is the norm induced by the inner product is given by*

$$\hat{v}_N(k) = \sum_{n=0}^{N-1} \hat{\beta}_n \phi_n(k) \ ; \ \hat{\beta}_n = \frac{\langle v, \phi_n \rangle}{||\phi_n||^2} \tag{4.37}$$

The corresponding minimum value $\hat{J}$ *of* J *is given by*

$$\hat{J} = \sqrt{||v||^2 - \sum_{m=0}^{N-1} |\hat{\beta}_m|^2 \ ||\phi_m||^2} \tag{4.38}$$

In order to prove this result, first observe that

$$J^2 = ||v - v_N||^2 = \langle v - v_N, v - v_N \rangle$$

Then after expansion

$$\begin{aligned} J^2 &= \langle v, v \rangle - \langle v, v_N \rangle - \langle v_N, v \rangle + \langle v_N, v_N \rangle \\ &= ||v||^2 - \langle v, v_N \rangle - \langle v_N, v \rangle + \sum_{m=0}^{N-1} |\beta_m|^2 \ ||\phi_m||^2 \end{aligned}$$

Now

$$\langle v, v_N \rangle = \sum_{m=0}^{N-1} \beta_m \langle v, \phi_m \rangle \ ; \ \langle v_N, v \rangle = \overline{\langle v, v_N \rangle} = \sum_{m=0}^{N-1} \overline{\beta_m} \ \overline{\langle v, \phi_m \rangle}$$

and so if we add and subtract the term

$$\sum_{m=0}^{N-1} \frac{|\langle v, \phi_m \rangle|^2}{||\phi_m||^2} = \sum_{m=0}^{N-1} \frac{\langle v, \phi_m \rangle}{||\phi_m||} \frac{\overline{\langle v, \phi_m \rangle}}{||\phi_m||}$$

in J^2, we get

$$\begin{aligned} J^2 = ||v||^2 &- \sum_{m=0}^{N-1} \frac{|\langle v, \phi_m \rangle|^2}{||\phi_m||^2} \\ &+ \sum_{m=0}^{N-1} \frac{\langle v, \phi_m \rangle}{||\phi_m||} \frac{\overline{\langle v, \phi_m \rangle}}{||\phi_m||} - \beta_m \langle v, \phi_m \rangle - \overline{\beta_m} \ \overline{\langle v, \phi_m \rangle} + |\beta_m|^2 \ ||\phi_m||^2 \end{aligned}$$

That is

$$J^2 = ||v||^2 - \sum_{m=0}^{N-1} \frac{|\langle v, \phi_m \rangle|^2}{||\phi_m||^2}$$
$$+ \sum_{m=0}^{N-1} \left\{ \frac{\langle v, \phi_m \rangle}{||\phi_m||} - \beta_m ||\phi_m|| \right\} \left\{ \overline{\frac{\langle v, \phi_m \rangle}{||\phi_m||} - \beta_m ||\phi_m||} \right\}$$

Since only the last term is dependent on β_m, we get the result as stated in Theorem 4.1.7.

Notice that the form of the cost J in (4.36) is determined by the inner product which in turn determines the orthogonality of the set $\{\phi_n; 0 \le n \le N-1\}$. In particular, if

$$\langle x, y \rangle = \sum_{k=0}^{L-1} \lambda_k x(k) \overline{y(k)} \;;\; \lambda_k > 0 \text{ for all } k$$

for some set of positive parameters $\{\lambda_k\}$, then

$$||v - v_N|| = \sqrt{\sum_{k=0}^{L-1} \lambda_k |v(k) - v_N(k)|^2}$$

This result is used as the basis for many applications of signal optimization. One key advantage of this result is that if the orthogonal set $\{\phi_n; 0 \le n \le N-1\}$ of N signals is augmented with a signal ϕ_N to produce an orthogonal set $\{\phi_n; 0 \le n \le N\}$ of $N+1$ signals, then the first N optimal coefficients $\{\hat{\beta}_n; 0 \le n \le N-1\}$ in the new optimal estimate

$$\hat{v}_{N+1}(k) = \sum_{m=0}^{N} \hat{\beta}_m \phi_m(k)$$

are the *same* as the N parameters in the optimal estimate $\hat{v}_N$ in (4.35); that is, we only need to calculate one new coefficient

$$\hat{\beta}_N = \frac{\langle v, \phi_N \rangle}{||\phi_N||^2}$$

Example 4.1.9. (a) Consider the two signals $\{\phi_0, \phi_1\}$ where

$$\phi_0 = \{1, 1, 1, 1\} \;;\; \phi_1 = \{2, -3, 2, -1\}$$

which are orthogonal with respect to the inner product

$$\langle x, y \rangle = \sum_{k=0}^{3} x(k) \overline{y(k)}$$

Then the signal $v_2 = \beta_0 \phi_0 + \beta_1 \phi_1$ which minimizes

$$||v - v_2|| = \sqrt{\sum_{k=0}^{3} |v(k) - v_2(k)|^2}$$

when $v = \{-2, 6, 4, 1\}$ is given by $\hat{v}_2 = \hat{\beta}_0\phi_0 + \hat{\beta}_1\phi_1$ where

$$\hat{\beta}_0 = \frac{\langle v, \phi_0\rangle}{\langle \phi_0, \phi_0\rangle} = \frac{9}{4} \; ; \; \hat{\beta}_1 = \frac{\langle v, \phi_1\rangle}{\langle \phi_1, \phi_1\rangle} = -\frac{5}{6}$$

The corresponding minimum value $\hat{J}$ is given by

$$\hat{J} = \sqrt{||v||^2 - \left\{|\hat{\beta}_0|^2||\phi_0||^2 + |\hat{\beta}_1|^2||\phi_1||^2\right\}}$$
$$= \sqrt{57 - \left\{\left(\frac{81}{16}\right)(4) + \left(\frac{25}{36}\right)(18)\right\}} = \sqrt{32.75}$$

(b) Now define

$$\phi_2 = \{1, 3, 1, -5\}$$

Then the set $\{\phi_0, \phi_1, \phi_2\}$ where $\{\phi_0, \phi_1\}$ are defined in part (a) is orthogonal with respect to the given inner product. Hence the signal $v_3 = \beta_0\phi_0 + \beta_1\phi_1 + \beta_2\phi_2$ which minimizes

$$J = ||v - v_3|| = \sqrt{\sum_{k=0}^{3} |v(k) - v_2(k)|^2} \tag{4.39}$$

is given by the signal

$$\hat{v}_3 = \hat{\beta}_0\phi_0 + \hat{\beta}_1\phi_1 + \hat{\beta}_2\phi_2$$

where

$$\hat{\beta}_0 = \frac{9}{4} \; ; \; \hat{\beta}_1 = -\frac{5}{6} \; ; \; \hat{\beta}_2 = \frac{\langle v, \phi_2\rangle}{\langle \phi_2, \phi_2\rangle} = \frac{5}{12}$$

The corresponding minimum value $\hat{J}$ is given by

$$\hat{J} = \sqrt{||v||^2 - \left\{|\hat{\beta}_0|^2||\phi_0||^2 + |\hat{\beta}_1|^2||\phi_1||^2 + |\hat{\beta}_2|^2||\phi_2||^2\right\}}$$
$$= \sqrt{57 - \left\{\left(\frac{81}{16}\right)(4) + \left(\frac{25}{36}\right)(18) + \left(\frac{25}{144}\right)(36)\right\}} = \sqrt{26.50}$$

which is less than $\hat{J}$ in (a). (The theory says that it cannot be greater.)

(c) Now suppose we consider the new inner product

$$\langle x, y\rangle = \sum_{k=0}^{3} \lambda_k x(k)\overline{y(k)}$$

where $\{\lambda_0 = \lambda_3 = 1;\ \lambda_1 = \lambda_2 = 2\}$. Then $\{\psi_0, \psi_2\}$ where

$$\psi_0 = \{1, 1, 1, 1\}\ ;\ \psi_1 = \{2, -1.5, 1, -1\}$$

form an orthogonal set. The signal $\hat{v}_2 = \hat{\gamma}_0\psi_0 + \hat{\gamma}_1\psi_1$ which minimizes

$$J = ||v - v_2|| = \sqrt{\sum_{k=0}^{3} \lambda_k |v(k) - v_2(k)|^2} \tag{4.40}$$

where v is defined in part (a) is given by

$$\hat{\gamma}_0 = \frac{\langle v, \psi_0\rangle}{\langle \psi_0, \psi_0\rangle} = \frac{19}{6}\ ;\ \hat{\gamma}_1 = \frac{\langle v, \psi_1\rangle}{\langle \psi_1, \psi_1\rangle} = -\frac{30}{23}$$

The corresponding minimum value $\hat{J}$ is given by

$$\begin{aligned}\hat{J} &= \sqrt{||v||^2 - \{|\hat{\gamma}_0|^2||\psi_0||^2 + |\hat{\gamma}_1|^2||\psi_1||^2\}}\\ &= \sqrt{109 - \left\{\left(\frac{361}{36}\right)(6) + \left(\frac{900}{529}\right)(11.5)\right\}} = \sqrt{29.26}\end{aligned}$$

Notice that the solutions in (a) and (c) *cannot* be compared since a different inner product means that we are minimizing different errors in (4.39) and (4.40).

Construction of Orthonormal Signals

Suppose v and w are two *independent* signals, and that the signal r is linearly dependent on v and w; that is

$$r = \gamma v + \delta w \tag{4.41}$$

for some constants $\{\gamma, \delta\}$. We now present a procedure for finding two *orthonormal* signals v_N and w_N such that *any* signal r in (4.41) can also be expressed in the form

$$r = \gamma_N v_N + \delta_N w_N \tag{4.42}$$

for some constants $\{\gamma_N, \delta_N\}$.

To begin, since $v \neq 0$, define

$$v_N \triangleq \frac{v}{||v||} \tag{4.43}$$

which implies $||v_N|| = 1$, and then define a signal y by

$$y = w - \alpha v_N \tag{4.44}$$

for some scalar α. Now

$$\langle y, v_N \rangle = \langle w - \alpha v_N, v_N \rangle = \langle w, v_N \rangle - \alpha \langle v_N, v_N \rangle = \langle w, v_N \rangle - \alpha$$

Hence $\langle y, v_N \rangle = 0$ if the scalar α is given by

$$\alpha = \langle w, v_N \rangle \tag{4.45}$$

Therefore the set of signals $\{v_N, w_N\}$ defined by

$$v_N \triangleq \frac{v}{||v||} \tag{4.46}$$
$$w_N \triangleq \frac{y}{||y||} \; ; \; y \triangleq w - \langle w, v_N \rangle v_N$$

are *orthonormal*. It then follows from (4.42) and (4.18) that

$$\begin{aligned}\langle r, v_N \rangle &= \langle \gamma_N v_N + \delta_N w_N, v_N \rangle \\ &= \gamma_N \langle v_N, v_N \rangle + \delta_N \langle w_N, v_N \rangle = \gamma_N\end{aligned}$$

Likewise, it follows that $\delta_N = \langle r, w_N \rangle$, so that (4.42) can also be expressed in the form

$$r = \langle r, v_N \rangle v_N + \langle r, w_N \rangle w_N \tag{4.47}$$

We have the following result.

Theorem 4.1.8. *Consider a set of bounded independent complex valued signals $\mathcal{S} = \{f_1, f_2, ..., f_p\}$ where $f_k : \mathcal{T} \to \mathcal{C}$, and suppose $\langle \cdot, \cdot \rangle$ is any inner product defined on $\mathcal{S}$. Then the set of signals $\mathcal{S}_N = \{\phi_1, \phi_2, ..., \phi_p\}$ defined by*

$$\phi_1 \triangleq \frac{f_1}{||f_1||} \tag{4.48}$$

and for $m = 2, 3, \ldots, p$

$$\phi_m \triangleq \frac{y_m}{||y_m||} \; ; \; y_m = f_m - \sum_{k=1}^{m-1} \langle f_m, \phi_k \rangle \phi_k \tag{4.49}$$

is a orthonormal set with respect to the given inner product.

Any signal v which is a linear combination of the *independent* signals $\{f_k\}$ in $\mathcal{S}$ can therefore be expressed in terms of the *orthonormal* signals $\{\phi_k\}$ in $\mathcal{S}_N$ in the form

$$f(k) = \sum_{m=1}^{p} < f, \phi_m > \phi_m(k) \ ; \ k \in \mathcal{Z}$$

Example 4.1.10. Consider the three digital signals $\{f_1, f_2, f_3\}$ as defined for all $-2 \leq k \leq 2$ by

$$f_1(k) = 1 \ ; \ f_2(k) = k \ ; \ f_3(k) = k^2$$

and the inner product

$$\langle f, g \rangle = \sum_{k=-2}^{2} f(k)\overline{g(k)}$$

Then

$$||f_1||^2 = \sum_{k=-2}^{2} |f(k)|^2 = 5$$

and so from (4.48)

$$\phi_1(k) = \frac{1}{\sqrt{5}}$$

Also from (4.49)

$$y_2 = f_2 - \langle f_2, \phi_1 \rangle \phi_1$$

where

$$\langle f_2, \phi_1 \rangle = (-2)(\frac{1}{\sqrt{5}}) + (-1)(\frac{1}{\sqrt{5}}) + 0 + (1)(\frac{1}{\sqrt{5}}) + (2)(\frac{1}{\sqrt{5}}) = 0$$

Hence $y_2 = f_2$, and so

$$\phi_2(k) = \frac{y_2(k)}{||y_2||} = \frac{k}{\sqrt{10}}$$

Finally,

$$y_3 = f_3 - \langle f_3, \phi_1 \rangle \phi_1 - \langle f_3, \phi_2 \rangle \phi_2$$

where

$$\langle f_3, \phi_1 \rangle = \frac{10}{\sqrt{5}} \ ; \ \langle f_3, \phi_2 \rangle = 0$$

and so

$$y_3(k) = k^2 - \left(\frac{10}{\sqrt{5}}\right)\frac{1}{\sqrt{5}} = k^2 - 2$$

Since $||y_3|| = \sqrt{14}$, we have

$$\phi_3(k) = \frac{k^2 - 2}{\sqrt{14}}$$

4.1.4 Signal Bounds

A signal model is said to be in *input-output form* if the output signal y is expressed directly in terms of the input signal u in the form of either an nth order difference equation, or a linear convolution. In a *state space model*, the output y is expressed indirectly in terms of the input u since the model also involves the components x_m of the state vector $\boldsymbol{x}$.

In this section, signal norms and inner products are used to derive a magnitude bounds on the output signal y in terms of a norm on the input signal u. These results are particularly useful in Chapter 5 for both scaling a digital filter, and analysing the errors as a result of a finite wordlength implementation of a digital filter.

Convolution

Signal norms and inner products can be used for obtaining bounds on the magnitude of a signal y which is the response of a difference equation in terms of a norm of the input signal u. In this way, some information about the solution is available even though the precise form of the input signal is unknown. To begin, consider the nth order difference equation

$$y(k+n) = \sum_{p=0}^{n-1} a_p y(k+p) + \sum_{p=0}^{n} b_p u(k+p) \tag{4.50}$$

Then as shown in Chapter 2, the complete response y for $k \geq 0$ can also be expressed in the form

$$y(k) = y_i(k) + \sum_{m=0}^{k} h(k-m)u(m) \tag{4.51}$$

where y_i is the zero input response, and h is the unit impulse response. Therefore

$$|y(k)| \leq |y_i(k)| + |\sum_{m=0}^{k} h(k-m)u(m)| \leq |y_i(k)| + (\sum_{m=0}^{k} |h(k-m)|)\ ||u||_\infty$$

where $||\cdot||_\infty$ is the *infinity norm* defined by (4.10). Now

$$|\sum_{m=0}^{k} h(k-m)| \leq \sum_{p=0}^{k} |h(p)| = ||h||_1$$

where $||\cdot||_1$ is the 1-norm defined by (4.11). Now by definition of the infinity norm, $|y_i(k)| \leq ||y_i||_\infty$. Hence we conclude that

$$||y||_\infty \leq ||y_i||_\infty + ||h||_1 \, ||u||_\infty \tag{4.52}$$

Alternatively, from (4.51), we have

$$|y(k)| \leq |y_i(k)| + ||h||_\infty (\sum_{m=0}^{k} |u(m)|)$$

which gives a second norm inequality condition

$$||y||_\infty \leq ||y_i||_\infty + ||h||_\infty \, ||u||_1 \tag{4.53}$$

In order to obtain a third norm inequality, observe that the *linear convolution* in (4.51) is actually an *inner product*; that is

$$\sum_{m=0}^{k} h(k-m)u(m) = \langle g, u \rangle \triangleq \sum_{m=0}^{k} g(m)u(m)$$

where $\{g(m) = h(k-m);\ 1 \leq m \leq k\}$. Then since $||g||_2 = ||h||_2$, we conclude from Theorem 4.1.4 that

$$|\sum_{m=0}^{k} h(k-m)u(m)| \leq ||h||_2 \, ||u||_2$$

and hence we have the norm inequality

$$||y||_\infty \leq ||y_i||_\infty + ||h||_2 \, ||u||_2 \tag{4.54}$$

We summarize these results as follows.

Theorem 4.1.9. *Upper bounds on the infinity norm of the response y of the linear time invariant difference equation (4.50) are given in terms of norms on the unit impulse response h and the input signal u by:*

$$||y||_\infty \leq ||y_i||_\infty + L$$

where y_i is the zero input response and L is given by any one of the following expressions:

$$L = ||h||_1 \, ||u||_\infty \ ; \ L = ||h||_2 \, ||u||_2 \ ; \ L = ||h||_\infty \, ||u||_1$$

The usefulness of these bounds depends on which signal norms are available, or can be calculated.

First Order Equation: Consider a *first order* (i.e. $n = 1$) difference equation (4.50). Then $\{h(0) = b_1, h(1) = a_0 b_1 + b_0\}$ and $\{h(k) = h(1)a_0^{k-1}, k \geq 1\}$, and so for $|a_0| < 1$

$$||h||_1 = |b_1| + \frac{|a_0 b_1 + b_0|}{1 - |a_0|} \; ; \; ||h||_2 = \sqrt{b_1^2 + \frac{(a_0 b_1 + b_0)^2}{1 - a_0^2}}$$
$$||h||_\infty = \max\{|b_1|, |a_0 b_1 + b_0|\}$$

Second Order Equation: For a *second order* (i.e. $n = 2$) difference equation (4.50), we have $\{h(0) = b_2, h(1) = a_1 b_2 + b_1, h(2) = b_0 + a_1 h(1) + a_0 b_2\}$. Then if $\{\lambda_1, \lambda_2\}$ are the roots of the characteristic equation

$$\lambda^2 - a_1 \lambda - a_0 = 0$$

we have: for $k \geq 1$

$$h(k) = \begin{cases} (\beta_1 + k\beta_2)\lambda_1^{k-1} & ; \; \lambda_1 = \lambda_2 \\ \beta_1 \lambda_1^{k-1} + \beta_2 \lambda_2^{k-1} & ; \; \lambda_1 \neq \lambda_2 \end{cases}$$

where the coefficients $\{\beta_1, \beta_2\}$ are given in terms of $\{h(1), h(2)\}$ (and consequently in terms of the coefficients $\{a_1, a_0, b_2, b_1, b_0\}$).

Assuming the difference equation is asymptotically stable (i.e. $|\lambda_k| < 1, \; k = 1, 2$), we then conclude for $\lambda_1 \neq \lambda_2$ that

$$\begin{aligned} ||h||_1 &= |b_2| + \sum_{k=1}^{\infty} |\beta_1 \lambda_1^{k-1} + \beta_2 \lambda_2^{k-1}| \\ &\leq |b_2| + |\beta_1| \sum_{k=1}^{\infty} |\lambda_1|^{k-1} + |\beta_2| \sum_{k=1}^{\infty} |\lambda_2|^{k-1} \end{aligned}$$

That is,

$$||h||_1 \leq |b_2| + \frac{|\beta_1|}{1 - |\lambda_1|} + \frac{|\beta_2|}{1 - |\lambda_2|} \tag{4.55}$$

Another bound is given directly from (4.50). That is, for $n = 2$

$$\begin{aligned} |h(k+2)| \leq\; & |a_1||h(k+1)| + |a_0||h(k)| \\ & + |b_0||\delta(k)| + |b_1||\delta(k+1)| + |b_2||\delta(k+2)| \end{aligned}$$

where δ is a discrete unit impulse signal, and so

$$\sum_{k=0}^{\infty} |h(k+2)| \leq |a_1| \sum_{k=0}^{\infty} |h(k+1)| + |a_0| \sum_{k=0}^{\infty} |h(k)| + |b_0| \sum_{k=0}^{\infty} |\delta(k)| \\ + |b_1| \sum_{k=0}^{\infty} |\delta(k+1)| + |b_2| \sum_{k=0}^{\infty} |\delta(k+2)|$$

Now define $L_1 \triangleq ||h||_1$. Then the above inequality gives

$$L_1 - |h(0)| - |h(1)| \leq |a_1|\{L_1 - |h(0)|\} + |a_0| L_1 + |b_0|$$

Then provided $|a_0| + |a_1| < 1$, we conclude

$$||h||_1 \leq \frac{|b_0| + |h(0)| + |h(1)|}{1 - |a_0| - |a_1|} \tag{4.56}$$
$$|h(0)| = |b_2|, \; |h(1)| = |a_1 b_2 + b_1|$$

Note that asymptotic stability of (4.50) does *not* implies $1 - |a_0| - |a_1| > 0$, so the bound in (4.56) is not always applicable.

The norm bounds in (4.55) and (4.56) are not necessarily the same. We also have

$$||h||_\infty = \max_k \{|h(0)|, |\beta_1 \lambda_1^{k-1} + \beta_2 \lambda_2^{k-1}|\} \\ \leq \max\{|b_2|, |\beta_1| + |\beta_2|\} \tag{4.57}$$

Whereas only *upper bounds* are available for both the 1-norm and the infinity norm, an *exact expression* is available for the 2-norm. Specifically, for $\lambda_1 \neq \lambda_2$

$$||h||_2^2 = h^2(0) + \sum_{k=1}^{\infty} \{\beta_1^2 \lambda_1^{2(k-1)} + 2\beta_1\beta_2(\lambda_1\lambda_2)^{k-1} + \beta_2^2 \lambda_2^{2(k-1)}\}$$

and so since $\{|\lambda_k| < 1; k = 1, 2\}$, we have that

$$||h||_2 = \sqrt{h^2(0) + \frac{\beta_1^2}{1-\lambda_1^2} + \frac{2\beta_1\beta_2}{1-\lambda_1\lambda_2} + \frac{\beta_2^2}{1-\lambda_2^2}} \tag{4.58}$$

That is, even though $\{\beta_i, \lambda_j\}$ are (possibly complex) functions of the coefficients $\{a_i, b_j\}$, an exact (real) expression for $||h||_2$ is available.

Example 4.1.11. Consider the difference equation

$$y(k+2) = -0.1y(k+1) + 0.2y(k) - u(k+1) + 0.13u(k)$$

which in (4.50) corresponds to

$$a_1 = -0.1,\ a_0 = 0.2,\ b_2 = 0,\ b_1 = -1,\ b_0 = 0.13$$

the unit impulse response h is given by $\{h(0) = 0,\ h(1) = -1,\ h(2) = 0.23\}$, and more generally

$$h(k) = \beta_1(-0.5)^{k-1} + \beta_2(0.4)^{k-1}\ ;\ k \geq 1$$

where

$$-1 = \beta_1 + \beta_2\ ;\ 0.23 = -0.5\beta_1 + 0.4\beta_2$$

gives $\{\beta_1 = -0.7,\ \beta_2 = -0.3\}$. Therefore from (4.55)

$$||h||_1 \leq \frac{0.7}{1-0.5} + \frac{0.3}{1-0.4} = 1.9$$

whereas from (4.56) since $1 - |a_0| - |a_1| = 0.7 > 0$

$$||h||_1 \leq \frac{0.13+1}{1-0.1-0.2} = 1.61$$

Extending the derivations of the second order filter to an nth order filter, we obtain the following results.

Theorem 4.1.10. *(i) An upper bound on the 1-norm of the unit impulse response of an asymptotically stable difference equation (4.50) is given by*

$$||h||_1 \leq \frac{|b_0| + \sum_{k=0}^{n-1}|h(k)|}{1 - \sum_{k=0}^{n}|a_k|} \tag{4.59}$$

(ii) Suppose no two characteristic roots are equal. Then an upper bound on the infinity norm of the unit impulse response of an asymptotically stable difference equation (4.50) is given by

$$||h||_\infty \leq \max_{1\leq k\leq n-1}\{|h(k)|, \sum_{k=1}^{n}|\beta_n|\} \tag{4.60}$$

where

$$h(k) = \sum_{j=1}^{n}\beta_j\lambda_j^{k-1}$$

An exact expression for the 2-norm $||h||_2$ of the nth order difference equation (4.50) is developed later in Theorem 4.1.11 based on a state space representation of (4.50).

State Space Representation

Consider the state space representation

$$\begin{aligned} \boldsymbol{x}(k+1) &= \boldsymbol{F}\boldsymbol{x}(k) + \boldsymbol{g}u(k) \ ; \ \boldsymbol{x}(k) \in \mathcal{R}^n \\ y(k) &= \boldsymbol{c}^T\boldsymbol{x}(k) + du(k) \end{aligned} \tag{4.61}$$

of the nth order difference equation (4.50). In order to obtain bounds for each of the components $x_m(k)$ of the state vector $\boldsymbol{x}(k)$, we define the n-vectors $\{\boldsymbol{c}_m;\ 1 \leq m \leq n\}$ by

$$\boldsymbol{c}_m^T = [0 \ \ 0 \ \ \ldots \ \ 0 \ \ 1 \ \ 0 \ \ \ldots \ \ 0] \tag{4.62}$$

where the 1 appears in the mth component of $\boldsymbol{c}_m$, and all other components are zero. This implies $x_m(k) = \boldsymbol{c}_m^T\boldsymbol{x}(k)$. Consequently, using the results of Chapter 2, we have from (4.61) that

$$\begin{aligned} x_m(k) &= x_{mi}(k) + \sum_{p=0}^{k} h_m(k-p)u(k) \\ y(k) &= y_i(k) + \sum_{p=0}^{k} h(k-p)u(k) \end{aligned}$$

where the initial condition responses $\{x_{mi}, y_i\}$ are given by

$$x_{mi}(k) = \boldsymbol{c}_m^T\boldsymbol{F}^{k-1}\boldsymbol{x}(0) \ ; \ y_i(k) = \boldsymbol{c}^T\boldsymbol{F}^{k-1}\boldsymbol{x}(0) \tag{4.63}$$

and the unit impulse response h from the input signal u to the output signal y, and the unit impulse response h_m from the input signal u to the state component x_m are given by

$$h_m(k) = \begin{cases} 0 & ; \ k \leq 0 \\ \boldsymbol{c}_m^T\boldsymbol{F}^{k-1}\boldsymbol{g} & ; \ k \geq 1 \end{cases} \ ; \ h(k) = \begin{cases} 0 & ; \ k < 0 \\ d & ; \ k = 0 \\ \boldsymbol{c}^T\boldsymbol{F}^{k-1}\boldsymbol{g} & ; \ k \geq 1 \end{cases} \tag{4.64}$$

In this state space formaulation, the close relationship between the zero input responses and the unit impulse responses are evident. From Theorem 4.1.9, we conclude that

$$||x_m||_\infty \leq ||x_{mi}||_\infty + ||h_m||_1 \; ||u||_\infty$$
$$||x_m||_\infty \leq ||x_{mi}||_\infty + ||h_m||_2 \; ||u||_2$$
$$||x_m||_\infty \leq ||x_{mi}||_\infty + ||h_m||_\infty \; ||u||_1$$

We now develop an expression for the 2-norms.

To begin, from (4.64), we have

$$h_m^2(0) = 0 \; ; \; h_m^2(k) = \boldsymbol{c}_m^T \boldsymbol{F}^{k-1} \boldsymbol{g}\boldsymbol{g}^T (\boldsymbol{F}^{k-1})^T \boldsymbol{c}_m \; , \; k \geq 1$$

so that

$$||h_m||_2^2 = \sum_{k=0}^{\infty} h_m^2(k) = \boldsymbol{c}_m^T \boldsymbol{K} \boldsymbol{c}_m \; ; \; \boldsymbol{K} \triangleq \sum_{k=1}^{\infty} \boldsymbol{F}^{k-1} \boldsymbol{g}\boldsymbol{g}^T (\boldsymbol{F}^{k-1})^T \qquad (4.65)$$

provided the infinite summation in (4.65) leads to a matrix $\boldsymbol{K}$ whose components are all bounded.

We do not formally prove the result here, However it can be shown that this infinite summation exists if the difference equation (4.50) is asymptotically stable when is equivalent to the condition that all the eigenvalues λ_k of $\boldsymbol{F}$ in the state space representation (4.61) satisfy the condition $|\lambda_k| < 1$.

Given that $\boldsymbol{K}$ in (4.65) exists, we then conclude that

$$\boldsymbol{F}\boldsymbol{K}\boldsymbol{F}^T + \boldsymbol{g}\boldsymbol{g}^T = \sum_{k=1}^{\infty} \boldsymbol{F}^k \boldsymbol{g}\boldsymbol{g}^T (\boldsymbol{F}^k)^T + \boldsymbol{g}\boldsymbol{g}^T = \boldsymbol{K}$$

Furthermore

$$\boldsymbol{K}^T = \sum_{k=1}^{\infty} [\boldsymbol{F}^{k-1} \boldsymbol{g}\boldsymbol{g}^T (\boldsymbol{F}^{k-1})^T]^T = \sum_{k=1}^{\infty} \boldsymbol{F}^{k-1} \boldsymbol{g}\boldsymbol{g}^T (\boldsymbol{F}^{k-1})^T$$

and so $\boldsymbol{K}^T = \boldsymbol{K}$. Any (square) matrix $\boldsymbol{K}$ which has this property is said to be *symmetric*. We summarize these results as follows.

Theorem 4.1.11. *The 2-norms* $\{||h_m||_2, \; ||h||_2\}$ *of the unit impulse responses* $\{h_m, \; h\}$ *of the state component* x_m *and the output* y *respectively of an asymptotically state space representation (4.61) are given by*

$$||h_m||_2 = \sqrt{\boldsymbol{c}_m^T \boldsymbol{K} \boldsymbol{c}_m} \; ; \; ||h||_2 = \sqrt{\boldsymbol{c}^T \boldsymbol{K} \boldsymbol{c}}$$

where $\boldsymbol{c}_m$ *is defined by (4.62), and* $\boldsymbol{K}$ *is the solution of the linear algebraic equation*

$$\boldsymbol{K} = \boldsymbol{F}\boldsymbol{K}\boldsymbol{F}^T + \boldsymbol{g}\boldsymbol{g}^T \qquad (4.66)$$

As mentioned earlier, an important signal processing application of scaling methods is in the design of a digital filter (or for that matter any signal processing algorithm) for which the precise form of the input signal u is unknown; but instead only some overall bound $||u||_1$, $||u||_2$ or $||u||_\infty$ is apriori available. Given such information, a magnitude bound $||x_m||_\infty$ can then be calculated for each component x_m of a state space representation (4.61) of a digital filter $H(z) = d + \boldsymbol{c}^T(z\boldsymbol{I} - \boldsymbol{F})^{-1}\boldsymbol{g}$. That is, from Theorem 4.1.9, if $x_{m,i}$ is the initial condition response of x_m, and h_m is the unit impulse response from the input u to the filter state component x_m, then

$$||x_m||_\infty \leq ||x_{m,i}||_\infty + L_\infty$$

where

$$L_\infty = ||h_m||_1||u||_\infty \quad \text{or} \quad L_\infty = ||h_m||_2||u||_2 \quad \text{or} \quad L_\infty = ||h_m||_\infty||u||_1$$

Clearly, the particular norm $||h_m||$ that must be calculated depends on which norm $||u||$ of the input u is available. However, as already noted, an exact expression is only available for $||h_m||_2$. If either $||h_m||_1$ or $||h_m||_\infty$ is required, then only (generally conservative) analytical estimates are available; otherwise $||h_m||_1$ and $||h_m||_\infty$ can be estimated by numerical computation; that is, for some 'sufficiently large' integer N

$$||h_m||_1 \simeq \sum_{k=0}^{N} |h_m(k)| \; ; \; ||h_m||_\infty \simeq \max_{0 \leq k \leq N} |h_m(k)|$$

Unfortunately, for narrow bandwidth filters, the unit impulse responses $\{h_m\}$ only slowly decay to zero, and so it is not apriori obvious how many $(= N)$ points should be used to compute these estimates.

However once the $\{h_m\}$ are calculated for one particular state space filter structure, scaling can be implemented to bound the magnitudes of the internal signals. That is, if x_m are the state components of $\boldsymbol{x}$ and $\{s_m > 0\}$ are scaling parameters, then

$$z_m = s_m x_m \; ; \; 1 \leq m \leq n$$

implies that the state $\boldsymbol{z}$ where

$$\boldsymbol{z} = \boldsymbol{S}\boldsymbol{x} \; ; \; \boldsymbol{S} = diag\{s_1, s_2, \ \ldots \ , s_n\}$$

results in the scaled state space filter structure

$$\begin{aligned} \boldsymbol{z}(k+1) &= \boldsymbol{F}_s\boldsymbol{z}(k) + \boldsymbol{g}_s u(k) \\ y(k) &= \boldsymbol{c}_s^T\boldsymbol{z}(k) + du(k) \end{aligned}$$

where

$$\boldsymbol{F}_s = \boldsymbol{S}\boldsymbol{F}\boldsymbol{S}^{-1} \; ; \; \boldsymbol{g}_s = \boldsymbol{S}\boldsymbol{g} \; ; \; \boldsymbol{c}_s^T = \boldsymbol{c}^T\boldsymbol{S}^{-1}$$

We now conclude this section with some further comments on the (exact) calculation of the 2-norm $||h_m||_2$. Normally the calculation of such norms via the linear algebraic equation (4.66) requires a numerical solution for which numerically efficient routines (including in MATLAB) are available. However, in some cases, analytical solutions are also possible. In particular, since $\boldsymbol{K}$ is symmetric, there are only $0.5n(n+1)$ *different* components of $\boldsymbol{K}$. The linear algebraic equation (4.66) can the be reduced to a system of linear equations of the form

$$\boldsymbol{M}\boldsymbol{\ell} = \gamma$$

where $\boldsymbol{M}$ is a square matrix of dimension $0.5n(n+1)$, and $\boldsymbol{\ell}$ is a $0.5n(n+1)$ dimensional vector whose components are the $0.5n(n+1)$ different components of the symmetric matrix $\boldsymbol{K}$.

Example 4.1.12. (i) Consider the second order difference equation

$$y(k+2) = a_1 y(k+1) + a_0 y(k) + b_0 u(k) + b_1 u(k+1)$$

and the phase variable state space representation (4.61) where

$$\boldsymbol{F} = \begin{bmatrix} 0 & 1 \\ a_0 & a_1 \end{bmatrix} \; ; \; \boldsymbol{g} = \begin{bmatrix} 0 \\ 1 \end{bmatrix} \; ; \; \boldsymbol{c} = \begin{bmatrix} b_0 \\ b_1 \end{bmatrix} \; ; \; d = 0$$

This equation is asymptotically stable, and hence both eigenvalues of $\boldsymbol{F}$ have modulus less than unity if and only if

$$|a_0| < 1 \; ; \; |a_1| < 1 - a_0$$

Then since $\boldsymbol{K} = \boldsymbol{K}^T$, the solution $\boldsymbol{K}$ of (4.66) is of the form

$$\boldsymbol{K} = \begin{bmatrix} k_{11} & k_{12} \\ k_{12} & k_{22} \end{bmatrix}$$

After substituting into (4.66), we have

$$\boldsymbol{F}\boldsymbol{K}\boldsymbol{F}^T + \boldsymbol{g}\boldsymbol{g}^T = \begin{bmatrix} k_{22} & a_0 k_{12} + a_1 k_{22} \\ a_0 k_{12} + a_1 k_{22} & a_0^2 k_{11} + 2a_0 a_1 k_{12} + a_1^2 k_{22} + 1 \end{bmatrix}$$

which gives the $0.5n(n+1) = 3$ equations

$$k_{11} = k_{22} \; ; \; k_{12} = a_0 k_{12} + a_1 k_{22}$$
$$k_{22} = a_0^2 k_1 + 2a_0 a_1 k_{12} + a_1^2 k_{22} + 1$$

Hence $\boldsymbol{K}$ is given by

$$\boldsymbol{K} = \beta \begin{bmatrix} 1 & a_1(1-a_0)^{-1} \\ a_1(1-a_0)^{-1} & 1 \end{bmatrix}$$

$$\beta = \frac{1}{1 - a_0^2 - a_1^2 - 2a_0 a_1^2 (1-a_0)^{-1}}$$

(Note that asymptotic stability implies $1 - a_0 \neq 0$ and β finite.)

The 2-norm of the unit impulse response h is therefore given by

$$||h||_2 = \sqrt{\boldsymbol{c}^T \boldsymbol{K} \boldsymbol{c}}$$

where

$$\boldsymbol{c}^T \boldsymbol{K} \boldsymbol{c} = \beta[b_0^2 + \frac{2b_0 b_1 a_1}{1-a_0} + b_1^2]$$

(ii) Suppose both eigenvalues $\{\lambda_1, \lambda_2\}$ of $\boldsymbol{F}$ are real and unequal. Then another state space represenatation is given by

$$\boldsymbol{F} = \begin{bmatrix} \lambda_1 & 0 \\ 0 & \lambda_2 \end{bmatrix} \;;\; \boldsymbol{g} = \begin{bmatrix} 1 \\ 1 \end{bmatrix} \;;\; \boldsymbol{c} = \begin{bmatrix} \gamma_1 \\ \gamma_2 \end{bmatrix} \;;\; d = 0$$

where

$$\frac{b_0 + b_1 z}{z^2 - a_1 z - a_0} = \frac{\gamma_1}{z - \lambda_1} + \frac{\gamma_2}{z - \lambda_2}$$

Then

$$\boldsymbol{F}\boldsymbol{K}\boldsymbol{F}^T + \boldsymbol{g}\boldsymbol{g}^T = \begin{bmatrix} \lambda_1^2 k_{11} + 1 & \lambda_1\lambda_2 k_{12} + 1 \\ \lambda_1\lambda_2 k_{12} + 1 & \lambda_2^2 k_{22} + 1 \end{bmatrix}$$

which implies $\boldsymbol{K}$ is given by

$$\boldsymbol{K} = \begin{bmatrix} (1-\lambda_1^2)^{-1} & (1-\lambda_1\lambda_2)^{-1} \\ (1-\lambda_1\lambda_2)^{-1} & (1-\lambda_2^2)^{-1} \end{bmatrix}$$

(iii) Suppose both eigenvalues $\{\lambda_1, \lambda_1\}$ of $\boldsymbol{F}$ are real and equal. Then another state space represenatation is given by

$$\boldsymbol{F} = \begin{bmatrix} \lambda_1 & 1 \\ 0 & \lambda_1 \end{bmatrix} \;;\; \boldsymbol{g} = \begin{bmatrix} 0 \\ 1 \end{bmatrix} \;;\; \boldsymbol{c} = \begin{bmatrix} \gamma_1 \\ \gamma_2 \end{bmatrix} \;;\; d = 0$$

Then

$$\boldsymbol{F}\boldsymbol{K}\boldsymbol{F}^T + \boldsymbol{g}\boldsymbol{g}^T = \begin{bmatrix} \lambda_1^2 k_{11} + 2\lambda_1 k_{12} + k_{22} & \lambda_1^2 k_{12} + \lambda_1 k_{22} \\ \lambda_1^2 k_{12} + \lambda_1 k_{22} & \lambda_1^2 k_{22} + 1 \end{bmatrix}$$

which implies $\boldsymbol{K}$ is given by

$$\boldsymbol{K} = \beta \begin{bmatrix} 1 + \lambda_1^2 & \lambda_1(1-\lambda_1^2) \\ \lambda_1(1-\lambda_1^2) & (1-\lambda_1^2)^2 \end{bmatrix} \;;\; \beta = \frac{1}{(1-\lambda_1)^3}$$

4.2 Discrete Fourier Transform

We have seen in section 3.1.3 that if $\{\phi_n : 0 \leq n \leq N-1\}$ is a set of orthogonal signals (i.e. $\langle \phi_n, \phi_m \rangle = 0$ for $n \neq m$), then for any signal v which is a linear combination of the $\{\phi_n\}$; that is

$$v(k) = \sum_{m=0}^{N-1} \beta_m \phi_m(k) \tag{4.67}$$

we have

$$\beta_m = \frac{\langle v, \phi_m \rangle}{\langle \phi_m, \phi_m \rangle} \tag{4.68}$$

In particular, if the set $\{\phi_n\}$ is also *orthonormal* (i.e. $\langle \phi_n, \phi_n \rangle = 1$ for n), then $\beta_m = \langle v, \phi_m \rangle$.

We have also seen that one important advantage of having an orthogonal representation (when compared to a non-orthogonal representation) is that it is relatively easy to compute the *least norm approximation* of a given signal in terms of such a representation. that is, given *any* signal v (which is not necessarily a linear combination of the orthogonal signals $\{\phi_n\}$), then the 2-norm $||v - \sum_{m=0}^{N-1} \beta_m \phi_m||_2$ is minimized when the coefficients β_m are given by (4.68).

In this section, we show that an infinite length *periodic signal* can be conveniently represented by an orthonormal set of complex exponential signals. Furthermore, once it is shown how any *finite length* signal can be represented by a periodic signal, it will follow that any finite length signal can also be represented by a set of orthonormal signals. Such a representation of both periodic and a finite length signals is known as the *Discrete Fourier Transform* representation.

4.2.1 Periodic Signal

For any positive integers $\{N, N_0\}$, consider the set of complex valued periodic signals $\mathcal{S}_{N,N_0}$ as defined by

$$\mathcal{S}_{N,N_0} \triangleq \{\phi_{n,N_0} \; ; \; 0 \leq n \leq N-1\} \tag{4.69}$$

where $\phi_{n,N_0} : \mathcal{Z} \to \mathcal{C}$ for each $n \in \mathcal{Z}$ is defined by

$$\phi_{n,N_0}(k) \triangleq exp\{\frac{j2\pi nk}{N_0}\} \; ; \; k \in \mathcal{Z} \tag{4.70}$$

Example 4.2.1. (a) When $N_0 = 3$ and $N = 5$, (4.69) and (4.70) define the five periodic signals $\{\phi_{0,3}, \phi_{1,3}, \phi_{2,3}, \phi_{3,3}, \phi_{4,3}\}$ by:

$$\phi_{0,3}(k) = 1 \ ; \ \phi_{1,3}(k) = exp\{\frac{j2\pi k}{3}\} \ ; \ \phi_{2,3}(k) = exp\{\frac{j4\pi k}{3}\}$$

and

$$\phi_{3,3}(k) = exp\{\frac{j6\pi k}{3}\} = 1$$
$$\phi_{4,3}(k) = exp\{\frac{j8\pi k}{3}\} = exp\{\frac{j2\pi k}{3}\} = \phi_{1,3}(k)$$

(b) The component values $\phi_{1,3}(k)$ for $0 \leq k \leq 4$ are:

$$\phi_{1,3}(0) = 1 \ ; \ \phi_{1,3}(1) = exp\{\frac{j2\pi}{3}\} \ ; \ \phi_{1,3}(2) = exp\{\frac{j4\pi}{3}\}$$

and

$$\phi_{1,3}(3) = exp\{\frac{j6\pi}{3}\} = 1 \ ; \ \phi_{1,3}(4) = exp\{\frac{j8\pi}{3}\} = exp\{\frac{j2\pi}{3}\} = \phi_{1,3}(1)$$

where

$$exp\{\frac{j2\pi}{3}\} = -\frac{1}{2} + j\frac{\sqrt{3}}{2} \ ; \ exp\{\frac{j4\pi}{3}\} = -\frac{1}{2} - j\frac{\sqrt{3}}{2}$$

(c) The component values $\phi_{2,3}(k)$ for $0 \leq k \leq 4$ are:

$$\phi_{2,3}(0) = 1 \ ; \ \phi_{2,3}(1) = exp\{\frac{j4\pi}{3}\} \ ; \ \phi_{2,3}(2) = exp\{\frac{j8\pi}{3}\} = exp\{\frac{j2\pi}{3}\}$$

and

$$\phi_{2,3}(3) = exp\{\frac{j12\pi}{3}\} = 1 \ ; \ \phi_{2,3}(4) = exp\{\frac{j16\pi}{3}\} = exp\{\frac{j4\pi}{3}\} = \phi_{2,3}(1)$$

More generally, given any integer $N_0 > 0$, we have the following properties of the signal ϕ_{n,N_0} defined by (4.70):

- For all integers n, k

  $$[\phi_{n,N_0}(k)]^{N_0} = 1$$

  and hence $\phi_{n,N_0}(k)$ for each $k \in \mathcal{Z}$ is an N_0th root of unity. The roots of unity when $N_0 = 5$ are illustrated in Figure 4.4.
- For all integers m, $exp\{j2\pi m\} = 1$ which implies that

  $$\phi_{n,N_0}(k) = \phi_{n,N_0}(k + N_0) \tag{4.71}$$

  Hence all signals $\{\phi_{n,N_0}\}$ in the set $\mathcal{S}_{N_0}$ are periodic with period *at most* N_0. However since there may still be a *smaller* integer N_1 such that

  $$\phi_{n,N_0}(k) = \phi_{n,N_0}(k + N_1)$$

  and so a particular signal $\{\phi_{n,N_0}\}$ may in fact have period $N_1 < N_0$.

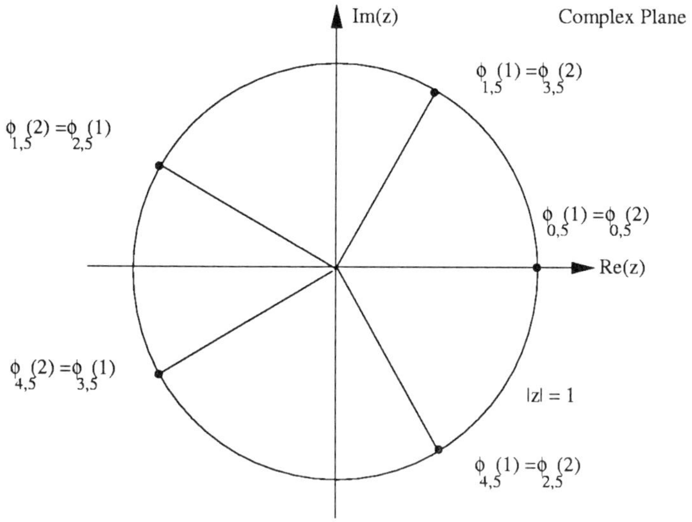

Fig. 4.4. Complex roots of unity

- For all integers n, k

$$\phi_{n,N_0}(k) = \phi_{n+N_0,N_0}(k) \tag{4.72}$$

  Hence there are *at most* N_0 *different* signals $\{\phi_{n,N_0}\}$ in the set $\mathcal{S}_{N,N_0}$ in (4.69). Furthermore, since there are *only* N_0 different signals for all $N \geq N_0$, and so there is no need to consider $N > N_0$.

We have the following result.

Theorem 4.2.1. *(i) For $N \geq N_0$, the set $\mathcal{S}_{N,N_0}$ in (4.69) with ϕ_{n,N_0} given by (4.70) is the same as the set*

$$\mathcal{S}_{N_0} \triangleq \{\phi_{n,N_0} \ ; \ 0 \leq n \leq N_0 - 1\} \ ; \ \phi_{n,N_0}(k) = exp\{\frac{j2\pi nk}{N_0}\} \tag{4.73}$$

(ii) For each $0 \leq n \leq N_0 - 1$, the signal ϕ_{n,N_0} is periodic of period P_{n,N_0} where P_{n,N_0} is of the form

$$P_{n,N_0} = \frac{N_0}{m} \quad \text{for some integer} \quad m > 0 \tag{4.74}$$

Example 4.2.2. Suppose $N_0 = 6$. Then the six signals in the set $\mathcal{S}_6$ in (4.73) are given by:

$$\phi_{0,6}(k) = 1 \ ; \ \phi_{1,6}(k) = exp\{\frac{j\pi k}{3}\} \ ; \ \phi_{2,6}(k) = exp\{\frac{j2\pi k}{3}\}$$

$$\phi_{3,6}(k) = exp\{j\pi k\} \ ; \ \phi_{4,6}(k) = exp\{\frac{j4\pi k}{3}\} \ ; \ \phi_{5,6}(k) = exp\{\frac{j5\pi k}{3}\}$$

Since by definition, the period of a periodic signal $s : \mathcal{Z} \to \mathcal{C}$ is the *smallest* integer P such that $s(k) = s(k+P)$ for all $n \in \mathcal{Z}$, we have that: the signal $\phi_{0,6}$ has period $P_{0,6} = 1$, the signal $\phi_{1,6}$ has period $P_{1,6} = 6$, the signal $\phi_{2,6}$ has period $P_{2,6} = 3$, the signal $\phi_{3,6}$ has period $P_{3,6} = 2$, the signal $\phi_{4,6}$ has period $P_{4,6} = 3$, and the signal $\phi_{5,6}$ has period $P_{5,6} = 6$. All periods $P_{n,6}$ are of the form (4.74).

Also

$$\phi_{6,6}(k) = 1 = \phi_{0,6}(k)$$

$$\phi_{7,6}(k) = exp\{\frac{j7\pi n}{3}\} = exp\{\frac{j\pi n}{3}\} = \phi_{1,6}(k)$$

and in general (as in (4.72) with $N_0 = 6$)

$$\phi_{n,6}(k) = \phi_{n+6,6}(k) \ ; \ n \in \mathcal{Z}$$

That is, the set $\mathcal{S}_6$ is the same as the set $\mathcal{S}_{N,6}$ in (4.69) for all $N \geq 6$.

Now, consider the *inner product* between the periodic signals ϕ_{n,N_0} and ϕ_{ℓ,N_0} in (4.73) in the set $\mathcal{S}_{N_0}$ as defined by

$$\langle \phi_{n,N_0}, \ \phi_{\ell,N_0} \rangle \triangleq \frac{1}{N_0} \sum_{k=0}^{N_0-1} \phi_{n,N_0}(k)\overline{\phi_{\ell,N_0}(k)} \tag{4.75}$$

$$= \frac{1}{N_0} \sum_{k=0}^{N_0-1} exp\{\frac{j2\pi(n-\ell)k}{N_0}\} \tag{4.76}$$

Then since it can be shown that

$$\sum_{k=0}^{N_0-1} exp\{\frac{j2\pi mk}{N_0}\} = \begin{cases} N_0 : m \text{ an integer multiple of } N_0 \\ 0 \ ; \text{ otherwise} \end{cases} \tag{4.77}$$

we conclude that: for $0 \leq n, \ell \leq N_0 - 1$

$$\langle \phi_{n,N_0}, \phi_{\ell,N_0} \rangle = \begin{cases} 1 : n = \ell \\ 0 \ ; n \neq \ell \end{cases}$$

That is, $\mathcal{S}_{N_0}$ in (4.73) is an *orthonormal* set of N_0 signals.

Example 4.2.3. To illustrate the result (4.77), when $N_0 = 6$ and $m = 2$, we have

$$\sum_{k=0}^{5} exp\{\frac{j4\pi k}{6}\} = 1 + exp\{\frac{j4\pi}{6}\} + exp\{\frac{j8\pi}{6}\} + exp\{\frac{j12\pi}{6}\}$$
$$+exp\{\frac{j16\pi}{6}\} + exp\{\frac{j20\pi}{6}\}$$
$$= 1 + exp\{\frac{j2\pi}{3}\} + exp\{\frac{j4\pi}{3}\} + 1 + exp\{\frac{j2\pi}{3}\} + exp\{\frac{j4\pi}{3}\}$$
$$= 0$$

and when $m = 12$

$$\sum_{k=0}^{5} exp\{\frac{j24\pi k}{6}\} = \sum_{k=0}^{5} exp\{j4\pi k\} = \sum_{k=0}^{5} 1 = 6$$

Now from (4.67) and (4.68), since $\mathcal{S}_{N_0}$ in (4.73) is an othonormal set, any signal $\tilde{v}$ which is a linear combination of the periodic signals $\{\phi_{n,N_0}\}$ can be written in the form

$$\tilde{v}(k) = \sum_{n=0}^{N_0-1} \beta_n \phi_{n,N_0}(k) \ ; \ \beta_n = \langle \tilde{v}, \phi_{n,N_0} \rangle \tag{4.78}$$

Conversely, it can be shown that *any* signal $\tilde{v}$ which is *periodic of period* N_0 can be written in the form (4.78).

Now define

$$\tilde{V}(n) \triangleq N_0 \beta_n$$

so that from (4.78)

$$\tilde{v}(k) = \sum_{n=0}^{N_0-1} \frac{\tilde{V}(n)}{N_0} \phi_{n,N_0}(k) \ ; \ \langle \tilde{v}, \phi_{n,N_0} \rangle = \frac{\tilde{V}(n)}{N_0} \tag{4.79}$$

Therefore from (4.75) and (4.79), we have the following result.

Theorem 4.2.2. *(i) Any periodic digital signal $\tilde{v}$ of period N_0 can be represented in the form*

$$\tilde{v}(k) = \frac{1}{N_0} \sum_{n=0}^{N_0-1} \tilde{V}(n) exp\{\frac{j2\pi nk}{N_0}\} \tag{4.80}$$

where the components $\tilde{V}(n)$ for $0 \leq n \leq N_0 - 1$ are given by

$$\tilde{V}(n) = \sum_{k=0}^{N_0-1} \tilde{v}(k) exp\{-\frac{j2\pi nk}{N_0}\} \tag{4.81}$$

(ii) Given any integer $p \neq 0$, then any periodic digital signal $\tilde{v}$ of period N_0 can also be represented in the form

$$\tilde{v}(k) = \frac{1}{pN_0} \sum_{n=0}^{pN_0-1} \tilde{V}(n) exp\{\frac{j2\pi nk}{pN_0}\} \tag{4.82}$$

where the components $\tilde{V}(n)$ for $0 \leq n \leq pN_0 - 1$ are given by

$$\tilde{V}(n) = \sum_{k=0}^{pN_0-1} \tilde{v}(k) exp\{-\frac{j2\pi nk}{pN_0}\} \tag{4.83}$$

(iii) Consider any periodic signal $\tilde{v}$ of period N_0. Then

$$\tilde{v}(k) = \frac{1}{pN_0} \sum_{n=0}^{pN_0-1} \left[\sum_{m=0}^{N_1-1} \tilde{v}(m) exp\{-\frac{j2\pi nm}{N_1}\} \right] exp\{\frac{j2\pi nk}{pN_0}\}$$

and

$$\tilde{V}(n) = \sum_{k=0}^{N_2-1} \left[\frac{1}{pN_0} \sum_{r=0}^{pN_0-1} \tilde{V}(r) exp\{\frac{j2\pi rk}{pN_0}\} \right] exp\{-\frac{j2\pi nk}{N_2}\}$$

if and only if $N_1 = qN_0$ and $N_2 = \ell N_0$ for some integers $q \neq 0, \ell \neq 0$.

Periodic Extension

Even though the components of the signal $\tilde{V}$ in (4.81) are only defined over the finite range $0 \leq n \leq N_0 - 1$, the definition of $\tilde{V}(n)$ can be extended over *all* integers $\mathcal{Z}$ by means of a *periodic extension*; that is, for any $n \in \mathcal{Z}$, we define $\tilde{V}(n)$ by

$$\tilde{V}(n + mN_0) = \tilde{V}(n) \ ; \ 0 \leq n \leq N_0 - 1, m \in \mathcal{Z} \tag{4.84}$$

The resulting periodic signal $\tilde{V}$ is of period N_0 or less. The one-to-one correspondence between the signals $\tilde{v}$ and $\tilde{V}$ requires that the periods of both signals are known.

Example 4.2.4. Consider the periodic signal $\tilde{v}$ of period N_0 defined by

$$\tilde{v}(k) = \begin{cases} 1 \ ; \ k = mN_0 \text{ for any } m \in \mathcal{Z} \\ 0 \ ; \ \text{otherwise} \end{cases}$$

Then $\tilde{V}$ in (4.81) and (4.84) is given by

$$\tilde{V}(n) = \sum_{k=0}^{N_0-1} \tilde{v}(k) exp\{-\frac{j2\pi nk}{N_0}\} = 1 \; ; \; 0 \leq n \leq N_0 - 1$$

$$\tilde{V}(n + pN_0) = \tilde{V}(n) \; ; \; 0 \leq n \leq N_0 - 1 \text{ for all integers } p$$

Hence $\tilde{V}(n) = 1$ for all integers n, and so $\tilde{V}$ has period 1.

Given the periodic signal $\tilde{V}$, the periodic signal $\tilde{v}$ may be recovered using (4.80); that is

$$\tilde{v}(k) = \frac{1}{N_0} \sum_{n=0}^{N_0-1} 1. \; exp\{\frac{j2\pi nk}{N_0}\} = \begin{cases} 1 \; ; \; k = mN_0 \text{ for any } m \in \mathcal{Z} \\ 0 \; ; \; \text{otherwise} \end{cases}$$

In particular, note that the *same* N_0 (or an integer multiple of N_0) must be used to determine $\tilde{v}$ from $\tilde{V}$, and *not* the period of $\tilde{V}$ (which in this example is 1).

We have the following definition.

The periodic signal $\tilde{V}$ defined in (4.81) and (4.84) is called the Discrete Fourier Transform (DFT) representation of the periodic signal $\tilde{v}$, and we write

$$\tilde{V} = DFT(\tilde{v}) \tag{4.85}$$

The periodic signal $\tilde{v}$ in (4.80) is called the Inverse Discrete Fourier Transform (IDFT) representation of the periodic signal $\tilde{V}$, and we write

$$\tilde{v} = IDFT(\tilde{V}) \tag{4.86}$$

Example 4.2.5. Consider a digital periodic signal $\tilde{v} : \mathcal{Z} \to \mathcal{R}$ of period $N_0 = 3$ where

$$\tilde{v}(0) = 2, \tilde{v}(1) = -1, \tilde{v}(2) = 3 \; .$$

Then in (4.73), we consider the 3 orthonormal periodic signals $\{\phi_{0,3}, \phi_{1,3}, \phi_{2,3}\}$ given by

$$\phi_{0,3}(k) = 1 \; ; \; \phi_{1,3}(k) = exp\{\frac{j2\pi k}{3}\} \; ; \; \phi_{2,3}(k) = exp\{\frac{j4\pi k}{3}\}$$

Each signal $\phi_{n,3}$ is uniquely defined by the three values $\{\phi_{n,3}(k); k = 0, 1, 2\}$. The signal $\phi_{0,3}$ has period $P_{0,3} = 1$, the signal $\phi_{1,3}$ has period $P_{1,3} = 3$ and the signal $\phi_{2,3}$ has period $P_{2,3} = 3$.

From (4.80), the signal $\tilde{v}$ of period 3 then has the digital Fourier transform representation $\tilde{v} = IDFT(\tilde{V})$ where

$$\tilde{v}(k) = \sum_{n=0}^{2} < \tilde{v}, \phi_{n,3} > \phi_{n,3}(k) = \frac{1}{3}\sum_{n=0}^{2} \tilde{V}(n) exp\{\frac{j2\pi nk}{3}\}$$

and from (4.81) for $n = 0, 1, 2$

$$\begin{aligned} \tilde{V}(n) &= 3\langle \tilde{v}, \phi_{n,3}\rangle = \sum_{k=0}^{2} \tilde{v}(k) exp\{-\frac{j2\pi nk}{3}\} \\ &= 2 - exp\{-\frac{j2\pi n}{3}\} + 3exp\{-\frac{j4\pi n}{3}\} \end{aligned}$$

Then

$$\begin{aligned} &\tilde{V}(0) = 2 - 1 + 3 = 4 \\ &\tilde{V}(1) = 2 - exp\{-\frac{j2\pi}{3}\} + 3exp\{-\frac{j4\pi}{3}\} = 1 + j2\sqrt{3} \\ &\tilde{V}(2) = 2 - exp\{-\frac{j4\pi}{3}\} + 3exp\{-\frac{j8\pi}{3}\} = 1 - j2\sqrt{3} = \overline{\tilde{V}(1)} \end{aligned}$$

This definition of $\tilde{V}(n)$ over the set $\mathcal{Z}$ of integers is given by *periodic extension*; that is

$$\tilde{V}(n + 3m) = \tilde{V}(n) \; ; \; n = 0, 1, 2 \; ; \; m \in \mathcal{Z}$$

That is, the given periodic signal $\tilde{v}$ of period 3 with values $\{\tilde{v}(0) = 2, \tilde{v}(1) = -1, \tilde{v}(2) = 3\}$ can be equivalently represented by the periodic signal $\tilde{V}$ with values $\{\tilde{V}(0) = 4, \tilde{V}(1) = 1 + j2\sqrt{3}, \tilde{V}(2) = 1 - j2\sqrt{3}\}$. In this case, $\tilde{V}$ has the same period (=3) as $\tilde{v}$, but this is not always the case.

Since

$$1 + j2\sqrt{3} = \sqrt{13} exp\{j\theta\} \; ; \; \tan\theta = 2\sqrt{3}$$

we have

$$\begin{aligned} \tilde{v}(k) &= \frac{1}{3}[4 + \sqrt{13} exp\{j(\frac{2\pi k}{3} + \theta)\} + \sqrt{13} exp\{j(\frac{4\pi k}{3} - \theta)\}] \\ &= \frac{1}{3}[4 + \sqrt{13} exp\{j(\frac{2\pi k}{3} + \theta)\} + \sqrt{13} exp\{-j(\frac{2\pi k}{3} + \theta)\}] \end{aligned}$$

That is, the periodic signal $\tilde{v}$ can also be represented in the form

$$\tilde{v}(k) = \frac{1}{3}[4 + 2\sqrt{13}\cos(\frac{2\pi k}{3} + \theta)]$$

Example 4.2.6. Consider the periodic signal $\tilde{g}$ of period N_0 defined by

$$\tilde{g}(k) = \cos\frac{2\pi k}{N_0} = \frac{1}{2}\left[exp\{\frac{j2\pi k}{N_0}\} + exp\{-\frac{j2\pi k}{N_0}\}\right]$$

Then from (4.81) we have that $\tilde{G} = DFT(\tilde{g})$ is given by

$$\begin{aligned}
\tilde{G}(n) &= \sum_{k=0}^{N_0-1}\frac{1}{2}\left[exp\{\frac{j2\pi k}{N_0}\} + exp\{-\frac{j2\pi k}{N_0}\}\right]exp\{-\frac{j2\pi nk}{N_0}\} \\
&= \frac{1}{2}\sum_{k=0}^{N_0-1}\left[exp\{\frac{j2\pi k(1-n)}{N_0}\} + exp\{-\frac{j2\pi k(1+n)}{N_0}\}\right] \\
&= \frac{1}{2}\sum_{k=0}^{N_0-1}\left[exp\{\frac{j2\pi(1-n)}{N_0}\}\right]^k + \frac{1}{2}\sum_{n=0}^{N_0-1}\left[exp\{-\frac{j2\pi(1+n)}{N_0}\}\right]^k
\end{aligned}$$

Then using the fact that

$$\sum_{k=0}^{N_0-1}\left[exp\{\frac{j2\pi q}{N_0}\}\right]^k = \begin{cases} N_0 \ ; \text{ if } q = \ell N_0 \text{ for some } \ell \in \mathcal{Z} \\ 0 \ ; \text{ otherwise} \end{cases}$$

we conclude with $q = 1 - n$ and $q = 1 + n$ that

$$\tilde{G}(n) = \begin{cases} 0.5N_0 \ ; & n = 1, N_0 - 1 \\ 0 \ ; & n = 0;\ 2 \le n \le N_0 - 2 \\ \tilde{G}(n + mN_0) \ ; & 0 \le n \le N_0 - 1 \text{ for all integers } m \end{cases}$$

The periodic function $\tilde{g}$ and $\tilde{G} = DFT(\tilde{g})$ are illustrated in Figure 4.5 for $N_0 = 8$.

Similarly, for the periodic signal

$$\tilde{f}(k) = \sin\frac{2\pi k}{N_0} = \frac{1}{2j}\left[exp\{\frac{2\pi k}{N_0}\} - exp\{-\frac{2\pi k}{N_0}\}\right]$$

we have that $\tilde{F} = DFT(\tilde{f})$ is given by

$$\tilde{F}(n) = \begin{cases} -j0.5N_0 \ ; & n = 1 \\ 0 \ ; & n = 0;\ 2 \le n \le N_0 - 2 \\ j0.5N_0 \ ; & n = N_0 - 1 \\ \tilde{F}(n + mN_0) \ ; & 0 \le n \le N_0 - 1 \text{ for all integers } m \end{cases}$$

Least Squares Estimation

Consider the signal y given by

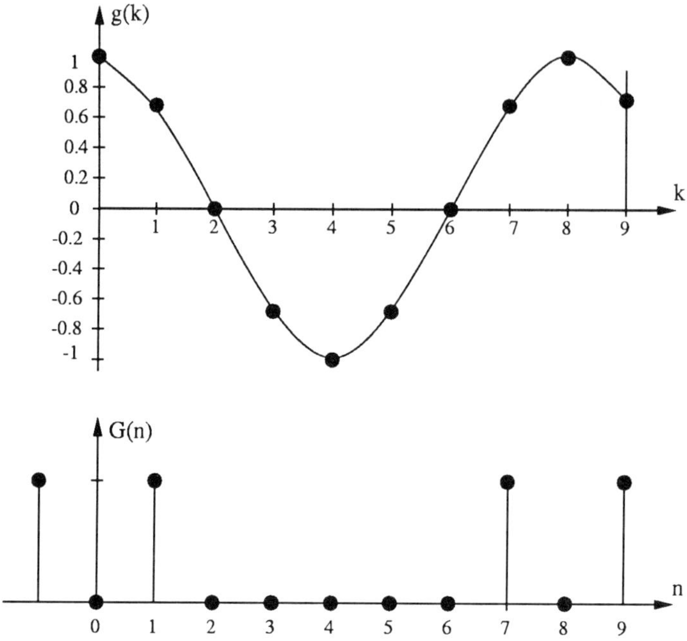

Fig. 4.5. Signal g and the DFT representation G

$$y(k) = \alpha_0 \cos k\Omega_0 + \beta_0 \sin k\Omega_0 \tag{4.87}$$

Then y is a periodic signal (of period $2M$) if and only if: for some integers q, N

$$\Omega_0 = \frac{q\pi}{N}$$

in which case it follows directly from (4.87) that

$$\alpha_0 = \frac{1}{N} \sum_{k=0}^{2N-1} y(k) \cos \frac{\pi k q}{N} \ ; \ \beta_0 = \frac{1}{N} \sum_{k=0}^{2N-1} y(k) \sin \frac{\pi k q}{N} \tag{4.88}$$

These expressions for $\{\alpha_0, \beta_0\}$ can also be obtained directly from the DFT. That is, from example 4.2.6, $Y = DFT(y)$ for $0 \le n \le 2N-1$ is given by

$$Y(n) = \begin{cases} N(\alpha_0 - j\beta_0) & ; \ n = 1 \\ 0 & ; \ n = 0; \ 2 \le n \le 2N-2 \\ N(\alpha_0 + j\beta_0) & ; \ n = 2N-1 \end{cases}$$

where

$$Y(n) = \sum_{k=0}^{2N-1} y(k) exp\{-\frac{j2\pi nkq}{2N}\}$$

Then $Y(1)$ gives the formula in (4.88).

Since the two signals $\{\ exp\{j\pi kq/N\},\ exp\{-j\pi kq/N\}\ \}$ of period $2N$ are orthogonal (in fact orthonormal), Theorem 4.1.7 asserts the following: Given *any* signal y, then the signal $\hat{y}$ of the form

$$\hat{y}(k) = \hat{\beta}_1 exp\{\frac{j\pi kq}{N}\} + \hat{\beta}_2 exp\{-\frac{j\pi kq}{N}\}$$

which *minimizes the sum of squares*

$$||y - \hat{y}||_2 = \sqrt{\frac{1}{2N}\sum_{k=0}^{2N-1}[y(k) - \hat{y}(k)]^2}$$

is given by (the inner products)

$$\hat{\beta}_1 = \frac{1}{2N}\sum_{k=0}^{2N-1} y(k)exp\{\frac{j\pi kq}{N}\}\ ;\ \hat{\beta}_2 = \frac{1}{2N}\sum_{k=0}^{2N-1} y(k)exp\{-\frac{j\pi kq}{N}\}$$

Equivalently, we have

$$\hat{y}(k) = \hat{\alpha}_0 \cos\frac{k\pi q}{N} + \hat{\beta}_0 \sin\frac{k\pi q}{N}$$

where

$$\hat{\alpha}_0 = \frac{1}{N}\sum_{k=0}^{2N-1} y(k)\cos\frac{\pi kq}{N}\ ;\ \hat{\beta}_0 = \frac{1}{N}\sum_{k=0}^{2N-1} y(k)\sin\frac{\pi kq}{N}$$

The result presented later in example 4.3.8 is more general in that it provides the optimal least squares estimates $\{\alpha_0, \beta_0\}$ in (4.163) when the number M of data points $\{y(k);\ 0 \leq k \leq M-1\}$ is *not* equal to an integer multiple of the frequency of $\hat{y}$.

4.2.2 Finite Length Signal

In the previous section, we considered the discrete Fourier transform representation of a *periodic* signal. In this section, we extend this representation to the case of bounded *nonperiodic* signals defined on a *finite domain.* The possibility of this extension is based on being able to define a new signal which is as a periodic extension of the given signal. Specifically, consider a (bounded) signal v where $v\ :\ \mathcal{T}_L \to \mathcal{C}$ and where $\mathcal{T}_L$ is a finite subset of Z given by

$$\mathcal{T}_L \triangleq \{n \in Z\ :\ 0 \leq n \leq L-1\} \tag{4.89}$$

Periodic Extension

The signal v can be *extended* to a *periodic* signal $\tilde{v} : \mathcal{Z} \to \mathcal{C}$ by means of a *periodic extension* ; that is, for any $m \in \mathcal{Z}$, define

$$\tilde{v}(k + mL) = v(k) \quad ; \quad 0 \leq k \leq L - 1 \tag{4.90}$$

Conversely, a finite length signal $v : \mathcal{T}_L \to \mathcal{C}$ can be *recovered* from a periodic signal $\tilde{v}$ by selecting the first L components of v; that is, we can define

$$v(k) \triangleq \tilde{v}(k) \; ; \; 0 \leq k \leq L - 1 \tag{4.91}$$

We have the following definition.

Consider any bounded finite length signal $v : \mathcal{T}_L \to \mathcal{C}$ of length L where $\mathcal{T}_L$ is defined by (4.89), and define the periodic extension $\tilde{v}$ of v by (4.90). Let $\tilde{V}$ denote the discrete Fourier transform of $\tilde{v}$; that is,

$$\tilde{V} = DFT(\tilde{v}) \; ; \; \tilde{v} = IDFT(\tilde{V})$$

and define V to be the first L components of $\tilde{V}$; that is,

$$V(n) \triangleq \tilde{V}(n) \; ; \; 0 \leq n \leq L - 1 \tag{4.92}$$

Then V is called the discrete Fourier transform of v, and v is called the inverse discrete Fourier transform of V. We then write

$$V = DFT(v) \quad ; \quad v = IDFT(V) \tag{4.93}$$

Example 4.2.7. Consider the digital signal $v : \mathcal{T}_8 \to \mathcal{R}$ where $\mathcal{T}_8 = \{n \in \mathcal{Z} : 0 \leq n \leq 7\}$ defined by

$$v(k) = a^k \quad ; \quad 0 \leq k \leq 7$$

Then from (4.91), the periodic extension $\tilde{v} : \mathcal{Z} \to \mathcal{C}$ of v is defined by

$$\tilde{v}(k + 8m) = v(k) \; ; \; 0 \leq k \leq 7, m \in \mathcal{Z}$$

Then $\tilde{V} = DFT(\tilde{v})$ is given by

$$\tilde{V}(n) = \sum_{k=0}^{7} a^k exp\{-\frac{j2\pi nk}{8}\}$$

When $\beta_n \neq 1$, we have

$$\tilde{V}(n) = \frac{1 - \beta_n^8}{1 - \beta_n} \; ; \; \beta_n \triangleq aexp\{-\frac{j2\pi n}{8}\}$$

and when $\beta_n = 1$, $\tilde{V}(n) = 8$. Conversely, given $\tilde{V}$, we have that $\tilde{v} = IDFT(\tilde{V})$ is given by

$$\tilde{v}(k) = \frac{1}{7}\sum_{n=0}^{7} \tilde{V}(n)exp\{\frac{j2\pi nk}{8}\}$$

The discrete Fourier transform $V = DFT(v)$ is given by the signal V of length 8 by

$$V(n) = \tilde{V}(n) \ ; \ 0 \le n \le 7$$

4.2.3 Properties of the DFT

By means of the concept of periodic extension, we seen that we are able to treat any finite length signal of length N as if it were a periodic signal of period N. In this section, we develop some fundamental properties of the DFT and the IDFT that apply equally to both finite length and periodic signals. However we first need to introduce some definitions.

Time-Shifting

If v is a periodic signal of period N, then the periodic signal p_M of period N is said to be a *time-shifted* version of v of shift length M when: for some integer M

$$p_M(k) \stackrel{\Delta}{=} v(k - M) \ ; \ k \in \mathcal{Z} \tag{4.94}$$

It then follows that since v has period N, then for any integer ℓ

$$p_M(k) = v(\ell N + k - M)$$

In particular, for all $0 \le k \le N - 1$, an integer ℓ can always be selected such that $0 \le \ell N + k - M \le N - 1$.

Example 4.2.8. Suppose v is a periodic signal of period $N = 5$, and the shift length $M = 3$. then the periodic signal p_3 of period 5 is defined by

$$p_3(0) = v(-3) = v(2) \ ; \ p_3(1) = v(3) \ ; \ p_3(2) = v(4)$$
$$p_3(3) = v(0) \ ; \ p_3(4) = v(1)$$

Likewise, if $M = 7$, then the periodic signal p_7 of period 5 is given by

$$p_7(0) = v(-7) = v(3) \ ; \ p_7(1) = v(4) \ ; \ p_7(2) = v(0)$$
$$p_7(3) = v(1) \ ; \ p_7(4) = v(2)$$

and if $M = 2$, then

$$p_2(0) = v(-2) = v(3) \; ; \; p_2(1) = v(4) \; ; \; p_2(2) = v(0)$$
$$p_2(3) = v(1) \; ; \; p_2(4) = v(2)$$

and if $M = -2$, then

$$p_{-2}(0) = v(2) \; ; \; p_{-2}(1) = v(3) \; ; \; p_{-2}(2) = v(4)$$
$$p_{-2}(3) = v(0) \; ; \; p_{-2}(4) = v(1)$$

Hence the time-shifted periodic signals p_7 and p_2 are identical, and the time-shifted periodic signals p_{-2} and p_3 are identical.

More generally, given a periodic signal v of period N, then the time-shifted periodic signals p_{M_1} and p_{M_2} are identical if: for some integer ℓ, $M_1 = M_2 + \ell N$. Consequently, when defining a time-shifted signal p_M of shift length M in (4.94), it is sufficient to restrict the integer M to lie in the range $0 \le M \le N-1$. We now define what is meant by a *time-shifted finite length signal.*

Suppose the N values of a *finite length signal* v are given for some integer Q by $\{v(Q), v(Q+1), \;.... \; v(Q+N-1)\}$, then the *time-shifted* version p_M of v of shift length M for $0 \le M \le N-1$ is defined by

$$p_M(k) \triangleq \begin{cases} v(k+N-M) & ; \; Q \le k \le M-1 \\ v(k-M) & ; \; Q+M \le k \le Q+N-1 \end{cases} \tag{4.95}$$

For example, if $Q = 0$, $N = 6$ and $M = 2$, then

$$p_2(k) \triangleq \begin{cases} v(k+4) \; ; \; 0 \le k \le 1 \\ v(k-2) \; ; \; 2 \le k \le 5 \end{cases}$$

Given a finite length signal v of length N, then the time-shifted signal p_M for any integer M is defined such that p_{M_1} and p_{M_2} are identical if: for some integer ℓ, $M_1 = M_2 + \ell N$.

Time-Reversal

Given a *periodic* signal v of period N defined on the integers $\mathcal{Z}$, the periodic signal v_- of period N is said to be a *time-reversal* of v when

$$v_-(k) \triangleq v(-k) \; ; \; k \in \mathcal{Z} \tag{4.96}$$

For example, if v has period 5, then v_- is uniquely defined by the 5 values:

$$\{v_-(0) = v(0),\ v_-(1) = v(-1) = v(4),\ v_-(2) = v(3),\ v_-(3) = v(2),\ v_-(4) = v(1)\}$$

Likewise, if v is a *finite-length* signal v of length N which for some integer Q is defined by $\{v(Q), v(Q+1), \ldots\ v(Q+N-1)\}$, then the *time-reversal* v_- of v is defined by

$$v_-(Q+k) \triangleq \begin{cases} v(Q)\ ;\ k = 0 \\ v(Q+N-k)\ ;\ 1 \leq k \leq N-1 \end{cases} \tag{4.97}$$

Even and Odd Signals

A signal v of period N, or finite length of length N, is said to be an *even signal* if v is equal to its time-reversal; that is, if $v = v_-$. A signal v of period N, or finite length of length N, is said to be an *odd signal* if v is equal to the negative of its time-reversal; that is, if $v = -v_-$.

Given any real valued signal v, we can define two real valued signals $\{v_e,\ v_o\}$ by

$$v_e(k) = 0.5[v(k) + v(-k)]\ ;\ v_o(k) = 0.5[v(k) - v(-k)] \tag{4.98}$$

Since for all k

$$v_e(k) = v_e(-k)\ ;\ v_o(k) = -v_o(-k)$$

the signal v_e is an *even signal*, and the signal v_o is an *odd signal*. Hence any real valued signal v can be expressed in the form

$$v(k) = v_e(k) + v_o(k) \tag{4.99}$$

for some (nonunique) *real valued even signal* v_e, and some (nonunique) *real valued odd signal* v_o.

Real and Complex Signals

Given that $\overline{w}$ denotes the complex conjugate of a number w, a signal v is said to be an *real signal* if $v(k) = \overline{v(k)}$ for all integers k. A signal v is said to be a *purely imaginary signal* if $v(k) = -\overline{v(k)}$ for all integers k.

Given any complex signal w, we can define a *real signal* w_r and a *purely imaginary signal* w_i by

$$w_r(k) = 0.5[w(k) + \overline{w(k)}]\ ;\ w_i(k) = 0.5[w(k) - \overline{w(k)}] \tag{4.100}$$

Hence any complex signal w can be expressed in the form

$$w(k) = w_r(k) + w_i(k) \tag{4.101}$$

for some (nonunique) *real valued signal* w_r, and some (nonunique) *purely imaginary valued signal* w_i. We now have the following result.

Theorem 4.2.3. *Suppose v and w are periodic (or finite length) digital signals of period (or length) N_0 defined on the integers $\mathcal{Z}$ such that*

$$V = DFT(v) \; ; \; W = DFT(w) \tag{4.102}$$

Then:

1. *(Linearity) For any constants a, b*

   $$aV + bW = DFT(av + bw)$$

2. *(Time shifting) Let the signal p_M be the time-shifted version of the signal v of shift-length M. Then $P_M = DFT(p_M)$ is given by*

   $$P_M(n) = exp\{-\frac{j2\pi nM}{N_0}\}V(n)$$

3. *(Time reversal) Let the signal v_- denote the time reversal of the signal v. Then $V_- = DFT(v_-)$ is given by*

   $$V_-(n) = V(-n)$$

4. *(Conjugation) Suppose the signal v^* is defined by $v^*(k) \triangleq \overline{v(k)}$. Then $Y = DFT(v^*)$ is given by*

   $$Y(n) = \overline{V(-n)}$$

5. *(Real signal) Suppose v is a real valued signal. Then $V = DFT(v)$ is given by*

   $$V(n) = \overline{V(-n)}$$

6. *(Purely imaginary signal) Suppose w is a purely imaginary valued signal. Then $W = DFT(w)$ is given by*

   $$W(n) = -\overline{W(-n)}$$

7. *(Even and real signal) Suppose v is an even real valued signal. Then $V = DFT(v)$ is given by*

   $$V(n) = V(-n)$$

   is real for all n.
8. *(Odd and real signal) Suppose w is an odd real valued signal. Then $W = DFT(w)$ is given by*

   $$W(n) = -W(-n)$$

   is purely imaginary for all n.

9. *(Parseval's relation)*

$$\frac{1}{N_0}\sum_{k=0}^{N_0-1}|v(k)|^2 = \sum_{n=0}^{N_0-1}|V(n)|^2$$

We now present a (partial) proof of each of the results of Theorem 4.2.3. The proof of part 1 follows directly from the definition of DFT. For part 2, we have

$$\begin{aligned}
P_M(n) &= \sum_{k=0}^{N_0-1} p_M(k)exp\{-\frac{j2\pi nk}{N_0}\} \\
&= \sum_{k=0}^{M-1} v(k+N_0-M)exp\{-\frac{j2\pi nk}{N_0}\} \\
&\quad + \sum_{k=M}^{N_0-1} v(k-M)exp\{-\frac{j2\pi nk}{N_0}\} \\
&= \sum_{\ell=N_0-M}^{N_0-1} v(\ell)exp\{-\frac{j2\pi n(\ell+M-N)}{N_0}\} \\
&\quad + \sum_{\ell=0}^{N_0-M-1} v(\ell)exp\{-\frac{j2\pi n(\ell+M)}{N_0}\} \\
&= exp\{-\frac{j2\pi n(M-N_0)}{N_0}\}\sum_{\ell=N_0-M}^{N_0-1} v(\ell)exp\{-\frac{j2\pi n\ell}{N_0}\} \\
&\quad + exp\{-\frac{j2\pi nM}{N_0}\}\sum_{\ell=0}^{N_0-M-1} v(\ell)exp\{-\frac{j2\pi n\ell}{N_0}\}
\end{aligned}$$

Hence since $exp\{-j2\pi n(M-N_0)/N_0\} = exp\{-j2\pi n\ell/N_0\}$, we have the result.

For part 3, we have

$$\begin{aligned}
V_-(n) &= \sum_{k=0}^{N_0-1} v(-k)exp\{-\frac{j2\pi nk}{N_0}\} \\
&= \sum_{k=0}^{N_0-1} v(-k)exp\{\frac{j2\pi n(-k)}{N_0}\}.\ exp\{\frac{j2\pi N_0}{N_0}\} \\
&= \sum_{k=0}^{N_0-1} v(N_0-k)exp\{\frac{j2\pi n(N_0-k)}{N_0}\}
\end{aligned}$$

$$= \sum_{\ell=0}^{N_0-1} v(\ell)exp\{\frac{j2\pi n\ell}{N_0}\}$$

$$= \sum_{\ell=0}^{N_0-1} v(\ell)exp\{-\frac{j2\pi(-n)\ell}{N_0}\} = V(-n)$$

For part 4, we have

$$Y(n) = \sum_{k=0}^{N_0-1} v^*(k)exp\{-\frac{j2\pi nk}{N_0}\}$$

$$= \sum_{k=0}^{N_0-1} \overline{v(k)}\ \overline{exp\{-\frac{j2\pi nk}{N_0}\}} = \sum_{k=0}^{N_0-1} \overline{v(k)exp\{\frac{j2\pi nk}{N_0}\}}$$

$$= \overline{\sum_{k=0}^{N_0-1} v(k)exp\{-\frac{j2\pi(-n)k}{N_0}\}} = \overline{V(-n)}$$

Part 5 follows directly from part 4, since if v is real, then $v^*(k) = v(k)$ for all k. Part 6 also follows directly from part 4, since if v is purely imaginary, then $v^*(k) = -v(k)$ for all k. Part 7 follows from part 3, and part 8 follows from part 6.

Part 9 follows directly from Theorem 4.1.6 for the particular orthonormal signals $\{\phi_n(k) = exp\{j2\pi nk/N_0\}$.

4.2.4 Fast Fourier Transform

A particular advantage of the DFT representation lies in the fact that there is a numerically efficient procedure known as the *Fast Fourier Transform* or FFT which is available for its computation. Once a particular FFT algorithm is developed, we then go on and show how such a procedure can be used to efficiently compute the linear convolution of two signals. Since a linear convolution is necessary to compute the response of a linear digital filter, the availability of the FFT algorithm is consequently very important for the real time implementation of a digital filter.

Computational Complexity of Algorithms

Since the Fourier transform of a real signal is generally *complex*, complex arithmetic (commonly available on many hardware devices) is required. A multiplication

$$(\lambda_1 + j\beta_1)(\lambda_2 + j\beta_2) = \lambda_1\lambda_2 - \beta_1\beta_2 + j(\lambda_1\beta_2 + \beta_1\lambda_2)$$

of two complex numbers requires 4 real multiplications and 2 real additions, while a complex addition

$$(\lambda_1 + j\beta_1) + (\lambda_2 + j\beta_2) = \lambda_1 + \lambda_2 + j(\beta_1 + \beta_2)$$

requires 2 real additions. The number of additions is generally of the same order as the number of multiplications, and so for the purposes of comparing different algorithms, we often simply measure the *computational complexity* of a signal processing algorithm by the number of multiplications.

The discrete Fourier transform V of a signal v of length N is given by

$$V(n) = \sum_{k=0}^{N-1} v(k)\phi_{n,N}(-k) \quad ; \quad 0 \leq n \leq N-1 \tag{4.103}$$

$$\phi_{n,N}(r) = exp\{\frac{j2\pi nr}{N}\}$$

Similarly, the inverse discrete Fourier transform is given by

$$v(k) = \frac{1}{N}\sum_{n=0}^{N-1} V(n)\phi_{k,N}(n) \quad ; \quad 0 \leq k \leq N-1$$

Since $\phi_{n,N}(0) = 1$ for all n, N, it can be seen that: for each integer n, $V(n)$ can be obtained using $N-1$ complex multiplications. Hence overall, the number of complex multiplications $\tilde{T}_N$ required to compute V from v (and compute v from V) is given by

$$\tilde{T}_N = N(N-1) \tag{4.104}$$

We now develop a number of so called "fast" algorithms in which the required number T_N of multiplications is less than $\tilde{T}_N$. As we shall see, the particular expression for T_N will depend on how the length N of the signal is factored.

Product of N_1 and N_2

Suppose the signal length N is factored according to

$$N = N_1.N_2$$

for some integers N_1, N_2 which are both greater than unity. Then $V(n)$ in (4.103) can be expressed in the form

$$V(n) = \sum_{p=0}^{N_1-1}\left\{\sum_{r=0}^{N_2-1} v(N_1r+p)\phi_{n,N}(-N_1r-p)\right\}$$

Now from the definition of $\phi_{n,N}$ in (4.103), we have that

$$\phi_{n,N}(-N_1r-p) = \phi_{n,N}(-p)\phi_{n,N}(-N_1r) \tag{4.105}$$

Hence V can be written in the form

$$V(n) = \sum_{r=0}^{N_2-1} v(N_1 r)\phi_{n,N}(-N_1 r) + \sum_{p=1}^{N_1-1} \left\{ \phi_{n,N}(-p) \sum_{r=0}^{N_2-1} v(N_1 r + p)\phi_{n,N}(-N_1 r) \right\} \tag{4.106}$$

That is, V is now in the form of a total of N_1 N_2-point DFTs. In particular, suppose T_{N_2} complex multiplications are required to compute each N_2-point DFT. Then the total number T_N of complex multiplications required to compute $V(n)$ for $0 \leq n \leq N-1$ is given by

$$T_N = N_1 T_{N_2} + N(N_1 - 1) \tag{4.107}$$

where the $N(N_1 - 1)$ multiplications come from the multiplications in (4.106) by the coefficients $\phi_{n,N}(-p)$. This factorization can lead to a reduction in computational complexity.

Example 4.2.9. (i) Suppose $N = 12 = N_1.N_2$ where $N_1 = 3$ and $N_2 = 4$. Then from (4.104)

$$\tilde{T}_{12} = 12(12-1) = 132$$

However from (4.107)

$$T_{12} = 3T_4 + 12(2) = 3T_4 + 24$$

Then if we take $T_4 = \tilde{T}_4 = 12$, we conclude that

$$T_{12} = 60 < 0.5\tilde{T}_{12}$$

(ii) Now suppose $N = 4 = N_1.N_2$ where $N_1 = N_2 = 2$. Then from (4.104), $\tilde{T}_4 = 12$. However from (4.107)

$$T_4 = 2T_2 + 4(2-1) = 2T_2 + 4$$

Now a 2-point DFT is of the form

$$\sum_{r=0}^{1} h(r)\phi_{n,2}(-r) = \begin{cases} h(0) + h(1) \ ; \ n = 0 \\ h(0) - h(1) \ ; \ n = 1 \end{cases}$$

and so can be accomplished with two additions and *no multiplications*; that is, $T_2 = 0$. Hence

$$T_4 = 4 < \tilde{T}_4 = 12$$

If in (i) we use $T_4 = 4$ rather than $T_4 = 12$, we then have

$$T_{12} = 36 \simeq 0.25\tilde{T}_{12}$$

(iii) Now suppose $N = 8$. Then from (4.107) with $N_1 = 2, N_2 = 4$, we have

$$T_8 = 2T_4 + 8(2-1) = 2T_4 + 8 = 16$$

Similarly with $\{N = 16; N_1 = 2, N_2 = 8\}$ and $\{N = 32; N_1 = 2, N_2 = 16\}$, we have

$$T_{16} = 2T_8 + 16(2-1) = 2T_8 + 16 = 48 \ ; \ T_{32} = 2T_{16} + 32 = 128$$

More generally, it follows that $S(k) \triangleq T_{2k}$ is given by

$$S(k+1) = 2S(k) + u(k), \ S(1) = 0 \ ; \ u(k) = 2^{k+1}$$

This is a first order difference equation for S, and so the general solution is given by

$$S(p) = 2^{p-1}S(1) + \sum_{m=1}^{p} h(p-m)u(m) \ ; \ p \geq 1$$

where the unit impulse response h is given by

$$h(0) = 0 \ ; \ h(k) = 2^{k-1}, \ k \geq 1$$

Hence

$$S(p) = \sum_{m=1}^{p-1} 2^{p-m-1}.2^{m+1} = 2^p(p-1)$$

We therefore have the following result.

Theorem 4.2.4. *Suppose the length N of a signal v is given by*

$$N = 2^p$$

for some integer $p \geq 2$. Then the number T_N of complex multiplications necessary to compute $V = DFT(v)$ is given by

$$T_N = N\log_2(N/2)$$

The following result is established in exercise 4.29.

Theorem 4.2.5. *(i) Suppose the length N of a signal v is given by*

$$N = 3^q$$

for some integer $q \geq 2$. Then the number T_N of complex multiplications necessary to compute $V = DFT(v)$ is given by

$$T_N = 2(N/3) + 2N\log_3(N/3)$$

(ii) Suppose

$$N = \ell^q$$

for some integer $q \geq 2$ and prime number ℓ. Then the number T_N of complex multiplications necessary to compute $V = DFT(v)$ is given by

$$T_N = (\ell - 1)(N/\ell) + \ell N\log_\ell(N/\ell)$$

The explicit form of the FFT algorithm when the length N is a power of 2 is now considered in more detail.

Power of 2 Algorithm

Consider the computation of the discrete Fourier transform V of a finite length signal v of length N. Suppose N is *even*, then from (4.81), $V = DFT(v)$ can be written in the form

$$V(n) = \sum_{r=0}^{N/2-1} v(2r)\phi_{n,N}(-2r) + \sum_{r=0}^{N/2-1} v(2r+1)\phi_{n,N}(-2r-1)$$

But from (4.103)

$$\phi_{n,N}(-2r-1) = \phi_{n,N}(-1)\phi_{n,N}(-2r) \; ; \; \phi_{n,N}(-2r) = \phi_{n,N/2}(-r)$$

and so we have *STAGE #1* of the *Fast Fourier Transform* (FFT) algorithm.

•*STAGE #1*:

Suppose N is even. Then for $0 \leq n \leq N-1$

$$V(n) = V_0(n) + \phi_{n,N}(-1)V_1(n) \tag{4.108}$$

where

$$V_0(n) \triangleq \sum_{r=0}^{N/2-1} v(2r)\phi_{n,N/2}(-r) \tag{4.109}$$

$$V_1(n) \triangleq \sum_{r=0}^{N/2-1} v(2r+1)\phi_{n,N/2}(-r)$$

That is, V_0 is the $N/2$ DFT of the $N/2$ *even* points $\{v(2r)\ ;\ 0 \le r \le N/2-1\}$ of the signal v, and V_1 is the $N/2$ DFT of the $N/2$ odd points $\{v(2r+1)\ ;\ 0 \le r \le N/2-1\}$ of the signal v.

Direct substitution into both V_0 and V_1 reveals that: for $0 \le n \le N/2-1$,

$$V_0(n+N/2) = V_0(n)\ ;\ \ V_1(n+N/2) = V_1(n)$$

Consequently, even though $V(n)$ in (4.108) needs to be computed for $0 \le n \le N-1$, both V_0 and V_1 in (4.109) need only be computed for $0 \le n \le N/2-1$.

•*STAGE #2*:

Suppose N is divisible by $4 = 2^2$. Then for $0 \le n \le N/2-1$, we similarly conclude that

$$\begin{aligned} V_0(n) &= V_{00}(n) + \phi_{n,N/2}(-1)V_{10}(n) \\ V_1(n) &= V_{01}(n) + \phi_{n,N/2}(-1)V_{11}(n) \end{aligned} \tag{4.110}$$

where

$$\begin{aligned} V_{00}(n) &\triangleq \sum_{r=0}^{N/4-1} v(4r)\phi_{n,N/4}(-r) \\ V_{10}(n) &\triangleq \sum_{r=0}^{N/4-1} v(4r+2)\phi_{n,N/4}(-r) \\ V_{01}(n) &\triangleq \sum_{r=0}^{N/4-1} v(4r+1)\phi_{n,N/4}(-r) \\ V_{11}(n) &\triangleq \sum_{r=0}^{N/4-1} v(4r+3)\phi_{n,N/4}(-r) \end{aligned} \tag{4.111}$$

That is, $\{V_{00}, V_{10}, V_{01}, V_{11}\}$ are respectively the $N/4$ point DFTs of the finite length signals $\{v(4r); 0 \le r \le N/4-1\}$, $\{v(4r+1); 0 \le r \le N/4-1\}$, $\{v(4r+2); 0 \le r \le N/4-1\}$ and $\{v(4r+3); 0 \le r \le N/4-1\}$.

Also, it follows that

$$V_{ij}(n+N/4) = V_{ij}(n) \text{ for } 0 \le n \le N/4-1$$

Consequently, even though $V_0(n)$ and $V_1(n)$ in (4.110) needs to be computed for $0 \le n \le N/2-1$, the four signals $\{V_{00}, V10, V_{01}, V_{11}\}$ need only be evaluated for $0 \le n \le N/4-1$.

•*STAGE #3*:

Suppose N is divisible by $8 = 2^3$. Then for $0 \le n \le N/4-1$, we again conclude that

$$\begin{aligned} V_{00}(n) &= V_{000}(n) + \phi_{n,N/4}(-1)V_{100}(n) \\ V_{10}(n) &= V_{010}(n) + \phi_{n,N/4}(-1)V_{110}(n) \\ V_{01}(n) &= V_{001}(n) + \phi_{n,N/4}(-1)V_{101}(n) \\ V_{11}(n) &= V_{011}(n) + \phi_{n,N/4}(-1)V_{111}(n) \end{aligned} \tag{4.112}$$

where

$$\begin{aligned} V_{000}(n) &\triangleq \sum_{r=0}^{N/8-1} v(8r)\phi_{n,N/8}(-r) \\ V_{100}(n) &\triangleq \sum_{r=0}^{N/8-1} v(8r+4)\phi_{n,N/8}(-r) \\ V_{010}(n) &\triangleq \sum_{r=0}^{N/8-1} v(8r+2)\phi_{n,N/8}(-r) \\ V_{110}(n) &\triangleq \sum_{r=0}^{N/8-1} v(8r+6)\phi_{n,N/8}(-r) \\ V_{001}(n) &\triangleq \sum_{r=0}^{N/8-1} v(8r+1)\phi_{n,N/8}(-r) \\ V_{101}(n) &\triangleq \sum_{r=0}^{N/8-1} v(8r+5)\phi_{n,N/8}(-r) \\ V_{011}(n) &\triangleq \sum_{r=0}^{N/8-1} v(8r+3)\phi_{n,N/8}(-r) \\ V_{111}(n) &\triangleq \sum_{r=0}^{N/8-1} v(8r+7)\phi_{n,N/8}(-r) \end{aligned} \tag{4.113}$$

That is, $\{V_{pqr}\}$ are respectively the $N/8$ point DFTs of finite length signals of length $N/8$.

Also, it follows that

$$V_{pqr}(n + N/8) = V_{pqr}(n) \text{ for } 0 \leq n \leq N/8 - 1$$

Consequently, even though $\{V_{00}, V_{10}, V_{01}, V_{11}\}$ in (4.112) needs to be computed for $0 \leq n \leq N/4-1$, the eight signals $\{V_{pqr}(n)\}$ need only be evaluated for $0 \leq n \leq N/8 - 1$.

•*STAGE #p*:

If $N = 2^p$ for some integer $p \geq 2$, then the final pth stage requires the computation of 2^{p-1} 2-point DFTs. However as we have already seen, a 2-point DFT can be accomplished with only two additions and no multiplications.

Computational Requirements: Based on the above discussions, we make the following conclusions about the FFT algorithm when N is a power of 2:

- Given the two signals $\{V_m(n)\ ;\ 0 \leq n \leq N-1\}$ for $m = 0, 1$, then STAGE #1 involves N complex multiplications by $\phi_{n,N}(-1)$ to compute $V(n)$ for $0 \leq n \leq N-1$.
- Given the four signals $\{V_{m\ell}(n)\ ;\ 0 \leq n \leq N/2-1\}$ for $m = 0, 1$ and $\ell = 0, 1$, STAGE #2 involves $2(N/2) = N$ complex multiplications by $\phi_{n,N/2}(-1)$ to compute both $\{V_0(n), V_1(n)\}$ for $0 \leq n \leq N/2-1$.
- Similarly, given the eight signals $\{V_{m\ell p}(n)\ ;\ 0 \leq n \leq N/8-1\}$ for $m = 0, 1$, $\ell = 0, 1$ and $p = 0, 1$, STAGE #3, involves $4(N/4) = N$ complex multiplications by $\phi_{n,N/4}(-1)$ to compute $\{V_{00}(n), V_{01}(n), V_{10}(n), V_{11}(n)\}$ for $0 \leq n \leq N/4-1$.

We deduce the following result.

Theorem 4.2.6. *(FFT Algorithm)*

If a finite length signal v has N points where $N = 2^p$ for some integer $p \geq 2$, then the computation of $V = DFT(v)$ can be decomposed into p stages. STAGE #m for $1 \leq m \leq p-1$ requires:

- *$L \times (N/L) = N$ complex multiplications by $\phi_{n,N/L}(-1)$ where $L = 2^{m-1}$, and*
- *The determination of $2L = 2^m$ different DFTs each of length N/L.*

The last STAGE #p requires the evaluation of 2^{p-1} different 2-point DFTs each of which can be computed with only two additions.

Hence the overall computation of V using the FFT algorithm can therefore be achieved using a total of T_N complex multiplications, or $4T_N$ real multiplications, where

$$T_N = N\log_2(N/2)$$

More general FFT algorithms for computing the Fourier transform of a signal whose length N is expressed as a product of the powers of prime numbers can also be derived.

Example 4.2.10. Consider the calculation of

$$V(n) = \sum_{k=0}^{7} v(k)\phi_{n,8}(-k)\ ;\ 0 \leq n \leq 7$$

A straightforward calculation appears to require 8 complex multiplications to compute $V(n)$ for each value of n. That is, without any further consideration, it appears that a total of $\tilde{T}_8 = 64$ complex multiplications are required.

However, since $\phi_{0,8}(-n) = 1$ for all n, it follows that $V(0)$ can be computed without any multiplications (only 7 additions). Also, since $\phi_{0,n}(-k) = 1$ for all integers n and k, $V(n)$ can be arranged as

$$V(n) = v(0) + \sum_{k=1}^{7} v(k)\phi_{n,8}(-k) \ ; \ 1 \leq n \leq 7$$

so that only 7 complex multiplications are required for each value of n; that is, after further consideration, it is evident that only a total of $7 \times 7 = 49$ complex multiplications in all are required.

However if we arrange the calculation according to the *FFT procedure*, at most 16 complex multiplications are required. Thus:

•*STAGE #1*:

$$V(n) = V_0(n) + \phi_{n,8}(-1)V_1(n) \ ; \ 0 \leq n \leq 7$$

where

$$V_0(n) = \sum_{r=0}^{3} v(2r)\phi_{n,4}(-r) \ ; \ V_1(n) = \sum_{r=0}^{3} v(2r+1)\phi_{n,4}(-r)$$

However since $V_0(n) = V_0(n+4)$ and $V_1(n) = V_1(n+4)$, both $\{V_0(n), V_1(n)\}$ need only be evaluated for $0 \leq n \leq 3$.

•*STAGE #2*: For $0 \leq n \leq 3$

$$V_0(n) = V_{00}(n) + \phi_{n,4}(-1)V_{10}(n) \ ; \ V_1(n) = V_{01}(n) + \phi_{n,4}(-1)V_{11}(n)$$

where

$$V_{00}(n) = \sum_{r=0}^{1} v(4r)\phi_{n,2}(-r) \ ; \ V_{10}(n) = \sum_{r=0}^{1} v(4r+2)\phi_{n,2}(-r)$$

$$V_{01}(n) = \sum_{r=0}^{1} v(4r+1)\phi_{n,2}(-r) \ ; \ V_{11}(n) = \sum_{r=0}^{1} v(4r+3)\phi_{n,2}(-r)$$

Since $V_{ij}(n) = V_{ij}(n+2)$ for $i = 0, 1$ and $j = 0, 1$, we only need compute $V_{ij}(n)$ for $0 \leq n \leq 1$.

Now since $\phi_{n,2}(-r) = \pm 1$ for all n and r, $\{V_{00}(n), V_{10}(n), V_{01}(n), V_{11}(n)\}$ can all be computed without any multiplications. Hence in STAGE #2, only 2 complex multiplications by $\phi_{n,4}(-1)$ are required to compute $\{V_0(n), V_1(n)\}$ from $\{V_{00}(n), V_{10}(n), V_{01}(n), V_{11}(n)\}$ for each value of n in the range $0 \leq n \leq$

3. Then in STAGE #1, only 1 complex multiplication by $\phi_{n,8}(-1)$ is required to compute $\{V(n)\}$ from the $\{V_0(n), V_1(n)\}$ for each value of n in the range $0 \leq n \leq 7$.

Hence overall, a total of

$$(4 \times 2) + (8 \times 1) = 16 = N\log_2(N/2) \ ; \ N = 8$$

complex multiplications are required. (If it is also recognised that $\phi_{0,m}(-r) = 1$ for all m, r then only 6 complex multiplications are required in both STAGE #2 and STAGE #1, so that now only

$$(N-2)\log_2(N/2) = 12 \ ; \ N = 8$$

complex multiplications are actually required.)

4.2.5 Linear Convolution via FFT

The *linear convolution* $y = w * v$ of two signals w and v was defined in Chapter 2 by

$$y(k) = \sum_{m=0}^{k} w(k-m)v(m) \tag{4.114}$$

In particular, if w is the unit impulse response of a digital filter with input signal v, then for zero initial consitions, y is the output signal. An efficient implementation of this algorithm is therefore of interest.

We first consider the computational complexity required to calculate y directly from this formula, and then we consider a DFT based procedure. To begin, suppose w and v are two *finite length* signals of length N as given by $\{w(k);\ 0 \leq k \leq N-1\}$ and $\{v(k);\ 0 \leq k \leq N-1\}$. Then the *linear convolution* $z_{2N-1} = w * v$ of w and v is given by (4.114). However by the definitions of the signals w and v, it follows that $y(k) = 0$ for $k < 0$ and $k \geq 2N-1$; that is, y is a *finite length* signal of length $2N-1$ given by $\{y(k);\ 0 \leq k \leq 2N-2\}$ where

$$y(k) = \begin{cases} \sum_{m=0}^{k} w(k-m)v(m) \ ; \ 0 \leq k \leq N-1 \\ \sum_{m=k-N+1}^{N-1} w(k-m)v(m) \ ; \ N \leq k \leq 2N-2 \end{cases} \tag{4.115}$$

Example 4.2.11. (a) Suppose $N = 3$. Then from (4.115) with $N-1 = 2$ and $2N-2 = 4$, we first have

$$\begin{aligned} y(0) &= w(0)v(0) \\ y(1) &= w(1)v(0) + w(0)v(1) \\ y(2) &= w(2)v(0) + w(1)v(1) + w(0)v(2) \end{aligned}$$

and then

$$\begin{aligned} y(3) &= w(2)v(1) + w(1)v(2) \\ y(4) &= w(2)v(2) \end{aligned}$$

Hence in order to compute $\{y(k);\ 0 \leq k \leq 4\}$, it takes a total of n_1 *real* multiplications and n_2 *real* additions where

$$n_1 = (1+2+3) + (2+1) = 9 \ ; \ n_2 = (1+2) + 1 = 4$$

(b) Suppose $N = 4$. Then from (4.115) with $N - 1 = 3$ and $2N - 2 = 6$, we have

$$\begin{aligned} y(0) &= w(0)v(0) \\ y(1) &= w(1)v(0) + w(0)v(1) \\ y(2) &= w(2)v(0) + w(1)v(1) + w(0)v(2) \\ y(3) &= w(3)v(0) + w(2)v(1) + w(1)v(2) + w(0)v(3) \end{aligned}$$

and then

$$\begin{aligned} y(4) &= w(3)v(1) + w(2)v(2) + w(1)v(3) \\ y(5) &= w(3)v(2) + w(2)v(3) \\ y(6) &= w(3)v(3) \end{aligned}$$

Hence to compute $\{y(k);\ 0 \leq k \leq 6\}$, it takes a total of n_1 real multiplications and n_2 real additions where

$$\begin{aligned} n_1 &= (1+2+3+4) + (3+2+1) = 16 \\ n_2 &= (1+2+3) + (2+1) = 9 \end{aligned}$$

More generally, it can be seen that in order to compute the linear convolution $y = w * v$ in (4.115) by *direct multiplication*, a total number of n_1 *real* multiplications, and n_2 *real* additions are required where

$$n_1 = \sum_{m=1}^{N} m + \sum_{m=1}^{N-1} m \ ; \ n_2 = \sum_{m=1}^{N-1} m + \sum_{m=1}^{N-2} m$$

That is

$$n_1 = \frac{N(N+1)}{2} + \frac{(N-1)N}{2} = N^2 \tag{4.116}$$

and

$$n_2 = \frac{(N-1)N}{2} + \frac{(N-2)(N-1)}{2} = (N-1)^2 \tag{4.117}$$

The aim now is to show that the *computational complexity* can be *reduced* by using a *DFT Based Procedure* to compute the linear convolution of two finite length signals w and v of length N *provided* each DFT is computed using the FFT algorithm. In particular, whereas the number of real multiplications to compute the linear convolution $y = w * v$ by direct calculation is N^2, we will now see that by using the FFT algorithm, only the order of $16N\log_2 N$ real multiplications are required.

We begin the discussion by defining the *circular convolution* of two periodic (or finite length) signals, and then go on to show how a DFT based procedure can reduce the computational complexity of a circular convolution. We then show how a *linear convolution* of two signals can be expressed as the circular convolution of two related signals which finally then gives a *DFT based procedure for linear convolution.*

Circular Convolution

The *circular convolution* $y = w \circ v$ of two signals w, v of length (or period) N is defined by

$$y(k) = \sum_{m=0}^{N-1} w(k-m)v(m) \ ; \ 0 \le k \le N-1 \tag{4.118}$$

Note that in (4.118), the components of w with negative arguments; that is, $\{w(p);\ -N+1 \le p \le -1\}$ are interpreted in the context of a *time-shifting* of w. Also, observe that since $y = w \circ v$ is given by

$$y(k) = \sum_{m=0}^{k} w(k-m)v(m) + \sum_{m=k+1}^{N-1} w(k-m)v(m)$$

the circular convolution $w \circ v$ is *not* generally equal to the *linear convolution* $w * v$.

Example 4.2.12. Given the signal $\{v(k);\ 0 \le k \le 2\}$ and the signal $\{w(k);\ 0 \le k \le 3\}$, the linear convolution $y = w * v$ is given by

$$\begin{aligned}
y(0) &= v(0)w(0) \\
y(1) &= v(0)w(1) + v(1)w(0) \\
y(2) &= v(0)w(2) + v(1)w(1) + v(2)w(0) \\
y(3) &= v(0)w(3) + v(1)w(2) + v(2)w(1) \\
y(4) &= v(1)w(3) + v(2)w(2) \\
y(5) &= v(2)w(3)
\end{aligned}$$

and so $y(k)$ is only defined for $0 \leq k \leq 5$.

Now define the signals v_a and w_a of length 5 by

$$v_a(k) = \begin{cases} v(k) & ; \ 0 \leq k \leq 2 \\ 0 & ; \ 2 \leq k \leq 5 \end{cases} \ ; \ w_a(k) = \begin{cases} w(k) & ; \ 0 \leq k \leq 3 \\ 0 & ; \ 4 \leq k \leq 5 \end{cases}$$

Then since

$$w_a(-1) = w_a(4) = 0 \ ; \ w_a(-2) = w_a(3) = 0 \ ; \ v_a(3) = v_a(4) = v_a(5) = 0$$

the component $y(0)$ can also be expressed in the form

$$y(0) = \sum_{m=0}^{5} v_a(m) w_a(-m)$$

More generally, we have the following result.

Theorem 4.2.7. *Given a signal $\{v(k);\ 0 \leq k \leq N_v - 1\}$ of length (or period) N_v, and a signal $\{w(k);\ 0 \leq k \leq N_w - 1\}$ of length (or period) N_w, define the signals v_a and w_a of length (or period) N_0 where*

$$N_0 = N_v + N_w - 1$$

by

$$v_a(k) = \begin{cases} v(k) & ; \ 0 \leq k \leq N_v - 1 \\ 0 & ; \ N_v \leq k \leq N_0 - 1 \end{cases}$$

$$w_a(k) = \begin{cases} w(k) & ; \ 0 \leq k \leq N_w - 1 \\ 0 & ; \ N_w \leq k \leq N_0 - 1 \end{cases}$$

Then the linear convolution $y = w * v$ in (4.114) is given by the circular convolution of v_a and w_a; that is, $y = w_a \circ v_a$ or

$$y(k) = \sum_{m=0}^{N_0 - 1} v_a(m) w_a(k - m) \tag{4.119}$$

We also have the next result.

Theorem 4.2.8. *Let y be the circular convolution of the two signals v, w of length (or period) N. Then:*

- $y = v \circ w = w \circ v$

- $$Y = DFT(y)$$
  is given by
  $$Y(n) = V(n)W(n) \quad ; \quad 0 \leq n \leq N-1$$

The first part of this result follows from (4.118) after introducing the change of variable $p = k - m$. That is

$$\sum_{m=0}^{N_0-1} w(k-m)v(m) = \sum_{\ell=k}^{k-N_0+1} w(\ell)v(k-\ell)$$

But by periodic extension (or periodicity), we have for any integer p that

$$\sum_{\ell=k}^{k-N_0+1} w(\ell)v(k-\ell) = \sum_{\ell=p}^{p-N_0+1} w(\ell)v(k-\ell)$$

The result follows with $p = 0$.

To prove the last part of Theorem 4.2.8, we have

$$Q(n) \triangleq V(n)W(n)exp\{\frac{j2\pi nk}{N_0}\}$$

is given by

$$\left[\sum_{m=0}^{N_0-1} v(m)exp\{-\frac{j2\pi nm}{N_0}\}\right]\left[\sum_{p=0}^{N_0-1} w(p)exp\{-\frac{j2\pi np}{N_0}\}\right] exp\{\frac{j2\pi nk}{N_0}\}$$

Hence

$$\begin{aligned}\sum_{n=0}^{N_0-1} Q(n) &= \sum_{m=0}^{N_0-1} v(m)\left[\sum_{p=0}^{N_0-1} w(p)\sum_{n=0}^{N_0-1} exp\{-\frac{j2\pi n(m+p-k)}{N_0}\}\right] \\ &= \sum_{m=0}^{N_0-1} v(m)w(k-m)\end{aligned}$$

where the last step follows by carrying out the summation over n using (4.77).

Example 4.2.13. In order to illustrate the result for the circular convolution in Theorem 4.2.8, consider a signal v of length 4 defined by

$$v(0) = 1 \;,\; v(1) = 2 \;;\; v(2) = 2 \;;\; v(3) = 1$$

Then

$$\begin{aligned} V(n) &= \sum_{k=0}^{3} v(k)exp\{-\frac{j2\pi nk}{4}\} \\ &= 1 + 2exp\{-\frac{j\pi n}{2}\} + 2exp\{-j\pi n\} + exp\{-\frac{j3\pi n}{2}\} \\ &= 1 + 2(-1)^n + (-j)^n[2 + (-1)^n] \end{aligned}$$

and hence

$$V(0) = 6 \ ; \ V(1) = -1 - j \ ; \ V(2) = 0 \ ; \ V(3) = -1 + j$$

Now suppose $w = v$. Then $y = v \circ w$ is given by

$$y(k) = \sum_{m=0}^{N_0 - 1} v(m)w(k - m) \ ; \ N_0 = 4 \tag{4.120}$$

can be calculated in two ways.

(a) Using discrete Fourier transforms

$$\tilde{Y}(n) = \tilde{V}(n)\tilde{W}(n) = [\tilde{V}(n)]^2$$

and so

$$\begin{aligned} 4y(k) &= \sum_{n=0}^{3} [V(n)]^2 exp\{\frac{j2\pi nk}{4}\} \\ &= 6^2 + (-1 - j)^2 exp\{\frac{j\pi k}{2}\} + 0 + (-1 + j)^2 exp\{\frac{j3\pi k}{2}\} \\ &= 36 + 2j^{k+1}[1 - (-1)^k] \end{aligned}$$

Therefore

$$y(0) = 9 \ ; \ y(1) = 8 \ ; \ y(2) = 9 \ ; \ y(3) = 10$$

(b) Directly from (4.120), we have

$$y(k) = \sum_{m=0}^{3} v(m)v(k - m) \ ; \ 0 \leq k \leq 3$$

and so

$$\begin{aligned} y(0) &= \sum_{m=0}^{3} v(m)v(-m) \\ &= [v(0)]^2 + v(1)v(-1) + v(2)v(-2) + v(3)v(-3) \\ &= [v(0)]^2 + v(1)v(3) + [v(2)]^2 + v(3)v(1) \\ &= 1 + 2 + 4 + 2 = 9 \end{aligned}$$

where we have used the periodic property: $v(-m) = v(4-m)$. Similarly

$$\begin{aligned}
y(1) &= v(0)v(1) + v(1)v(0) + v(2)v(3) + v(3)v(2) \\
&= 2 + 2 + 2 + 2 = 8 \\
y(2) &= v(0)v(2) + [v(1)]^2 + v(2)v(0) + [v(3)]^2 \\
&= 2 + 4 + 2 + 1 = 9 \\
y(3) &= v(0)v(3) + v(1)v(2) + v(2)v(1) + v(3)v(0) \\
&= 1 + 4 + 4 + 1 = 10
\end{aligned}$$

We now consider the computation of a *circular convolution* by means of an FFT algorithm. To begin, consider the special case when the signals v and w are both of the same length N. Then in Theorem 4.2.7, $N_0 = 2N-1$. Now

$$V(n) = \sum_{m=0}^{2N-2} v_a(m)\phi_{n,2N-1}(-m) \quad ; \quad 0 \le n \le 2N-1 \tag{4.121}$$

$$W(n) = \sum_{r=0}^{2N-2} w_a(r)\phi_{n,2N-1}(-r) \quad ; \quad 0 \le n \le 2N-1$$

Also

$$\phi_{n,L}(r) \triangleq exp\{\frac{j2\pi nr}{L}\} \tag{4.122}$$

so that

$$\phi^p_{n,L}(r) = (exp\{\frac{j2\pi nr}{L}\})^p = exp\{\frac{j2\pi nrp}{L}\} \tag{4.123}$$

Other properties of $\phi_{n,L}$ which follow from the properties of the complex exponential signal are:

$$\phi_{n,L}(p)\ \phi_{n,L}(q) = \phi_{n,L}(p+q) \tag{4.124}$$

$$\sum_{m=0}^{L-1} \phi^{\ell}_{n,L}(m) = \begin{cases} L \ ; \ \ell = kL, \ k \text{ integer} \\ 0 \ ; \ \text{otherwise} \end{cases}$$

Continuing, the signal $\{W(n)V(n)\}$ is then given by

$$W(n)V(n) = \sum_{m=0}^{2N-2} \sum_{r=0}^{2N-2} v_a(m)w_a(r)\phi_{n,2N-1}(-m-r) \triangleq Y(n) \tag{4.125}$$

and so $y = IDFT(Y)$ is given by

$$
\begin{aligned}
y(k) &= \frac{1}{2N-1} \sum_{n=1}^{2N-2} W(n)V(k)\phi_{k,2N-1}(n) \\
&= \frac{1}{2N-1} \sum_{m=0}^{2N-2} v_a(m) \sum_{r=0}^{2N-2} w_a(r) \left[\sum_{k=0}^{2N-2} \phi_{n-m-r,2N-1}(k) \right]
\end{aligned}
$$

But from (4.124)

$$
\frac{1}{2N-1} \sum_{k=0}^{2N-2} \phi_{n-m-r,2N-1}(k) = \begin{cases} 1 \,;\; r = n - m + l(2N-1) \\ 0 \,;\; \text{otherwise} \end{cases}
$$

for any integer l, and so

$$
v(k) = \sum_{m=0}^{2N-2} v_a(m) w_a(k-m)
$$

We summarise this result as follows.

Theorem 4.2.9. *Consider the linear convolution $y = w * v$ of the finite length signals $\{w(k);\ 0 \le k \le N-1\}$ and $\{v(k);\ 0 \le k \le N-1\}$ of length N, and define the discrete Fourier transforms*

$$
W = DFT(w_a) \;;\; V = DFT(v_a)
$$

of the finite length signals w_a and v_a of length $2N-1$.

Then $Y = DFT(y)$ is given by

$$
Y(n) = W(n)V(n)
$$

and

$$
y(k) = \frac{1}{2N-1} \sum_{n=0}^{2N-2} Y(n)\phi_{k,2N-1}(n) \;;\; 0 \le k \le 2N-2
$$

Example 4.2.14. Suppose v is a signal of length 2, and w is a signal of length 3. Then $N_0 \triangleq gcd(2,3) = 6$, and so the circular convolution y of v and w is given by

$$
y(k) = \sum_{m=0}^{5} v(m)w(k-m)
$$

For example

$$
\begin{aligned}
y(2) &= v(0)w(2) + v(1)w(1) + v(2)w(0) \\
&\quad + v(3)w(-1) + v(4)w(-2) + v(5)w(-3)
\end{aligned}
$$

where

$$w(-1) = w(2) \; ; \; w(-2) = w(1) \; ; \; w(-3) = w(0)$$

The calculation of the circular convolution via (4.118) requires a total of T_1 real multiplications where

$$T_1 = N_0^2 \tag{4.126}$$

However as shown in Theorem 4.2.3, the signal $\tilde{y}$ may also be computed via a *DFT based procedure* as follows:

- Calculate $V = DFT(v)$ and $W = DFT(w)$
- Calculate $Y(n) = V(n)W(n)$ for $0 \leq n \leq N_0 - 1$, and finally
- Calculate $y(k)$ for $0 \leq k \leq N_0 - 1$ by

$$y(k) = \sum_{n=0}^{N_0-1} Y(n)exp\{\frac{j2\pi nk}{N_0}\}$$

Now suppose $N_0 = 2^p$ for some integer $p > 0$. Then as shown in Theorem 4.2.6:

- The calculation of V and W each generally requires $4N_0\log_2(0.5N_0)$ real multiplications
- The calculation of Y generally requires $4N_0$ real multiplications, and finally
- The calculation of y from $\tilde{Y}$ generally requires $4N_0\log_2(0.5N_0)$ real multiplications.

Hence overall, the FFT based procedure for calculating the circular convolution $\tilde{y}$ requires a total of T_2 real multiplications where

$$T_2 = 4N_0 + 12N_0\log_2(0.5N_0) \tag{4.127}$$

Then from (4.126) and (4.127), we have that

$$\frac{T_1}{T_2} = \frac{0.25N_0}{1 + 3N_0\log_2(0.5N_0)}$$

Hence the FFT procedure has less multiplications (i.e. $T_2 < T_1$), and so is more numerically efficient, when

$$N_0 < 2^{r_o} \; ; \; r_0 = \frac{0.25N_0 + 2}{3}$$

or $N_0 > 64$. For example, when $N_0 = 32$,

$$\frac{T_1}{T_2} = \frac{8}{1 + 3\log_2(16)} = \frac{8}{13} \approx 0.6$$

when $N_0 = 64$

$$\frac{T_1}{T_2} = \frac{16}{1 + 3\log_2(32)} = \frac{16}{16} = 1$$

and when $N = 1024$

$$\frac{T_1}{T_2} = \frac{256}{1 + 3\log_2(512)} = \frac{256}{28} \approx 9$$

DFT Based Procedure

As a consequence of the result in Theorem 4.2.9, the *linear convolution* $y = w * v$ of two signals w and v of length N may be computed by the *circular convolution* $y = w_a \circ v_a$ of two related signals $\{w_a, v_a\}$ of length $2N-1$ defined in Theorem 4.2.7 as follows:

- Compute $W = DFT(w_a)$ and $V = DFT(v_a)$.
- Compute $\{W(n)V(n)\ ;\ 0 \leq k \leq 2N-2\}$.
- Compute $z = IDFT(Z)$ where $Z(n) = W(n)V(n)$.

Given that w and v are both of length N, and the linear convolution $w * v$ is of length $2N-1$, $w * v$ could also be achieved if the DFTs were computed on the basis of L points for any $L \geq 2N-1$. For example, if $2N-1 = 63$, then since $64 = 2^6$, it is more convenient to choose $L = 64$.

The *DFT Based Procedure* requires the calculation of two $2N-1$ points DFTs (as in *Step 1*), and the calculation of one IDFT (as in *Step 3*). Normally, the number of complex multiplications n_f is the same for both the DFT and the IDFT. When L is a power of 2, then

$$n_f = (L-2)\log_2(\frac{L}{2})$$

However half the values of w_a and v_a are zero, so that both DFTs in *Step 1* only require $n_f/2$ multiplications.

As Step 2 requires $2N-1$ complex multiplications, we have overall that the linear convolution $w * v$ of two signals w and v of length N which is computed via the *DFT Based Procedure* requires a total of $n_{2,m}$ *real* multiplications where

$$n_{2,m} = 4(n_f/2 + n_f/2 + n_f + 1 + 2N - 1) = 8n_f + 2N \tag{4.128}$$

The savings in computational complexity *only* come about as a result of the availability of the special *Fast Fourier Transform* algorithms. That is, since $L = 2N$ is a power of 2, then using a Fast Fourier Transform algorithm, we only require $n_f = (2N-2)\log_2(N)$ rather than $n_f = (2N-1)^2$ multiplications. Consequently, there will be a *reduction* in the number of real multiplications using the *DFT Based Procedure* compared with *direct computation* of the linear convolution if in (4.116), (4.128), we have $n_{1,m} < n_{2,m}$; that is, if

$$8n_f + 2N = 8(2N-2)\log_2(N) + 2N < N^2 \tag{4.129}$$

The inequality (4.129) is satisfied for $N > 128$. For example, when $N = 1024 = 2^{10}$, then

$$8(2N-2)\log_2(N) + 2N \approx 16\log_2 N + 2N = 165,648$$

and $N^2 = 1,048,576$, so in this case, the number of multiplications using the *DFT Based Procedure* is reduced to less than 1/6 of the number of multiplications required using the *direct method.*

For large values of N, the DFT based procedure for linear convolution requires the order of $16\log_2(N)$ multiplications whereas direct computation requires N^2 multiplications. Hence significant savings can be made when N is large and/or many linear convolutions need to be computed.

Segmentation and Overlap

When the length of one signal (say w) is much larger than the length of the other signal v, then it can be more numerically convenient to segment w. For example, w may be the unit impulse response of a FIR filter with a large number of coefficients, or a (large) finite length approximation of the unit impulse reponse of an IIR filter.

To begin, suppose the length N_w of the signal w is given by

$$N_w = ML$$

for some integers M, L. (If necessary, zeros may be appended to the end of w to get appropriate values for M, L.) Then the signal w can be expressed in the form

$$w(k) = \sum_{p=0}^{M-1} w_p(k) \tag{4.130}$$

where the signal w_p of length L is defined by

$$w_p(k) \triangleq \begin{cases} w(k) & ; \ pL \le k \le (p+1)L - 1 \\ 0 & ; \ \text{otherwise} \end{cases} \tag{4.131}$$

That is, the signal w is the *concatenation* of the *segmented signals* w_p.

Then from (4.114) and (4.130)

$$y(k) = \sum_{p=0}^{M-1} y_p(k) \tag{4.132}$$

where the signal y_p of length $L + N_v - 1$ is defined by

$$y_p(k) = \begin{cases} \sum_{m=pL}^{k} v(m) w_p(k-m) & ; \ pL \le k \le (p+1)L + N_v - 1 \\ 0 & ; \ \text{otherwise} \end{cases}$$

Now after a change of variable $\ell = k - pL$, and then a further change of variable $r = m - pL$, we have for: $0 \le \ell \le L + N_v - 2$

$$\begin{aligned} y_p(\ell + pL) &= \sum_{m=pL}^{\ell+pL} v(m) w_p(\ell + pL - m) \\ &= \sum_{r=0}^{\ell} v(r + pL) w_p(\ell - r) \end{aligned}$$

Therefore if we define

$$y_{p,p}(\ell) \triangleq y_p(\ell + pL) \ ; \ v_p(r) \triangleq v(r + pL) \tag{4.133}$$

we have

$$y_{p,p}(\ell) = \sum_{r=0}^{\ell} v_p(r) w_p(\ell - r) \ ; \ 0 \le \ell \le L + N_v - 2$$

The signal $y_{p,p}$, which is the linear convolution of the signals v_p and w_p, can also be expressed as a *circular convolution* of the form

$$y_{p,p}(\ell) = \sum_{r=0}^{L+N_v-2} \tilde{v}_p(r) \tilde{w}_p(\ell - r)$$

which can be calculated using an FFT based procedure. The complete response y in (4.132) is obtained by calculating $y_{p,p}$ (and hence y_p) for $0 \le p \le M - 1$. *However* forming y from the components y_p is *not* simply a matter of concatenating the components y_p, but an *overlap-and-add* proceedure is required.

Overlap-and Save procedure: In order to see what needs to be done, observe that the signal $y_{0,0}$ is given by $\{y_{0,0}(\ell); 0 \le \ell \le N_v + L - 2\}$, and so y_0 is given by $\{y_0(\ell); 0 \le \ell \le N_v + L - 2\}$. Next the signal $y_{1,1}$ is given by $\{y_{1,1}(\ell); 0 \le \ell \le N_v + L - 2\}$, and so y_1 is given by $\{y_1(\ell); L \le \ell \le N_v + 2L - 2\}$. That is, the domains of the signals y_0 and y_1 *overlap*. This implies $y^{(1)} \triangleq y_0 + y_1$ is given by

$$y^{(1)}(k) = \begin{cases} y_0(k) & ; \ 0 \le k \le L - 1 \\ y_0(k) + y_1(k) & ; \ L \le k \le N_v + L - 2 \\ y_1(k) & ; \ N_v + L - 1 \le k \le N_v + 2L - 2 \end{cases}$$

Similarly, $y^{(2)} \triangleq y^{(1)} + y_2$ is given by the *overlap-and-add* of $y^{(1)}$ and y_2.

In conclusion, the signal y is given by

$$y = y^{(p)} \triangleq y^{(p-1)} + y_p$$

where with $y^{(0)} = y_0$

$$y^{(n)} = y^{(n-1)} + y_n \ ; \ 1 \le n \le p$$

4.3 Least Squares Estimation

Many problems in signal estimation can be formulated in the context of a problem in linear algebra whereby one seeks a real n-vector $\boldsymbol{x}$ so as to minimizes the 2-norm of

$$\mathcal{J}(\boldsymbol{x}) \triangleq ||\boldsymbol{H}\boldsymbol{x} - \boldsymbol{f}||_2 \tag{4.134}$$

where $\boldsymbol{H}$ is a given $m \times n$ real matrix, and $\boldsymbol{f}$ is a given real m-vector. The value $\boldsymbol{x}_{min}$ of $\boldsymbol{x}$ in (4.134) which minimizes $\mathcal{J}(\boldsymbol{x})$ is known as the *least squares solution.*

In order to find the least squares solution to this linear algebra problem, we first need to develop some preliminary results based on the derivative of a scalar function $g(\boldsymbol{x})$ of $\boldsymbol{x}$ with respect to the vector $\boldsymbol{x}$ as follows. The *derivative of a scalar function $g(\boldsymbol{x})$ with respect to the real n-vector $\boldsymbol{x}$* where $\boldsymbol{x}^T = [x_1\ x_2 \cdots x_n]^T$ is an n-vector defined by

$$\frac{dg(\boldsymbol{x})}{d\boldsymbol{x}} \triangleq \left[\frac{\partial g(\boldsymbol{x})}{\partial x_1}\ \ \frac{\partial g(\boldsymbol{x})}{\partial x_2}\ \cdots\ \frac{\partial g(\boldsymbol{x})}{\partial x_n}\right]^T$$

Based on this definition, it can be shown that for any constant n-vector $\boldsymbol{h}$ and any constant n-square matrix $\boldsymbol{Q}$ that:

$$\frac{d}{d\boldsymbol{x}}(\boldsymbol{h}^T\boldsymbol{x}) = \frac{d}{d\boldsymbol{x}}(\boldsymbol{x}^T\boldsymbol{h}) = \boldsymbol{h}\ ;\ \frac{d}{d\boldsymbol{x}}(\boldsymbol{x}^T\boldsymbol{Q}\boldsymbol{x}) = (\boldsymbol{Q} + \boldsymbol{Q}^T)\boldsymbol{x} \tag{4.135}$$

Example 4.3.1. When $n = 2$, we have $\boldsymbol{h}^T\boldsymbol{x} = h_1x_1 + h_2x_2$, and so

$$\frac{d}{d\boldsymbol{x}}(\boldsymbol{h}^T\boldsymbol{x}) = \begin{bmatrix} \frac{\partial}{\partial x_1}(\boldsymbol{h}^T\boldsymbol{x}) \\ \frac{\partial}{\partial x_2}(\boldsymbol{h}^T\boldsymbol{x}) \end{bmatrix} = \begin{bmatrix} h_1 \\ h_2 \end{bmatrix} = \boldsymbol{h}$$

Also, when $n = 2$, and

$$\boldsymbol{Q} = \begin{bmatrix} q_{11} & q_{12} \\ q_{21} & q_{22} \end{bmatrix}$$

then

$$\boldsymbol{x}^T\boldsymbol{Q}\boldsymbol{x} = q_{11}x_1^2 + (q_{12} + q_{21})x_1x_2 + q_{22}x_2^2$$

and so

$$\begin{aligned} \frac{d}{d\boldsymbol{x}}(\boldsymbol{x}^T\boldsymbol{Q}\boldsymbol{x}) &= \begin{bmatrix} \frac{\partial}{\partial x_1}(\boldsymbol{x}^T\boldsymbol{Q}\boldsymbol{x}) \\ \frac{\partial}{\partial x_2}(\boldsymbol{x}^T\boldsymbol{Q}\boldsymbol{x}) \end{bmatrix} = \begin{bmatrix} 2q_{11}x_1 + (q_{12} + q_{21})x_2 \\ (q_{12} + q_{21})x_1 + 2q_{22}x_2 \end{bmatrix} \\ &= \begin{bmatrix} 2q_{11} & q_{12} + q_{21} \\ q_{12} + q_{21} & 2q_{22} \end{bmatrix} \begin{bmatrix} x_1 \\ x_2 \end{bmatrix} = (\boldsymbol{Q} + \boldsymbol{Q}^T)\boldsymbol{x} \end{aligned}$$

Now from (4.134), define

$$\begin{aligned} g(\boldsymbol{x}) &\triangleq \mathcal{J}^2(\boldsymbol{x}) = (\boldsymbol{H}\boldsymbol{x} - \boldsymbol{f})^T(\boldsymbol{H}\boldsymbol{x} - \boldsymbol{f}) \\ &= \boldsymbol{x}^T\boldsymbol{H}^T\boldsymbol{H}\boldsymbol{x} - \boldsymbol{x}^T\boldsymbol{H}^T\boldsymbol{f} - \boldsymbol{f}^T\boldsymbol{H}\boldsymbol{x} + \boldsymbol{f}\boldsymbol{f}^T \end{aligned}$$

so that the value of $\boldsymbol{x}$ which minimizes $\mathcal{J}(\boldsymbol{x})$ also minimizes $g(\boldsymbol{x})$. Since $(\boldsymbol{H}^T\boldsymbol{H})^T = \boldsymbol{H}^T\boldsymbol{H}$ and $(\boldsymbol{f}^T\boldsymbol{H})^T = \boldsymbol{H}^T\boldsymbol{f}$, we have using (4.135) that

$$\frac{dg(\boldsymbol{x})}{d\boldsymbol{x}} = 2\boldsymbol{H}^T\boldsymbol{H}\boldsymbol{x} - 2\boldsymbol{H}^T\boldsymbol{f} \tag{4.136}$$

By a fundamental theorem of calculus, any value of $\boldsymbol{x}$ which minimizes $g(\boldsymbol{x})$ occurs when

$$\frac{dg(\boldsymbol{x})}{d\boldsymbol{x}} = 0$$

Furthermore, since $g(\boldsymbol{x})$ is a nonnegative quadratic function in the components of $\boldsymbol{x}$, *all* such values of $\boldsymbol{x}$ minimize $g(\boldsymbol{x}) \triangleq \mathcal{J}^2(\boldsymbol{x})$. We then have the following result.

Theorem 4.3.1. *Suppose $\boldsymbol{H}$ is a real $m \times n$ matrix with full column rank n, and $\boldsymbol{f}$ is a real m-vector. Then $\mathcal{J}(\boldsymbol{x})$ in (4.134) is minimized uniquely by $\boldsymbol{x} = \boldsymbol{x}_{min}$ where*

$$\boldsymbol{x}_{min} = (\boldsymbol{H}^T\boldsymbol{H})^{-1}\boldsymbol{H}^T\boldsymbol{f}$$

with

$$\mathcal{J}(\boldsymbol{x}_{min}) = ||\boldsymbol{H}(\boldsymbol{H}^T\boldsymbol{H})^{-1}\boldsymbol{H}^T\boldsymbol{f} - \boldsymbol{f}||_2$$

Note that if an $m \times n$ matrix $\boldsymbol{H}$ does *not* have full column rank n, then the n-square matrix $\boldsymbol{H}^T\boldsymbol{H}$ is *not* invertible. However, it can still be shown that $\mathcal{J}(\boldsymbol{x})$ in (4.134) *always* has *at least one* minimizing solution irrespective of the rank of $\boldsymbol{H}$ and the particular m-vector $\boldsymbol{f}$. In our applications of least squares analysis to signal processing problems, we will assume that $\boldsymbol{H}$ *has* full column rank, but for completeness we now state the general result.

Theorem 4.3.2. *Suppose $\boldsymbol{H}$ is any real $m \times n$ matrix, and $\boldsymbol{f}$ is a real m-vector. Then all values $\boldsymbol{x}_{min}$ which minimize $\mathcal{J}(\boldsymbol{x})$ in (4.134) are of the form*

$$\boldsymbol{x}_{min} = \boldsymbol{H}^+\boldsymbol{f}$$

where $\boldsymbol{H}^+$ is a pseudo inverse of $\boldsymbol{H}$. When $\boldsymbol{H}$ has full column rank n, then

$$\boldsymbol{H}^+ = (\boldsymbol{H}^T\boldsymbol{H})^{-1}\boldsymbol{H}^T$$

Example 4.3.2. (a) Consider the problem of finding a 1-vector (i.e. a scalar) x which minimizes $\mathcal{J}(x) = ||\boldsymbol{H}_1 x - \boldsymbol{f}_1||_2$ where

$$\boldsymbol{H}_1 = \begin{bmatrix} 1 \\ 1 \\ 0 \end{bmatrix} \quad ; \quad \boldsymbol{f}_1 = \begin{bmatrix} 2 \\ 3 \\ -1 \end{bmatrix}$$

To solve this problem, observe that $\boldsymbol{H}_1$ has full column rank 1, and $\boldsymbol{H}_1^T \boldsymbol{H}_1 = 2$ with $(\boldsymbol{H}_1^T \boldsymbol{H}_1)^{-1} = 0.5$. Therefore using Theorem 4.3.1, the unique minimizing solution is given by

$$x_{min} = (\boldsymbol{H}_1^T \boldsymbol{H}_1)^{-1} \boldsymbol{H}_1^T \boldsymbol{f}_1 = 0.5[1 \ \ 1 \ \ 0] \begin{bmatrix} 2 \\ 3 \\ -1 \end{bmatrix} = 2.5$$

Also

$$J(x_{min}) = \left\| \begin{bmatrix} 1 \\ 1 \\ 0 \end{bmatrix} 2.5 - \begin{bmatrix} 2 \\ 3 \\ -1 \end{bmatrix} \right\|_2 = \left\| \begin{bmatrix} 0.5 \\ -0.5 \\ 1 \end{bmatrix} \right\|_2 = \sqrt{1.5}$$

(b) Suppose now that

$$\boldsymbol{H}_2 = \begin{bmatrix} 1 & 0 \\ 1 & 1 \\ 0 & 1 \end{bmatrix} \quad ; \quad \boldsymbol{f}_2 = \begin{bmatrix} 2 \\ 3 \\ -1 \end{bmatrix}$$

Since $\boldsymbol{H}_2$ has full column rank 2, we have

$$(\boldsymbol{H}_2^T \boldsymbol{H}_2)^{-1} = \begin{bmatrix} 2 & 1 \\ 1 & 2 \end{bmatrix}^{-1} = \frac{1}{3} \begin{bmatrix} 2 & -1 \\ -1 & 2 \end{bmatrix}$$

and so by Theorem 4.3.1, the unique minimizing least squares solution which minimizes $||\boldsymbol{H}_2 \boldsymbol{x} - \boldsymbol{f}_2||_2$ is given by

$$\boldsymbol{x}_{min} = (\boldsymbol{H}_2^T \boldsymbol{H}_2)^{-1} \boldsymbol{H}_2^T \boldsymbol{f}_2 = \frac{1}{3} \begin{bmatrix} 2 & -1 \\ -1 & 2 \end{bmatrix} \begin{bmatrix} 1 & 1 & 0 \\ 0 & 1 & 1 \end{bmatrix} \begin{bmatrix} 2 \\ 3 \\ -1 \end{bmatrix} = \frac{1}{3} \begin{bmatrix} 8 \\ -1 \end{bmatrix}$$

Also

$$\mathcal{J}(\boldsymbol{x}_{min}) = ||\frac{1}{3} \begin{bmatrix} 1 & 0 \\ 1 & 1 \\ 0 & 1 \end{bmatrix} \begin{bmatrix} 8 \\ -1 \end{bmatrix} - \begin{bmatrix} 2 \\ 3 \\ -1 \end{bmatrix} ||_2 = \frac{2\sqrt{3}}{3}$$

4.3.1 Linear Phase FIR Filter Design

Consider a FIR digital filter as defined by the frequency response function

$$H(z) = b_n + b_{n-1}z^{-1} + \ \dots\ b_1 z^{-n+1} + b_0 z^{-n} \tag{4.137}$$

As described in Theorem 3.2.1, there are four basic types of *linear phase* FIR filters depending on the relationship between the filter coefficients $\{b_\ell\}$. The corresponding magnitude response functions $A_\ell(\omega)$ for $\ell = 1, 2, 3, 4$ of the linear phase FIR filter is then either a sum of sine or cosine functions of the frequency ω.

Given the desired magnitudes $\{A_d(\omega_k) : \ 1 \le k \le M\}$ at some given set of given frequencies $\{\omega_k : \ 1 \le k \le M\}$, the *least squares filter design problem* we now address is; Find the $n+1$ filter coefficients $\{b_p : \ 0 \le p \le n\}$ which minimize the sum of squares $\mathcal{J}_M^2$ of the magnitude errors

$$\mathcal{J}_M^2 \triangleq \sum_{k=1}^{M} e^2(\omega_k) \ ; \ e(\omega_k) \triangleq A_d(\omega_k) - A_\ell(\omega_k) \tag{4.138}$$

In order to solve this problem, define the M-vectors $\{\boldsymbol{\varepsilon}_M, \boldsymbol{f}_M\}$ by

$$\boldsymbol{\varepsilon}_M \triangleq \begin{bmatrix} e(\omega_1) \\ e(\omega_2) \\ \cdot \\ \cdot \\ \cdot \\ e(\omega_M) \end{bmatrix} \ ; \ \boldsymbol{f}_M \triangleq \begin{bmatrix} A_d(\omega_1) \\ A_d(\omega_2) \\ \cdot \\ \cdot \\ \cdot \\ A_d(\omega_M) \end{bmatrix} \tag{4.139}$$

Also with reference to Theorem 3.2.1, define the vector $\boldsymbol{x}_{M,\ell}$ and the matrix $\boldsymbol{H}_{M,\ell}$ for each filter type corresponding to $\ell = 1, 2, 3, 4$ as follows:

- *Type 1 Filter* (n even; $N \triangleq 0.5n$)

In this case

$$b_m = b_{n-m} \ ; \ m = 0, 1, 2, \ \dots\ , N$$

and define the $(N+1)$-vector $\boldsymbol{x}_{M,1}$ and the $M \times (N+1)$ matrix $\boldsymbol{H}_{M,1}$ by

$$\boldsymbol{H}_{M,1} = \begin{bmatrix} 2\cos N\omega_1 & 2\cos(N-1)\omega_1 & \cdot\cdot & 2\cos\omega_1 & 1 \\ 2\cos N\omega_2 & 2\cos(N-1)\omega_2 & \cdot\cdot & 2\cos\omega_2 & 1 \\ \cdot & \cdot & \cdot\cdot & \cdot & \cdot \\ \cdot & \cdot & \cdot\cdot & \cdot & \cdot \\ 2\cos N\omega_M & 2\cos(N-1)\omega_M & \cdot\cdot & 2\cos\omega_M & 1 \end{bmatrix} \tag{4.140}$$

$$\boldsymbol{x}_{M,1}^T = \begin{bmatrix} b_0 \ b_1 \ \cdot\cdot \ b_N \end{bmatrix}$$

- *Type 2 Filter* (n odd; $N \triangleq 0.5(n-1)$)

In this case

$$b_m = b_{n-m} \; ; \; m = 0, 1, 2, \; \ldots \; , N$$

and define the $(N+1)$-vector $\boldsymbol{x}_{M,2}$ and the $M \times (N+1)$ matrix $\boldsymbol{H}_{M,2}$

$$\boldsymbol{H}_{M,2} = \begin{bmatrix} 2\cos(N+\frac{1}{2})\omega_1 & 2\cos(N-\frac{1}{2})\omega_1 & \cdot\cdot & 2\cos\frac{1}{2}\omega_1 \\ 2\cos(N+\frac{1}{2})\omega_2 & 2\cos(N-\frac{1}{2})\omega_2 & \cdot\cdot & 2\cos\frac{1}{2}\omega_2 \\ \cdot & \cdot & \cdot\cdot & \cdot \\ \cdot & \cdot & \cdot\cdot & \cdot \\ 2\cos(N+\frac{1}{2})\omega_M & 2\cos(N-\frac{1}{2})\omega_M & \cdot\cdot & 2\cos\frac{1}{2}\omega_M \end{bmatrix} \tag{4.141}$$

$$\boldsymbol{x}_{M,2}^T = \left[b_0 \; b_1 \; \cdot\cdot \; b_N \right]$$

- *Type 3 Filter* (n even; $N \triangleq 0.5n$)

In this case

$$b_N = 0; \; b_m = -b_{n-m} \; ; \; m = 0, 1, 2, \; \ldots \; , N$$

and define the N-vector $\boldsymbol{x}_{M,3}$ and the $M \times N$ matrix $\boldsymbol{H}_{M,3}$ by

$$\boldsymbol{H}_{M,3} = \begin{bmatrix} 2\sin N\omega_1 & 2\sin(N-1)\omega_1 & \cdots & 2\sin\omega_1 \\ 2\sin N\omega_2 & 2\sin(N-1)\omega_2 & \cdots & 2\sin\omega_2 \\ \cdot & \cdot & \cdots & \\ \cdot & \cdot & \cdots & \\ 2\sin N\omega_M & 2\sin(N-1)\omega_M & \cdots & 2\sin\omega_M \end{bmatrix} \tag{4.142}$$

$$\boldsymbol{x}_{M,3}^T = \left[b_0 \; b_1 \; \cdot\cdot \; b_{N-1} \right]$$

- *Type 4 Filter* (n odd; $N \triangleq 0.5(n-1)$)

In this case

$$b_m = -b_{n-m} \; ; \; m = 0, 1, 2, \; \ldots \; , N$$

and define the $(N+1)$-vector $\boldsymbol{x}_{M,4}$ and the $M \times (N+1)$ matrix $\boldsymbol{H}_{M,4}$ by

$$\boldsymbol{H}_{M,4} = \begin{bmatrix} 2\sin(N+\frac{1}{2})\omega_1 & 2\sin(N-\frac{1}{2})\omega_1 & \cdot\cdot & 2\sin\frac{1}{2}\omega_1 \\ 2\sin(N+\frac{1}{2})\omega_2 & 2\sin(N-\frac{1}{2})\omega_2 & \cdot\cdot & 2\sin\frac{1}{2}\omega_2 \\ \cdot & \cdot & \cdot\cdot & \cdot \\ \cdot & \cdot & \cdot\cdot & \cdot \\ 2\sin(N+\frac{1}{2})\omega_M & 2\sin(N-\frac{1}{2})\omega_M & \cdot\cdot & 2\sin\frac{1}{2}\omega_M \end{bmatrix} \tag{4.143}$$

$$\boldsymbol{x}_{M,4}^T = \left[b_0 \; b_1 \; \cdot\cdot \; b_N \right]$$

Then $\mathcal{J}_M$ in (4.138) for each type of linear phase FIR filtercan be written in the form

$$\mathcal{J}_M = ||\boldsymbol{\epsilon}_M||_2 = ||\boldsymbol{H}_{M,\ell}\boldsymbol{x}_{M,\ell} - \boldsymbol{f}_M||_2$$

From Theorem 4.3.1, we then have the following result.

Theorem 4.3.3. *Suppose $\boldsymbol{H}_{M,\ell}$ as defined in (4.140) - (4.143) has full column rank. Then the FIR filter coefficients $(\boldsymbol{x}_{M,\ell})_{min}$ which minimize the sum of squares $\mathcal{J}_M$ in (4.138) is given by*

$$(\boldsymbol{x}_{M,\ell})_{min} = (\boldsymbol{H}_{M,\ell}^T\boldsymbol{H}_{M,\ell})^{-1}\boldsymbol{H}_{M,\ell}^T\boldsymbol{f}_M \tag{4.144}$$

Note that since $\boldsymbol{H}_{M,\ell}$ involves only the desired magnitude response data, an appropriate choice can always be made so as to guarantee that $\boldsymbol{H}_{M,\ell}$ indeed has full column rank $n+1$.

Example 4.3.3. (a) Consider the 5 coefficient linear phase FIR filter with frequency response function H defined by

$$H(z) = b_4 + b_3z^{-1} + b_2z^{-2} + b_1z^{-3} + b_0z^{-4} \tag{4.145}$$

Then for a *Type 1 Filter* ($n = 4, N = 2$), we have $\{b_4 = b_0\ ;\ b_3 = b_1\}$, and

$$\boldsymbol{x}_{M,1} = \begin{bmatrix} b_0 \\ b_1 \\ b_2 \end{bmatrix} \ ; \ \boldsymbol{H}_{M,1} = \begin{bmatrix} 2\cos 2\omega_1 & 2\cos\omega_1 & 1 \\ 2\cos 2\omega_2 & 2\cos\omega_2 & 1 \\ \cdot & \cdot & \cdot \\ \cdot & \cdot & \cdot \\ 2\cos 2\omega_M & 2\cos\omega_M & 1 \end{bmatrix}$$

Hence at least 3 data points are required to uniquely determine $\{b_0, b_1, b_2\}$.

For a *Type 3 Filter* ($n = 4, N = 2$), we have $\{b_2 = 0;\ b_4 = -b_0\ ;\ b_3 = -b_1\}$, and

$$\boldsymbol{x}_{M,3} = \begin{bmatrix} b_0 \\ b_1 \end{bmatrix} \ ; \ \boldsymbol{H}_{M,3} = \begin{bmatrix} 2\sin 2\omega_1 & 2\sin\omega_1 \\ 2\sin 2\omega_2 & 2\sin\omega_2 \\ \cdot & \cdot \\ \cdot & \cdot \\ 2\sin 2\omega_M & 2\sin\omega_M \end{bmatrix}$$

Hence at least 2 data points are required to uniquely determine $\{b_0, b_1\}$.

(b) Now consider the 6 coefficient linear phase FIR filter with frequency response function H defined by

$$H(z) = b_5 + b_4z^{-1} + b_3z^{-2} + b_2z^{-3} + b_1z^{-4} + b_0z^{-5} \tag{4.146}$$

Then for a *Type 2 Filter* ($n = 5, N = 2$), we have $\{b_5 = b_0\ ;\ b_4 = b_1\ ;\ b_3 = b_2\}$, and

$$\boldsymbol{x}_{M,2} = \begin{bmatrix} b_0 \\ b_1 \\ b_2 \end{bmatrix} \; ; \; \boldsymbol{H}_{M,2} = \begin{bmatrix} 2\cos 2.5\omega_1 & 2\cos 1.5\omega_1 & 2\cos 0.5\omega_1 \\ 2\cos 2.5\omega_2 & 2\cos 1.5\omega_2 & 2\cos 0.5\omega_2 \\ \cdot & \cdot & \cdot \\ \cdot & \cdot & \cdot \\ 2\cos 2.5\omega_M & 2\cos 1.5\omega_M & 2\cos 0.5\omega_M \end{bmatrix}$$

Hence at least 3 data points are required to uniquely determine $\{b_0, b_1, b_2\}$.

For a *Type 4 Filter* ($n = 5, N = 2$), we have $\{b_5 = -b_0 \; ; \; b_4 = -b_1 \; ; \; b_3 = -b_2\}$, and

$$\boldsymbol{x}_{M,4} = \begin{bmatrix} b_0 \\ b_1 \\ b_2 \end{bmatrix} \; ; \; \boldsymbol{H}_{M,4} = \begin{bmatrix} 2\sin 2.5\omega_1 & 2\sin 1.5\omega_1 & 2\sin 0.5\omega_1 \\ 2\sin 2.5\omega_2 & 2\sin 1.5\omega_2 & 2\sin 0.5\omega_2 \\ \cdot & \cdot & \cdot \\ \cdot & \cdot & \cdot \\ 2\sin 2.5\omega_M & 2\sin 1.5\omega_M & 2\sin 0.5\omega_M \end{bmatrix}$$

Hence again at least 3 data points are required to uniquely determine $\{b_0, b_1, b_2\}$.

Selection of Reference Frequencies $\{\omega_k : 1 \le k \le M\}$

In order to determine the particular linear phase FIR filter coefficients $\{b_0, b_1, \ldots, b_n\}$ which then determine via Theorem 3.2.1, the corresponding magnitude response $A_\ell(\omega)$ for all frequencies ω, it is necessary to select a sufficiently large number of desired magnitudes $\{A_d(\omega_k)\}$ at particular fequencies $\{\omega_k\}$. In order to guarantee a solution via Theorem 4.3.3, the number M of desired magnitudes must be greater than the number M_0 of independent coefficients where

$$M_0 = \begin{cases} 0.5(n+2) & ; \text{Type 1} \quad \text{(n even)} \\ 0.5(n+1) & ; \text{Type 2} \quad \text{(n odd)} \\ 0.5n & ; \text{Type 3} \quad \text{(n even)} \\ 0.5(n+1) & ; \text{Type 4} \quad \text{(n odd)} \end{cases}$$

In addition, the desired magnitudes must be selected such that the relevant matrix $\boldsymbol{H}_{M,\ell}$ has full column rank. This condition is not a problem in practice, and furthermore, since by Theorem 3.2.1, $A_\ell(\omega)$ is periodic of period 2π, but symmetric about $\omega = \pi$. Hence it is only necessary to select the frequencies $\{\omega_k\}$ in the range $0 \le \omega_k < \pi$ for all k. If a *low pass* linear phase FIR filter is required, then as discussed in section 2.6.3, either a Type 1 or Type 2 filter must be used. A *high pass* characteristic can be obtained from either a Type 3 or Type 4 filter.

Consider an ideal low pass digital filter with bandwidth ω_c. If such a characteristic is to be approximated, then it is necessary to select a number of values $A_d(\omega)$ such that

$$A_d(\omega_k) \simeq \begin{cases} 1 \; ; \; 1 \le k \le P, \; \omega_k < \omega_c \\ 0 \; ; \; P+1 \le k \le M, \; \omega_k > \omega_c \end{cases}$$

However the particular values chosen can have a significant effect on the characteristic of the final design $A_\ell(\omega)$.

In particular, following the discussion of the Gibbs phenomenon in section 3.1.2, it is known that any attempt to approximate a *discontinuity* (such as the transition from $A_d(\omega) = 1$ to $A_d(\omega) = 0$ at $\omega = \omega_c$) can result in a large oscillatory behaviour in $A_\ell(\omega)$ near $\omega = \omega_c$. Consequently, a number of points should be selected along a transition band from $\omega = \omega_c - d$ to $\omega = \omega_c + d$ for some $d > 0$.

4.3.2 Input-Output Signal Model

A digital signal y which is expressed as the output of an nth order difference equation with input signal u is said to have an *input-output signal model.* We now consider a problem of signal estimation when the input-output signal model is FIR.

FIR Model

Consider the system with input signal u and output signal y having the ideal FIR signal model

$$y(k) = h_1 u(k-1) + h_2 u(k-2) + \ \dots \ h_p u(k-p) \tag{4.147}$$

This system can for example arise as a model for multipath propogation in a communication system. In this case $h_1 u(k-1)$ is the primary signal while the other components $\{h_m u(k-m);\ m \geq 2\}$ are interfering signals. The primary (or direct path) signal component is $h_1 u(k-1)$, and corresponds to the generated signal $u(k)$ being received one sample period after transmission with an attenuation factor h_1. The interfering signal component $h_2 u(k-2) + \dots\ h_p u(k-p)$ in the received signal $y(k)$ arrives via indirect paths as a result multiple reflections and attenuations of $u(k)$. In order to be able to compensate for multipath propogation, it is necessary to have estimates the values of the coefficients $\{h_1, h_2, \ \dots \ h_p\}$.

Consider conducting an experiment in which the input data $\{u(k) :\ 0 \leq k \leq M-p\}$ and the output data points $\{y(k) :\ 1 \leq k \leq M\}$ are known. The aim is then to find the coefficients $\{h_1, h_2, \ \dots \ h_p\}$ which minimize the sum of squares

$$\mathcal{J}_M^2 \triangleq \sum_{k=1}^{M} e^2(k) \ ; \ e(k) \triangleq y(k) - \sum_{\ell=1}^{p} h_\ell u(k-\ell) \tag{4.148}$$

This formulation of the problem recognises the fact that even if an ideal FIR model as in (4.147) exists, measurement errors in recording both the input and output signals will mean that the error signal e in (4.148) will be nonzero. In some other applications where an ideal FIR model does not exist, the purpose of the exercise may be simply to find the "best" least squares model of the form (4.147) that "fits" the given data.

If we define the M-vectors $\{\boldsymbol{\varepsilon}_M, \boldsymbol{f}_M\}$, the p-vector $\boldsymbol{x}_M$ and the $M \times p$ matrix $\boldsymbol{H}_M$ by

$$\boldsymbol{\varepsilon}_M \triangleq \begin{bmatrix} e(1) \\ e(2) \\ \cdot \\ \cdot \\ \cdot \\ e(M) \end{bmatrix} ; \boldsymbol{f}_M \triangleq \begin{bmatrix} y(1) \\ y(2) \\ \cdot \\ \cdot \\ \cdot \\ y(M) \end{bmatrix} ; \boldsymbol{x}_M \triangleq \begin{bmatrix} h_1 \\ h_2 \\ \cdot \\ \cdot \\ h_p \end{bmatrix}$$

$$\boldsymbol{H}_M \triangleq \begin{bmatrix} u(0) & u(-1) & \cdot\cdot & u(1-p) \\ u(1) & u(0) & \cdot\cdot & u(2-p) \\ \cdot & & & \\ \cdot & & & \\ \cdot & & & \\ u(M-1) & u(M-2) & \cdot\cdot & u(M-p) \end{bmatrix} \tag{4.149}$$

then $\mathcal{J}_M$ in (4.148) can be written in the form

$$\mathcal{J}_M = ||\boldsymbol{\varepsilon}_M||_2 = ||\boldsymbol{H}_M \boldsymbol{x}_M - \boldsymbol{f}_M||_2$$

From Theorem 4.3.1, we then have the following result.

Theorem 4.3.4. *Suppose $M \geq p$ and $\boldsymbol{H}_M$ in (4.149) has full column rank p. Then the FIR model coefficients $\boldsymbol{x}_{M,min} = [h_1 \ h_2 \ \ldots \ h_p]^T$ which minimize the sum of squares $\mathcal{J}_M$ in (4.148) is given by*

$$\boldsymbol{x}_{M,min} = (\boldsymbol{H}_M^T \boldsymbol{H}_M)^{-1} \boldsymbol{H}_M^T \boldsymbol{f}_M \tag{4.150}$$

Note that since the matrix $\boldsymbol{H}_M$ involves only the input signal u, an appropriate choice for this signal can guarantee that $\boldsymbol{H}_M$ indeed has full rank p.

Example 4.3.4. (a) Consider the FIR signal model

$$y(k) = h_1 u(k-1) + h_2 u(k-2) + h_3 u(k-3)$$

Then

$$\boldsymbol{H}_M = \begin{bmatrix} u(0) & u(-1) & u(-2) \\ u(1) & u(0) & u(-1) \\ \cdot & \cdot\cdot & \\ \cdot & \cdot\cdot & \\ \cdot & \cdot\cdot & \\ u(M-1) & u(M-2) & u(M-3) \end{bmatrix} ; \boldsymbol{f}_M = \begin{bmatrix} y(1) \\ y(2) \\ \cdot \\ \cdot \\ \cdot \\ y(M) \end{bmatrix}$$

$$\boldsymbol{x}_M^T \triangleq \begin{bmatrix} h_1 & h_2 & h_3 \end{bmatrix}$$

Since the 3-square matrix $\boldsymbol{H}_M \boldsymbol{H}_M^T$ is symmetric, there are only $0.5(3)(4) = 6$ different components. If ℓ_{ij} is the ijth component of $\boldsymbol{H}_M \boldsymbol{H}_M^T$, then $\ell_{ij} = \ell_{ji}$ for all i, j, and

$$\ell_{11} = \sum_{k=0}^{M-1} u^2(k) \; ; \; \ell_{12} = \sum_{k=0}^{M-1} u(k)u(k-1) \; ; \; \ell_{13} = \sum_{k=0}^{M-1} u(k)u(k-2)$$

$$\ell_{22} = \sum_{k=1}^{M-2} u^2(k) \; ; \; \ell_{23} = \sum_{k=-1}^{M-2} u(k)u(k-1) \; ; \; \ell_{33} = \sum_{k=-2}^{M-3} u(k)u(k-2)$$

(b) A *linear phase constraint* on the FIR signal model can be incorporated into the least squares signal analysis problem. For example, the model

$$y(k) = h_1 u(k-1) + h_2 u(k-2) + h_1 u(k-3)$$

has *linear phase*, and in order to estimate the coefficients $\{h_1, h_2\}$, we define the error signal e by

$$e(k) \triangleq y(k) - h_1 v(k-1) - h_2 u(k-2) \; ; \; v(k) \triangleq u(k) + u(k-2)$$

The best least squares parameters $\{h_1, h_2\}$ that minimize

$$\mathcal{J}_M^2 = \sum_{k=1}^{M} e^2(k)$$

is given by (4.150) where now

$$\boldsymbol{H}_M = \begin{bmatrix} v(0) & u(-1) \\ v(1) & u(0) \\ \cdot & \cdot \\ \cdot & \cdot \\ \cdot & \cdot \\ v(M-1) & u(M-2) \end{bmatrix} \; ; \; \boldsymbol{f}_M = \begin{bmatrix} y(1) \\ y(2) \\ \cdot \\ \cdot \\ \cdot \\ y(M) \end{bmatrix} \; ; \; \boldsymbol{x}_M \triangleq \begin{bmatrix} h_1 \\ h_2 \end{bmatrix}$$

IIR Model

Consider now the signal y which is the unit impulse response of the ideal IIR signal model

$$y(k+n) = \sum_{\ell=0}^{n-1} a_\ell y(k+\ell) + u(k+n) \tag{4.151}$$

As discussed in Chapter 1, this signal model is used in speech compression to represent (typically) 20 milliseconds of speech, and can be derived based on a physical model of the human vocal tract. In order to be able to compress or code speech in an efficient manner, it is necessary to first obtain estimates the values of the model coefficients $\{a_0, a_1, \ldots a_{n-1}\}$.

The unit impulse response is given by the initial condition response

$$y(k) = \sum_{\ell=0}^{n-1} a_\ell y(k-n+\ell) \ ; \ k \geq 1$$

Then given the output signal data $\{y(k) : -(n-1) \leq k \leq M\}$, the least squares problem is to find the coefficients $\{a_0, a_1, \ \dots \ a_{n-1}\}$, which minimize the sum of squares

$$\mathcal{J}_M^2 \triangleq \sum_{k=1}^{M} e^2(k) \ ; \ e(k) \triangleq y(k) - \sum_{\ell=0}^{n-1} a_\ell y(k-n+\ell) \tag{4.152}$$

This formulation recognises the fact that even if an ideal IIR model exists, measurement errors in recording the signal y will mean that the error signal e in (4.152) will be nonzero.

In other applications where an ideal IIR model does not exist, the aim is again to find the "best" least squares IIR model of the form (4.151) that "fits" the given data. If we define the M-vectors $\{\boldsymbol{\varepsilon}_M, \boldsymbol{f}_M\}$, the n-vector $\boldsymbol{x}_M$ and the $M \times n$ matrix $\boldsymbol{H}_M$ by

$$\boldsymbol{\varepsilon}_M = \begin{bmatrix} e(1) \\ e(2) \\ \cdot \\ \cdot \\ \cdot \\ e(M) \end{bmatrix} \ ; \ \boldsymbol{f}_M = \begin{bmatrix} y(1) \\ y(2) \\ \cdot \\ \cdot \\ \cdot \\ y(M) \end{bmatrix} \ ; \ \boldsymbol{x}_M = \begin{bmatrix} a_0 \\ a_1 \\ \cdot \\ \cdot \\ a_{n-1} \end{bmatrix} \tag{4.153}$$

$$\boldsymbol{H}_M = \begin{bmatrix} y(-n+1) & y(-n+2) & \cdot\cdot & y(1) & y(0) \\ y(-n+2) & y(-n+3) & \cdot\cdot & y(2) & y(1) \\ \cdot & & & & \\ \cdot & & & & \\ \cdot & & & & \\ y(-n+M) & y(-n+M+1) & \cdot\cdot & y(M) & y(M-1) \end{bmatrix}$$

then $\mathcal{J}_M$ in (4.152) can be written in the form

$$\mathcal{J}_M = ||\boldsymbol{\varepsilon}_M||_2 = ||\boldsymbol{H}_M \boldsymbol{x}_M - \boldsymbol{f}_M||_2$$

From Theorem 4.3.1, we then have the following result.

Theorem 4.3.5. *Suppose $M \geq n$ and $\boldsymbol{H}_M$ in (4.153) has full column rank n. Then the IIR model coefficients $\boldsymbol{x}_{M,min} = [a_0 \ a_1 \ \dots \ a_{n-1}]^T$ which minimize the sum of squares $\mathcal{J}_M$ in (4.152) is given by*

$$\boldsymbol{x}_{M,min} = (\boldsymbol{H}_M^T \boldsymbol{H}_M)^{-1} \boldsymbol{H}_M^T \boldsymbol{f}_M \tag{4.154}$$

Note that $\boldsymbol{H}_M$ involves the measurement data signal y so that unlike the case of the FIR least squares problem, the rank condition on $\boldsymbol{H}_M$ *cannot* be guaranteed.

Example 4.3.5. (a) Consider the unit impulse response h of the IIR model

$$y(k+2) = a_1 y(k+1) + a_0 y(k) + u(k+2)$$

Then we have that h is given by the initial condition response

$$h(k+2) = a_1 h(k+1) + a_0 h(k) \ ; \ k \geq 1$$

where $\{h(0) = 1, \ h(1) = a_1\}$ which then gives $y(2) = a_1^2 + a_0$. However since we are implicitly assuming that the the data y has measurement errors, it is not appropriate to simply use the two values $\{h(0), \ h(1)\}$ to estimate $\{a_0, \ a_0\}$. We therefore instead use the least squares approach as described in Theorem 4.3.5.

For this second order IIR model, we have from (4.153) that

$$\boldsymbol{H}_M = \begin{bmatrix} y(-1) & y(0) \\ y(0) & y(1) \\ \cdot & \cdot \\ \cdot & \cdot \\ \cdot & \cdot \\ y(M-2) & y(M-1) \end{bmatrix} \ ; \ \boldsymbol{f}_M = \begin{bmatrix} y(1) \\ y(2) \\ \cdot \\ \cdot \\ \cdot \\ y(M) \end{bmatrix} \ ; \ \boldsymbol{x}_M = \begin{bmatrix} a_0 \\ a_1 \end{bmatrix}$$

and so the coefficients $\{\ell_{ij}\}$ of the 2-square matrix $\boldsymbol{H}_M^T \boldsymbol{H}_M$ are given by

$$\ell_{11} = \sum_{k=-1}^{M-2} y^2(k) \ ; \ \ell_{12} = \ell_{21} = \sum_{k=-1}^{M-2} y(k) y(k+1)$$

$$\ell_{22} = \sum_{k=0}^{M-1} y^2(k)$$

(b) Consider now the IIR model

$$y(k+2) = a_1 y(k+1) + a_0 y(k) + b_1 u(k+1) + b_0 u(k)$$

where now both input and output data is available. Again we seek estimates of the parameters $\{a_1, \ a_0, \ b_1, \ b_0\}$ which minimize

$$\mathcal{J}_M^2 \triangleq \sum_{k=1}^{M} e^2(k)$$

where

$$e(k) \triangleq y(k) - a_1 y(k-1) - a_0 y(k-2) - b_1 u(k-1) - b_0 u(k-2)$$

The solution is given by (4.154) where now

$$\boldsymbol{H}_M = \begin{bmatrix} y(-1) & y(0) & u(-1) & u(0) \\ y(0) & y(1) & u(0) & u(1) \\ \cdot & \cdot & \cdot & \cdot \\ \cdot & \cdot & \cdot & \cdot \\ \cdot & \cdot & \cdot & \cdot \\ y(M-2) & y(M-1) & u(M-2) & u(M-1) \end{bmatrix} ; \ \boldsymbol{f}_M = \begin{bmatrix} y(1) \\ y(2) \\ \cdot \\ \cdot \\ \cdot \\ y(M) \end{bmatrix}$$

$$\boldsymbol{x}_M^T = \begin{bmatrix} a_0 \ a_1 \ b_0 \ b_1 \end{bmatrix}$$

4.3.3 State Space Signal Model

A digital signal y can also be expressed as the initial condition response of the state space signal model

$$\begin{aligned} \boldsymbol{x}(k+1) &= \boldsymbol{F}\boldsymbol{x}(k) \ ; \ \boldsymbol{x}(0) \in \mathcal{R}^n \\ y(k) &= \boldsymbol{h}^T \boldsymbol{x}(k) \end{aligned} \tag{4.155}$$

where the initial state $\boldsymbol{x}(0)$ determines the parameters of the signal. A least squares estimate of the initial state will then provide a least squares estimate of the signal parameters.

Example 4.3.6. (i) Suppose an object is travelling at *constant velocity* v_0. Then the position $d(t)$ at time t is given by

$$d(t) = v_0 t + d_0$$

where d_0 is the initial position measured with respect to some reference at time $t = 0$. Define

$$x_1(k) \triangleq d(kT) \ ; \ x_2(k) \triangleq v_0$$

where T is constant. Then since the velocity v_0 is constant, we have

$$\begin{aligned} x_2(k+1) &= x_2(k) = v_0 \\ x_1(k+1) &= (k+1)Tv_0 + d_0 = x_1(k) + Tv_0 = x_1(k) + Tx_2(k) \end{aligned}$$

That is, the state space representation with measurement $y(k) = d(kT)$ is given by

$$\begin{aligned} \boldsymbol{x}(k+1) &= \boldsymbol{F}\boldsymbol{x}(k) \\ y(k) &= \boldsymbol{h}^T \boldsymbol{x}(k) \end{aligned} \tag{4.156}$$

where

$$\boldsymbol{F} = \begin{bmatrix} 1 & T \\ 0 & 1 \end{bmatrix} \; ; \; \boldsymbol{x}(k) = \begin{bmatrix} x_1(k) \\ x_2(k) \end{bmatrix} \; ; \; \boldsymbol{x}(0) = \begin{bmatrix} d_0 \\ v_0 \end{bmatrix} \; ; \; \boldsymbol{h} = \begin{bmatrix} 1 \\ 0 \end{bmatrix} \tag{4.157}$$

Thus the initial position d_0 and the constant velocity v_0 are determined by the choice of the initial state $\boldsymbol{x}(0)$.

(ii) Suppose now the object is travelling at *constant acceleration* α_0. Then the velocity $v(t)$ and position $d(t)$ at time t are given by

$$v(t) = \alpha_0 t + v_0 \; ; \; d(t) = 0.5\alpha_0 t^2 + v_0 t + d_0$$

where $\{v_0, d_0\}$ are respectively the velocity and position at time $t = 0$. Define

$$x_1(k) \triangleq d(kT) \; ; \; x_2(k) \triangleq v(kT) \; ; \; x_3(k) \triangleq \alpha_0$$

Then since the acceleration α_0 is constant, we have

$$\begin{aligned}
x_3(k+1) &= x_3(k) = \alpha_0 \\
x_2(k+1) &= (k+1)T\alpha_0 + v_0 = x_2(k) + T\alpha_0 = x_2(k) + Tx_3(k) \\
x_1(k+1) &= 0.5(k+1)^2\alpha_0 T^2 + (k+1)v_0 T + d_0 = x_1(k) + Tx_2(k) + 0.5T^2 x_3(k)
\end{aligned}$$

That is, the state space signal model with measurement $y(k) = d(kT)$ is given by

$$\begin{aligned}
\boldsymbol{x}(k+1) &= \boldsymbol{F}\boldsymbol{x}(k) \\
y(k) &= \boldsymbol{h}^T \boldsymbol{x}(k)
\end{aligned}$$

where

$$\boldsymbol{F} = \begin{bmatrix} 1 & T & 0.5T^2 \\ 0 & 1 & T \\ 0 & 0 & 1 \end{bmatrix} \; ; \; \boldsymbol{x}(k) = \begin{bmatrix} x_1(k) \\ x_2(k) \\ x_3(k) \end{bmatrix} \; ; \; \boldsymbol{x}(0) = \begin{bmatrix} d_0 \\ v_0 \\ \alpha_0 \end{bmatrix} \; ; \; \boldsymbol{h} = \begin{bmatrix} 1 \\ 0 \\ 0 \end{bmatrix}$$

Thus the initial position d_0, the initial velocity v_0, and the constant acceleration α_0 are determined by the choice of the initial state $\boldsymbol{x}(0)$.

Example 4.3.7. Consider a discrete time sinusoidal signal y given by

$$y(k) = \alpha_0 \cos k\Omega_0 + \beta_0 \sin k\Omega_0 \quad ; \quad \Omega_0 \neq 0 \tag{4.158}$$

which as earlier shown in example 2.2.4 is given by the solution of the second order difference equation

$$y(k+2) = 2\cos \Omega_0^2 y(k+1) - y(k) \; ; \; k \geq 0$$

where

$$y(0) = \alpha_0 \ ; \ y(1) = \alpha_0 \cos \Omega_0 + \beta_0 \sin \Omega_0$$

From Theorem 2.5.1 (or equivalently, Theorem 2.5.4 with $n = 2$), an equivalent state space representation is given by

$$\begin{aligned} \boldsymbol{x}(k+1) &= \boldsymbol{F}\boldsymbol{x}(k) \\ y(k) &= \boldsymbol{h}^T \boldsymbol{x} \end{aligned}$$

where

$$\boldsymbol{F} = \begin{bmatrix} \cos \Omega_0 & \sin \Omega_0 \\ -\sin \Omega_0 & \cos \Omega_0 \end{bmatrix} \ ; \ \boldsymbol{h} = \begin{bmatrix} 1 \\ 0 \end{bmatrix} \ ; \ \boldsymbol{x}(0) = \begin{bmatrix} \alpha_0 \\ \beta_0 \end{bmatrix} \tag{4.159}$$

Hence assuming the frequency Ω_0 is fixed, the amplitude and the phase of the sinusoidal signal y are determined by the choice of the initial state $\boldsymbol{x}(0)$.

Given the data $\{y(k) \ : \ 0 \leq k \leq M-1\}$ and the model coefficients $\{\boldsymbol{F}, \boldsymbol{h}\}$, we now consider the problem of finding the initial condition $\boldsymbol{x}(0)$ which minimize the sum of squares

$$\mathcal{J}_M^2 \triangleq \sum_{k=0}^{M-1} e^2(k) \ ; \ e(k) \triangleq y(k) - \boldsymbol{h}^T \boldsymbol{F}^k \boldsymbol{x}(0) \tag{4.160}$$

Once again, this least squares formulation recognises the fact that even if an ideal model as in (4.155) exists, measurement errors in recording the signal y will mean that the error signal e in (4.160) will be nonzero. In other applications where an ideal model does not exist, the purpose of the exercise is again to find the "best" least squares model of the form (4.155) that "fits" the given data.

If we define the M-vectors $\{\boldsymbol{\varepsilon}_M, \boldsymbol{f}_M\}$, the n-vector $\boldsymbol{x}_M = \boldsymbol{x}(0)$, and the $M \times n$ matrix $\boldsymbol{H}_M$ by

$$\boldsymbol{\varepsilon}_M = \begin{bmatrix} e(0) \\ e(1) \\ \cdot \\ \cdot \\ \cdot \\ e(M-1) \end{bmatrix} \ ; \ \boldsymbol{f}_M = \begin{bmatrix} y(0) \\ y(1) \\ \cdot \\ \cdot \\ \cdot \\ y(M-1) \end{bmatrix} \tag{4.161}$$

$$\boldsymbol{H}_M = \begin{bmatrix} \boldsymbol{h}^T \\ \boldsymbol{h}^T \boldsymbol{F} \\ \cdot \\ \cdot \\ \cdot \\ \boldsymbol{h}^T \boldsymbol{F}^{M-1} \end{bmatrix}$$

then $\mathcal{J}_M$ in (4.160) can be written in the form

$$\mathcal{J}_M = ||\boldsymbol{\varepsilon}_M||_2 = ||\boldsymbol{H}_M\boldsymbol{x}_M - \boldsymbol{f}_M||_2 \; ; \; \boldsymbol{x}_M = \boldsymbol{x}(0)$$

From Theorem 4.3.1, we then have the following result.

Theorem 4.3.6. *Suppose $M \geq n$ and $\boldsymbol{H}_M$ in (4.161) has full column rank n. Then the initial state $\boldsymbol{x}_{M,min} = \boldsymbol{x}(0)$ which minimizes the sum of squares $\mathcal{J}_M$ in (4.160) is given by*

$$\boldsymbol{x}_{M,min} = (\boldsymbol{H}_M^T\boldsymbol{H}_M)^{-1}\boldsymbol{H}_M^T\boldsymbol{f}_M \tag{4.162}$$

Note that provided there is sufficient data (ie $M \geq n$), a signal model can always be selected such that $\boldsymbol{H}_M$ has full column rank n.

Example 4.3.8. We now consider the problem of obtaining least squares estimates $\{\hat{\alpha}_0, \hat{\beta}_0\}$ of the parameters $\{\alpha_0, \beta_0\}$ based on the ideal signal model

$$y(k) = \alpha_0 \sin k\Omega_0 + \beta_0 \cos k\Omega_0$$

given the frequency $\Omega_0 \neq 0$. The model is ideal in the sense that we assume the presence of measurement errors.

In the case of the sinusoidal signal model (4.158) in which $\{\boldsymbol{F}, \boldsymbol{h}, \boldsymbol{x}(0)\}$ are given by (4.159), it follows that

$$\boldsymbol{F}^2 = \begin{bmatrix} \cos 2\Omega_0 & \sin 2\Omega_0 \\ -\sin 2\Omega_0 & \cos 2\Omega_0 \end{bmatrix}$$

and more generally

$$\boldsymbol{F}^m = \begin{bmatrix} \cos m\Omega_0 & \sin m\Omega_0 \\ -\sin m\Omega_0 & \cos m\Omega_0 \end{bmatrix} \; ; \; m \geq 1$$

so that in this case with $n = 2$, the $M \times 2$ matrix $\boldsymbol{H}_M$ in (4.161) is given by

$$\boldsymbol{H}_M = \begin{bmatrix} 1 & 0 \\ \cos \Omega_0 & \sin \Omega_0 \\ \cos 2\Omega_0 & \sin 2\Omega_0 \\ \cdot & \cdot \\ \cdot & \cdot \\ \cdot & \cdot \\ \cos(M-1)\Omega_0 & \sin(M-1)\Omega_0 \end{bmatrix} \; ; \; \boldsymbol{f}_M = \begin{bmatrix} y(0) \\ y(1) \\ y(2) \\ \cdot \\ \cdot \\ \cdot \\ y(M-1) \end{bmatrix}$$

which implies

$$\boldsymbol{H}_M\boldsymbol{H}_M^T = \begin{bmatrix} \ell_{11} & \ell_{12} \\ \ell_{12} & \ell_{22} \end{bmatrix}$$

where

$$\ell_{11} = \sum_{k=0}^{M-1} \cos^2 k\Omega_0 \ ; \ \ell_{12} = 0.5 \sum_{k=0}^{M-1} \sin 2k\Omega_0 \ ; \ \ell_{22} = \sum_{k=0}^{M-1} \sin^2 k\Omega_0$$

Therefore from (4.162), provide $d \stackrel{\Delta}{=} \ell_{11}\ell_{22} - \ell_{12}^2 \neq 0$ where

$$d = (\sum_{k=0}^{M-1} \cos^2 k\Omega_0)(\sum_{k=0}^{M-1} \sin^2 k\Omega_0) - 0.25(\sum_{k=0}^{M-1} \sin 2k\Omega_0)^2$$

it follows that $\hat{\boldsymbol{x}}_M = [\hat{\alpha}_0 \ \ \hat{\beta}_0]^T$ is given by

$$\hat{\alpha}_0 = d^{-1}\gamma_0 \ ; \ \hat{\alpha}_0 = d^{-1}\gamma_1 \tag{4.163}$$

where

$$\gamma_0 = (\sum_{k=0}^{M-1} \sin^2 k\Omega_0)(\sum_{k=0}^{M-1} y(k) \cos k\Omega_0)$$
$$-0.5(\sum_{k=0}^{M-1} \sin 2k\Omega_0)(\sum_{k=0}^{M-1} y(k) \sin k\Omega_0)$$
$$\gamma_1 = -0.5(\sum_{k=0}^{M-1} \sin 2k\Omega_0)(\sum_{k=0}^{M-1} y(k) \cos k\Omega_0)$$
$$+(\sum_{k=0}^{M-1} \cos^2 k\Omega_0)(\sum_{k=0}^{M-1} y(k) \sin k\Omega_0)$$

In particular, if the frequency Ω_0 is of the form

$$\Omega_0 = q\left(\frac{\pi}{M}\right) \ ; \ q \text{ an integer}$$

then it follows that

$$\sum_{k=0}^{M-1} \sin 2k\Omega_0 = 0 \ ; \ \sum_{k=0}^{M-1} \sin^2 k\Omega_0 = \sum_{k=0}^{M-1} \cos^2 k\Omega_0 = 0.5M$$

Then in (4.163), we have

$$\hat{\alpha}_0 = \frac{2}{M} \sum_{k=0}^{M-1} y(k) \cos k\Omega_0 \ ; \ \hat{\beta}_0 = \frac{2}{M} \sum_{k=0}^{M-1} y(k) \sin k\Omega_0 \tag{4.164}$$

This result can be extended to the case of obtaining least squares estimates of the amplitudes $\{\alpha_m, \beta_m; \ 0 \leq m \leq L\}$ of the periodic signal

$$y(k) = \sum_{m=0}^{L} \{\alpha_n \cos k\Omega_m + \beta_n \sin k\Omega_m\}$$

given the frequencies $\{\Omega_m; \ 0 \leq m \leq L\}$.

Example 4.3.9. We now consider the problem of obtaining a least squares estimate $\hat{v}_0$ of the velocity of an aircraft which is assumed to be moving at a constant (but unknown) velocity v_0 based on measurements of the position y. This problem is related to radar processing as described in Chapter 1.

For an object moving with constant velocity, the signal model is given by (4.155) with $\{\boldsymbol{F}, \boldsymbol{h}, \boldsymbol{x}(0)$ given by (4.157). It follows that

$$\boldsymbol{F}^k = \begin{bmatrix} 1 & kT \\ 0 & 1 \end{bmatrix}$$

and so the $M \times 2$ matrix $\boldsymbol{H}_M$ in (4.161) is given by

$$\boldsymbol{H}_M = \begin{bmatrix} 1 & 0 \\ 0 & T \\ \cdot & \cdot \\ \cdot & \cdot \\ \cdot & \cdot \\ 1 & T^{M-1} \end{bmatrix} ; \ \boldsymbol{f}_M = \begin{bmatrix} y(0) \\ y(1) \\ \cdot \\ \cdot \\ \cdot \\ y(M-1) \end{bmatrix}$$

which implies

$$\boldsymbol{H}_M \boldsymbol{H}_M^T = \begin{bmatrix} \ell_{11} & \ell_{12} \\ \ell_{12} & \ell_{22} \end{bmatrix}$$

where

$$\ell_{11} = M \ ; \ \ell_{12} = \sum_{k=1}^{M-1} T^k \ ; \ \ell_{22} = \sum_{k=1}^{M-1} T^{2k}$$

Hence

$$\begin{bmatrix} \hat{d}_0 \\ \hat{v}_0 \end{bmatrix} = d^{-1} \begin{bmatrix} \ell_{22} & -\ell_{12} \\ -\ell_{12} & \ell_{11} \end{bmatrix} \begin{bmatrix} \sum_{k=1}^{M-1} y(k) \\ \sum_{k=1}^{M-1} T^k y(k) \end{bmatrix}$$

where

$$d = M \sum_{k=1}^{M-1} T^{2k} - (\sum_{k=1}^{M-1} T^k)^2$$

and so the velocity estimate $\hat{v}_0$ is given by

$$\hat{v}_0 = d^{-1} \left\{ - \left(\sum_{k=1}^{M-1} T^k)(\sum_{k=1}^{M-1} y(k) \right) + M \sum_{k=1}^{M-1} T^k y(k) \right\}$$

Based on the position measurements $y(k) = d(kT)$ (with assumed errors) at times $t = kT$, a least squares estimate $\hat{d}_0$ of the initial position $d(0)$ is given by

$$\hat{d}_0 = d^{-1} \left\{ \left(\sum_{k=1}^{M-1} T^{2k})(\sum_{k=1}^{M-1} y(k) \right) - \left(\sum_{k=1}^{M-1} T^k)(\sum_{k=1}^{M-1} T^k y(k) \right) \right\}$$

4.3.4 Recursive Estimation

We have developed a least squares algorithm that depends explicitly on the number M of measurements which finds the vector $\hat{\boldsymbol{x}}$ which minimizes $\mathcal{J}_M$ as defined by

$$\mathcal{J}_M = ||\boldsymbol{H}_M \boldsymbol{x}_M - \boldsymbol{f}_M||_2 \tag{4.165}$$

In particular, assuming $\boldsymbol{H}_M$ has full column rank, then the least squares estimate $\hat{\boldsymbol{x}}$ is given by

$$\hat{\boldsymbol{x}}_M = (\boldsymbol{H}_M^T \boldsymbol{H}_M)^{-1} \boldsymbol{H}_M^T \boldsymbol{f}_M \tag{4.166}$$

If an *additional data point* becomes available, then provided $\boldsymbol{H}_{M+1}$ has full column rank, a new estimate $\hat{\boldsymbol{x}}_{M+1}$ that minimizes $\mathcal{J}_{M+1}$ is given by

$$\hat{\boldsymbol{x}}_{M+1} = (\boldsymbol{H}_{M+1}^T \boldsymbol{H}_{M+1})^{-1} \boldsymbol{H}_{M+1}^T \boldsymbol{f}_{M+1} \tag{4.167}$$

As it stands, the calculation of $\hat{\boldsymbol{x}}_{M+1}$ requires the calculation of another matrix inverse. However in a real time signal processing application, matrix inversion can be both too time consuming and too inaccurate. Therefore in this section, we develop a *recursive algorithm* which updates the current estimate $\hat{\boldsymbol{x}}_M$ to the new estimate $\hat{\boldsymbol{x}}_{M+1}$ *without the need to compute a matrix inverse.*

In particular, suppose $\hat{\boldsymbol{x}}_M$ is a p-vector, $\boldsymbol{H}_M$ is an $M \times p$ matrix, and $\boldsymbol{f}_M$ is an M-vector. Then for all the applications considered in sections 3.2.1 - 3.2.3, the $(M+1) \times p$ matrix $\boldsymbol{H}_{M+1}$ and the $(M+1)$-vector $\boldsymbol{f}_{M+1}$ can be written in the form

$$\boldsymbol{H}_{M+1} = \begin{bmatrix} \boldsymbol{H}_M \\ \boldsymbol{v}^T(M) \end{bmatrix} \ ; \ \boldsymbol{f}_{M+1} = \begin{bmatrix} \boldsymbol{f}_M \\ z(M) \end{bmatrix}$$

where $\boldsymbol{v}_M$ is a p-vector and z_M is a scalar. In this case, given that $\boldsymbol{H}_M$ has full column rank p, then $\boldsymbol{H}_{M+1}$ will also have full column rank p. That is, if the p-square matrix $\boldsymbol{H}_M^T \boldsymbol{H}_M$ is invertible then so too is the p-square matrix $\boldsymbol{H}_{M+1}^T \boldsymbol{H}_{M+1}$. We then have the following result.

Theorem 4.3.7. *(i) Suppose*

$$\boldsymbol{H}_{k+1} = \begin{bmatrix} \boldsymbol{H}_k \\ \boldsymbol{v}^T(k) \end{bmatrix} \ ; \ \boldsymbol{f}_{k+1} = \begin{bmatrix} \boldsymbol{f}_k \\ z(k) \end{bmatrix} \tag{4.168}$$

where $\boldsymbol{H}_M$ has full column rank p. Then $\boldsymbol{H}_k$ has full column rank for all $k \geq M$.

(ii) When $\boldsymbol{H}_k$ has full column rank, define

$$\boldsymbol{Q}(k) \triangleq (\boldsymbol{H}_k^T \boldsymbol{H}_k)^{-1} \ ; \ k \geq M \tag{4.169}$$

Then the optimal least squares estimate $\hat{\boldsymbol{x}}_k$ that minimizes $||\boldsymbol{H}_k \boldsymbol{x}_k - \boldsymbol{f}_k||_2$ for $k \geq M$ is given by

$$\hat{\boldsymbol{x}}_{k+1} = \hat{\boldsymbol{x}}_k + \boldsymbol{g}(k)[z(k) - \boldsymbol{v}^T(k)\hat{\boldsymbol{x}}_k] \tag{4.170}$$

where $\boldsymbol{g}(k)$ *and* $\boldsymbol{Q}(k)$ *are given recursively for* $k \geq M$ *by*

$$\begin{aligned} \boldsymbol{g}(k) &= \frac{\boldsymbol{Q}(k)\boldsymbol{v}(k)}{1 + \boldsymbol{v}^T(k)\boldsymbol{Q}(k)\boldsymbol{v}(k)} \\ \boldsymbol{Q}(k+1) &= \boldsymbol{Q}(k) - \frac{\boldsymbol{Q}(k)\boldsymbol{v}(k)\boldsymbol{v}^T(k)\boldsymbol{Q}(k)}{1 + \boldsymbol{v}^T(k)\boldsymbol{Q}(k)\boldsymbol{v}(k)} \end{aligned} \tag{4.171}$$

In order to prove this result, observe from (4.168) that

$$\begin{aligned} \boldsymbol{H}_{k+1}^T\boldsymbol{H}_{k+1} &= [\boldsymbol{H}_k^T \;\; \boldsymbol{v}(k)] \begin{bmatrix} \boldsymbol{H}_k \\ \boldsymbol{v}^T(k) \end{bmatrix} = \boldsymbol{H}_k^T\boldsymbol{H}_k + \boldsymbol{v}(k)\boldsymbol{v}^T(k) \\ \boldsymbol{H}_{k+1}^T\boldsymbol{f}_{k+1} &= [\boldsymbol{H}_k^T \;\; \boldsymbol{v}(k)] \begin{bmatrix} \boldsymbol{f}_k \\ z(k) \end{bmatrix} = \boldsymbol{H}_k^T\boldsymbol{f}_k + \boldsymbol{v}(k)z(k) \end{aligned}$$

Hence

$$(\boldsymbol{H}_{k+1}^T\boldsymbol{H}_{k+1})^{-1} = [\boldsymbol{H}_k^T\boldsymbol{H}_k + \boldsymbol{v}(k)\boldsymbol{v}^T(k)]^{-1}$$

An expression for the inverse of $\boldsymbol{H}_{k+1}^T\boldsymbol{H}_{k+1}$ in terms of the inverse of $\boldsymbol{H}_k^T\boldsymbol{H}_k$ can then be determined from what is called the *matrix inversion formula.* We therefore postpone the proof of Theorem 4.3.7, and now consider the following result.

Theorem 4.3.8. *Suppose* $\boldsymbol{A}$ *is an n-square invertible matrix, and* $\{\boldsymbol{b}, \boldsymbol{c}\}$ *are n-vectors. Then* $\boldsymbol{A} + \boldsymbol{b}\boldsymbol{c}^T$ *is invertible if and only if* $1 + \boldsymbol{c}^T\boldsymbol{A}^{-1}\boldsymbol{b} \neq 0$. *Furthermore, if* $\boldsymbol{A} + \boldsymbol{b}\boldsymbol{c}^T$ *is invertible, then*

$$(\boldsymbol{A} + \boldsymbol{b}\boldsymbol{c}^T)^{-1} = \boldsymbol{A}^{-1} - \frac{\boldsymbol{A}^{-1}\boldsymbol{b}\boldsymbol{c}^T\boldsymbol{A}^{-1}}{1 + \boldsymbol{c}^T\boldsymbol{A}^{-1}\boldsymbol{b}}$$

In order to prove Theorem 4.3.8, observe that since $\boldsymbol{A}^{-1}$ is assumed to exist

$$\boldsymbol{A} + \boldsymbol{b}\boldsymbol{c}^T = \boldsymbol{A}(\boldsymbol{I}_n + \boldsymbol{A}^{-1}\boldsymbol{b}\boldsymbol{c}^T)$$

Then using the properties of the determinant $det[\boldsymbol{M}]$ of a matrix $\boldsymbol{M}$, we have

$$\begin{aligned} det[\boldsymbol{A} + \boldsymbol{b}\boldsymbol{c}^T] &= det[\boldsymbol{A}(\boldsymbol{I}_n + \boldsymbol{A}^{-1}\boldsymbol{b}\boldsymbol{c}^T)] = det[\boldsymbol{A}]\; det[\boldsymbol{I}_n + \boldsymbol{A}^{-1}\boldsymbol{b}\boldsymbol{c}^T] \\ &= det[\boldsymbol{A}]\; det[(\boldsymbol{I}_n + \boldsymbol{A}^{-1}\boldsymbol{b}\boldsymbol{c}^T)^T] \\ &= det[\boldsymbol{A}]\; det[(1 + \boldsymbol{c}^T\boldsymbol{A}^{-1}\boldsymbol{b})\boldsymbol{I}_n] \\ &= (1 + \boldsymbol{c}^T\boldsymbol{A}^{-1}\boldsymbol{b})det[\boldsymbol{A}] \end{aligned}$$

Therefore since $det[\boldsymbol{A}] \neq 0$, we have that $det[\boldsymbol{A} + \boldsymbol{b}\boldsymbol{c}^T] \neq 0$ if and only if $1 + \boldsymbol{c}^T \boldsymbol{A}^{-1} \boldsymbol{b} \neq 0$. Finally

$$(\boldsymbol{A} + \boldsymbol{b}\boldsymbol{c}^T)\left(\boldsymbol{A}^{-1} - \frac{\boldsymbol{A}^{-1}\boldsymbol{b}\boldsymbol{c}^T\boldsymbol{A}^{-1}}{1 + \boldsymbol{c}^T\boldsymbol{A}^{-1}\boldsymbol{b}}\right)$$
$$= \boldsymbol{I}_n - \frac{\boldsymbol{b}\boldsymbol{c}^T\boldsymbol{A}^{-1}}{1 + \boldsymbol{c}^T\boldsymbol{A}^{-1}\boldsymbol{b}} + \boldsymbol{b}\boldsymbol{c}^T\boldsymbol{A}^{-1} - \frac{\boldsymbol{b}(\boldsymbol{c}^T\boldsymbol{A}^{-1}\boldsymbol{b})\boldsymbol{c}^T\boldsymbol{A}^{-1}}{1 + \boldsymbol{c}^T\boldsymbol{A}^{-1}\boldsymbol{b}} = \boldsymbol{I}_n$$

Example 4.3.10. Suppose

$$\boldsymbol{A} = \begin{bmatrix} 1 & 0 & 0 \\ 0 & 1 & 0 \\ 0 & 0 & 1 \end{bmatrix} = \boldsymbol{I}_3 \ ; \ \boldsymbol{b} = \begin{bmatrix} 3 \\ -2 \\ 1 \end{bmatrix} \ ; \ \boldsymbol{c} = \begin{bmatrix} -1 \\ 2 \\ 0 \end{bmatrix}$$

Then $1 + \boldsymbol{c}^T\boldsymbol{A}^{-1}\boldsymbol{b} = 1/6$ and so $(\boldsymbol{A} + \boldsymbol{b}\boldsymbol{c}^T)^{-1}$ exists. Furthermore

$$\boldsymbol{K} \triangleq \boldsymbol{A} + \boldsymbol{b}\boldsymbol{c}^T = \begin{bmatrix} -2 & 6 & 0 \\ 2 & -3 & 0 \\ -1 & 2 & 1 \end{bmatrix}$$

and so from Theorem 4.3.8, the inverse of $\boldsymbol{K}$ is given by

$$\boldsymbol{K}^{-1} = \begin{bmatrix} 1 & 0 & 0 \\ 0 & 1 & 0 \\ 0 & 0 & 1 \end{bmatrix} + \frac{1}{6}\begin{bmatrix} -2 & 6 & 0 \\ 2 & -3 & 0 \\ -1 & 2 & 1 \end{bmatrix} = \frac{1}{6}\begin{bmatrix} 3 & 6 & 0 \\ 2 & 2 & 0 \\ -1 & 2 & 6 \end{bmatrix}$$

Now, let us return to the proof of Theorem 4.3.7. With $\boldsymbol{Q}(k)$ as defined by Theorem 4.3.7, we have using the matrix inversion formula in Theorem 4.3.8 that

$$\boldsymbol{Q}(k+1) = \boldsymbol{Q}(k) - \frac{\boldsymbol{Q}(k)\boldsymbol{v}(k)\boldsymbol{v}^T(k)\boldsymbol{Q}(k)}{1 + \boldsymbol{v}^T(k)\boldsymbol{Q}(k)\boldsymbol{v}(k)}$$

and so from (4.167)

$$\begin{aligned}
\hat{\boldsymbol{x}}_{k+1} &= \boldsymbol{Q}(k+1)[\boldsymbol{H}_k^T \boldsymbol{f}_k + \boldsymbol{v}(k)z(k)] \\
&= [\boldsymbol{Q}(k) - \frac{\boldsymbol{Q}(k)\boldsymbol{v}(k)\boldsymbol{v}^T(k)\boldsymbol{Q}(k)}{1 + \boldsymbol{v}^T(k)\boldsymbol{Q}(k)\boldsymbol{v}(k)}][\boldsymbol{H}_k^T \boldsymbol{f}_k + \boldsymbol{v}(k)z(k)] \\
&= \boldsymbol{Q}(k)\boldsymbol{H}_k^T \boldsymbol{f}_k - \frac{\boldsymbol{Q}(k)\boldsymbol{v}(k)\boldsymbol{v}^T(k)\boldsymbol{Q}(k)\boldsymbol{H}_k^T \boldsymbol{f}_k}{1 + \boldsymbol{v}^T(k)\boldsymbol{Q}(k)\boldsymbol{v}(k)} \\
&\quad + \left(\boldsymbol{Q}(k)\boldsymbol{v}(k) - \frac{\boldsymbol{Q}(k)\boldsymbol{v}(k)\boldsymbol{v}^T(k)\boldsymbol{Q}(k)\boldsymbol{v}(k)}{1 + \boldsymbol{v}^T(k)\boldsymbol{Q}(k)\boldsymbol{v}(k)}\right) z(k) \\
&= \hat{\boldsymbol{x}}_k - \boldsymbol{g}(k)\boldsymbol{v}^T(k)\hat{\boldsymbol{x}}_k + \frac{\boldsymbol{Q}(k)\boldsymbol{v}(k)}{1 + \boldsymbol{v}^T(k)\boldsymbol{Q}(k)\boldsymbol{v}(k)} z(k)
\end{aligned}$$

which then gives the result of Theorem 4.3.7. Observe that the calculation in (4.171) of $\boldsymbol{Q}(k+1)$ from $\boldsymbol{Q}(k)$ (and hence $\hat{\boldsymbol{x}}_{k+1}$ from $\hat{\boldsymbol{x}}_k$) involves matrix multiplication and addition, but *no matrix inversion*.

4.3.5 Applications of Recursive Estimation

We now consider some applications of the recursive least squares algorithm (4.170), (4.171) in Theorem 4.3.7 by identifying, in each case, the appropriate signals z in (4.170) and v in (4.171).

Linear Phase FIR Model

For a type 1 (i.e. $N = 0.5n$ with n even) linear phase FIR filter, we have from (4.140) that

$$\boldsymbol{v}^T(k) \triangleq \left[2\cos N\omega_{k+1}\ 2\cos(N-1)\omega_{k+1}\ \cdots\ 2\cos\omega_{k+1}\ 1\right]$$
$$z(k) \triangleq A_d(\omega_{k+1})$$

Then given the optimal FIR parameters

$$\hat{\boldsymbol{x}}_k^T \triangleq [b_0\ \ b_1\ \ \ldots\ \ b_N]$$

based on the amplitude values $\{A_d(\omega_m)\ :\ 1 \leq m \leq k+1\}$, a new set of optimal parameters $\hat{\boldsymbol{x}}_{k+1}$ based on the new data point $\{\omega_{k+2}, A_d(\omega_{k+2})\}$ can be obtained without computing a new matrix inverse. (Similar results can also be obtained for the other types of linear phase FIR filters.)

FIR Model Parameters

A recursive least squares algorithm can also be developed in order to update the optimal parameter vector

$$\hat{\boldsymbol{x}}_k^T \triangleq [h_1\ \ h_2\ \ \ldots\ \ h_p]$$

of the FIR model (4.147). In particular, in terms of Theorem 4.3.7, we have from (4.149) that

$$\boldsymbol{v}^T(k) \triangleq \left[u(k)\ u(k-1)\ \cdots\ u(k-p+1)\right]$$
$$z(k) \triangleq y(k+1)$$

The resulting least squares recursive algorithm is used as the basis for linear predictive speech coding as described in the Introduction.

IIR Model Parameters

In this case, for the IIR model (4.151), we have from (4.153) that

$$\begin{aligned} \boldsymbol{v}^T(k) &\triangleq \left[\, y(-n+k+1)\ y(-n+k+2) \cdots y(k+1)\ y(k)\,\right] \\ z(k) &\triangleq y(k+1) \end{aligned}$$

Signal Parameters

Using (4.161), it follows for the nth order state space signal model (4.155) that

$$\boldsymbol{v}^T(k) \triangleq \boldsymbol{h}^T \boldsymbol{F}^k \ ; \ z(k) \triangleq y(k) \tag{4.172}$$

The recursive least squares state space signal model algorithm is then given by

$$\hat{\boldsymbol{x}}_{k+1} = \hat{\boldsymbol{x}}_k + \boldsymbol{g}(k)[y(k) - \boldsymbol{h}^T \boldsymbol{F}^k \hat{\boldsymbol{x}}_k] \tag{4.173}$$

where $\hat{\boldsymbol{x}}_k$ is the *least squares estimate* of the *initial state* $\boldsymbol{x}(0)$ based on the data $\{y(m);\ m \leq k\}$ that is available up and to and including time k.

Now the signal model (4.161) implies

$$\boldsymbol{x}(k) = \boldsymbol{F}^k \boldsymbol{x}(0) \ ; \ y(k) = \boldsymbol{h}^T \boldsymbol{F}^k \boldsymbol{x}(0)$$

and so it follows that the *least squares estimate* $\hat{\boldsymbol{x}}(k)$ of the state $\boldsymbol{x}(k)$ and the least squares estimate $\hat{y}(k)$ of the output $y(k)$ based on the data $\{y(m);\ m \leq k\}$ that is available up and to and including time k are given by

$$\hat{\boldsymbol{x}}(k) = \boldsymbol{F}^k \hat{\boldsymbol{x}}_k \ ; \ \hat{y}(k) = \boldsymbol{h}^T \boldsymbol{F}^k \hat{\boldsymbol{x}}_k \tag{4.174}$$

From (4.173), we have

$$\boldsymbol{F}^{k+1} \hat{\boldsymbol{x}}_{k+1} = \boldsymbol{F}^{k+1} \hat{\boldsymbol{x}}_k + \boldsymbol{F}^{k+1} \boldsymbol{g}(k)[y(k) - \boldsymbol{h}^T \boldsymbol{F}^k \hat{\boldsymbol{x}}_k]$$

which from (4.174) implies

$$\hat{\boldsymbol{x}}(k+1) = \boldsymbol{F} \hat{\boldsymbol{x}}(k) + \boldsymbol{\ell}(k)[y(k) - \hat{y}(k)] \ ; \ \boldsymbol{\ell}(k) = \boldsymbol{F}^{k+1} \boldsymbol{g}(k)$$

Now from Theorem 4.3.7 and (4.172), we have

$$\boldsymbol{F}.\boldsymbol{F}^k \boldsymbol{g}(k) = \frac{\boldsymbol{F}.\boldsymbol{F}^k \boldsymbol{Q}(k)(\boldsymbol{F}^k)^T \boldsymbol{h}}{1 + \boldsymbol{h}^T \boldsymbol{F}^k \boldsymbol{Q}(k)(\boldsymbol{F}^k)^T \boldsymbol{h}}$$

and so if we define

$$\boldsymbol{P}(k) \triangleq \boldsymbol{F}^k \boldsymbol{Q}(k)(\boldsymbol{F}^k)^T$$

we get the following result.

Theorem 4.3.9. *The best least squares signal estimate* $\hat{y}$ *of the signal* y *which is defined by the signal model*

$$\begin{aligned} \boldsymbol{x}(k+1) &\triangleq \boldsymbol{F}\boldsymbol{x}(k) \ ; \ \boldsymbol{x}(k) \in \mathcal{R}^n \\ y(k) &\triangleq \boldsymbol{h}^T\boldsymbol{x}(k) \end{aligned}$$

for some unknown initial state $\boldsymbol{x}(0)$ *is given recursively by*

$$\begin{aligned} \hat{\boldsymbol{x}}(k+1) &\triangleq \boldsymbol{F}\hat{\boldsymbol{x}}(k) + \boldsymbol{\ell}(k)[y(k) - \hat{y}(k)] \ ; \ k \geq M \\ \hat{y}(k) &\triangleq \boldsymbol{h}^T\hat{\boldsymbol{x}}(k) \end{aligned}$$

where the estimator gain $\{\boldsymbol{\ell}(k);\ k \geq M\}$ *is given by*

$$\begin{aligned} \boldsymbol{\ell}(k) &= \frac{\boldsymbol{F}\boldsymbol{P}(k)\boldsymbol{h}}{1 + \boldsymbol{h}^T\boldsymbol{P}(k)\boldsymbol{h}} \\ \boldsymbol{P}(k+1) &= \boldsymbol{F}\boldsymbol{P}(k)\boldsymbol{F}^T - (1 + \boldsymbol{h}^T\boldsymbol{P}(k)\boldsymbol{h})\boldsymbol{\ell}(k)\boldsymbol{\ell}^T(k) \end{aligned}$$

where

$$\boldsymbol{P}(M) = \boldsymbol{F}^M(\boldsymbol{H}_M^T\boldsymbol{H}_M)^{-1}(\boldsymbol{F}^M)^T$$

Note that for a valid signal model of order n, the $M \times n$ matrix $\boldsymbol{H}_M$ will have full rank when $M = n$. Once $(\boldsymbol{H}_n^T\boldsymbol{H}_n)^{-1}$ is computed (which initiates the algorithm), no further matrix inverses are required to be computed during the implementation of the recursive algorithm.

Example 4.3.11. Consider the state space signal model (4.156), (4.157) which models the position $x_1(k)$ and velocity $x_2(k)$ of an object (such as an aircraft). The aircraft which at time $k = 0$ was at an unknown distance d_0 from a radar sensor is assumed to be travelling at a constant but unknown velocity v_0. Then

$$\begin{aligned} x_1(k+1) &= x_1(k) + Tx_2(k) \ ; \ x_1(0) = D \\ x_2(k+1) &= x_2(k) \ ; \ x_2(k) = V \end{aligned}$$

If we assume time $t = kT$ that the radar signal provides a measurement $y(k)$ of the distance $x_1(k)$ at time k, then

$$y(k) = x_1(k) + \varepsilon(k)$$

where $\varepsilon(k)$ is a measurement error at time k. The corresponding recursive least squares estimation algorithm then provides estimates $\{\hat{y}(k), \hat{x}_1(k), \hat{x}_2(k)\}$ as follows:

$$\begin{aligned}\hat{x}_1(k+1) &= \hat{x}_1(k) + T\hat{x}_2(k) + \ell_1(k)[y(k) - \hat{x}_1(k)]\\ \hat{x}_2(k+1) &= \hat{x}_2(k) + \ell_2(k)[y(k) - \hat{x}_1(k)]\\ \hat{y}(k) &= \hat{x}_1(k)\end{aligned}$$

where $\boldsymbol{\ell}^T(k) = [\ell_1(k) \;\; \ell_2(k)]$ is given from Theorem 4.3.7 with

$$\boldsymbol{F} = \begin{bmatrix} 1 & T \\ 0 & 1 \end{bmatrix} \; ; \; \boldsymbol{h} = \begin{bmatrix} 1 \\ 0 \end{bmatrix} \tag{4.175}$$

As previously noted, the reason for implementing this algorithm is because the actual measurement y is uncertain. As well, the model may also be uncertain in that the model only approximates the aircraft motion because the velocity $x_2(k)$ may actually *not* be constant for all k. The recursive algorithm can correct for small errors in both the measurement and the signal model.

Another estimation algorithm is also possible. Specifically, suppose now both the position $x_1(k)$ and velocity $x_2(k)$ are measured (with error) so that now we can assume that

$$y(k) = x_1(k) + Tx_2(k) + \varepsilon(k)$$

is available from measurements. The corresponding recursive least squares estimation algorithm which provides estimates $\{\hat{y}(k), \hat{x}_1(k), \hat{x}_2(k)\}$ is now given by

$$\begin{aligned}\hat{x}_1(k+1) &= \hat{x}_1(k) + T\hat{x}_2(k) + \ell_1(k)[y(k) - \hat{x}_1(k) - T\hat{x}_2(k)]\\ \hat{x}_2(k+1) &= \hat{x}_2(k) + \ell_2(k)[y(k) - \hat{x}_1(k) - T\hat{x}_2(k)]\\ \hat{y}(k) &= \hat{x}_1(k) + T\hat{x}_2(k)\end{aligned}$$

where $\boldsymbol{\ell}^T(k) = [\ell_1(k) \;\; \ell_2(k)]$ is given from Theorem 4.3.7 with $\boldsymbol{F}$ as in (4.175), but now with

$$\boldsymbol{h}^T = [1 \; T]$$

This latter algorithm can be arranged into the form

$$\begin{aligned}\hat{x}_1(k+1) &= (1 - \ell_1(k))\hat{x}_1(k) + T(1 - \ell_1(k))\hat{x}_2(k) + \ell_1(k)y(k)\\ \hat{x}_2(k+1) &= -\ell_2(k)\hat{x}_1(k) + (1 - \ell_2(k)T)\hat{x}_2(k) + \ell_2(k)y(k)\end{aligned}$$

If we let $\{\ell_1(k) = \alpha, \ell_2(k) = \beta T^{-1}, y(k) = z(k+1)\}$, this recursive algorithm can be seen to correspond to the α-β radar tracking algorithm described in the Introduction.

Summary

This Chapter introduced some fundamental properties of signals; specifically, length, boundedness, independence and periodicity. We then provided measures for the size (i.e. norm) and orthogonality of both nonperiodic and periodic signals. In particular, we showed how both a finite length signal and a periodic signal v can be expressed as an *orthogonal representation*; that is, as a linear combination of a set of orthogonal signals. One key advantage of an orthogonal representation is that once it is known that a signal has such a representation, then the coefficients in the representation are computed from the inner product between the given signal and the orthogonal functions. Furthermore, given any signal w, the particular signal v which minimises the least squares error between v and w when v is restricted to be a linear combination of orthogonal functions $\{\phi_\ell\}$ was shown to be directly obtained from the inner products between w and the $\{\phi_\ell\}$.

Furthermore, given any signal w, the particular signal v which minimizes the least squares error between v and w when v is restricted to be a linear combination of a set of orthogonal signals $\{\phi_\ell\}$ was shown to be directly obtained from the inner products between v and $\{\phi_\ell\}$. One particular orthogonal (in fact orthonormal) representation in the form of complex exponential signals resulted in the discrete Fourier transform (DFT) representation of either a periodic or finite length signal. Efficient algorithms known as the fast Fourier transform (FFT) was developed for the computation of the DFT. For a signal of length (or period) N, the FFT required the order of $N\log_2 N$ multiplications as compared with the order of N^2 which is required for direct calculation. One key application for the FFT was the computation of the linear convolution between two signals.

The norm and inner product of signals were used to develop bounds on signals which are the response of linear difference equations in terms of the bounds on the input to such equations. A general problem of least squares signal estimation was presented in the form of the minimization of a 2-norm. Subsequently, the algebraic solution of this problem was used to solve a number of problems in signal estimation including fitting the magnitude response of a FIR filter to some desired specifications, and the estimation of signal parameters. In particular, an efficient recursive least squares algorithm was developed which permit real-time applications of the least squares solution without the need to compute a matrix inverse.

Exercises

4.1 Determine which of the following signals v where $k \in \mathcal{Z}$ are periodic, and determine their periods ?

$$(a)\ v(k) = \sin 0.2k \quad (b)\ v(k) = \cos 10\pi k \quad (c)\ v(k) = 2\sin 0.2\pi k + \cos\frac{3\pi k}{7}$$

$$(d)\ v(k) = \cos\frac{3\pi k}{7} \quad (e)\ v(k) = exp\{j0.4\pi k\} \quad (f)\ v(k) = \sin(0.2\pi k + 0.72)$$

4.2 Suppose the analog signal $f(t) = \sin 0.6\pi t$ is sampled at the time instants $t = kT; k \in \mathcal{Z}$ resulting in the digital signal $v(k) = f(kT)$.

What is the condition on the sampling period T in order that v is periodic, and what is the period of v in terms of T ?

4.3 (i) Find the largest sampling period T at which the analog signal $f(t) = \sin 100t$ can be sampled in order that the digital signal $v(k) = f(kT); k \in \mathcal{Z}$ is periodic. What is the period of v ?

(ii) Repeat (i) for the analog signals: $f(t) = \cos 100\pi t$ and $f(t) = \sin 5t$.

4.4 Which of the infinite summations:

$$\sum_{n=-\infty}^{\infty} |v(n)|^2 \ ; \quad \sum_{n=-\infty}^{0} |v(n)|^2 \ ; \quad \sum_{n=0}^{\infty} |v(n)|^2$$

are finite with respect to the following signals:

$$(a)\ v(k) = a^k \ ; \ (b)\ v(k) = a^k \sin 2\pi bk \ ; \ (c)\ v(k) = a^{|k|}\cos\frac{3bk}{5}$$

and (where possible) give the range of values $\{a, b\}$ for convergence.

4.5 (i) Define the digital signal

$$f(k) = (0.75)^k \ ; \ k \geq 0 \ \text{ and } \ f(k) = 0 \ ; \ k < 0$$

where $k \in \mathcal{Z}$, and compute the appropriate norms $||f||_p$ for $p = 1, 2, \infty$.

(ii) Repeat (i) for the signal

$$f(k) = 3\sin(2\pi k + \pi/4) \ ; \ k \in \mathcal{Z}$$

4.6 Consider the digital signals $\{f, g, h\}$ of length 3 as defined by:

(a) $\{f(0) = 1, f(1) = 0, f(2) = 0\}$
(b) $\{g(0) = 1, g(1) = j, g(2) = 0\}$ and
(c) $\{h(0) = 1, h(1) = -3, h(2) = 2 + j2\}$.

(i) Find the signals norms and the angles between all signals with respect to the inner product defined by

$$\langle x, y\rangle = x(0)\overline{y(0)} + x(1)\overline{y(1)} + x(2)\overline{y(2)}$$

(ii) Repeat (i) with respect to the inner product defined by

$$\langle x, y\rangle = x(0)\overline{y(0)} + 2x(1)\overline{y(1)} + 4x(2)\overline{y(2)}$$

4.7 Find the 1-norm and the 2-norm of the periodic signals in exercise 4.1.

4.8 Consider the following real scalar valued digital signals:

(a) $f(n) = (0.5)^{n-1}$

(b) $g(n) = (-0.5)^n$

(c) $h(n) = \begin{cases} 1 \ ; n = 0,\ 2 \\ -1 \ ; n = 1,\ 3 \\ 0 \ ; n \geq 4 \end{cases}$

defined on the domain $\mathcal{T} = \mathcal{Z}_+$.

(i) Compute the norms $||\cdot||_\infty$, $||\cdot||_1$ and $||\cdot||_2$ for f, g and h.
(ii) Define the inner product between two signals x and y by

$$\langle x,\ y\rangle = \sum_{n=0}^{\infty} z(n)\overline{y(n)}$$

Compute $\langle f,\ g\rangle$, $\langle f,\ h\rangle$ and $\langle g,\ h\rangle$.
(iii) Repeat part (ii) using the inner product

$$\langle x,\ y\rangle = \sum_{n=0}^{\infty} (0.2)^n z(n)\overline{y(n)}$$

4.9 (i) Show that the digital signals $\{f_0(n) = 1,\ f_1(n) = k,\ f_2(n) = k^2\}$ are linearly independent on the interval $[0, 2] \subset \mathcal{Z}$
(ii) Consider the signal

$$v(k) = \alpha_0 f_0(k) + \alpha_1 f_1(k) + \alpha_2 f_2(k)$$

Find the coefficients $\{\alpha_0, \alpha_1, \alpha_2\}$ such that $v(k) = w(k)$ for $k = 0, 1, 2$ where

$$w(k) = \sin 0.25\pi k$$

4.10 (i) Approximate the digital signal $v(k) = \sin 0.25\pi k$ using:

(a) Walsh signals $\{w_{0,1}, w_{1,1}\}$ of length $2^1 = 2$ defined on $[0, 1] \subset \mathcal{Z}$, and
(b) Walsh signals $\{w_{0,2}, w_{1,2}, w_{2,2}, w_{3,2}\}$ of length $2^2 = 4$ defined on $[0, 1, 2, 3] \subset \mathcal{Z}$

(ii) Repeat (i) with respect to the digital signal $v(k) = k$.

4.11 (i) Find the scalar α such that the finite length digital signals $\{\phi_0, \phi_1\}$ are orthogonal on $[-1, 1] \subset \mathcal{Z}$ where

$$\phi_0(k) = 1 \ ; \ \phi_1(k) = 1 - \alpha k$$

with respect to the inner product

$$\langle x, y\rangle = \sum_{n=0}^{L-1} x(n)\overline{y(n)}$$

(ii) Repeat (i) on the interval $[0, 2] \subset \mathcal{Z}$
(iii) Repeat (i) in terms of the signals

$$\phi_0(k) = 1 \ ; \ \phi_1(k) = 1 - \alpha(0.5)^k$$

4.12 (i) Find the best approximation v of the finite length digital signal w on the interval $[-1, 1] \subset \mathcal{Z}$ where

$$v(k) = \beta_0\phi_0(k) + \beta_1\phi_1(k)$$

in terms of the orthogonal signals

$$\phi_0(k) = 1 \ ; \ \phi_1(k) = 1 - \alpha k$$

with respect to the inner product

$$\langle x, y\rangle = \sum_{n=0}^{L-1} x(n)\overline{y(n)}$$

on the interval $[-1, 1]$ for some α (to be determined). What is the minimum approximation error ?

(ii) Repeat (i) with respect to the inner product

$$\langle x, y\rangle = \sum_{n=0}^{L-1} \lambda_n x(n)\overline{y(n)}$$

where $\{\lambda_{-1} = 1 \ ; \ \lambda_0 = 4 \ ; \ \lambda_1 = 1\}$.

4.13 (i) Consider the signals $\{f_1, f_2, f_3\}$ defined on the interval $-2 \leq k \leq 2$ where

$$f_1(k) = 1 \ ; \ f_2(k) = k \ ; \ f_3(k) = k^2$$

and define an inner product by

$$\langle f, g\rangle = \sum_{k=-2}^{2} \lambda_k f(k)\overline{g(k)}$$

where $\{\lambda_{-2} = \lambda_2 = 0.5 \ ; \ \lambda_{-1} = \lambda_0 = \lambda_1 = 1\}$.

(ii) Show that any signal u which is a linear combination of the signals $\{f_k\}$ can also be written as a linear combination of the orthonormal set $\{\phi_0, \phi_1, \phi_2\}$ of signals where

$$\phi_0(k) = 0.5 \;\; ; \;\; \phi_1(k) = \frac{k}{\sqrt{6}} \;\; ; \;\; \phi_2(k) = \frac{k^2 - 1.5}{3}$$

4.14 (i) Find the approximation v of the digital signal $\{g(k) = \sin 0.25\pi k : 0 \le k \le 3\}$ of the form

$$v(k) = \beta_0\phi_0(k) + \beta_1\phi_1(k)$$

which minimizes $\sum_{n=0}^{3}[v(n) - g(n)]^2$ where

$$(a) \;\; \phi_0 = \{1, 1, 1, 1\} \;\; ; \;\; \phi_1 = \{1, -1, -1, 1\}$$

and

$$(b) \;\; \phi_0 = \{1, -1, 1, -1\} \;\; ; \;\; \phi_1 = \{1, -1, -1, 1\}$$

In each case, calculate the approximation error.

(ii) Find another pair of signals which are orthogonal on the interval $[0, 3] \subset \mathcal{Z}$ which offer a better approximation to the signal g.

4.15 (i) Given the finite length signal u where $\{u(k) = 1;\ 0 \le k \le 7\}$. Show that $U = DFT(u)$ is given by

$$U(n) = \begin{cases} 8 \; ; \; n = 0 \\ 0 \; ; \; 1 \le n \le 7 \end{cases}$$

(ii) Given the finite length signal u where $\{u(k) = exp\{j2\pi Tk/N\};\ 0 \le k \le 7\}$ for some *non-integer* value T. Show that the DFT U of u is given by

$$U(n) = \frac{1 - exp\{j2\pi(T - n)\}}{1 - exp\{j0.5\pi(T - n)\}} \;\; ; \;\; 0 \le n \le 7$$

(iii) Suppose that T in (ii) is now an *integer* in the range $0 \le T \le 7$. Show that $U(n) = 0$ for all n.

(iv) Suppose now T in (ii) is an integer which is *not* in the range $0 \le T \le 7$. What now is $U = DFT(u)$?

4.16 Suppose the values $x(k)$ of a finite length signal x over the range $0 \le k \le 7$ are given by $x(k) \in \{0, 1, 1, 1, 1, 1, 0, 0\}$. Show that $X = DFT(x)$ is given by

$$X(n) = 2\cos 0.5\pi n + 2\cos 0.25\pi n + 1 \;\; ; \;\; 0 \le n \le 7$$

4.17 (i) Use the linearity property of the DFT to determine the 16 point DFT Y of the signal y defined by

$$y(k) = 2 + 6.2^{-k} - 4\sin 0.25\pi k$$

(ii) Find an analytical expression for the 16 point DFT Y of the signal y defined by

$$y(k) = 2\cos 0.25\pi(k-2)$$

4.18 (i) An 8-point $U = DFT(u)$ of a real signal u was evaluated at particular points as given by

$$U(n) = \begin{cases} 0 & ;\ n = 3,5 \\ 3 & ;\ n = 6 \\ 0.5 + j & ;\ n = 7 \end{cases}$$

Use the properties of the DFT to find as many other values as possible.

(ii) A 7-point $U = DFT(u)$ of a real signal u is given by

$$U(n) \in \{3, 2exp\{-j\pi/3\}, -exp\{-j\pi/4\}, 4exp\{-j\pi/8\}, \ldots\}$$

Sketch the magnitude and phase of the complete DFT.

4.19 Determine the DFT representation for each of the following digital signals of period $N_0 = 4$ where

(a) $f(k) = \begin{cases} 1\ ;\ k = 0,2 \\ -1\ ;\ k = 1,3 \end{cases}$

(b) $f(k) = \begin{cases} 1\ ;\ k = 0,3 \\ -1\ ;\ k = 1,2 \end{cases}$

(c) $f(k) = \begin{cases} 1\ ;\ k = 0,1 \\ -1\ ;\ k = 2,3 \end{cases}$

4.20 Consider the periodic signal v of period 2, and the periodic signal w of period 3 as defined by

$$v(0) = 1\ ;\ v(1) = -2\ ;\ v(k) = v(k+2),\ k \in \mathcal{Z}$$
$$w(0) = -2\ ;\ w(1) = 1\ ;\ w(2) = -2\ ;\ w(k) = w(k+3),\ k \in \mathcal{Z}$$

(i) Find $DFT(v+w)$ of the signal $v + w$

(ii) Find $DFT(vw)$ of the signal vw defined by $vw(k) = v(k)w(k)$.

(iii) Using the DFT, find the circular convolution of v and w.

4.21 Determine the DFT for each of the following signals:

$$(a)\ \ v_1(k) = \sin\frac{2\pi(k-1)}{3} + \cos\frac{3\pi k}{7}\ ;\ k \in \mathcal{Z}$$
$$(b)\ \ v_2(k) = 2\sin\frac{3\pi k}{4}\sin\frac{2\pi k}{7}\ ;\ k \in \mathcal{Z}$$
$$(c)\ \ v_3(k) = 0.25^k\ ;\ \ 0 \le k \le 3\ ;\ k \in [0,3]$$

4.22 Suppose

$$V(n) = \begin{cases} exp\{-j0.8n\pi\}\ ; n = 0,1,2 \\ exp\{j0.8n\pi\}\ ; n = 3,4 \\ 0\ ; n = 5 \end{cases}$$

is the DFT of a periodic signal v of period 5.
(i) Show that $V(n) = \overline{V(5-n)}$.
(ii) Compute the signal v.

4.23 (i) Compute $F_m = DFT(f_m)$ for each of the following signals of length 9:

$$\text{(a)}\ f_1(k) = \begin{cases} 2\ ; 0 \le k \le 4 \\ 0\ ; 5 \le k \le 8 \end{cases}$$

$$\text{(b)}\ f_2(k) = \begin{cases} 1\ ; 0 \le k \le 4 \\ -1\ ; 5 \le k \le 8 \end{cases}$$

$$\text{(c)}\ f_3(k) = \begin{cases} 2\ ; 0 \le k \le 2 \\ 0\ ; 3 \le k \le 8 \end{cases}$$

$$\text{(d)}\ f_4(k) = \begin{cases} 1\ ; 0 \le k \le 2 \\ -1\ ; 3 \le k \le 8 \end{cases}$$

(ii) What are the key differences between: F_1 and F_2, F_3 and F_4, and F_2 and F_4?

4.24 (i) Suppose $V = DFT(v)$ is given by

$$V(n) = 2exp\{\frac{j2\pi n}{5}\}\ ;\ \ 0 \le n \le 4$$

Compute the corresponding periodic signal v.
(ii) Suppose v is a periodic signal of period N_0 such that

$$v(n) = -v(2-n)$$

Show that $V(0) = 0$.
(iii) Suppose v is a periodic signal of period N_0 such that

$$v(n) = v(3-n)$$

Show that $V(0.5N_0) = 0$.

4.25 (i) Develop an FFT algorithm based on $n = 3^3$ points. Extend this result to the case when $N = 3^p$ for any integer $p \geq 3$.

(ii) How many multiplications are required in the general case ?

4.26 Suppose w_N and u_M are finite length signals of length N and M respectively, and z_{N+M-1} is their linear convolution; that is

$$z_{N+M-1} = w_N * u_M$$

(i) How many multiplications are required to compute z_{N+M-1} using the direct method ?

(ii) How many multiplications are required to compute z_{N+M-1} using an FFT based procedure when: (a) $N + M - 1 = 2^p$, and (b) $N + M - 1 = 3^p$ for some integer p ?

4.27 Consider the digital convolution

$$y(k) = \sum_{m=0}^{k} h(k-m)u(m)$$

where h is the unit impulse response of a linear time invariant system such that $h(m) = 0$ for all $m \geq 99$, and also that the input u is such that $u(m) = 0$ for all $m \geq 99$.

(i) How many multiplications are required to calculate all non zero values of $y(k)$?

(ii) How many real multiplications would be required to calculate all non zero values of $y(k)$ using the DFT method given that an N-point FFT requires $24N\log_2(2N)$ real multiplications ?

4.28 [Under certain conditions which relate to the existence of a particular integral, the DFT developed for a signal defined on a finite domain can be extended to the limiting case of a *nonperiodic infinite length signal.* In this situation, the resulting transforms have a different form than the DFT and IDFT, and so are more commonly referred to as the *Discrete-Time Fourier Transform* (DTFT).]

The one-sided Discrete-Time Fourier transform H_+ of the infinite length signal $h : \mathcal{Z}_+ \to \mathcal{C}$ (provided it exists) is defined by

$$H_+(\exp\{j\omega\}) \triangleq \sum_{k=0}^{\infty} h(k)exp\{-j\omega k\}$$

which can be written in the form

$$H_+ = DTFT_+(h)$$

(i) Compute H_+ where $h(k) = a^k$.

(ii) Let $\omega_n \triangleq 2\pi nN^{-1}$, and define

$$H(\omega_n) \triangleq \lim_{N\to\infty} \sum_{k=0}^{N-1} h(k)exp\{-j\omega_n k\}$$

Show that provided $H(\omega_n)$ exists, then

$$h(k) = \lim_{N\to\infty} \sum_{n=0}^{N-1} H(\omega_n)\frac{exp\{j\omega_n k\}}{N}$$

(iii) Use the fact that $N^{-1} = (2\pi)^{-1}(\omega_n - \omega_{n-1})$ to show

$$h(k) = \frac{1}{2\pi} \lim_{N\to\infty} \sum_{n=0}^{N-1} H(\omega_n)exp\{j\omega_n k\}(\omega_n - \omega_{n-1})$$

Hence conclude that (in a well-defined sense)

$$h(k) = \frac{1}{2\pi}\int_0^\infty H_+(\exp\{j\omega\})exp\{j\omega k\}d\omega$$

which can be written in the form

$$h = IDTFT_+(H_+)$$

[*Observe* that

$$H_+(z) = \sum_{k=0}^{\infty} h(k)z^{-k}$$

and so when h is the unit impulse response of an asymptotically stable digital filter, $H_+(exp\{j\omega\})$ is in fact the frequency reponse function of the digital filter.]

4.29 Let T_m denote the number of complex multiplications required to compute an m-point DFT.

(i) Show $T_3 = 2$.

(ii) Show

$$T_9 = 3T_3 + 9(2) \quad ; \quad T_{27} = 3T_9 + 27(2)$$

(iii) When $N = 3^p$, show

$$T_N = 2.3^{p-1} + 2.3^p(p-1)$$

(iv) Suppose

$$N = P.Q \quad ; \quad P = 2^p, \; Q = 3^q$$

Show

$$T_N = PT_Q + 2P\log_2(P) \quad ; \quad T_Q = 2(Q/3) + 2Q\log_3(Q/3)$$

4.30 Define two finite length signals h and g of length 6 by

$$h(k) \in \{1,2,0,0,2,1\} \;\; ; \;\; g(k) \in \{2,2,2,2,1,1\}$$

(i) Compute the circular convolution of h and g
(ii) Compute the linear convolution of h and g

4.31 Consider two finite length signals v and w defined for $0 \leq k \leq 4$ by

$$v(k) = \cos\frac{k\pi}{3} \;\; ; \;\; w(k) = \sin\frac{k\pi}{3}$$

(i) Compute the circular convolution of v and w
(ii) Compute the linear convolution of v and w

4.32 Consider a finite length signal u of length 4 defined by

$$u(0) = 1 \; , \; u(1) = 2 \; , \; u(2) = 2 \; , \; u(3) = 1$$

Show that $U = DFT(u)$ is given by

$$U(n) = \begin{cases} 6 & ; \; n = 0 \\ \sqrt{2}exp\{j1.25\pi\} & ; \; n = 1 \\ 0 & ; \; n = 2 \\ \sqrt{2}exp\{j0.75\pi\} & ; \; n = 3 \end{cases}$$

4.33 Suppose a signal y is defined by

$$y(k) = \begin{cases} 3 \; ; & 0 \leq k \leq 2M \\ 0 \; ; & k \geq 2M+1 \end{cases}$$

Show that the $Y = DFT(y)$ is given by

$$Y(n) = 3\frac{\sin(M+0.5)\omega_n}{\sin 0.5\omega_n}\; exp\{-jM\omega_n\} \;\; ; \;\; \omega_n = \frac{2\pi n}{2M+1}$$

4.34 Find the best linear phase FIR filter

$$H(z) = b_1 + b_0 z^{-1}$$

that matches an amplitude function A_d at points where $\{A_d(0.1) = 1; \; A_d(0.2) = 0.9\}$ in the sense that

$$[H_d(0.1) - A_d(0.1)]^2 + [H_d(0.2) - A_d(0.2)]^2$$

is minimized where H_d is the magnitude function of the linear phase FIR filter.

4.35 Define the 2×3 matrix $\boldsymbol{H}$ and the 3-vector $\boldsymbol{f}$ by

$$\boldsymbol{H} = \begin{bmatrix} 2 & 2 \\ 1 & \alpha \\ 0 & 0 \end{bmatrix} \quad ; \quad \boldsymbol{f} = \begin{bmatrix} 1 \\ 0 \\ -1 \end{bmatrix}$$

(i) Find the vector(s) $\boldsymbol{x} \in \mathcal{R}^2$ which minimizes $g(\boldsymbol{x}) \triangleq ||\boldsymbol{H}\boldsymbol{x} - \boldsymbol{f}||_2$ when $\alpha \neq 1$.

(ii) Repeat (i) when $\alpha = 1$.

4.36 Consider the second order IIR model defined by $e(k) = 0$ where

$$e(k) = y(k) - a_1 y(k-1) - a_0 y(k-2) - b_0 u(k) - b_1 u(k-1) - b_2 u(k-2)$$

(i) Find conditions on the parameters $\{b_0, b_1, b_2\}$ such that the model is all-pass ?

(ii) Given the data $\{y(k), u(k);\ 0 \leq k \leq 4\}$, formulate the problem of finding the optimal set of parameters for the all-pass model which minimizes

$$\sum_{k=k_0}^{k_1} e^2(k)$$

for appropriately defined integers $\{k_0, k_1\}$.

4.37 (i) Formulate the problem of obtaining the least squares parameter estimates $\{\hat{\gamma}_0, \hat{\alpha}_0, \hat{\beta}_0\}$ of the parameters $\{\gamma_0, \alpha_0, \beta_0\}$ of the ideal signal model

$$y(k) = \gamma_0 + \alpha_0 \sin k\Omega_0 + \beta_0 \cos \Omega_0$$

based on the given data $\{y(k);\ 0 \leq k \leq 9\}$.

(ii) Simplify the result in (i) when $\Omega_0 = 0.1q\pi$ for some integer q.

(iii) Repeat the least squares development with respect to the model

$$y(k) = \gamma_0 + \sum_{m=0}^{1} (\alpha_m \sin k\Omega_m + \beta_m \cos \Omega_m)$$

(iv) Simplify the result in (iii) when $\Omega_0 = 0.1q\pi$ and $\Omega_1 = 0.1p\pi$ for some integers $\{q, p\}$.

4.38 (i) Formulate the problem for optimizing the choice of coefficients $\{c_0, c_1, c_2\}$ so as to best fit the function

$$f(x) = c_0 + c_1 x + c_2 x^2$$

to the data points $\{x_k, f(x_k)\}$

(ii) Relate the solution in (i) to the problem of estimating the position, velocity and acceleration of an object that is known to be moving at a constant (but unknown) rate of change of acceleration.

4.39 Consider an ideal low pass digital filter whose magnitude response is given by

$$A(\omega) = \begin{cases} 1 \; ; \; 0 \leq \omega < 0.5\pi \\ 0 \; ; \; 0.5\pi < \omega < \pi \end{cases}$$

The aim is to derive a linear phase FIR filter $H(z)$ with 8 coefficients whose magnitude response approximates $A(\omega)$ at the discrete frequencies $\{\omega_k; \; 0 \leq k \leq 7\}$.

(i) Consider a uniform frequency separation where $\omega_k = k\pi/8; \; 0 \leq k \leq 7\}$ and

$$A(\omega_k) = \begin{cases} 1 \; ; \; 0 \leq k \leq 5 \\ 0 \; ; \; 6 \leq k \leq 7 \end{cases}$$

With the help of MATLAB, find the optimal least squares lowpass FIR filter when: (a) $N = 12$ and (b) $N = 13$

(ii) Repeat (i) when

$$A(\omega_k) = \begin{cases} 1 & ; \; 0 \leq k \leq 3 \\ 0.75 & ; \; k = 4 \\ 0.40 & ; \; k = 5 \\ 0 & ; \; 6 \leq k \leq 7 \end{cases}$$

(iii) Now consider nonuniform frequency separation where ω_k belong to the set

$$\omega_k \in \{0, \pi/10, \pi/5, 3\pi/10, 6\pi/10, 7\pi/10, 8\pi/10, 9\pi/10\}$$

With the help of MATLAB, find the optimal least squares lowpass FIR filter when: (a) $N = 12$ and (b) $N = 13$ using the magnitude specifications in (i), part (a).

(iv) Repeat (iii) using the magnitude specifications in (i), part (b).

(v) Plot all your results and compare the solutions.

5. Finite Wordlength IIR Filter Implementation

The difference equation

$$\begin{aligned} y(k+n) &= a_{n-1}y(k+n-1) + \ \dots\ + a_1 y(k+1) + a_0 y(k) \\ &\quad + b_0 u(k) + b_1 u(k+1) + \ \dots\ + b_n u(k+n) \end{aligned} \tag{5.1}$$

or its state space equivalent

$$\begin{aligned} \boldsymbol{x}(k+1) &= \boldsymbol{F}\boldsymbol{x}(k) + \boldsymbol{g}u(k) \quad ; \quad \boldsymbol{x}(k) \in \mathcal{R}^n \\ y(k) &= \boldsymbol{h}^T \boldsymbol{x}(k) + du(k) \end{aligned} \tag{5.2}$$

which are both representations of a digital filter with frequency response function $H(z)$ given by

$$\begin{aligned} H(z) &= \frac{b_0 + b_1 z + \ \dots\ + b_n z^n}{z^n - a_{n-1}z^{n-1} - \ \dots\ - a_1 z - a_0} \\ &= d + \boldsymbol{c}^T (s\boldsymbol{I}_{\mathbf{n}} - \boldsymbol{F})^{-1}\boldsymbol{g} \end{aligned} \tag{5.3}$$

must be realized using either a general purpose digital computer, or special purpose digital hardware. Then the coefficients $\{a_p, b_q\}$, or the components of the matrices $\{\boldsymbol{F}, \boldsymbol{g}, \boldsymbol{c}, d\}$, and the results of the computation at each discrete time step k must be stored in binary form in registers of finite length. Since speed and accuracy are important considerations, the *finite wordlength* nature of these calculations cannot be ignored. The *significance* of the finite register length depends very much on the numerical sensitivity of the filtering algorithm. Narrow band digital filters can be particularly sensitive to numerical errors, and sometimes even small arithmetic errors can lead to numerical instability problems if suitable precautions are not taken. This section considers the effects of finite wordlength on the numerical realization of a digital filter.

5.1 Arithmetic Format

Numbers in digital computers are represented as binary representations consisting of strings of bits known as (binary) *words* where the *wordlength* is the number of bits in the word. The most common arithmetic representations which are supported by standard processors are *fixed point* and *floating point* arithmetic. The *dynamic range* of the arithmetic representation is the total numerical range of numbers that can be represented, and the *resolution* of the arithmetic representation is the difference between consecutively represented numerical values. For fixed point arithmetic, the resolution is constant, but in floating point arithmetic, the resolution increases as the magnitude of the number increases. As a consequence, for the same number of bits, a floating point format has a greater dynamic range than a fixed point format, but at the expense of an increasingly larger resolution between successive numbers.

When two binary words are added or multiplied, the result may well exceed the allowable dynamic range. If this range is exceeded, then we say that *overflow* has occurred, and a numerical error results. If overflow has *not* occurred, but the numerical result cannot be exactly represented within the precision available, then again a numerical error results. In this case, we say that *underflow* has occurred. If either overflow or underflow occurs after an addition or multiplication, then the implementation must select a numerical result from the allowable values that *can* be represented. This selection process is referred to as *quantization*.

5.1.1 Fixed Point

The fixed point method for representing a number w with a wordlength of n $(= 1+p+t)$ bits assigns 1 bit for the sign, p bits for the integer, and t bits for the fraction. We then say that w is in the representative class $[p,t]$, and write

$$w \approx [p,t] \tag{5.4}$$

Each particular word w has the binary form

$$a_p a_{p-1} ... a_1 a_0 \;\Delta\; a_{-1} a_{-2} ... a_{-t} \tag{5.5}$$

where a_p is the sign bit with 0 for a positive and 1 for a negative number. The string $a_{p-1}a_{p-2}...a_1a_0$ represents the magnitude of the integer part, and the string $a_{-1}a_{-2}...a_{-t}$ represents the (unsigned) fractional part.

Applications in fixed point arithmetic are usually based on the assumption that once the location of the binary point is decided, then this location remains *fixed*. However, the operations of multiplication and addition do not depend on the location of the binary point, and the position is left to the interpretation of the user in the implementation of fixed point algorithms. The two extreme positions for the binary point are either: (i) at the *left but one* (corresponding to $p = 0$ in (5.5)) in which case a_0 is the sign bit, and w is a *signed fraction*, or (ii) at the *right* in which case there is no fractional part, and w is a *signed integer*.

In signal processing applications, the value of a number w in fixed point arithmetic is normally represented in the *two's complement format.* In this representation, the value of a positive number (ie $a_p = 0$) has the value

$$2^{p-1}a_{p-1} + 2^{p-2}a_{p-2} + ... + 2a_1 + a_0 + 2^{-1}a_{-1} + ... + 2^{-t}a_{-t} \qquad (5.6)$$

For a *negative* number (i.e. $a_p = 1$), the two's complement number w has the value determined by

$$-(2^{p-1}\bar{a}_{p-1}+2^{p-2}\bar{a}_{p-2}+...+2\bar{a}_1+\bar{a}_0+2^{-1}\bar{a}_{-1}+...+2^{-t}\bar{a}_{-t})-1-2^{-t} \qquad (5.7)$$

The *dynamic range* $\mathcal{D}$ of the fixed point representation in (5.5) is given by $\mathcal{D} = 2^{p+1}$, and the *resolution* $\mathcal{R} = 2^{-t}$ between any two successive numbers is *constant.* If the magnitude of the sum or product of two numbers is larger that the maximum number 2^{p-1} that can be represented, then *overflow* has occurred.

A two's complement representation has a characteristic known as the *sign extension property.* That is, if a number w has a correct $[p,t]$ representation, then w also has a $[q,t]$ representation for any $q > p$ if the sign bit a_p is repeated $q-p$ times. Conversely, any number w with a $[q,t]$ representation in which the sign bit a_q is repeated $q-p$ times, also has a $[p,t]$ representation. For example, the integer -6 has the $[4,0]$ representation (=1010), and the $[7,0]$ representation (=1111010).

Table 5.1. Fixed point addition

	One's complement	Two's complement	
6	0110	0110	
<u>−6</u>	<u>1001</u>	<u>1010</u>	
0	(0) 1111	(1) 0000	(ignore carry no overflow)
4	0100	0100	
<u>−1</u>	<u>1110</u>	<u>1111</u>	
3	(1) 0010	(1) 0011	(ignore carry no overflow)
4	0100	0100	
<u>+6</u>	<u>0110</u>	<u>0110</u>	
10	1010	(0) 1010	(overflow)

In *addition* (or subtraction), *sign errors* can arise if overflow occurs and the carry is ignored. An error can arise in the *sign* of the result because an overflow of the bits representing the magnitude of the result will change the sign of the result unless the answer is interpreted to extra precision. For example, 4+6 (=01010) is correct to 5 bits in Table 5.1, but the 4 bit interpretation (= 1010), which is -6 in two's complement arithmetic, is incorrect. Overflow which results from addition can be detected by observing the carry into the sign bit position, and the carry out of the sign bit position. If both carries are equal, then no overflow has occurred. For example, in the addition of $4 + 6$ in Table 3.1, the carry into the sign bit position is 1 and the carry out of the sign bit position is 0, and so overflow has occurred.

A property of two's complement arithmetic is: provided the *end result* of an addition is known *not* to overflow, then any *intermediate* overflow that may occur in reaching the end result is of no consequence. For example, the calculation $(4 + 6) - (7)$ to 4 bit precision has an intermediate overflow due to the sum $4+6$. However, if -7 (= 1001) is added to the 4 bit result of $4+6$ (= 1010), then the correct 4 bit answer (= 0011) is obtained.

The *multiplication* of two's complement numbers can be (inefficiently) obtained using an *add-and-shift algorithm* as follows:

1. Make negative operands positive; initialize result to zero.
2. Test the least significant bit (=LSB) of multiplier for 0 or 1.
3. If 1, add multiplicand to the result, then execute step 4; else directly execute step 4.
4. Shift multiplicand left one bit; go to step 2.

However a more *efficient* procedure known as *Booth's algorithm* which is often the basis for the hardware multiplication of two fixed point two's complement numbers is described as follows:

1. Initialize result to zero.
2. Test the LSB of multiplier for 0 or 1.
3. If 1, subtract multiplicand from product ignoring carry, then execute step 4; else directly execute step 4.
4. Shift product right one bit, repeating MSB.
5. If no more multiplier bits terminate algorithm; else test transition of next multiplier bit.
6. If 0 to 1 transition, subtract multiplicand from most significant bits of product ignoring carry, then execute step 4; if 1 to 0 transition, add multiplicand to most significant bits of product ignoring any carry; else directly execute step 4.

Example 5.1.1. The 4 bit multiplication of -3 (= 1101) by 5 (= 0101) using Booth's algorithm is illustrated in Table 5.2. Using the sign extension property for two's complement arithmetic, the FINAL result is equivalent to 10001 or -15.

Table 5.2. Illustration of Booth's algorithm

multiplicand	(-3)	1101
multiplier	(5)	0101
LSB 1 so subtract		0000
	[1]	0011
intermediate product		0011
shift right repeating MSB		00011
transition $(1 \rightarrow 0)$ *so add*	[0]	1101
intermediate product		11101
shift right repeating MSB		111101
transition $(0 \rightarrow 1)$ *so subtract*	[1]	0011
*intermediate product (***ignore** *carry*		001001
shift right repeating MSB		0001001
transition $(1 \rightarrow 0)$ *so add*	[0]	1101
FINAL result		1110001

Several different algorithms are used for the (software) *division* of two's complement numbers. The simplest method tests the signs of both the divisor and dividend to determine the signs of the quotient and remainder. The divisor and dividend are then both made positive and the magnitudes of the numbers are divided by a repeated subtraction of the divisor from the dividend. The number of times the divisor is subtracted without a borrow is the *quotient*, and the difference after the last successive subtraction is the *remainder*.

Other algorithms for division only lead to a small improvement in execution time when compared to the relative large improvement obtainable in multiplication using Booth's algorithm. For purposes of obtaining a *fast* implementation, any division that appears in a numerical algorithm should be converted where possible to a multiplication.

Fixed Point Quantization Characteristics

When two fixed point numbers are added, the *magnitude* of the result may exceed the available wordlength. For example, if the result is confined to be a *fraction*, then the sum of 0.5 and 0.75 cannot be represented, and so *overflow* has occurred. In terms of the sign extension property, overflow *has* occurred if the most significant bit is *not* equal to the sign bit. For example, the 6 bit number -15 (=110001) can be reduced to the 5 bit number 10001 without error. However, if the 5 bit number 10001 is reduced to 4 bits, then an overflow occurs.

Overflow: If *overflow is detected*, then the result of the addition of the two words can be set at the *maximum value* consistent with the correct sign. That is, if the quantized sum $Q[w]$ cannot exceed unity magnitude, then set

$$Q[w] = \begin{cases} 1 \ ; \ w \geq 1 \\ w \ ; \ -1 \leq w \leq 1 \\ -1 \ ; \ w \leq -1 \end{cases}$$

This overflow characteristic (which is referred to as *saturation arithmetic*) is illustrated in Figure 5.1(a). If *overflow is not detected* (which in a twos complement fixed point representation means that the most significant bit is *not* equal to the sign bit), and the result is used as is, then the result will not have the correct sign. The resulting overflow characteristic will then be defined by that which appears in Figure 5.1(b).

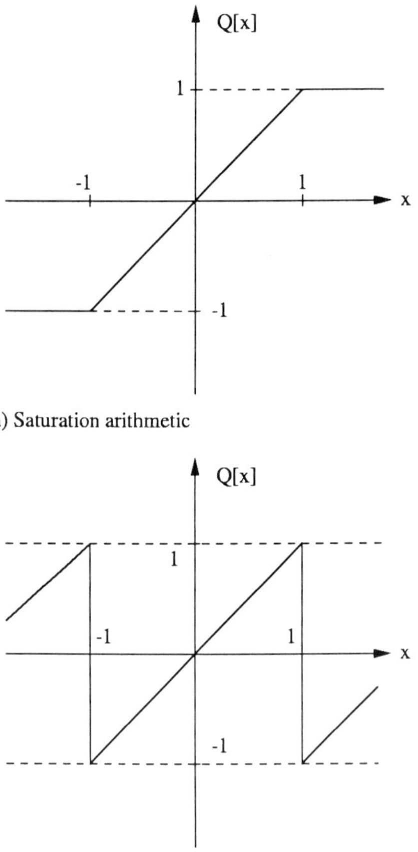

(a) Saturation arithmetic

(b) Overflow undetected

Fig. 5.1. Overflow characteristics

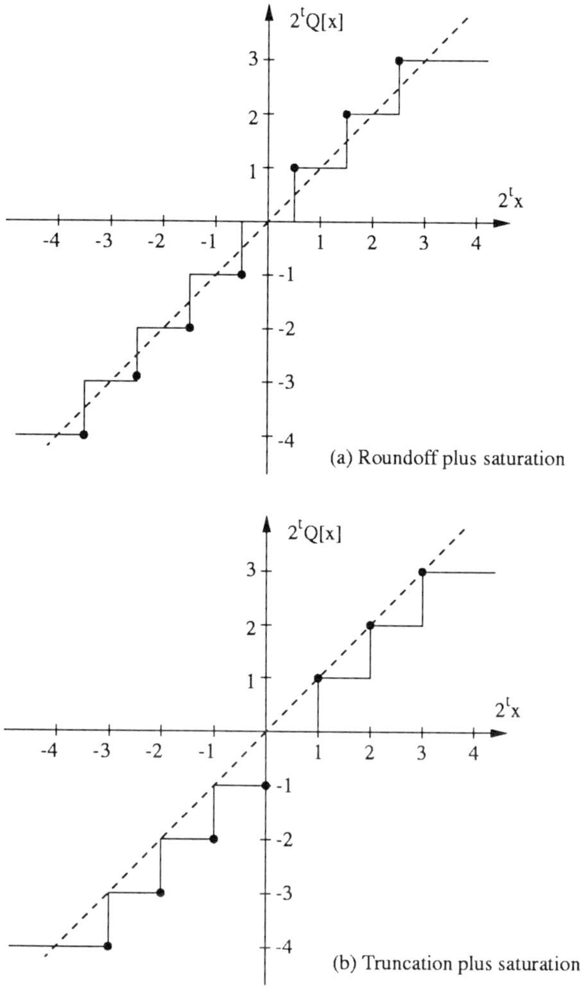

Fig. 5.2. Quantization plus saturation arithmetic

Underflow: When two fixed point numbers having t bits in each fractional representation are *added*, the fractional representation of the sum also only requires t fractional bits. However, the *product* of two such numbers generally requires $2t$ bits for an exact representation of the fraction. If only t bits are available, then *underflow* can occur in which case, an appropriate *quantization* $Q[\,\cdot\,]$ must be implemented. The two possible quantization characteristics for two's complement arithmetic are *roundoff* and *truncation*. Roundoff and truncation together with saturation arithmetic for a two's complement $[2,0]$ representation has one of the combined nonlinear characteristics illustrated in Figure 5.2.

Underflow Using Truncation: Ignoring the least significant bit(s) in a two's complement representation is equivalent to implementing *truncation* $Q_T[\cdot]$. For example, for $t = 1$ fractional bits in the representation of $Q_T[\cdot]$, we have

$$y \approx 0_\Delta 01 \ (= 0.25) \ \text{ implies } \ Q_T[y] \approx 0_\Delta 0 \ (= 0)$$

so that $y - Q_T[y] = +0.25$, and

$$y \approx 1_\Delta 01 \ (= -0.75) \ \text{ implies } \ Q_T[y] \approx 1_\Delta 0 \ (= -1.0)$$

so that again $y - Q_T[y] = +0.25$. The *truncation quantization error* $y - Q_T[y]$ will therefore be *non-negative* for both positive *and* negative values of y.

If a discrete signal y is quantized using *truncation quantization*, then the error signal

$$e(k) \stackrel{\Delta}{=} y(k) - Q_T[y(k)] \geq 0 \ \text{ for all } k$$

Therefore the *average error* $\overline{e}_N$ over the N values $1 \leq k \leq N$ as defined by

$$\overline{e}_N \stackrel{\Delta}{=} \frac{1}{N} \sum_{k=1}^{N} e(k) \geq 0 \tag{5.8}$$

is always non-negative irrespective of the values of $y(k)$. We therefore say that truncation quantization introduces a *bias* into the quantized signal $Q_T[y]$.

Underflow Using Roundoff: The *roundoff* operation $Q_{RO}[\ \cdot\ ]$ selects $Q_{RO}[y]$ to be the nearest representation to y. However when y is exactly half way between two consecutive values of $Q_{RO}[y]$, the decision as to whether to round up or round down must be selected in advance. For example, in a 4 bit fractional representation, 0.09375 $(= 0_\Delta 00011)$ can either be rounded *up* to 0.125 $(= 0_\Delta 0010)$ or rounded *down* to 0.0625 $(= 0_\Delta 0001)$. This decision is determined by the hardware implementation.

Upward rounding can be implemented (in two's complement arithmetic) by adding 1 to the least significant bit, and then truncating the bits of the summation which have significance less than 2^{-t}. For example, with $t = 2$ in (5.5) for $Q_{RO}[y]$, suppose $y \approx 0_\Delta 0101$. Then

$$y + 0_\Delta 0001 = 0_\Delta 0110 \ \text{ which implies } \ Q_{RO}[y] = 0_\Delta 01$$

so that $y - Q_{RO}[y] = +0.25$. When $y \approx 1_\Delta 0111$ we have

$$y + 0_\Delta 0001 = 1_\Delta 1000 \ \text{ which implies } \ Q_{RO}[y] = 1_\Delta 10$$

so that $y - Q_{RO}[y] = -0.25$. The *roundoff quantization error* $y - Q_{RO}[y]$ always has the *same sign* as y; that is

$$y(y - Q_{RO}[y]) \geq 0$$

If a discrete signal y is quantized using *roundoff quantization*, then

$$e(k) \triangleq y(k) - Q_{RO}[y(k)] \begin{cases} \geq 0; & y(k) \geq 0 \\ \leq 0; & y(k) \leq 0 \end{cases}$$

In this case, the average error $\overline{e}_N$ in (5.8) may either be nonnegative or nonpositive. In a statistical sense, if the value $y(k)$ is equally likely to be positive or negative, then for large values of N, $\overline{e}_N$ is very likely to be zero. That is, any bias with roundoff quantization is therefore generally much smaller than the bias associated with truncation quantization.

Convergent Rounding: If either *rounding up* or *rounding down* is *always* implemented, then a *small bias* will be added to the signal, and even though this bias is much smaller than that which is introduced when *truncation* is implemented, sometimes even a small bias can cause problems in particular IIR filtering applications. An *improved rounding procedure* that overcomes this bias problem is known as *convergent rounding.* In this case, when a number is to be rounded lies at the midpoint range between two allowable number representations, it is rounded either higher or lower according to the following criteria. If the bit that will become the least significant bit of the rounded number is 1, then the number is rounded up, and if the least significant bit of the rounded number is 0, then the number is rounded down. Convergent rounding is not supported in hardware by all digital signal processors. However the analog devices ASP-21xx and Motorola DSP5600x families both support convergent rounding.

5.1.2 Floating Point

The floating point representation of a number w is defined by *two* words: the *mantissa* m (of $t+1$ bits) and the *exponent* exp (of $p+1$ bits) meaning that a total of $n = t + p + 2$ bits are required. Thus, for a floating point number, we write

$$w \approx [m_{\Delta} exp] \tag{5.9}$$

where

$$m \approx m_{0\Delta} m_{-1}, m_{-2} \ldots m_{-t} \;\; ; \;\; exp \approx e_{p\Delta} e_{p-1} \ldots e_1 e_0 \tag{5.10}$$

One possible floating point representation is to have the mantissa m as a *sign magnitude* fixed point fraction with sign bit m_0, and the exponent exp as a *sign magnitude* integer with sign bit e_p. In this case, it is usual to *normalize* the mantissa m such that

$$0.5 \leq |m| < 1 \tag{5.11}$$

In this representation, the value of the floating point word w in (5.9) is given by

$$(-1)^{m_0}[2^{-1} + 2^{-2}m_{-1} + 2^{-3}m_{-2} + \ldots + 2^{-(t+1)}m_{-t}] \times 2^{exp} \tag{5.12}$$

where

$$exp \triangleq (-1)^{e_p}[2^{p-1}e_{p-1} + 2^{p-2}e_{p-2} + \dots + 2e_1 + e_0] \tag{5.13}$$

For example, if $\{p = 2, t = 3\}$, then

$$2.5 = 0.625 \times 2^2 \approx [m_{\Delta} exp]$$

where

$$m \approx [0_{\Delta}010] \quad ; \quad exp \approx [0_{\Delta}10]$$

Similarly

$$-0.1875 = -0.75 \times 2^{-2} \approx [m_{\Delta} exp]$$

where

$$m \approx [1_{\Delta}100] \quad ; \quad exp \approx [1_{\Delta}10]$$

An alternative representation for the exponent exp is to remove the sign bit, and interpret the exponent as being *biased*; that is, interpret exp in (5.10) according to

$$exp \triangleq 2^p e_p + 2^{p-1}e_{p-1} + \dots + 2e_1 + e_0 - 2^{p-1}$$

rather than by (5.13). In this representation with $\{p = 2, exp = 2\}$ in the number 2.5 is represented by $exp \approx [1_{\Delta}00]$, and $exp = -2$ in the number -0.1875 is represented by $exp \approx [0_{\Delta}00]$

With a *signed magnitude* exponent format in (5.9), then in (5.10) we have that

$$exp \leq (exp)_{max} \triangleq 2^p - 1 \quad ; \quad exp \geq (exp)_{min} \triangleq -(2^p - 1)$$

whereas for the *biased* exponent format

$$(exp)_{max} \triangleq 2^{p+1} - 1 - 2^{p-1} \; ; \; (exp)_{min} \triangleq -2^{p-1}$$

The *dynamic range* $\mathcal{D}$ of the floating point representation (5.9), (5.10) is given (approximately) by

$$\mathcal{D} \simeq 2^{(exp)_{max}+1}$$

The resolution between two successive floating point numbers varies according to the *magnitude* of the numbers. The *minimum resolution* $\mathcal{R}_{min}$ is given by

$$\mathcal{R}_{min} \triangleq 2^{-(t+1)} \times 2^{(exp)_{min}}$$

while the *maximum resolution* $\mathcal{R}_{max}$ is given by

$$\mathcal{R}_{max} \triangleq 2^{-(t+1)} \times 2^{(exp)_{max}}$$

For the *same number* of $p+t$ bits, both floating and fixed point formats can only represent the same total of different numbers. However, a floating point number with p bits for the exponent and t bits for the mantissa will have a *larger* dynamic range than a fixed format having p bits for the integer, and t points for the fraction. Furthermore, the resolution of the floating point format *increases* with the magnitude of the number whereas the resolution of the fixed point number remains *constant*.

Example 5.1.2. Suppose $t = p = 2$. Then the numbers which can be represented in normalized signed magnitude mantissa floating point format are defined by

$$(\pm 1)\alpha \times 2^{\beta}$$

where $\alpha \in \{0.0, 0.5, 0.75\}$ and

$$\beta \in \begin{cases} \{3, 2, 1, 0, -1, -2, -3\} & ; \text{signed magnitude exponent} \\ \{5, 4, 3, 2, 1, 0, -1, -2\} & ; \text{biased exponent} \end{cases}$$

The dynamic range

$$\mathcal{D} \simeq \begin{cases} 2^4 & ; \text{signed magnitude exponent} \\ 2^6 & ; \text{biased exponent} \end{cases}$$

while the minimum resolution $\mathcal{R}_{min}$, and the maximum resolution $\mathcal{R}_{max}$, are given by

$$\mathcal{R}_{min} = \begin{cases} 2^{-6} & ; \text{signed magnitude exponent} \\ 2^{-5} & ; \text{biased exponent} \end{cases}$$

$$\mathcal{R}_{max} = \begin{cases} 2^{0} & ; \text{signed magnitude exponent} \\ 2^{2} & ; \text{biased exponent} \end{cases}$$

Note that a larger dynamic range is achieved using a biased exponent (rather than a sign magnitude exponent representation), but this advantage is countered by the fact that a biased exponent representation has a lower resolution.

The numbers that can be represented in a two's complement fixed point format are defined by $(\pm 1)\alpha + \beta$ where $\alpha \in \{0, 1, 2\}$ and $\beta \in \{0, 0.25, 0.75, \}$.

Arithmetic operations between floating point numbers are more complicated than fixed point, and take more time to complete. However, a general purpose computer is normally not designed to take advantage of the speed of fixed point arithmetic. Floating point addition or subtraction first requires an alignment of the exponents before the mantissa can be added or subtracted. The product of two floating point numbers is formed by a fixed point multiplication of the two mantissa, and an addition of the exponents. No alignment of exponents is required, but a normalization of the mantissa of the product may be necessary.

For example, consider the *addition* of two floating point numbers x and y each of $n = p+t+2$ bits as represented in (5.9), (5.10). That is, t bits for the mantissa magnitude plus one sign bit (normalized sign magnitude mantissa format) and $p+1$ bits for the exponent. Assume a *signed magnitude exponent* representation, and write

$$\begin{aligned} x &= a \times 2^b \ ; \ 0.5 \leq |a| < 1 \ , \ |b| \leq 2^p - 1 \\ y &= c \times 2^d \ ; \ 0.5 \leq |c| < 1 \ , \ |d| \leq 2^p - 1 \end{aligned} \tag{5.14}$$

Suppose

$$d = b + r \ ; \ r \geq 0$$

then

$$x + y = (a2^{-r} + c) \times 2^d$$

Hence, *after* mantissa normalization, but *before* mantissa quantization

$$x + y = m \times 2^{exp} \ ; \ 0.5 \leq |m| < 1 \tag{5.15}$$

where for some integer k

$$|m| = |s|2^{k-1} \ ; \quad s \triangleq a2^{-r} + c \ ; \quad exp \triangleq d - (k-1) \tag{5.16}$$

The bounds on the magnitudes $|a|$ and $|c|$ in (5.14) and $r \geq 0$ guarantee $0 \leq |s| < 2$ in (5.9). Consequently, some choice for the integer $0 \leq k \leq 1$ will guarantee $|m| < 1$. Since the mantissa m must be normalized according to $0.5 \leq |m| < 1$, the required integer k in (5.16) satisfies

$$2^{-k} \leq |s| < 2^{-(k-1)}$$

A *necessary* condition for *overflow* to occur in floating point addition is $k = 0$. If overflow *does* occur, then saturation arithmetic is implemented whereby $|x + y|$ is fixed at the maximum allowable value consistent with the correct sign for $x + y$.

Floating Point Quantization Characteristics

Unlike fixed point arithmetic, *underflow* quantization can occur in *floating point addition*. Specifically, let $Q_{RO}[m]$ denote the fixed point *roundoff* quantization of the (normalized) *mantissa* m of the finite wordlength sum $x + y$, and let $FL[x + y]$ denote the *complete* floating point quantization of $x + y$. Then

$$FL[x + y] = Q_{RO}[m] \times 2^{exp} \tag{5.17}$$

where

$$Q_{RO}[m] = m - \varepsilon_a \tag{5.18}$$

Since $|m|$ is *normalized* as in (5.11), a t bit representation of $|m|$ implies

$$|\varepsilon_a| < 2^{-(t+1)} \tag{5.19}$$

Consider now the multiplication ($w = xy$) of two normalized floating point numbers x and y defined in (5.14). Since both $|a|$ and $|c|$ are bounded between 0.5 and 1, $|ac|$ is bounded between 0.25 and 1. Therefore, *before quantization* of the mantissa m in the product but *after normalization*, we have

$$xy = m \times 2^{exp} \ ; \ 0.5 \leq |m| < 1$$

where for a signed magnitude exponent representation

$$|m| = \begin{cases} 2|ac| \ , \ e = b + d - 1 & ; 0.25 \leq |ac| < 0.5 \\ |ac| \ , \ e = b + d & ; 0.5 \leq |ac| < 1 \end{cases}$$

Hence overflow or underflow can therefore occur in floating point multiplication. In particular, after roundoff quantization of the mantissa

$$Q_{RO}[m] = m - \varepsilon_m \ ; \ |\varepsilon_m| < 2^{-(t+1)} \tag{5.20}$$

and

$$FL[xy] = Q_{RO}[m] \times 2^{exp}$$

Then

$$xy = FL[xy] + \varepsilon_m \times 2^{exp} \tag{5.21}$$

and so the multiplicative error δ is defined by

$$FL[xy] \stackrel{\Delta}{=} (xy)(1 - \delta) \ ; \ \delta = \frac{\varepsilon_m}{m} \tag{5.22}$$

with

$$0 \leq |\delta| < 2^{-t}$$

Consider the floating point realization of the *ideal* first order finite wordlength (FWL) digital filter

$$y(k+1) = a_0 y(k) + b_0 u(k) \tag{5.23}$$

as described by

$$\hat{y}(k+1) = FL[FL[\hat{a}_0 \hat{y}(k)] + FL[\hat{b}_0 \hat{u}(k)]] \tag{5.24}$$

where $FL[\cdot]$ denotes 'floating point quantization'. At the outset, the ideal filter coefficients $\{a_0, b_0\}$ need to be quantized to $\{\hat{a}_0, \hat{b}_0\}$ consistent with the available floating point format. The input signal u may also require quantization to $\hat{u}$. (For example, the FWL representation $\hat{u}(k)$ of $u(k)$ may result from analog-to-digital conversion of the sampled value $u(k) \triangleq v(kT)$ of a continuous time signal $v(t)$ at the sampling time $t = kT$.)

During the processing of the FWL filter input signal $\hat{u}$ in (5.24), three separate arithmetic quantizations are required; one after the multiplication of $\hat{u}(k)$ by $\hat{b}_0$ which gives the result $FL[\hat{b}_0\hat{u}(k)]$, another after the multiplication of $\hat{y}(k)$ by $\hat{a}_0$ which gives the result $FL[\hat{a}_0\hat{y}(k)]$, and the last after the addition of $FL[\hat{a}_0\hat{y}(k)]$ to $FL[\hat{b}_0\hat{u}(k)]$ which gives the result $\hat{y}(k+1)$. If we then define the *multiplicative* floating point quantization errors $\{\delta_a(k), \delta_b(k)\}$ by

$$FL[\hat{a}_0\hat{y}(k)] = \hat{a}_0\hat{y}(k)(1 - \delta_a(k)) \quad ; \quad FL[\hat{b}_0\hat{u}(k)] = \hat{b}_0\hat{u}(k)(1 - \delta_b(k))$$

we have on substituting into (5.24) that

$$\begin{aligned} \hat{y}(k+1) &= FL[\hat{a}_0\hat{y}(k)(1 - \delta_a(k)) + \hat{b}_0\hat{u}(k)(1 - \delta_b(k))] \\ &= [\hat{a}_0\hat{y}(k)(1 - \delta_a(k)) + \hat{b}_0\hat{u}(k)(1 - \delta_b(k))][1 - \varepsilon(k)] \end{aligned} \qquad (5.25)$$

where $\varepsilon(k)$ is defined to be the floating point error resulting from *addition*.

After expansion, we therefore have

$$\hat{y}(k+1) = \alpha_0(k)\hat{y}(k) + \beta_0(k)\hat{u}(k) \qquad (5.26)$$

where

$$\alpha_0(k) \triangleq \hat{a}_0(1 - \delta_a(k))(1 - \varepsilon(k)) \ ; \ \beta_0(k) \triangleq \hat{b}_0(1 - \delta_b(k))(1 - \varepsilon(k)) \qquad (5.27)$$

The effect of the quantization errors $\{\delta_a(k), \delta_b(k), \varepsilon(k)\}$ is therefore seen to be equivalent to modifying the filter coefficients $\{\hat{a}_0, \hat{b}_0\}$ to $\{\alpha_0(k), \beta_0(k)\}$ respectively. Since such modifications occur at every time step k, the *accumulative effect* of such errors is difficult to analyse. We therefore restrict further consideration in this text to the consideration of underflow errors in a *fixed point* implementation.

5.2 First Order FWL Filter

We now begin the development of an analytical framework to enable the analysis of the effects of numerical quantization errors in a finite wordlength (FWL) fixed point implementation of a digital filter. Quantization errors occur in the filter coefficients, the input, the internal states and the output. If only *coefficient errors* are present, then filter *analysis* (but not necessarily filter *synthesis*) is relatively straight forward, and does not depend significantly on whether or not the arithmetic representation is a fixed point format or a floating point format. However as we have seen, the analysis of the *accumulative* effects of internal quantization errors in the filter states is impractical for a floating point implementation.

Fixed Point Realization

Consider again the ideal first order digital filter

$$y(k+1) = a_0 y(k) + b_0 u(k) \tag{5.28}$$

where u is the input signal, and y is the output signal. Finite precision fixed point arithmetic is also generally unable to *exactly* represent the coefficients a_0 and b_0. Also, the value of $y(k)$ in (5.28) can *never* be exactly represented for all discrete time instants k if a has a fractional component.

For example, suppose the coefficients a_0 and b_0, the input $u(k)$ and the initial condition $y(0)$ are all 8 bit fractions. Then $y(1)$ in (5.28) will in general have a 16 bit fractional part, $y(2)$ will have a 24 bit fractional part, and so on. Hence, given any particular wordlength, there will be some future time $m > 0$ at which the output value $y(m)$ *cannot* be exactly represented within the available wordlength.

A *finite wordlength* (FWL) *fixed point* realization of the first order filter (5.28) is described by

$$\begin{aligned} \hat{y}(k+1) &= \hat{a}_0 Q[\hat{y}(k)] + \hat{b}_0 \hat{u}(k) \\ Q[\hat{y}(k)] &= \hat{y}(k) - \varepsilon_Q(k) \end{aligned} \tag{5.29}$$

where $Q[\cdot]$ represents either *truncation plus saturation arithmetic*, or *roundoff plus saturation arithmetic*.

The first point to note is that since the arithmetic quantizer $Q[\cdot]$ is a *nonlinear* function, the FWL realization (5.29) is actually an example of a *nonlinear* first order difference equation. Consequently, if

$$\hat{y}_j(k+1) = \hat{a}_0 Q[\hat{y}_j(k)] + \hat{b}_0 \hat{u}_j(k) \quad ; \quad j = 1, 2$$

then it is *not* generally the case that the signal $\hat{y} = \hat{y}_1 + \hat{y}_2$ satisfies (5.29) when the input signal $\hat{u} = \hat{u}_1 + \hat{u}_2$. This nonlinear property makes analysis difficult.

Now, assume both the coefficients $\hat{a}_0$ and $\hat{b}_0$ in (5.29) have the same representation

$$\hat{a}_0, \hat{b}_0 \approx [p_0, t_0] \tag{5.30}$$

where as we recall from section 5.1.1, this notation means that p_0 bits represent the *integer* parts of the coefficients $\hat{a}_0$ and $\hat{b}_0$, and t_0 bits represent the fractional parts of $\hat{a}_0$ and $\hat{b}_0$. If we also assume that p_0 bits are sufficient to exactly represent the *integer* parts of the coefficients a_0 and b_0, then the magnitudes of the *coefficient errors* $a_0 - \hat{a}_0$ and $b_0 - \hat{b}_0$ are determined by the number of bits ($= t_0$) assigned to the *fractional* coefficient representation in (5.30).

Similarly, suppose the quantized value $Q[\hat{y}(k)]$ and the quantized input $\hat{u}(k)$ in (5.29) have the FWL representations

$$Q[\hat{y}(k)] \approx [p_1, t_1] \quad ; \quad \hat{u}(k) \approx [p_2, t_2] \tag{5.31}$$

Then from (5.30) and (5.31), it follows that the products in (5.29) have the *exact* representations:

$$\hat{a}_0 Q[\hat{y}(k)] \approx [p_0 + p_1, t_0 + t_1] \quad ; \quad \hat{b}_0 \hat{u}(k) \approx [p_0 + p_2, t_0 + t_2]$$

If (as is normally the case), we have that $\{p_2 \leq p_1 \ ; \ t_2 \leq t_1\}$ in (5.31), then $\hat{y}(k+1)$ in (5.29) has an *exact* representation

$$\hat{y}(k+1) \approx [p_0 + p_1 + 1, t_0 + t_1] \tag{5.32}$$

where one additional bit may be required in the accumulator to allow for the one addition.

If *all* the most significant $p_0 + 1$ bits of $\hat{y}(k+1)$ are *equal* to the sign bit, then by the *sign extension property* of two's complement fixed point arithmetic, $\hat{y}(k+1)$ also has the *exact* representation

$$\hat{y}(k+1) \approx [p_1, t_0 + t_1] \tag{5.33}$$

In this case, it follows from (5.31) that $\hat{y}(k+1)$ has *not overflowed.* The quantized state $Q[\hat{y}(k+1)]$ which results from quantizing $\hat{y}(k+1)$ to t_1 significant (fractional) bits, then has the required representation

$$Q[\hat{y}(k+1)] \approx [p_1, t_1]$$

which is consistent with $Q[\hat{y}(k)]$ in (5.31). The *quantization error signal* ε_Q defined by

$$\varepsilon_Q(k+1) \stackrel{\Delta}{=} \hat{y}(k+1) - Q[\hat{y}(k+1)] \tag{5.34}$$

in (5.29) also then has the representation

$$\varepsilon_Q(k+1) \approx [0, t_0 + t_1]$$

For *roundoff quantization*, the most significant t_1 fractional bits of $\varepsilon_Q(k)$ are all zero, and so all we can say about $\varepsilon_Q(k+1)$ is that

$$|\varepsilon_Q(k+1)| \leq 2^{-(t_1+1)} = 0.5 \times 2^{-t_1} \tag{5.35}$$

However for *truncation quantization*, only the most significant $t_1 - 1$ fractional bits of $\varepsilon_Q(k)$ are necessarily all zero, and so now all we can say is

$$|\varepsilon_Q(k+1)| \leq 2^{-t_1} \tag{5.36}$$

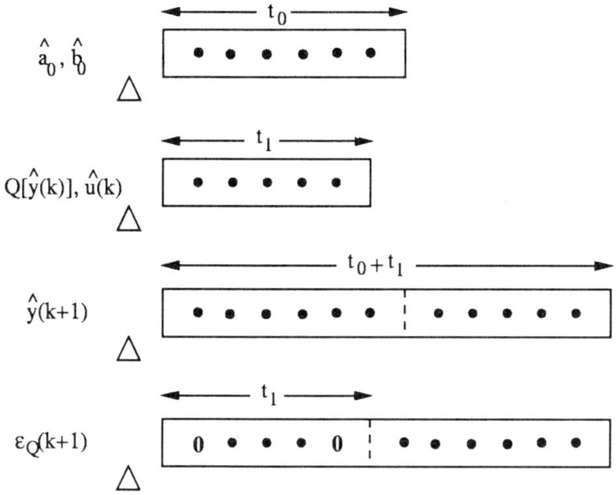

Fig. 5.3. Fixed point wordlengths

The wordlengths of the quantized coefficients, the state, and the state quantization error when $p_0 = p_1 = p_2 = 0$ (i.e. all pure fractional representations) are illustrated in Figure 5.3.

If all the most significant p_0+2 bits of $\hat{y}(k+1)$ in (5.32) are *not* equal, then overflow has occurred. In this case, if we assume $Q[w]$ represents the assignment of the *most significant* bits of w by implementing *saturation arithmetic*, then

$$Q[w] = \begin{cases} 2^{p_1} & ; w \geq 2^{p_1} \\ w - \varepsilon_Q & ; -2^{p_1} < w < 2^{p_1} \\ -2^{p_1} & ; w \leq -2^{p_1} \end{cases}$$

Hence if overflow *does* occur (and is detected), then all that will be known about the arithmetic error $\varepsilon_Q(k)$ (except perhaps knowing the sign) is that

$$|\varepsilon_Q(k)| \geq 2^{p_1}$$

Unless otherwise stated, we shall in future assume that $Q[\cdot]$ denotes *fixed point roundoff quantization incorporating saturation arithmetic.*

The first order FWL realization (5.29) employs signal *quantization before multiplication* by the coefficient $\hat{a}_0$. An alternative FWL realization of the ideal filter (5.28) which employs *quantization after multiplication* is given by

$$\hat{z}(k+1) = Q[\hat{a}_0\hat{z}(k) + \hat{b}_0\hat{u}(k)] \tag{5.37}$$

where now the quantization error $\nu_Q(k)$ is defined by

$$\nu_Q(k) \triangleq \hat{a}_0\hat{z}(k) + \hat{b}_0\hat{u}(k) - Q[\hat{a}_0\hat{z}(k) + \hat{b}_0\hat{u}(k)] \tag{5.38}$$

If the state $\hat{z}(k)$ in (5.37) has the *same* arithmetic representation as $Q[\hat{y}(k)]$ in (5.31); that is

$$\hat{z}(k) \approx [p_1, t_1] \tag{5.39}$$

then all multiplications in both (5.29) and (5.37) are $(p_0 + t_0) \times (p_1 + t_1)$. Furthermore, the accumulator in both implementations need only be $(p_0 + p_1 + 1) + (t_0 + t_1)$ bits wide for *exact* evaluation of the inner products on the right-hand side of both realizations. Therefore, the wordlength representation (5.39) for $\hat{z}(k)$ provides a fair basis with which to compare the relative numerical accuracies of the implementations (5.29) and (5.37).

Suppose numerical simulations are conducted for some input sequence $\{\hat{u}(k)\}$ using the realizations (5.29) and (5.37). Then after taking $Q[\cdot]$ of both sides of (5.29), we have

$$Q[\hat{y}(k+1)] = Q[\hat{a}_0 Q[\hat{y}(k)] + \hat{b}_0 \hat{u}(k)]$$

Hence it follows from (5.37) that the results of a simulation of (5.29) for the *same* input sequence is given by

$$\hat{z}(k) = Q[\hat{y}(k)]$$

provided the initial conditions $\hat{y}(0)$ and $\hat{z}(0)$ satisfy $\hat{z}(0) = Q[\hat{y}(0)]$. Conversely, the results of a simulation of (5.29) for some $\{\hat{u}(k)\}$ can be determined from a simulation of (5.37) provided $\hat{z}(0) = Q[\hat{y}(0)]$.

Note that in MATLAB, a number w can be rounded to a t_0 bit fractional representation via the instruction:

$$Q[w] = 2^{-t_0} * \text{round}[2^{t_0} * w]$$

where round$[x]$ rounds the number x to the nearest integer.

Example 5.2.1. Consider the ideal first order filter (5.28) when $\{b_0 = 0.375, a_0 = 0.625\}$. For $t_0 = 4$ in (5.30), $\{a_0 = \hat{a}_0 = 0.625\}$ and $\{b_0 = \hat{b}_0 = 0.375\}$ can be exactly represented. (Actually, a fractional wordlength of $t_0 = 3$ bits is sufficient.)

Suppose further that $t_1 = 2$ in (5.31). Then the unit step response (i.e. $u(k) = 1$ for $k \geq 0$) from a zero initial condition $y(0) = \hat{y}(0) = 0$ for the ideal and FWL cases (5.28) and (5.29) respectively are given by:

$$\text{Ideal} : \{0.3750,\ 0.6094,\ 0.7559,\ 0.8474,\ 0.9046,\ \ldots,\ 1.0000\}$$

$$\text{FWL} : \{0.3750,\ 0.5312,\ 0.6875,\ 0.8438,\ 0.8438,\ \ldots,\ 0.8438\}$$

Ideally, $y(k)$ approaches 1 in the limit as $k \to \infty$, but that a steady state numerical error of 0.1562 arises with the FWL implementation as a result of roundoff quantization even though there are no coefficient or input errors.

5.2.1 Quantization Errors

Let the output error e_y between the ideal (or infinite precision) filter output y and the finite precision filter output $\hat{y}$ be defined by

$$e_y(k) \triangleq y(k) - \hat{y}(k) \tag{5.40}$$

Then from (5.28) and (5.29), the error signal e_y satisfies the difference equation

$$e_y(k+1) = \hat{a}_0 e_y(k) + w_e(k) \tag{5.41}$$

where

$$w_e(k) \triangleq \hat{a}_0 \varepsilon_Q(k) + \hat{b}_0 \nabla u(k) + (\nabla a_0) y(k) + (\nabla b_0) u(k) \tag{5.42}$$

with the input and coefficient errors given by

$$\nabla u \triangleq u - \hat{u} \;\; ; \;\; \nabla a_0 \triangleq a_0 - \hat{a}_0, \;\; \nabla b_0 \triangleq b_0 - \hat{b}_0$$

In terms of bounds, we have

$$\begin{aligned} \|e_y\|_\infty \leq & \frac{1}{1-|\hat{a}_0|}\left\{|\hat{a}_0| \; \|\varepsilon_Q\|_\infty + |\hat{b}_0| \; \|\nabla u\|_\infty\right\} \\ & + \frac{1}{1-|\hat{a}_0|}\{|\nabla a_0| \; \|y\|_\infty + |\nabla b_0| \; \|u\|_\infty\} \end{aligned}$$

Also from (5.28)

$$\begin{aligned} \begin{bmatrix} e_y(k+1) \\ y(k+1) \end{bmatrix} = & \begin{bmatrix} \hat{a}_0 & \nabla a_0 \\ 0 & a_0 \end{bmatrix} \begin{bmatrix} e_y(k) \\ y(k) \end{bmatrix} + \begin{bmatrix} \nabla b_0 \\ b_0 \end{bmatrix} u(k) \\ & + \begin{bmatrix} \hat{a}_0 \\ 0 \end{bmatrix} \varepsilon_Q(k) + \begin{bmatrix} \hat{b}_0 \\ 0 \end{bmatrix} \nabla u(k) \end{aligned} \tag{5.43}$$

In principle, given the coefficients $\{a_0, b_0, \hat{a}_0, \hat{b}_0\}$, the internal quantization error signal ε_Q, and the input error signal ∇u, one can then solve these equations in order to determine the precise arithmetic error signal e_y that results. The calculation in (5.43) could be undertaken on a large wordlength computer in order to evaluate the performance of the FWL implementation for a particular input signal u. However, as we now illustrate, (5.43) is perhaps more useful for the purpose of analysing the relative significance of the input quantization errors, the coefficient quantization errors, and the internal quantization errors in $\hat{y}$.

The input quantization error signal ∇u and the internal quantization error signal ε_Q both depend on the input signal u, the coefficients $\{a_0, b_0\}$ and the coefficients errors $\{\nabla a_0, \nabla b_0\}$. However to a first approximation, we can look at the effects of input quantization, coefficient quantization and internal quantization as if the effects were independent.

Input Quantization Errors

The effect on the output error e_y as a result of the input quantization error ∇u in the absence of both coefficient and internal quantization errors is given by

$$e_y(k+1) = a_0 e_y(k) + b_0 \nabla u(k) \tag{5.44}$$

Hence the *error frequency response function* $E_{y\nabla u}(z)$ between the output error e_y and the input error ∇u due to input quantization is equal to the filter frequency response; that is

$$E_{y\nabla u}(z) = \frac{b_0}{z - a_0} \tag{5.45}$$

Coefficient Quantization Errors

From (5.43), the effect of coefficient errors on the output error e_y in the absence of both input and internal quantization errors is given by the equations

$$\begin{bmatrix} e_y(k+1) \\ y(k+1) \end{bmatrix} = \begin{bmatrix} \hat{a}_0 & \nabla a_0 \\ 0 & a_0 \end{bmatrix} \begin{bmatrix} e_y(k) \\ y(k) \end{bmatrix} + \begin{bmatrix} \nabla b_0 \\ b_0 \end{bmatrix} u(k)$$

$$e_y(k) = [1 \;\; 0] \begin{bmatrix} e_y(k) \\ y(k) \end{bmatrix}$$

The *error frequency response function* $E_{yu}(z)$ between the output error e_y and the input signal u is then given by

$$\begin{aligned} E_{yu}(z) &= [1 \;\; 0] \begin{bmatrix} z - \hat{a}_0 & -\nabla a_0 \\ 0 & z - a_0 \end{bmatrix}^{-1} \begin{bmatrix} \nabla b_0 \\ b_0 \end{bmatrix} \\ &= \frac{\nabla b_0}{z - \hat{a}_0} + \frac{(\nabla a_0) b_0}{(z - \hat{a}_0)(z - a_0)} \end{aligned} \tag{5.46}$$

For example, when $\{a_0 = 0.88, b_0 = 0.90\}$, the error magnitude response functions $|E_{yu}(e^{j\omega})|$ corresponding to wordlengths $t_0 = 4$ and $t_0 = 6$ are illustrated in Figure 5.4. This information may be used to estimate the relative importance of coefficients and input quantization errors. For example, suppose $\hat{u}$ is determined by an analog-to-digital conversion of u where $u(kT) = v(t = kT)$ for some analog signal v. Then ∇u is determined by the wordlength of the ADC while $\{\nabla a_0, \nabla b_0\}$ are determined by the wordlength of the coefficients $\{\hat{a}_0, \hat{b}_0\}$.

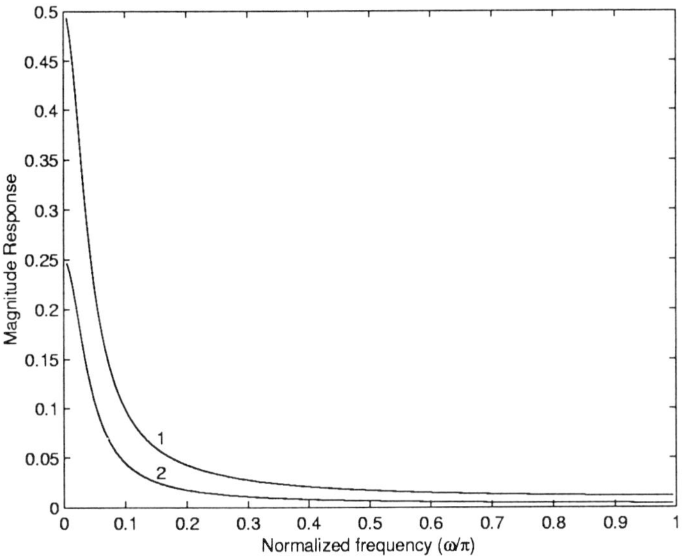

Fig. 5.4. Error frequency response function of first order filter for 1: $t_0 = 4$ and 2: $t_0 = 6$ bits

Internal Quantization Errors

From (5.43), the effect of the internal quantization errors ε_Q on the quantization error e_y in the absence of both input and coefficient errors is given by the equation

$$e_y(k+1) = a_0 e_y(k) + a_0 \varepsilon_Q(k) \tag{5.47}$$

Assuming $e_y(0) = 0$, we then have from section 4.3.3 that

$$||e_y||_\infty \leq L_e \tag{5.48}$$

where

$$\begin{aligned} L_e &= ||h_e||_\infty \; ||\varepsilon_Q||_1 \quad \text{or} \quad L_e = ||h_e||_2 \; ||\varepsilon_Q||_2 \quad \text{or} \\ L_e &= ||h_e||_1 \; ||\varepsilon_Q||_\infty \end{aligned} \tag{5.49}$$

where h_e is the unit impulse response of the error equation (5.47). The error frequency response $E_{y\varepsilon}$ in (5.47) is given by

$$E_{y\varepsilon}(z) = \frac{a_0}{z - a_0}$$

and

$$||h_e||_1 = \frac{|a_0|}{1-|a_0|} \quad ; \quad ||h_e||_2 = \frac{|a_0|}{1-a_0^2} \quad ; \quad ||h_e||_\infty = |a_0|$$

If we assume that there is no overflow, and quantization of $\hat{y}(k)$ to a t_1 bit fractional representation $Q[\hat{y}(k)]$, then

$$||\varepsilon_Q||_\infty \leq 0.5 \times 2^{-t_1} \text{ (roundoff) and } ||\varepsilon_Q||_\infty \leq 2^{-t_1} \text{ (truncation)}$$

with

$$L_e = \frac{|a_0|}{1-|a_0|} \, ||\varepsilon_Q||_\infty \tag{5.50}$$

Statistical Quantization Error Model

The quantization error signal ε_Q actually results from a nonlinear operation on the signal $\hat{y}$, and therefore depends explicitly on the particular input signal u. That is, without knowing the input and the initial condition $\hat{y}(0)$, all that is known about ε_Q is the infinity norm $||\varepsilon_Q||_\infty$ (provided there is no overflow) while the other norms $||\varepsilon_Q||_1$ and $||\varepsilon_Q||_2$ cannot be determined. However, provided the filter input signal u is "sufficiently exciting", then it has been found by simulation studies that a very good approximation for $||\varepsilon_Q||_2$ can be based on a *statistical model* for the signal ε_Q.

Specifically, if we assume quantization of $\hat{y}$ to a signal $Q[\hat{y}]$ having a t_1 bit fractional representation, then in the absence of overflow, then the statistical model assumes the quantization error signal ε_Q is uniformly distributed over an interval. For *roundoff quantization*, the interval is $[-0.5 \times 2^{-t_1}, 0.5 \times 2^{-t_1}]$ whereas for *truncation quantization*, the interval is $[0, 2^{-t_1}]$. (Recall from section 5.1.1 that the error ε_Q that results from truncation is always positive.)

The notion of "sufficiently exciting" does have a precise (but complicated) mathematical definition. In practice, it is enough that the input signal have a sufficiently large magnitude which varies sufficient over time. A constant input signal is not "sufficiently exciting" while a sinusoidal signal of sufficiently high frequency and sufficiently large magnitude can be "sufficiently exciting". Based on such a statistical model, it then follows that

$$||\varepsilon_Q||_2 = \sqrt{\frac{1}{12} 2^{-2t_1} + \bar{\varepsilon}_Q^2}$$

where the 'mean' $\bar{\varepsilon}_Q$ is given by

$$\bar{\varepsilon}_Q = \begin{cases} 0 & ; \text{ roundoff quantization} \\ 0.25 \times 2^{-t_1} & ; \text{ truncation quantization} \end{cases}$$

That is

$$||\varepsilon_Q||_2 = \begin{cases} 0.29 \times 2^{-t_1} & ; \text{ roundoff quantization} \\ 0.38 \times 2^{-t_1} & ; \text{ truncation quantization} \end{cases} \tag{5.51}$$

A bound L_e on the filter output error is then given by

$$L_e \leq ||h_e||_2 \; ||\varepsilon_Q||_2 = \frac{|a_0|}{\sqrt{1-a_0^2}} \; ||\varepsilon_Q||_2 \tag{5.52}$$

This bound is generally less conservative than the bound in (5.50). For example, suppose $a_0 = 0.8$. Then from (5.50) for roundoff quantization in the absence of overflow

$$||e_y||_\infty \leq 2 \times 2^{-t_1}$$

However from (5.52) for roundoff quantization in the absence of overflow, and using the statistical model for $||\varepsilon_Q||_2$, we have

$$||e_y||_\infty \leq 0.38 \times 2^{-t_1}$$

However note that whereas the higher bound for $||e_y||_\infty$ is known to be valid (but generally conservative), the lower bound is only an *estimate* based on a *statistical model*, and therefore is generally optimistic.

Internal Quantization Error Feedback

An alternative realization of (5.28) is given by

$$\begin{aligned} \hat{y}(k+1) &= \hat{a}_0 Q[\hat{y}(k)] + \hat{b}_0 \hat{u}(k) + \alpha_0 \varepsilon_Q(k) \\ Q[\hat{y}(k)] &= \hat{y}(k) - \varepsilon_Q(k) \end{aligned} \tag{5.53}$$

where α_0 is an *integer*. That is, the wordlength representations are given by

$$\begin{aligned} \hat{a}_0, \hat{b}_0 &\approx [p_0, t_0] \quad ; \quad \alpha_0 \approx [p_0, 0] \\ Q[\hat{y}(k)], \hat{u}(k) &\approx [p_1, t_1] \quad ; \quad \varepsilon_Q(k) \approx [0, t_1 + t_0] \end{aligned}$$

The inclusion of the term $\alpha_0 \varepsilon_Q(k)$ is referred to as *internal quantization error feedback*. Note that since $\varepsilon_Q(k)$ has a $t_0 + t_1$ bit fractional representation, α_0 must be an *integer*; otherwise, the product $\alpha_0 \varepsilon_Q(k)$ in (5.53) will *not* have a fractional wordlength representation of $t_0 + t_1$ bits, and therefore will be *wordlength inconsistent*.

Given $\hat{y}(k)$, the FWL realization in (5.53) consists of first rounding $\hat{y}(k)$ to $Q[\hat{y}(k)]$. Then $Q[\hat{y}(k)]$ is multiplied by $\hat{a}_0$ and $\hat{y}(k) - Q[\hat{y}(k)]$ is multiplied by the integer α_0. The FWL realization (5.53) can also be equivalently realized in the form

$$\hat{y}(k+1) = \alpha_0 \hat{y}(k) + \hat{\beta}_0 Q[\hat{y}(k)] + \hat{b}_0 \hat{u}(k)$$

where $\beta_0 = a_0 - \alpha_0$, and $\hat{\beta}_0$ is the FWL coefficient fractional representation of β_0. Since α_0 is an integer, it follows that $\hat{\beta}_0 = \hat{a}_0 - \alpha_0$. For example, if

$\{a_0 = 0.9, \alpha_0 = 1\}$, and all FWL coefficients are restricted to a 4 bit fractional representation, the $\{\beta_0 = -0.1, \hat{\beta}_0 = -0.125\}$.

Once again, since α_0 is an integer, $\alpha_0\hat{y}(k)$ has the same fractional wordlength representation as $\hat{y}(k)$, and therefore $\hat{y}(k)$ does not have to be quantized before multiplication by α_0.

For purposes of analysis (and not implementation), (5.53) can be written in the equivalent form

$$\hat{y}(k+1) = \hat{a}_0\hat{y}(k) + \hat{b}_0\hat{u}(k) + (\alpha_0 - \hat{a}_0)\varepsilon_Q(k) \tag{5.54}$$

and so when $\{a_0 = \hat{a}_0, b_0 = \hat{b}_0\}$ and $u = \hat{u}$, the only difference between the ideal (ie infinite precision) difference equation (5.28) and the FWL equivalent (5.54) is the input term $(\alpha_0 - \hat{a}_0)\varepsilon_Q(k)$.

Another way to understand the restriction that α_0 *must* be an integer is to observe that if α_0 could be freely chosen, then $\alpha_0 = \hat{a}_0$ $(= a_0)$ would imply that (5.54) becomes equivalent to (5.28); that is, infinite precision fractional arithmetic would be realized - which would be a contradiction. *Infinite precision fractional arithmetic* is only possible if $\hat{a}_0$ and $\hat{b}_0$ are both integers.

The effect on the output error e_y in the absence of both input and coefficient quantization errors as a result of the internal quantization errors ε_Q with *feedback* is given by

$$e_y(k+1) = a_0e_y(k) + (a_0 - \alpha_0)\varepsilon_Q(k) \tag{5.55}$$

That is, the *error frequency response function* $E_{y\varepsilon}(z)$ between the output error e_y and the quantization error ε_Q due to internal quantization is now given by

$$E_{y\varepsilon}(z) = \frac{a_0 - \alpha_0}{z - a_0} \tag{5.56}$$

Recall that the ideal first order filter (5.28) is asymptotically stable when $|a_0| < 1$. The in terms of minimizing the output error with respect to the internal quantization error, it follows for $|a_0| < 1$ that the optimal value α_{min} of the integer α_0 is given by

$$\alpha_{min} = \begin{cases} 1\,; & 0.5 \le a_0 < 1 \\ 0\,; & -0.5 \le ta_0 \le 0.5 \\ -1\,; & -1 < a_0 \le -0.5 \end{cases}$$

For example, if $a_0 = 0.99$ then $a_0 - \alpha_0 = 0.99$ when $\alpha_0 = 0$, and $a_0 - \alpha_0 = -0.01$ when $\alpha_0 = 1$. This means that a reduction in the dc gain of the error system function $E_{y\varepsilon}$ in (5.55) by a factor of 99 results when internal quantization error feedback is used. When $\alpha_0 = \pm 1$, the implementation of the quantization error feedback term $\alpha_0\varepsilon_Q(k)$ involves only one extra subtraction/addition.

5.2.2 Scaling

Scaling is concerned with the "large" signal behaviour, and so (at least to a first order approximation) we can ignor the effects of both coefficient and input quantization errors in the analysis. We first examine the problem of selecting a *constant* scaling factor, and then extend our considerations to *adaptive scaling.*

Constant Signal Scaling

Consider the *ideal* first order digital filter (5.28). Then if $\{a_0 = 0.5, b_0 = 1, y(0) = 1, u(0) = 2, u(1) = -0.5, u(2) = 0.5\}$, we ideally have

$$y(1) = 2.5,\ y(2) = 0.75,\ y(3) = 0.875 \tag{5.57}$$

However if overflow occurs when $|y(k)| \geq 1$, and *saturation arithmetic* is implemented according to

$$\hat{y}(k+1) = Q[a_0\hat{y}(k) + b_0u(k)]$$

where

$$Q[y] = \begin{cases} 1 \ ; \ y \geq 1 \\ y \ ; \ -1 \leq y \leq 1 \\ -1 \ ; \ y \leq -1 \end{cases} \tag{5.58}$$

then

$$\hat{y}(1) = Q[2.5] = 1,\ \hat{y}(2) = 0,\ \hat{y}(3) = 0.5$$

We now examine how overflow can be avoided by introducing a scaling factor.

For example, suppose simulation (or analysis) reveals for the *ideal calculation* in (5.28) that for a range of input signals u, we have $|y(k)| \leq L_0$ for all $k \geq 0$. Then if we define $z(k) \triangleq L_0^{-1}y(k)$, we have from (5.28) that

$$\begin{aligned} z(k+1) &= a_0z(k) + b_{0L}u(k) \ ; \quad b_{0L} = L_0^{-1}b_0 \\ y(k) &= L_0z(k) \end{aligned} \tag{5.59}$$

Now $|y(k)| \leq L_0$ implies $|z(k)| \leq 1$. Therefore the implementation

$$\hat{z}(k+1) = Q[a_0\hat{z}(k) + b_{0L}u(k)] \ ; \quad \hat{y}(k) = Q[L_0\hat{z}(k)]$$

with $Q[\cdot]$ defined by (5.58) which incorporates scaling by the scaling factor L_0 will not overflow in z, even though the output value y sometimes does. In particular, since z does not overflow, the values of y for which $|y(k)| \leq 1$ will always be correct. For example, if in (5.59), we have $\{a_0 = 0.5, b_0 = 1\}$ and $L_0 = 4$ (which implies $b_L = 4$), then $z(0) = 0.25$ implies $y(0) = 1$, and then $\{u(0) = 2, u(1) = -0.5, u(2) = 0.5\}$ implies

$$\hat{z}(1) = 0.625,\ \hat{z}(2) = 0.1875,\ \hat{z}(3) = 0.21875$$

which gives

$$\hat{y}(1) = Q[4 \times 0.625] = 1\ ,\ \hat{y}(2) = 4 \times 0.1875 = 0.75$$
$$\hat{y}(3) = 4 \times 0.21875 = 0.875$$

Hence after comparing with (5.57), we see that only the value $\hat{y}(1)$ is incorrect.

The practical problem however with scaling is to be able to find (either by numerical simulation or by analysis) an appropriate bound L_0 to use in (5.59). The *limitation of simulation* for determining a suitable value for L_0 is that not all possible input signals u can be simulated. The *limitation of analysis* is that generally the bound L_0 which is obtained is generally too conservative. In particular, assuming zero initial conditions, we have by Theorem 4.1.9 that

$$||y||_\infty \leq L_0$$

where possible bounds for L_0 are given by

$$L_0 = ||h||_1\ ||u||_\infty\ \ ;\ \ L_0 = ||h||_2\ ||u||_2\ \ ;\ \ L_0 = ||h||_\infty\ ||u||_1$$

where h is the unit impulse response of (5.28); that is, we have

$$||h||_1 = \frac{|b_0|}{1 - |a_0|}\ \ ;\ \ ||h||_2 = \frac{|b_0|}{\sqrt{1 - a_0^2}}\ \ ;\ \ ||h||_\infty = |b_0|$$

A problem if to large a value for L_0 is selected.

For example, suppose in (5.59), we choose $L_0 = 1000$ (rather than $L_0 = 4$). Then again with $\{a = 0.5, b = 1\}$, we have $b_L = 0.001$, so ideally $z(0) = 0.001$ corresponds to $y(0) = 1$. Then again with $\{u(0) = 2, u(1) = -0.5, u(2) = 0.5\}$, we get $\{z(1) = 0.0025, y(1) = 1000 \times z(1) = 2.5\}$ as before. However, if (say) $z(k)$ has a only a $[1, 4]$ bit fixed point roundoff representation, then $z(0) = 0.001$ would be quantized to $\hat{z}(0) = 0$, and then $\hat{y}(0) = Q[1000 \times \hat{z}(0)] = 0$. If the calculations are continued, then we would also get $\hat{y}(1) = \hat{y}(2) = 0$. Hence, in general terms, if the scaling factor L_0 is *increased*, then in order to avoid *underflow*, the fractional wordlength that is assigned to the signal z should also be *increased*. This example illustrates the point that a too conservative choice for a scaling factor L_0 can lead to an *increase* in the errors due to *underflow*.

Adaptive Scaling

Another way to reduce the possibility of overflow, while at the same time reducing the effects of internal quantization is to use *adaptive scaling*. We begin this development with consideration of an adaptively scaled first order difference equation.

Specifically, consider the difference equation where given $\{z(k), u(k)\}$

$$\begin{aligned} y(k) &= s^{-1}(k)z(k) \;\; ; \;\; p(k) = a_0 z(k) + s(k) b_0 u(k) \\ z(k+1) &= d(k+1)p(k) \end{aligned} \tag{5.60}$$

Then it follows that the signal y in (5.60) satisfies (5.28) if $s(k)$ is updated to $s(k+1)$ according to

$$s(k+1) = d(k+1)s(k) \tag{5.61}$$

Also, if in addition, $d(k+1)$ is selected such that

$$0.5 \leq |d(k+1)p(k)| < 1 \tag{5.62}$$

then $z(k+1)$ will lie in the range $0.5 \leq |z(k+1)| < 1$.

In particular, suppose $s(k)$ and $d(k)$ are both restricted to be a power of two of the form

$$d(k) = 2^{-\beta(k)} \;\; ; \;\; s(k) = 2^{-\gamma(k)} \tag{5.63}$$

for some *integers* $\{\beta(k), \gamma(k)\}$. Then (5.61) is equivalent to

$$\gamma(k+1) = \gamma(k) + \beta(k+1) \tag{5.64}$$

and (5.62) is satisfied if the integer $\beta(k+1)$ is chosen such that

$$2^{\beta(k+1)-1} \leq |p(k)| < 2^{\beta(k+1)}$$

We have the following adaptive scaling algorithm.

Adaptive First Order Scaling Algorithm

An adaptive scaling algorithm that is equivalent to the ideal first order difference equation (5.28) is provided as follows: Given $y(0)$, calculate $\{z(0), \gamma(0)\}$ such that: if $y(0) = 0$, then $\{z(0) = 0; \gamma(0) = 0\}$; else if $y(0) \neq 0$, then select $\gamma(0)$ such that

$$0.5 \leq |z(0)| < 1 \;\; ; \;\; z(0) = 2^{-\gamma(0)} y(0)$$

and then determine $p(0)$ from

$$p(0) = a_0 z(0) + 2^{-\gamma(0)} b_0 u(0)$$

Then for $k \geq 0$ given $\{z(k), u(k), \gamma(k)\}$:

- Calculate the integer $\beta(k+1)$ such that

$$2^{\beta(k+1)-1} \leq |p(k)| \leq 2^{\beta(k+1)}$$

- Update $\{z(k), \gamma(k)\}$ to $\{z(k+1), \gamma(k+1)\}$ by

$$z(k+1) = 2^{-\beta(k+1)} p(k) \quad ; \quad \gamma(k+1) = \gamma(k) + \beta(k+1)$$

- Update

$$\begin{aligned} p(k+1) &= a_0 z(k+1) + 2^{-\gamma(k+1)} b_0 u(k+1) \\ y(k+1) &= 2^{\gamma(k+1)} z(k+1) \end{aligned}$$

Since $\gamma(k)$ is an *integer*, the 'multiplications' in step 2 and step 3 can be realized by 'shifting operations' (which was the practical reason for imposing the restriction in (5.63) in the first place). The update to $\gamma(k+1)$ in step 2 then requires a further integer addition. The most time-consuming step of the algorithm is the determination of $\beta(k+1)$ in step 1 in order to guarantee $0.5 \leq |z(k+1)| < 1$. This operation (which may be accomplished either in hardware or software) is similiar to the alignment operation that occurs in the normalization of the mantissa in floating point arithmetic.

Even so since in practice both $\beta(k+1)$ and $z(k+1)$ are constrained, overflow is still possible although less likely. For example, suppose the integer $\beta(k+1)$ and the value $z(k+1)$ are restricted to lie within the range

$$-2 \leq \beta(k+1) \leq 2 \quad ; \quad 0.5 \leq |z(k+1)| < 1$$

Then if $\{\gamma(k) = 0, p(k) = 5\}$, the condition on $\beta(k+1)$ in step 1 cannot be satisfied. Instead, in practice, $\beta(k+1)$ would be set at the *maximum* value of $\beta(k+1) = 2$ which then gives $\{\gamma(k+1) = 2, z(k+1) = 1.25\}$ in step 3. This means that $z(k+1)$ would *overflow*.

If $\{\gamma(k) = 0, p(k) = 0.01\}$, the condition on $\beta(k+1)$ again cannot be satisfied. Instead, in practice, $\beta(k+1)$ would be set at the *minimum* value of $\beta(k+1) = -2$ which then gives $\{\gamma(k+1) = -2, z(k+1) = 0.04\}$ in step 3. This means that depending on the available fractional wordlength, the quantization of $z(k+1)$ could *underflow*, and thereby result in a significant loss of accuracy in values $\{z(m); m \geq k+1\}$.

Effect of Scaling on Quantization Errors

If we define $z(k) \triangleq L_0^{-1} y(k)$, then (5.28) can be written in the form (5.59) which has the FWL implementation

$$\begin{aligned} \hat{z}(k+1) &= \hat{a}_0 Q[\hat{z}(k)] + \hat{b}_{0L} \hat{u}(k) \\ Q[\hat{z}(k)] &= \hat{z}(k) - \eta_Q(k) \\ \hat{y}(k) &= L_0 \hat{z}(k) \end{aligned} \tag{5.65}$$

Here we assume that there is no coefficient error in implementing the scaling factor L_0. (For example, take L_0 to either be an integer or a power of 2.)

Now from (5.59) and (5.65), the output error $e_y(k) \triangleq y(k) - \hat{y}(k)$ is given by

$$e_y(k) = L_0\tilde{e}_y(k) \quad ; \quad \tilde{e}_y(k) \triangleq z(k) - \hat{z}(k)$$

where

$$\tilde{e}_y(k+1) = \hat{a}_0\tilde{e}_y(k) + \tilde{w}_e(k)$$

and

$$\tilde{w}_e(k) = (\hat{a}_0 - \alpha_0)\eta_Q(k) + \hat{b}_{0L}\nabla u(k) + (\nabla a_0)z(k) + (\nabla b_{0L})u(k)$$
$$\nabla b_{0L} \triangleq b_{0L} - \hat{b}_{0L} \quad ; \quad \nabla a_0 \triangleq a_0 - \hat{a}_0 \quad ; \quad \nabla u \triangleq u - \hat{u}$$

Hence

$$e_y(k+1) = \hat{a}_0 e_y(k) + w_{eL}(k) \tag{5.66}$$

where

$$\begin{aligned} w_{eL}(k) = {} & L_0(\hat{a}_0 - \alpha_0)\eta_Q(k) + L_0\hat{b}_{0L}\nabla u(k) \\ & + (\nabla a_0)y(k) + L_0(\nabla b_{0L})u(k) \end{aligned} \tag{5.67}$$

A comparison of (5.41), (5.42) which result from the implementation *without* scaling, and (5.66), (5.67) which result when a scaling factor L_0 is *included*, reveals the following points:

- Generally, $L_0\hat{b}_{0L} \neq \hat{b}_0$ and $L_0(\nabla b_{0L}) \neq \nabla b_0$ but the approximation is close. That is, the effect of both ∇u and u in both the scaled and unscaled versions are approximately the same.
- Without scaling, the term in w_e due to internal quantization error is $(\hat{a}_0 - \alpha_0)\varepsilon_Q$, whereas with scaling, the corresponding term w_{eL} due to internal quantization error is $L_0(\hat{a}_0 - \alpha_0)\eta_Q$. If $Q[\hat{y}(k)]$ in (5.53) and $Q[\hat{z}(k)]$ in (5.65) are assigned the *same* fractional wordlength of t_1 bits, then

$$|(\hat{a}_0 - \alpha_0)\varepsilon_Q(k)| \leq |\hat{a}_0 - \alpha_0| 2^{-(t_1+1)}$$
$$|(\hat{a}_0 - \alpha_0)\eta_Q(k)| \leq L_0|\hat{a}_0 - \alpha_0| 2^{-(t_1+1)}$$

  That is, *increasing* the scaling factor L_0 in order to prevent overflow leads to an *increase* in the output error e_y due to the effects of internal quantization error. Likewise, *decreasing* the scaling factor L_0 because overflow is less likely leads to an *decrease* in the output error e_y due to the effects of internal quantization error.

5.3 Second Order FWL Filter

Consider the FWL realization

$$\begin{aligned}\hat{y}(k+2) &= \hat{a}_1 Q[\hat{y}(k+1)] + \hat{a}_0 Q[\hat{y}(k)] + \hat{b}_0 \hat{u}(k) + \hat{b}_1 \hat{u}(k+1) \\ Q[\hat{y}(k)] &= \hat{y}(k) - \varepsilon_Q(k)\end{aligned} \tag{5.68}$$

of the ideal second order digital filter

$$y(k+2) = a_1 y(k+1) + a_0 y(k) + b_0 u(k) + b_1 u(k+1) \tag{5.69}$$

In this implementation, the past quantized outputs and inputs $\{Q[\hat{y}(k+1)], Q[\hat{y}(k)], \hat{u}(k+1), \hat{u}(k)\}$ must be available in order to be able to compute $\hat{y}(k+2)$. Then once $\hat{y}(k+2)$ is computed, the quantized value $Q[\hat{y}(k+2)]$ must be stored for future calculations along with $\{Q[\hat{y}(k+1)], \hat{u}(k+2), \hat{u}(k+1)\}$

Suppose all coefficients $\{\hat{a}_1, \hat{a}_0, \hat{b}_1, \hat{b}_0\}$ have the FWL coefficient representation $[p_0, t_0]$ as in (5.30). In addition, suppose $Q[\hat{y}(k)]$ has the fixed point representation $[p_1, t_1]$, and the quantized input $\hat{u}(k)$ has the fixed point representation $[p_2, t_2]$ with $\{p_2 \leq p_1; t_2 \leq t_1\}$. Then the output $\hat{y}(k+2)$ in (5.68) has the representation

$$\hat{y}(k) \approx [p_0 + p_1 + 2, t_0 + t_1]$$

where now two additional bits may be needed in the accumulator to allow for the three additions.

Alternative Realization: Another FWL fixed point realization of (5.69) in which quantization is implemented *after* all the multiplications have been completed is described by

$$\hat{z}(k+2) = Q[\hat{a}_1 \hat{z}(k+1) + \hat{a}_0 \hat{z}(k) + \hat{b}_0 \hat{u}(k) + \hat{b}_1 \hat{u}(k+1)] \tag{5.70}$$

However, if we take $Q[\cdot]$ of both sides of (5.68), we then see that $\hat{z} = Q[\hat{y}]$ is actually a solution of (5.70). However, there are other FWL realizations which are *not* equivalent. For example, the FWL implementation:

$$\hat{y}(k+2) = \hat{a}_1 Q[\hat{y}(k+1)] + Q[\hat{a}_0 Q[\hat{y}(k)] + \hat{b}_0 \hat{u}(k)] + \hat{b}_1 \hat{u}(k+1) \tag{5.71}$$

is different from (5.68), and consequently, will generally produce different numerical results.

5.3.1 Quantization Errors

Consider the FWL fixed point realization of the second order filter (5.69) as given by (5.68). Then it follows that the output error $e_y(k) \triangleq y(k) - \hat{y}(k)$ satisfies the difference equation

$$e_y(k+2) = \hat{a}_1 e_y(k+1) + \hat{a}_0 e_y(k) + w_e(k+1) \tag{5.72}$$

where

$$\begin{aligned} w_e(k+1) &= \hat{a}_1 \varepsilon_Q(k+1) + \hat{a}_0 \varepsilon_Q(k) + \hat{b}_0 \nabla u(k) + \hat{b}_1 \nabla u(k+1) \\ &\quad + (\nabla a_1) y(k+1) + (\nabla a_0) y(k) + (\nabla b_0) u(k) + (\nabla b_1) u(k+1) \end{aligned}$$

Hence in terms of magnitude bounds

$$||e_y||_\infty \le ||h_e||_1 \; ||w_e||_\infty$$

where h_e is the unit impulse response of (5.72) with respect to the input signal w_e, and

$$\begin{aligned} ||w_e||_\infty &\le (|\hat{a}_1| + |\hat{a}_0|)||\varepsilon_Q||_\infty + (|\hat{b}_1| + |\hat{b}_0|)||\nabla u||_\infty + \\ &\qquad (|\nabla a_1| + |\nabla a_0|)||y||_\infty + (|\nabla b_1| + |\nabla b_0|)||u||_\infty \\ &\le (|\hat{a}_1| + |\hat{a}_0|)||\varepsilon_Q||_\infty + (|\hat{b}_1| + |\hat{b}_0|)||\nabla u||_\infty + \\ &\qquad [|\nabla b_1| + |\nabla b_0| + ||h||_1(|\nabla a_1| + |\nabla a_0|)]||u||_\infty \end{aligned}$$

where from section 4.1.4 with $b_2 = 0$

$$||h_e||_1 \le \frac{|\hat{\beta}_1|}{1 - |\hat{\lambda}_1|} + \frac{|\hat{\beta}_2|}{1 - |\hat{\lambda}_2|} \quad ; \quad ||h||_1 \le \frac{|\beta_1|}{1 - |\lambda_1|} + \frac{|\beta_2|}{1 - |\lambda_2|}$$

with

$$\begin{aligned} \frac{z}{z^2 - \hat{a}_1 z - \hat{a}_0} &= \frac{\hat{\beta}_1}{1 - \hat{\lambda}_1} + \frac{\hat{\beta}_2}{1 - \hat{\lambda}_2} \quad ; \quad 1 > |\hat{\lambda}_1| \ne |\hat{\lambda}_2| < 1 \\ \frac{b_1 z + b_0}{z^2 - a_1 z - a_0} &= \frac{\beta_1}{1 - \lambda_1} + \frac{\beta_2}{1 - \lambda_2} \quad ; \quad 1 > |\lambda_1| \ne |\lambda_2| < 1 \end{aligned}$$

Also from (5.72), (5.73) and (5.69) can be written in the state space form

$$\begin{bmatrix} e_y(k+1) \\ e_y(k) \\ y(k+1) \\ y(k) \\ u(k) \end{bmatrix} = \begin{bmatrix} \hat{a}_1 & \hat{a}_0 & \nabla a_1 & \nabla a_0 & \nabla b_0 \\ 1 & 0 & 0 & 0 & 0 \\ 0 & 0 & a_1 & a_0 & b_0 \\ 0 & 0 & 1 & 0 & 0 \\ 0 & 0 & 0 & 0 & 0 \end{bmatrix} \begin{bmatrix} e_y(k) \\ e_y(k-1) \\ y(k) \\ y(k-1) \\ u(k-1) \end{bmatrix} + \begin{bmatrix} \nabla b_1 \\ 0 \\ b_1 \\ 0 \\ 1 \end{bmatrix} u(k) + \begin{bmatrix} 1 \\ 0 \\ 0 \\ 0 \\ 0 \end{bmatrix} \delta(k)$$

where

$$\delta(k) = \hat{a}_1\varepsilon_Q(k+1) + \hat{a}_0\varepsilon_Q(k) + \hat{b}_0\nabla u(k) + \hat{b}_1\nabla u(k+1)$$

The relative significance of the input errors, the coefficient errors, and the quantization errors in $\hat{y}$ can then be estimated as follows.

Input Quantization Errors

The effect of the input quantization error ∇u in the absence of both coefficient and internal quantization errors is described by the *error frequency response function* $E_{y\nabla u}$ between the output error e_y and the input quantization error ∇u; that is

$$E_{y\nabla u}(z) = \frac{b_1 z + b_0}{z^2 - a_1 z - a_0} \tag{5.73}$$

That is, as was the case of a first order filter, the error response function $E_{y\nabla u}$ is equal to the response function $H(z)$ of the digital filter.

Coefficient Quantization Errors

The effect of only coefficient errors is obtained by setting $\{\nabla u = 0;\ \varepsilon_Q = 0\}$. Then the *error frequency response function* $E_{yu}(z)$ between the output error e_y and the input signal u due to coefficient quantization is given by

$$E_{yu}(z) = \begin{bmatrix} 1\ 0\ 0\ 0\ 0 \end{bmatrix} (z\boldsymbol{I}_5 - \begin{bmatrix} \hat{a}_1 & \hat{a}_0 & \nabla a_1 & \nabla a_0 & \nabla b_0 \\ 1 & 0 & 0 & 0 & 0 \\ 0 & 0 & a_1 & a_0 & b_0 \\ 0 & 0 & 1 & 0 & 0 \\ 0 & 0 & 0 & 0 & 0 \end{bmatrix})^{-1} \begin{bmatrix} \nabla b_1 \\ 0 \\ b_1 \\ 0 \\ 1 \end{bmatrix} \tag{5.74}$$

which after expansion gives

$$E_{yu}(z) = \frac{(\nabla b_1)z + \nabla b_0}{z^2 - \hat{a}_1 z - \hat{a}_0} + \frac{((\nabla a_1)z + \nabla a_0)(b_1 z + b_0)}{(z^2 - \hat{a}_1 z - \hat{a}_0)(z^2 - a_1 z - a_0)} \tag{5.75}$$

In particular, suppose $\{a_1 = 1.8, a_0 = -0.81, b_1 = b_0 = 1\}$ so that $\{\nabla b_1 = \nabla b_0 = 0\}$. The error magnitude response functions $|E_{yu}(e^{j\omega})|$ for $t_0 = 10$ and $t_0 = 12$ bits are illustrated in Figure 5.5 .

Internal Quantization Errors

The effect of only the internal quantization error ε_Q is described by the difference equation

$$e_y(k+2) = a_1 e_y(k+1) + a_0 e_y(k) + a_1\varepsilon_Q(k+1) + a_0\varepsilon_Q(k)$$

The *error frequency response function* $E_{y\varepsilon}$ between the output error e_y and the quantization error ε_Q is therefore given by

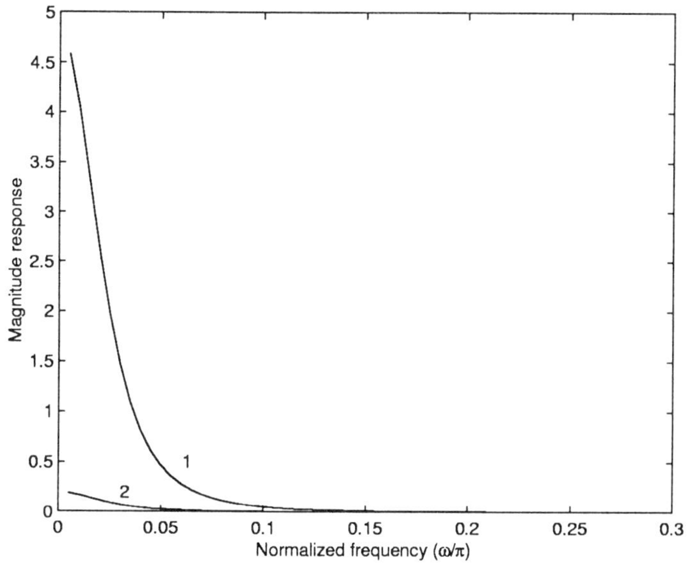

Fig. 5.5. Error frequency response functions of second order filter for 1: $t_0 = 10$ and 2: $t_0 = 12$ bits

$$E_{y\varepsilon}(z) = \frac{a_1 z + a_0}{z^2 - a_1 z - a_0} \tag{5.76}$$

Internal Quantization Error Feedback

An alternative FWL fixed point realization of the second order filter (5.69) is given by

$$\begin{aligned} \hat{y}(k+2) &= \hat{a}_1 Q[\hat{y}(k+1)] + \hat{a}_0 Q[\hat{y}(k)] + \hat{b}_1 \hat{u}(k+1) + \hat{b}_0 \hat{u}(k) + \\ &\qquad \alpha_1 \varepsilon_Q(k+1) + \alpha_0 \varepsilon_Q(k) \\ Q[\hat{y}(k)] &= \hat{y}(k) - \varepsilon_Q(k) \end{aligned} \tag{5.77}$$

where $\{\alpha_0, \alpha_1\}$ are both integers. That is, the wordlength representations are given by

$$\begin{aligned} \hat{a}_1, \hat{a}_0, \hat{b}_1, \hat{b}_0 &\approx [p_0, t_0] \quad ; \quad \alpha_1, \alpha_0 \approx [p_0, 0] \\ Q[\hat{y}(k)], \hat{u}(k) &\approx [p_1, t_1] \quad ; \quad \varepsilon_Q(k) \approx [0, t_1 + t_0] \end{aligned}$$

In this realization, there are two *internal quantization error feedback terms* $\{\alpha_1 \varepsilon_Q(k+1), \alpha_0 \varepsilon_Q(k)\}$. These feedback terms only effect the output error signal e_y with respect to the internal quantization error signal ε_Q.

Given $\{\hat{y}(k), \hat{y}(k+1)\}$, the FWL realization in (5.77) consists of: (i) rounding $\{\hat{y}(k), \hat{y}(k+1)\}$ to $\{Q[\hat{y}(k)], Q[\hat{y}(k+1)]\}$, (ii) multiplications by $\{\hat{a}_0, \hat{a}_1\}$ to give $\{\hat{a}_0 Q[\hat{y}(k)], \hat{a}_1 Q[\hat{y}(k+1)]\}$, and (iii) multiplications by the integers $\{\alpha_0, \alpha_1\}$ to give $\{\alpha_0(\hat{y}(k) - Q[\hat{y}(k)]), \alpha_1(\hat{y}(k+1) - Q[\hat{y}(k+1)])\}$. The FWL realization (5.77) can also be equivalently realized in the form

$$\begin{aligned}\hat{y}(k+2) = \alpha_1 \hat{y}(k+1) + \hat{\beta}_1 Q[\hat{y}(k+1)] + \alpha_0 \hat{y}(k) + \hat{\beta}_0 Q[\hat{y}(k)] \\ + \hat{b}_1 \hat{u}(k+1) + \hat{b}_0 \hat{u}(k)\end{aligned}$$

where $\{\beta_1 = a_1 - \alpha_1, \beta_0 = a_0 - \alpha_0\}$, and $\{\hat{\beta}_1 \hat{\beta}_0\}$ are the FWL coefficient fractional representation of $\{\beta_1, \beta_0\}$. Since $\{\alpha_1, \alpha_0\}$ are integers, it follows that $\{\hat{\beta}_1 = \hat{a}_1 - \alpha_1, \hat{\beta}_0 = \hat{a}_0 - \alpha_0\}$.

In particular, in the absence of both input and coefficient errors, we have

$$\begin{aligned}e_y(k+2) = a_1 e_y(k+1) + a_0 e_y(k) \\ +(\hat{a}_1 - \alpha_1)\varepsilon_Q(k+1) + (\hat{a}_0 - \alpha_0)\varepsilon_Q(k)\end{aligned} \tag{5.78}$$

and so the error frequency response $E_{y\varepsilon}(z)$ is now given by

$$E_{y\varepsilon}(z) = \frac{(a_1 - \alpha_1)z + (a_0 - \alpha_0)}{z^2 - a_1 z - a_0} \tag{5.79}$$

Recall that the second order filter (5.69) is *asymptotically stable* if $\{|a_0| < 1, |a_1| < |1 - a_0|\}$; that is, the filter coefficients $\{\hat{a}_0, \hat{a}_1\}$ can at most assume values in the ranges

$$-1 < a_0 < 1 \;\; ; \;\; -2 < a_1 < 2$$

Therefore in terms of minimizing the *bound* $(|\hat{a}_1 - \alpha_1| + |\hat{a}_0 - \alpha_0|)||\varepsilon_Q||_\infty$ as a result of the internal quantization error, it follows that the optimal values $\{\alpha_{0,min}, \alpha_{1,min}\}$ of the integers $\{\alpha_0, \alpha_1\}$ are given by

$$\alpha_{0,min} = \begin{cases} 1\;; & 0.5 \le a_0 < 1 \\ 0\;; & -0.5 \le a_0 \le 0.5 \\ -1\;; & -1 \le a_0 \le -0.5 \end{cases} \;\; ; \;\; \alpha_{1,min} = \begin{cases} 2\;; & 1.5 \le a_1 < 2 \\ 1\;; & 0.5 \le a_1 \le 1 \\ 0\;; & -0.5 \le a_1 \le 0.5 \\ -1\;; & -1 \le a_1 \le -0.5 \\ -2\;; & -2 < a_1 \le -1.5 \end{cases}$$

For example, for the (stable) filter defined by $\{a_0 = -0.99 \;;\; a_1 = 1.98\}$, we have

$$E_{y\varepsilon}(z) = \frac{g(z - 0.5)}{z^2 - 1.98z + 0.99}$$

where

$$g = \begin{cases} 1.98\;; & \alpha_0 = \alpha_1 = 0 \\ -0.02\;; & \alpha_0 = -1, \alpha_1 = 2 \end{cases}$$

This means a reduction in the dc gain of the error system function $E_{r\varepsilon}$ by a factor of 99 results when internal quantization error feedback is used.

More generally, we have that $||e_y||_\infty \leq L_e$ with bounds L_e given by (5.49) where now h_e is the unit impulse response of the error system $E_{y\varepsilon}(z)$ in (5.79). In particular, if there is no overflow, and $\hat{y}$ is rounded to $Q[\hat{y}]$ which has a t_1 bit fractional part, then

$$L_e = ||h_e||_1 ||\varepsilon_Q||_\infty \leq ||h_e||_1 \times 0.5 \times 2^{-t_1}$$

while if a statistical model is used to estimate L_e, then

$$L_e = ||h_e||_2 \times \sqrt{\frac{1}{12}} \times 2^{-t_1} = ||h_e||_2 \times 0.29 \times 2^{-t_1}$$

5.3.2 Scaling

If we define $z(k) \triangleq L_0^{-1} y(k)$, then (5.69) can be written in the form

$$z(k+2) = a_1 z(k+1) + a_0 z(k) + b_{0L} u(k) + b_{1L} u(k+1)$$

where $\{b_{mL} = L_0^{-1} b_m;\ m = 0, 1, 2\}$. Hence $|y(k) \leq L_0$ implies $|z(k) \leq 1$, and consequently, the implementation

$$\begin{aligned} \hat{z}(k+2) &= Q[a_1 \hat{z}(k+1) + a_0 \hat{z}(k) + b_{0L} u(k) + b_{1L} u(k+1)] \\ \hat{y}(k) &= L_0 \hat{z}(k) \end{aligned}$$

with $Q[\cdot]$ defined by (5.58) will not overflow in z.

Second Order Adaptive Scaling Algorithm

The difference equation (5.69) can also be expressed in the form

$$y(k) \triangleq s^{-1}(k) z(k)$$

where

$$z(k+1) = d(k+1) p(k) \quad ; \quad s(k+1) = d(k+1) s(k)$$

and

$$\begin{aligned} p(k+1) = &\ a_1 z(k+1) + b_1 s(k+1) u(k+1) \\ &+ d(k+1)[a_0 z(k) + b_0 s(k) u(k)] \end{aligned}$$

An *adaptive second order algorithm* is then described as follows. Given $\{y(m); m = -1, 0\}$, calculate $\{z(m), \gamma(m); m = -1, 0\}$ such that: if $y(m) = 0$, then $z(m) = \gamma(m) = 0$, else if $y(m) \neq 0$, then select $\gamma(m)$ such that

$$0.5 \le |z(m)| < 1 \ ; \ z(m) = 2^{-\gamma(m)} y(m), \ m = -1, 0$$
$$\beta(0) = \gamma(0) - \gamma(-1)$$

and then determine $p(0)$ from

$$p(0) = a_1 z(0) + 2^{-\gamma(0)} b_1 u(0) + 2^{-\beta(0)} [a_0 z(-1) + 2^{-\gamma(-1)} b_0 u(-1)]$$

Then for $k \ge 0$ given $\{z(k), u(k), u(k+1), \gamma(k)\}$:

- Calculate the integer $\beta(k+1)$ such that
$$2^{\beta(k+1)-1} \le |p(k)| \le 2^{\beta(k+1)}$$
- Update $\{z(k), \gamma(k)\}$ to $\{z(k+1), \gamma(k+1)\}$ by
$$z(k+1) = 2^{-\beta(k+1)} p(k) \ ; \ \ \gamma(k+1) = \gamma(k) + \beta(k+1)$$
- Update

$$p(k+1) = a_1 z(k+1) + b_1 2^{-\gamma(k+1)} u(k+1)$$
$$+ 2^{-\beta(k+1)} [a_0 z(k) + b_0 2^{-\gamma(k)} u(k)]$$
$$y(k+1) = 2^{\gamma(k+1)} z(k+1)$$

Example 5.3.1. Consider the ideal second order analog filter

$$y(k+2) = 1.8y(k+1) - 0.95y(k) + u(k+1)$$

for $k \ge 0$ when $y(-1) = y(0) = 0$ and $u(k) = 0.125$.

Four different responses are illustrated in Figure 5.6(a); (1) the response y above, (2) the magnitude constrained response as defined by

$$y(k+2) = Q[1.8y(k+1) - 0.95y(k) + u(k+1)]$$

where $Q[\cdot]$ denotes saturation arithmetic; that is

$$Q[w] = \begin{cases} 1 & ; \ w \ge 1 \\ w & ; \ |w| \le 1 \\ -1 & ; \ w \le -1 \end{cases}$$

(3) the dynamically scaled response under the constraint $0.5 \le |z(k)| < 1$ when $z(k) \ne 0$, and (4) the dynamically scaled response under the constraint $0.5 \le |z(k)| < 1$ when $z(k) \ne 0$, and the overflow constraint $|y(k)| \le 1$.

Note that the dynamically scaled response (3) under the constraint on $z(k)$ and the response (1) are identical. Also, response (4) gives the correct signal for $|y(k)| \le 1$. However response (2) is never correct once saturation first occurs at time $k = 5$.

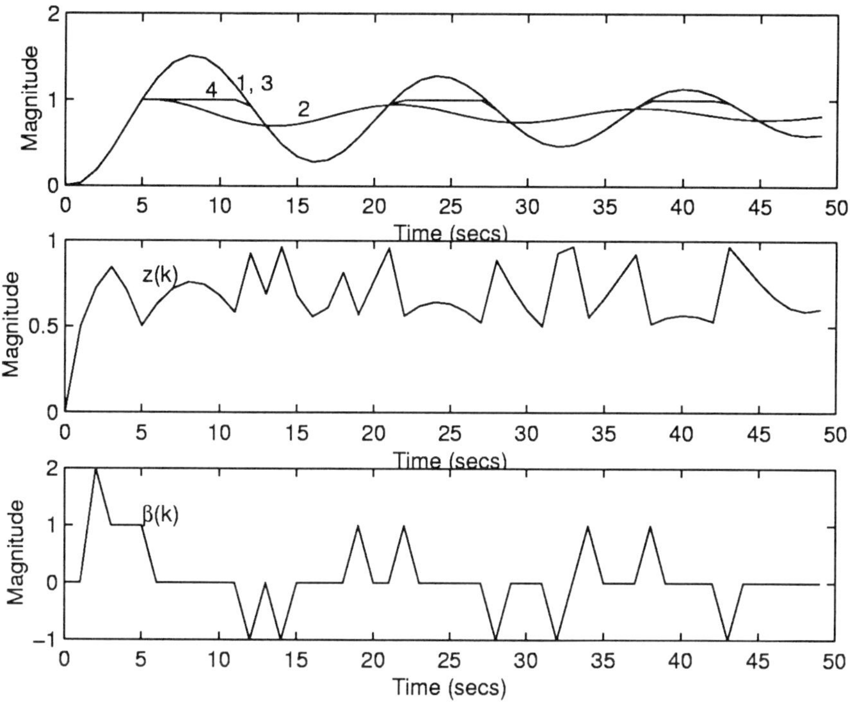

Fig. 5.6. Step responses of second order filter

The response in Figure 5.6(b) which illustrates the dynamically scaled signal z, and the response in Figure 5.6(c) which illustrates the dynamic scaling integer $\beta(k)$ both correspond to response 3 and response 4 in Figure 5.6(a)

Effect of Scaling

If we define $z(k) \stackrel{\Delta}{=} L_0^{-1}y(k)$, then (5.28) can be written in the form (5.59) which has the FWL implementation

$$\begin{aligned} \hat{z}(k+1) &= \hat{a}_1 Q[\hat{z}(k+1)] + \hat{a}_0 Q[\hat{z}(k)] + \hat{b}_{0L}\hat{u}(k) + \hat{b}_{1L}\hat{u}(k+1) \qquad (5.80) \\ Q[\hat{z}(k)] &= \hat{z}(k) - \eta_Q(k) \\ \hat{y}(k) &= L_0 \hat{z}(k) \end{aligned}$$

where we assume that there is no coefficient error in implementing the scaling factor L_0 which is possible if L_0 is an integer or a power of 2.

By using a similar argument to that used to analyse the effect of scaling on the output error for the FWL first order filter (5.65), it follows that

$$e_y(k+1) = \hat{a}_1 e_y(k+1) + \hat{a}_0 e_y(k) + w_{eL}(k) \tag{5.81}$$

where

$$\begin{aligned} w_{eL}(k) = {} & L_0(\hat{a}_1 - \alpha_1)\eta_Q(k+1) + L_0(\hat{a}_0 - \alpha)\eta_Q(k) + L_0\hat{b}_{1L}\nabla u(k+1) \\ & + L_0\hat{b}_{0L}\nabla u(k) + (\nabla a_1)y(k+1) + (\nabla a_0)y(k) \\ & + L_0(\nabla b_{0L})u(k+1) + L_0(\nabla b_{0L})u(k) \end{aligned}$$

$$\nabla b_{jL} \triangleq b_{jL} - \hat{b}_{jL} \;\; ; \;\; \nabla a_j \triangleq a_j - \hat{a}_j \;\; ; \;\; \nabla u \triangleq u - \hat{u} \tag{5.82}$$

As was the case for a first order filter, we again draw the conclusion that the main effect of scaling on the output error e_y is in respect of the component due to the *internal quantization error.* Specifically, *increasing* the scaling factor L_0 in order to prevent overflow leads to an *increase* in the output error e_y due to the effects of internal quantization error unless the fractional wordlength of the quantized signal $Q[\hat{z}(k)]$ is increased.

5.4 State Space FWL Filter

An ideal phase variable state space representation of the ideal second order digital filter (5.69) is given by

$$\begin{aligned} x_1(k+1) &= a_0 x_2(k) + b_0 u(k) \\ x_2(k+1) &= x_1(k) + a_1 x_2(k) + b_1 u(k) \\ y(k) &= x_2(k) \end{aligned} \tag{5.83}$$

One FWL phase variable state space realization is then given by

$$\begin{aligned} \hat{x}_1(k+1) &= \hat{a}_0 Q[\hat{x}_2(k)] + \hat{b}_0 \hat{u}(k) \\ \hat{x}_2(k+1) &= \hat{x}_1(k) + \hat{a}_1 Q[\hat{x}_2(k)] + \hat{b}_1 \hat{u}(k) \\ \hat{y}(k) &= \hat{x}_2(k) \end{aligned} \tag{5.84}$$

Then we have from (5.84) that

$$\begin{aligned} \hat{y}(k+2) &= \hat{x}_1(k+1) + a_1 Q[\hat{x}_2(k+1)] + \hat{b}_1 u(k+1) \\ &= \hat{a}_0 Q[\hat{x}_2(k)] + \hat{b}_0 u(k) + \hat{a}_1 Q[\hat{y}(k+1)] + \hat{b}_1 u(k+1) \end{aligned}$$

That is

$$\begin{aligned} \hat{y}(k+2) &= \hat{a}_0 Q[\hat{y}(k)] + \hat{a}_1 Q[\hat{y}(k+1)] + \hat{b}_0 \hat{u}(k) + \hat{b}_1 \hat{u}(k+1) \\ Q[\hat{y}(k)] &= \hat{y}(k) - \varepsilon_Q(k) \end{aligned}$$

which is the same as (5.68).

A second possible FWL phase variable state space realization of (5.83) is given by

$$\begin{aligned}\hat{x}_1(k+1) &= \hat{a}_0 Q[\hat{x}_2(k)] + \hat{b}_0 \hat{u}(k) \\ \hat{x}_2(k+1) &= \hat{x}_1(k) + \hat{a}_1 Q[\hat{x}_2(k)] + \hat{b}_1 \hat{u}(k) \\ \hat{z}(k) &= Q[\hat{x}_2(k)]\end{aligned} \tag{5.85}$$

The only difference between (5.84) and (5.85) is in the form of the output. Now we have

$$\begin{aligned}\hat{z}(k+2) &= Q[\ \hat{x}_1(k+1) + \hat{a}_1 Q[\hat{x}_2(k+1) + \hat{b}_1 \hat{u}(k+1)\ ] \\ &= Q[\ \hat{a}_0 Q[\hat{x}_2(k)] + \hat{b}_0 \hat{u}(k) + \hat{a}_1 Q[\hat{x}_2(k+1)] + \hat{b}_1 \hat{u}(k+1)\ ]\end{aligned}$$

That is

$$\hat{z}(k+2) = Q[\ \hat{a}_0 \hat{z}(k) + \hat{a}_1 \hat{z}(k+1) + \hat{b}_0 \hat{u}(k) + \hat{b}_1 \hat{u}(k+1)\ ]$$

which is the same as (5.70).

A third possible FWL phase variable state space realization of (5.83) is given by

$$\begin{aligned}\hat{x}_1(k+1) &= \hat{a}_0 Q[\hat{x}_2(k)] + \hat{b}_0 \hat{u}(k) \\ \hat{x}_2(k+1) &= Q[\hat{x}_1(k)] + \hat{a}_1 Q[\hat{x}_2(k)] + \hat{b}_1 \hat{u}(k) \\ \hat{q}(k) &= \hat{x}_2(k)\end{aligned} \tag{5.86}$$

Now the only difference between (5.84) and (5.86) is in the form of the first term in the expression for $\hat{x}_2(k+1)$. In this case, we have

$$\hat{q}(k+2) = \hat{a}_1 Q[\hat{q}(k+1)] + Q[\hat{a}_0 Q[\hat{q}(k)] + \hat{b}_0 \hat{u}(k)] + \hat{b}_1 \hat{u}(k+1) \tag{5.87}$$

which is the same as (5.71).

More generally, a second order state space representation of the ideal second order digital filter (5.69) is given by

$$\begin{aligned}x_1(k+1) &= f_{11} x_1(k) + f_{12} x_2(k) + g_1 u(k) \\ x_2(k+1) &= f_{21} x_1(k) + f_{22} x_2(k) + g_2 u(k) \\ y(k) &= h_1 x_1(k) + h_2 x_2(k)\end{aligned} \tag{5.88}$$

where from Theorem 2.5.1 in Chapter 2 with $d = 0$, we have

$$\begin{bmatrix} 1 & -a_1 \\ 0 & 1 \end{bmatrix} \begin{bmatrix} \boldsymbol{h}^T \boldsymbol{F} \boldsymbol{g} \\ \boldsymbol{h}^T \boldsymbol{g} \end{bmatrix} = \begin{bmatrix} b_0 \\ b_1 \end{bmatrix} \quad ; \quad det(\lambda \boldsymbol{I}_2 - \boldsymbol{F}) = \lambda^2 - a_1 \lambda - a_0$$

with

$$F = \begin{bmatrix} f_{11} & f_{12} \\ f_{21} & f_{22} \end{bmatrix} \quad ; \quad g = \begin{bmatrix} g_1 \\ g_2 \end{bmatrix} \quad ; \quad h = \begin{bmatrix} h_1 \\ h_2 \end{bmatrix}$$

A FWL realization of (5.88) is then given by

$$\begin{aligned} \hat{x}_1(k+1) &= \hat{f}_{11}Q[\hat{x}_1(k)] + \hat{f}_{12}Q\hat{x}_2(k)] + \hat{g}_1\hat{u}(k) \\ \hat{x}_2(k+1) &= \hat{f}_{21}Q\hat{x}_1(k)] + \hat{f}_{22}Q[\hat{x}_2(k)] + \hat{g}_2\hat{u}(k) \\ \hat{y}(k) &= \hat{h}_1Q[\hat{x}_1(k)] + \hat{h}_2Q[\hat{x}_2(k)] \end{aligned} \tag{5.89}$$

If each coefficient $\{\hat{f}_{ij}, \hat{g}_m, \hat{h}_p\}$ has a $[p_0, t_0]$ representation, and $\{Q[\hat{x}_1(k)], Q[\hat{x}_2(k)], \hat{u}(k)\}$ all have a $[p_1, t_1]$ representation, then all the products $\{\hat{g}_m\hat{u}(k), \hat{f}_{ij}Q[\hat{x}_p(k)]\}$ have a $[p_0 + p_1, t_0 + t_1]$ representation which implies $\{\hat{x}_1(k+1), \hat{x}_2(k+1), \hat{y}(k)\}$ have a $[p_0 + p_1 + 1, t_0 + t_1]$ representation. If the FWL output is also constrained to have a $[p_1, t_1]$ representation, then a further quantization as defined by

$$\hat{y}(k) = Q[\hat{h}_1Q[\hat{x}_1(k)] + \hat{h}_2Q[\hat{x}_2(k)]]$$

may be required.

Now each term $\hat{f}_{ij}Q[\hat{x}_p(k)]$ in (5.89) must have a $[p_0 + p_1, t_1 + t_0]$ representation, and so *normally* cannot be replaced by the term $\hat{f}_{ij}\hat{x}_p(k)$ which (normally) requires a $[p_0 + p_1, t_1 + 2t_0]$ representation. However such replacements *are* possible if any coefficient $\hat{f}_{ij}$ (or $\hat{h}_m$) is an *integer*. For example, if $\hat{f}_{21} = 1$, then $\hat{f}_{21}Q[\hat{x}_1(k)]$ has a $[p_1 + 1, t_1]$ representation, and $\hat{f}_{21}\hat{x}_1(k)$ has a $[p_1 + 1, t_1 + t_0]$ representation, and so both are acceptable.

In particular, in a phase variable realization (5.83) where

$$\{f_{11} = a_0, f_{12} = 0, g_1 = b_0;\ f_{21} = 1, f_{22} = a_1, g_2 = b_1;\ h_1 = 0, h_2 = 1\}$$

we can use the FWL realization (5.84) rather than (5.86), or instead of

$$\begin{aligned} \hat{x}_1(k+1) &= \hat{a}_0Q\hat{x}_2(k)] + \hat{b}_0\hat{u}(k) \\ \hat{x}_2(k+1) &= Q[\hat{x}_1(k)] + \hat{a}_1Q[\hat{x}_2(k)] + \hat{b}_1\hat{u}(k) \\ \hat{y}(k) &= Q[\hat{x}_2(k)] \end{aligned} \tag{5.90}$$

which would result from (5.89). As we have already indicated, the numerical characteristics of (5.84), (5.86) and (5.90) are *different*, and as one might expect, the implementation (5.84) will be shown to be generally more accurate. We leave further analysis of this problem until section 5.5.

High Order Realizations

For an n-vector $\boldsymbol{w}$ where $\boldsymbol{w}^T = [w_1\ w_2\ \ldots w_n\ ]$, let the quantized vector $Q[\boldsymbol{w}]$ be defined componentwise by

$$Q[\boldsymbol{w}] \triangleq [\ Q[w_1]\ \ Q[w_2]\ \ldots\ Q[w_n]\ ]^T$$

Then one possible FWL fixed point state space implementation of the ideal state space representation

$$\begin{aligned} \boldsymbol{x}(k+1) &= \boldsymbol{F}\boldsymbol{x}(k) + \boldsymbol{g}u(k) \\ y(k) &= \boldsymbol{h}^T\boldsymbol{x}(k) + du(k) \end{aligned} \tag{5.91}$$

of the nth order digital filter

$$y(k+n) = \sum_{p=0}^{n-1} a_p y(p) + \sum_{p=0}^{n} b_p u(p) \tag{5.92}$$

is described by

$$\begin{aligned} \hat{\boldsymbol{x}}(k+1) &= \hat{\boldsymbol{F}}Q[\hat{\boldsymbol{x}}(k)] + \hat{\boldsymbol{g}}\hat{u}(k) \\ Q[\hat{\boldsymbol{x}}(k)] &= \hat{\boldsymbol{x}}(k) - \boldsymbol{\varepsilon}_Q(k) \\ \hat{y}(k) &= \hat{\boldsymbol{h}}^T Q[\hat{\boldsymbol{x}}(k)] + \hat{d}\hat{u}(k) \end{aligned} \tag{5.93}$$

where $\boldsymbol{\varepsilon}_Q$ is the internal quantization error signal vector. Although, in principle, each component $Q[w_j]$ may be selected to be either truncation or rounding, in practice all components $Q[w_j]$ are implemented with the same form of quantization. Unless otherwise stated, we assume all such quantizations to be *roundoff quantization incorporating saturation arithmetic.*

Suppose all components $\{\hat{f}_{ij}\}$ of $\hat{\boldsymbol{F}}$, $\{\hat{g}_i\}$ of $\hat{\boldsymbol{g}}$, and $\{\hat{c}_k\}$ of $\hat{\boldsymbol{c}}$ as well as $\hat{d}$ have the FWL coefficient representation $[p_0, t_0]$ as in (5.30). In addition, suppose all the quantized state components $Q[\hat{x}_j(k)]$ of the n-vector $Q[\hat{\boldsymbol{x}}(k)]$ have the fixed point representation $[p_1, t_1]$, and the quantized input $\hat{u}(k)$ has the representation $[p_2, t_2]$ with $\{p_2 \le p_1\ ;\ t_2 \le t_1\}$. Then the output $\hat{y}(k)$ in (5.93) has the representation

$$\hat{y}(k) \approx [p_0 + p_1 + r_{add}, t_0 + t_1] \quad ; \quad r_{add} \triangleq \lceil \log_2(n) \rceil$$

where $\lceil h \rceil$ denotes the smallest integer which is greater than or equal to h.

Each update of $\hat{x}_j(k)$ to $\hat{x}_j(k+1)$ where $\hat{x}_j(k)$ is the jth component of $\hat{\boldsymbol{x}}(k)$ requires the evaluation of the *inner product*

$$\sum_{m=1}^{n} \hat{f}_{jm} Q[\hat{x}_m(k)] + \hat{g}_j \hat{u}(k)$$

Therefore, in order to be able to be guaranteed an *exact representation* for this inner product, we require $\hat{x}_j(k+1)$ to have the representation

$$\hat{x}_j(k+1) \approx [p_0 + p_1 + n_{add}, t_0 + t_1] \quad ; \quad n_{add} \triangleq \lceil \log_2(n+1) \rceil \geq r_{add}$$

A digital signal processor suitable for exact inner product evaluation of $\hat{x}_j(k+1)$ or $\hat{y}(k)$ requires *additional* accumulator precision of n_{add} bits. Many commercially available DSP chips already have extra precision. For example, the AT&T DSP16xx which supports 16×16 fixed point multiplication has an internal accumulator of 36 bits and so can compute exactly an inner product of $2^{36-32} = 16$ terms. Then if all terms are fractions, the most significant 4 bits of the accumulator represent the *integer part* of the sum. The Analog Devices ADSP-21xx also supports 16×16 fixed point multiplication, but has an internal accumulator of 40 bits, and so can add exactly 256 terms without any overflow.

Each product in the inner product of $\hat{x}_j(k+1)$ or $\hat{y}(k)$ can be evaluated using a $(p_0+t_0) \times (p_1+t_1)$ bit multiplier. When $p_1 \geq p_2$, the *most significant* bits of $\hat{u}(k)$ are increased with $p_1 - p_2$ *sign* bits since by the sign extension property of two's complement arithmetic, the numerical value $\hat{u}(k)$ is unaltered. For $t_1 \geq t_2$, the *fractional* representation of $\hat{u}(k)$ should be enlarged by adding $t_1 - t_2$ least significant fractional zeros. If *all* the most $p_0 + n_{add}$ bits of $\hat{x}_j(k+1)$ are *equal to the sign bit*, then overflow has *not* occurred, in which case, using the sign extension property of two's complement fixed point arithmetic

$$\hat{x}_j(k+1) \approx [p_1, t_0 + t_1]$$

For example, when

$$z^2 - a_1 z - a_0 = (z - \lambda_1)(z - \lambda_2)$$

with $\lambda_1 \neq \lambda_2$ both real, then a second order FWL state space implementation of (5.69) is given by (5.93) with

$$\hat{\boldsymbol{F}} = \begin{bmatrix} \hat{\lambda}_1 & 0 \\ 0 & \hat{\lambda}_2 \end{bmatrix} \quad ; \quad \hat{\boldsymbol{g}} = \begin{bmatrix} \hat{g}_1 \\ \hat{g}_2 \end{bmatrix} \quad ; \quad \hat{\boldsymbol{h}} = \begin{bmatrix} \hat{h}_1 \\ \hat{h}_2 \end{bmatrix} \quad ; \quad \hat{d} = 0$$

which on expansion corresponds to

$$\begin{aligned} \hat{x}_1(k+1) &= \hat{\lambda}_1 Q[\hat{x}_1(k)] + \hat{g}_1 \hat{u}(k) \quad ; \quad \hat{x}_2(k+1) = \hat{\lambda}_2 Q[\hat{x}_2(k)] + \hat{g}_2 \hat{u}(k) \\ \hat{y}(k) &= \hat{h}_1 Q[\hat{x}_1(k)] + \hat{h}_2 Q[\hat{x}_2(k)] \end{aligned}$$

However because of the nonlinear quantization functions $Q[\cdot]$, it is not possible in this case to derive a second order difference equation in $\hat{y}$ alone.

This situation is true more generally. That is, any FWL state space implementation (5.93) is *not* in general equivalent to a (nonlinear) nth order difference equation in the output signal $\hat{y}$. Furthermore, an nth order difference equation cannot in general be written in terms of (5.93). However by direct expansion, we can derive the following result.

Theorem 5.4.1. *The nth order FWL difference equation*

$$\hat{y}(k+n) = \sum_{p=0}^{n-1} \hat{a}_p Q[\hat{y}(p)] + \sum_{p=0}^{n} \hat{b}_p \hat{u}(p) \tag{5.94}$$

can be written as a FWL state space representation

$$\begin{aligned} \hat{\boldsymbol{x}}(k+1) &= \boldsymbol{J}_0 \hat{\boldsymbol{x}}(k) + \hat{\boldsymbol{M}} Q[\hat{\boldsymbol{x}}(k)] + \hat{\boldsymbol{g}} \hat{u}(k) \\ Q[\hat{\boldsymbol{x}}(k)] &= \hat{\boldsymbol{x}}(k) - \boldsymbol{\varepsilon}_Q(k) \\ \hat{y}(k) &= \hat{\boldsymbol{h}}_0^T \boldsymbol{x}(k) + \hat{d} \hat{u}(k) \end{aligned} \tag{5.95}$$

where

$$\hat{\boldsymbol{M}} = \begin{bmatrix} 0\ 0 \cdot \cdot 0 & \hat{a}_0 \\ 0\ 0 \cdot \cdot 0 & \hat{a}_1 \\ 0\ 0 \cdot \cdot 0 & \hat{a}_2 \\ \cdot\ \cdot\ \cdot\ \cdot\ \cdot & \cdot \\ \cdot\ \cdot\ \cdot\ \cdot\ \cdot & \cdot \\ 0\ 0 \cdot \cdot 0 & \hat{a}_{n-1} \end{bmatrix} \; ; \; \hat{\boldsymbol{g}} = \begin{bmatrix} \hat{b}_0 \\ \hat{b}_1 \\ \hat{b}_2 \\ \cdot \\ \cdot \\ \hat{b}_{n-1} \end{bmatrix} \; ; \; \hat{\boldsymbol{h}}_0 = \begin{bmatrix} 0 \\ 0 \\ 0 \\ \cdot \\ \cdot \\ 1 \end{bmatrix} \; ; \; \hat{d} = 0$$

and

$$\boldsymbol{J}_0 = \begin{bmatrix} 0\ 0 \cdot \cdot 0\ 0 \\ 1\ 0 \cdot \cdot 0\ 0 \\ 0\ 1\ 0 \cdot \cdot\ 0 \\ \cdot\ \cdot\ \cdot\ \cdot\ \cdot\ \cdot \\ \cdot\ \cdot\ \cdot\ \cdot\ \cdot\ \cdot \\ 0\ 0 \cdot \cdot 1\ 0 \end{bmatrix}$$

Since all components of $\boldsymbol{J}_0$ and $\boldsymbol{h}_0$ are either 0 or 1, the state space implementation (5.95) has no more hardware requirements than the input-output implementation (5.94).

5.4.1 Quantization Errors

We now analyse the effects of numerical errors on the *ideal* response (5.91) as a result of a FWL implementation which includes errors due to input quantization, coefficient quantization and internal quantization. The analysis of the first and second order equations in the previous two sections will be shown to be special cases of this more general approach.

Firstly, observe that (5.95) can be equivalently realized in the form

$$\begin{aligned} \hat{\boldsymbol{x}}(k+1) &= \hat{\boldsymbol{F}}Q[\hat{\boldsymbol{x}}(k)] + \boldsymbol{J}_0\boldsymbol{\varepsilon}_Q(k) + \hat{\boldsymbol{g}}\hat{u}(k) \\ Q[\hat{\boldsymbol{x}}(k)] &= \hat{\boldsymbol{x}}(k) - \boldsymbol{\varepsilon}_Q(k) \\ \hat{y}(k) &= \hat{\boldsymbol{h}}^T Q[\hat{\boldsymbol{x}}(k)] + \boldsymbol{h}_0^T\boldsymbol{\varepsilon}_Q(k) + \hat{d}\hat{u}(k) \end{aligned} \tag{5.96}$$

where $\{\hat{\boldsymbol{F}} = \hat{\boldsymbol{M}} + \boldsymbol{J}_0, \hat{\boldsymbol{h}} = \boldsymbol{h}_0\}$.

More generally, by extending the earlier arguments in section 5.3.1, it can be seen that (5.96) is a *wordlength consistent* FWL realization of (5.91) for *any* n-square matrix $\boldsymbol{J}_0$ and *any* n-vector $\boldsymbol{h}_0$ in which *all* components of $\{\boldsymbol{J}_0, \boldsymbol{h}_0\}$ are *integers*. The extra terms $\boldsymbol{J}_0\boldsymbol{\varepsilon}_Q(k)$ and $\boldsymbol{h}_0^T\boldsymbol{\varepsilon}_Q(k)$ in (5.96) are *quantization error feedback terms*. From the ideal state space representation (5.91) and the corresponding FWL representation (5.96), it then follows that the output error signal e_y and the state error signal $\boldsymbol{e}_x$ defined by

$$\boldsymbol{e}_x(k) \triangleq \boldsymbol{x}(k) - \hat{\boldsymbol{x}}(k) \quad ; \quad e_y(k) \triangleq y(k) - \hat{y}(k) \tag{5.97}$$

satisfies the equations

$$\begin{aligned} \boldsymbol{e}_x(k+1) &= \hat{\boldsymbol{F}}\boldsymbol{e}_x(k) + \hat{\boldsymbol{g}}\nabla u(k) + (\nabla\boldsymbol{F})\boldsymbol{x}(k) \\ &\quad +(\nabla\boldsymbol{g})u(k) + (\hat{\boldsymbol{F}} - \boldsymbol{J}_0)\boldsymbol{\varepsilon}_Q(k) \\ e_y(k) &= \hat{\boldsymbol{h}}^T\boldsymbol{e}_x(k) + (\nabla\boldsymbol{h})^T\boldsymbol{x}(k) \\ &\quad +\hat{d}\nabla u(k) + (\nabla d)u(k) + (\hat{\boldsymbol{h}} - \boldsymbol{h}_0)^T\boldsymbol{\varepsilon}_Q(k) \end{aligned} \tag{5.98}$$

where the input error signal $\nabla u \triangleq u - \hat{u}$, and the coefficient errors $\{\nabla\boldsymbol{F}, \nabla\boldsymbol{g}, \nabla\boldsymbol{h}, \nabla d\}$ are defined by

$$\nabla\boldsymbol{F} \triangleq \boldsymbol{F} - \hat{\boldsymbol{F}} \quad ; \quad \nabla\boldsymbol{g} \triangleq \boldsymbol{g} - \hat{\boldsymbol{g}} \quad ; \quad \nabla\boldsymbol{h} \triangleq \boldsymbol{h} - \hat{\boldsymbol{h}} \quad ; \quad \nabla d \triangleq d - \hat{d} \tag{5.99}$$

The equation for the *state error* $\boldsymbol{e}_x$ has *three forcing terms*: A signal $\hat{\boldsymbol{g}}\nabla u$ as a result of *input* quantization, a signal $(\nabla\boldsymbol{F})\boldsymbol{x} + (\nabla\boldsymbol{g})u$ as a result of *coefficient* errors in the system matrix $\boldsymbol{F}$ and the input matrix $\boldsymbol{g}$, and a signal $(\hat{\boldsymbol{F}} - \boldsymbol{J}_0)\boldsymbol{\varepsilon}_Q(k)$ as a result of *internal quantization errors.*

The *output error* signal e_y also has three terms: A signal $\hat{d}\nabla u$ as a result of input quantization, a signal $(\nabla\boldsymbol{h})^T\boldsymbol{x} + (\nabla d)u$ as a result of *coefficient* errors in the output matrix $\boldsymbol{h}$ and the output gain d, and and a signal $(\boldsymbol{h} - \boldsymbol{h}_0)^T\boldsymbol{\varepsilon}_Q(k)$ as a result of *internal quantization errors.*

Input Quantization Errors

From (5.98) and (5.91), the effect on the output error e_y due to the input quantization in the absence of both coefficient and internal quantization errors is described by the equations

$$\begin{bmatrix} e_x(k+1) \\ x(k+1) \end{bmatrix} = \begin{bmatrix} F & 0_{nn} \\ 0_{nn} & F \end{bmatrix} \begin{bmatrix} e_x(k) \\ x(k) \end{bmatrix} + \begin{bmatrix} g \\ 0_n \end{bmatrix} \nabla u(k)$$

$$e_y(k) = [h^T \; 0_n^T] \begin{bmatrix} e_x(k) \\ x(k) \end{bmatrix} + \hat{d} \nabla u(k) \tag{5.100}$$

where $\mathbf{0_{nn}}$ is an n-square matrix of zeros, and $\mathbf{0_n}$ is an n-vector of zeros. Hence the *error frequency function* $E_{y\nabla u}$ from the input error ∇u to the output error e_y is therefore given by

$$\begin{aligned} E_{y\nabla u}(z) &= d + [h^T \; 0_n^T] \begin{bmatrix} zI_n - F & 0_{nn} \\ 0_{nn} & zI_n - F \end{bmatrix}^{-1} \begin{bmatrix} g \\ 0_n \end{bmatrix} \\ &= d + h^T (zI_n - F)^{-1} g \end{aligned} \tag{5.101}$$

which is the same as the frequency response function $H(z)$ of the digital filter which is to be realized. Hence the corresponding unit impulse response $h_{y\nabla u}$ between the input error signal ∇u and the output error signal e_y is given by

$$h_{y\nabla u}(k) = \begin{cases} d & ; \; k = 0 \\ h^T F^{k-1} g \; ; & k \geq 1 \end{cases} \tag{5.102}$$

We then have the following result whose proof follows in a similar way to Theorem 5.4.3.

Theorem 5.4.2. *Assume all eigenvalues λ_k of F have modulus less than unity. Then the 2-norm $||h_{y\nabla u}||_2$ of the unit impulse response function $h_{y\nabla u}$ between the input error signal ∇u and the FWL output signal e_y as a result of only input quantization is given by*

$$||h_{y\nabla u}||_2^2 = d^2 + h^T K h \quad ; \quad K = FKF^T + gg$$

Coefficient Quantization Errors

From (5.98) and (5.91), the effect on the output error e_y due to coefficient quantization in the absence of both input and internal quantization errors is described by the equations

$$\begin{bmatrix} e_x(k+1) \\ x(k+1) \end{bmatrix} = \begin{bmatrix} \hat{F} & \nabla F \\ 0_{nn} & F \end{bmatrix} \begin{bmatrix} e_x(k) \\ x(k) \end{bmatrix} + \begin{bmatrix} \nabla g \\ g \end{bmatrix} u(k)$$

$$e_y(k) = [\hat{h}^T \; (\nabla h)^T] \begin{bmatrix} e_x(k) \\ x(k) \end{bmatrix} + (\nabla d) u(k) \tag{5.103}$$

The *error frequency function* E_{yu} from the input u to the output error e_y as a result of only coefficient quantization is therefore given by

$$E_{yu}(z) = \nabla d + [\hat{\boldsymbol{h}}^T \quad (\nabla \boldsymbol{h})^T] \begin{bmatrix} z\boldsymbol{I}_{\mathbf{n}} - \hat{\boldsymbol{F}} & -\nabla \boldsymbol{F} \\ \mathbf{0}_{\mathbf{nn}} & z\boldsymbol{I}_{\mathbf{n}} - \boldsymbol{F} \end{bmatrix}^{-1} \begin{bmatrix} \nabla \boldsymbol{g} \\ \boldsymbol{g} \end{bmatrix} \tag{5.104}$$

Now given three n-square matrices $\{\boldsymbol{A}, \boldsymbol{B}, \boldsymbol{D}\}$, the partitioned matrix

$$\begin{bmatrix} \boldsymbol{A} & \boldsymbol{B} \\ \mathbf{0}_{\mathbf{nn}} & \boldsymbol{D} \end{bmatrix}$$

is a $2n$-square matrix. Furthermore, if both $\boldsymbol{A}$ and $\boldsymbol{D}$ are invertible, then

$$\begin{bmatrix} \boldsymbol{A} & \boldsymbol{B} \\ \mathbf{0}_{\mathbf{nn}} & \boldsymbol{D} \end{bmatrix} \begin{bmatrix} \boldsymbol{A}^{-1} & -\boldsymbol{A}^{-1}\boldsymbol{B}\boldsymbol{D}^{-1} \\ \mathbf{0}_{\mathbf{nn}} & \boldsymbol{D}^{-1} \end{bmatrix} = \begin{bmatrix} \boldsymbol{I}_{\mathbf{n}} & -\boldsymbol{B}\boldsymbol{D}^{-1} + \boldsymbol{B}\boldsymbol{D} \\ \mathbf{0}_{\mathbf{nn}} & \boldsymbol{I}_{\mathbf{n}} \end{bmatrix}$$

where $\boldsymbol{I}_{\mathbf{n}}$ is the n-square identity matrix. Hence

$$\begin{bmatrix} \boldsymbol{A} & \boldsymbol{B} \\ \mathbf{0}_{\mathbf{nn}} & \boldsymbol{D} \end{bmatrix}^{-1} = \begin{bmatrix} \boldsymbol{A}^{-1} & -\boldsymbol{A}^{-1}\boldsymbol{B}\boldsymbol{D}^{-1} \\ \mathbf{0}_{\mathbf{nn}} & \boldsymbol{D}^{-1} \end{bmatrix} \tag{5.105}$$

This algebraic result when used in (5.104) gives

$$\begin{aligned} E_{yu}(z) = \nabla d &+ \hat{\boldsymbol{h}}^T (z\boldsymbol{I}_{\mathbf{n}} - \hat{\boldsymbol{F}})^{-1} \nabla \boldsymbol{g} + (\nabla \boldsymbol{h})^T (z\boldsymbol{I}_{\mathbf{n}} - \boldsymbol{F})^{-1} \boldsymbol{g} \\ &+ \boldsymbol{h}^T (z\boldsymbol{I}_{\mathbf{n}} - \hat{\boldsymbol{F}})^{-1} \nabla \boldsymbol{F} (z\boldsymbol{I}_{\mathbf{n}} - \boldsymbol{F})^{-1} \boldsymbol{g} \end{aligned} \tag{5.106}$$

The *unit impulse response* h_{yu} between the input u and the output error e_y as a result of only the coefficient quantization error is also given from (5.103) by

$$h_{yu}(k) = \begin{cases} \nabla d & ; \ k = 0 \\ \boldsymbol{h}_{yu}^T \boldsymbol{F}_{yu}^{k-1} \boldsymbol{g}_{yu} & ; \ k \geq 1 \end{cases} \tag{5.107}$$

where

$$\boldsymbol{F}_{yu} \triangleq \begin{bmatrix} \hat{\boldsymbol{F}} & \nabla \boldsymbol{F} \\ \mathbf{0}_{\mathbf{nn}} & \boldsymbol{F} \end{bmatrix} \ ; \ \boldsymbol{g}_{yu} \triangleq \begin{bmatrix} \nabla \boldsymbol{g} \\ \boldsymbol{g} \end{bmatrix} \ ; \ \boldsymbol{h}_{yu} \triangleq \begin{bmatrix} \boldsymbol{h} \\ \nabla \boldsymbol{h} \end{bmatrix} \tag{5.108}$$

We then have the following result.

Theorem 5.4.3. *Assume all eigenvalues $\hat{\lambda}_k$ of $\hat{\boldsymbol{F}}$ and λ_k of $\boldsymbol{F}$ have modulus less than unity. Then the 2-norm $||h_{yu}||_2$ of the unit impulse response function h_{yu} between the input signal u and the FWL output signal e_y as a result of only coefficient quantization is given by*

$$||h_{yu}||_2^2 = (\nabla d)^2 + \boldsymbol{h}_{yu}^T \boldsymbol{K}_{yu} \boldsymbol{h}_{yu} \ ; \ \boldsymbol{K}_{yu} = \boldsymbol{F}_{yu} \boldsymbol{K}_{yu} \boldsymbol{F}_{yu}^T + \boldsymbol{g}_{yu} \boldsymbol{g}_{yu}$$

where $\{\boldsymbol{F}_{yu}, \boldsymbol{g}_{yu}, \boldsymbol{h}_{yu}\}$ given by (5.108).

In order to establish this result, observe from (5.107) that

$$\begin{aligned} ||h_{yu}||_2^2 &= (\nabla d)^2 + \sum_{k=1}^{\infty} \boldsymbol{h}_{yu}^T \boldsymbol{F}_{yu}^{k-1} \boldsymbol{g}_{yu} \cdot \boldsymbol{g}_{yu}^T (\boldsymbol{F}_{yu}^T)^{k-1} \boldsymbol{h}_{yu} \\ &= (\nabla d)^2 + \boldsymbol{h}_{yu}^T \boldsymbol{K}_{yu} \boldsymbol{h}_{yu} \end{aligned}$$

where

$$\boldsymbol{K}_{yu} \triangleq \sum_{k=1}^{\infty} \boldsymbol{F}_{yu}^{k-1} \boldsymbol{g}_{yu} \boldsymbol{g}_{yu}^T (\boldsymbol{F}_{yu}^T)^{k-1}$$

The result then follows from Theorem 4.1.11.

Internal Quantization Errors

From (5.98), the state output error $\boldsymbol{e}_x$ and the output error e_y as a result of *internal quantization error feedback* but in the absence of both input and coefficient quantization are given in terms of the internal quantization errors $\boldsymbol{\varepsilon}_Q$ by

$$\begin{aligned} \boldsymbol{e}_x(k+1) &= \boldsymbol{F}\boldsymbol{e}_x(k) + (\boldsymbol{F} - \boldsymbol{J}_0)\boldsymbol{\varepsilon}_Q(k) \\ e_y(k) &= \boldsymbol{h}^T \boldsymbol{e}_x(k) + (\boldsymbol{h} - \boldsymbol{h}_0)^T \boldsymbol{\varepsilon}_Q(k) \end{aligned} \tag{5.109}$$

Now the internal quantization error $\boldsymbol{\varepsilon}_Q$ in (5.96) is in general an n-dimensional vector with components ε_{Qj} given by

$$\varepsilon_{Qj}(k) = \hat{x}_j(k) - Q[\hat{x}_j(k)] \ ; \ 1 \le j \le n \tag{5.110}$$

where $\hat{x}_j$ is the jth component of the filter state $\hat{\boldsymbol{x}}$. That is, $\varepsilon_{Qj}(k) = \boldsymbol{c}_j^T \boldsymbol{\varepsilon}_Q(k)$ where $\boldsymbol{c}_j$ is the n-dimensional vector whose components are all zero except the jth component which is unity; that is

$$\boldsymbol{c}_j^T = [0 \ \ 0 \ \ 0 \ \ \ldots \ \ 0 \ \ 1 \ \ 0 \ \ \ldots \ \ 0] \tag{5.111}$$

The error frequency response function $E_{r\varepsilon_j}$ between the jth internal quantization error component ε_{Qj} and the output error e_y is then given by

$$E_{y\varepsilon_j}(z) = (\boldsymbol{h} - \boldsymbol{h}_0)^T \boldsymbol{c}_j + \boldsymbol{h}^T (z\boldsymbol{I}_\mathbf{n} - \boldsymbol{F})^{-1} (\boldsymbol{F} - \boldsymbol{J}_0) \boldsymbol{c}_j \tag{5.112}$$

The corresponding *unit impulse response* $h_{r\varepsilon_j}$ between the jth internal quantization error component ε_{Qj} and the output error e_y as a result of only the internal quantization error is given from (5.112) by

$$h_{y\varepsilon_j}(k) = \begin{cases} (\boldsymbol{h} - \boldsymbol{h}_0)^T \boldsymbol{c}_j & ; \ k = 0 \\ \boldsymbol{h}^T \boldsymbol{F}^{k-1} (\boldsymbol{F} - \boldsymbol{J}_0) \boldsymbol{c}_j & ; \ k \ge 1 \end{cases} \tag{5.113}$$

We then have the following result.

Theorem 5.4.4. *Assume all eigenvalues λ_k of $\boldsymbol{F}$ have modulus less than unity. Then the 2-norm $||h_{y\varepsilon_j}||_2$ of the unit impulse response function $h_{y\varepsilon_j}$ between the jth internal quantization error signal component ε_{Qj} defined in (5.110) and the FWL output signal e_y as a result of only internal quantization is given by*

$$||h_{y\varepsilon_j}||_2^2 = \boldsymbol{c}_j^T \boldsymbol{Z} \boldsymbol{c}_j \tag{5.114}$$

where

$$\begin{aligned}\boldsymbol{Z} &= (\boldsymbol{h} - \boldsymbol{h}_0)(\boldsymbol{h} - \boldsymbol{h}_0)^T + (\boldsymbol{F} - \boldsymbol{J}_0)^T \boldsymbol{W} (\boldsymbol{F} - \boldsymbol{J}_0)\\ \boldsymbol{W} &= \boldsymbol{F}^T \boldsymbol{W} \boldsymbol{F} + \boldsymbol{h}\boldsymbol{h}^T\end{aligned}$$

In order to establish this result, observe from (5.113) that

$$\begin{aligned}||h_{y\varepsilon_j}||_2^2 &= \boldsymbol{c}_j^T (\boldsymbol{h} - \boldsymbol{h}_0)(\boldsymbol{h} - \boldsymbol{h}_0)^T \boldsymbol{c}_j \\ &\quad + \boldsymbol{c}_j^T (\boldsymbol{F} - \boldsymbol{J}_0)^T \sum_{k=1}^{\infty} (\boldsymbol{F}^T)^{k-1} \boldsymbol{h} \,.\, \boldsymbol{h}^T \boldsymbol{F}^{k-1} (\boldsymbol{F} - \boldsymbol{J}_0) \boldsymbol{c}_j \\ &= \boldsymbol{c}_j^T \left[(\boldsymbol{h} - \boldsymbol{h}_0)(\boldsymbol{h} - \boldsymbol{h}_0)^T + (\boldsymbol{F} - \boldsymbol{J}_0)^T \boldsymbol{W} (\boldsymbol{F} - \boldsymbol{J}_0) \right] \boldsymbol{c}_j\end{aligned}$$

where

$$\boldsymbol{W} = \sum_{k=1}^{\infty} (\boldsymbol{F}^T)^{k-1} \boldsymbol{h}\boldsymbol{h}^T (\boldsymbol{F}^{k-1}$$

The result then follows from Theorem 4.1.11.

If the aim is to make $||h_{y\varepsilon_j}||$ for $1 \le j \le n$ "small", then a "good" (but not necessarily minimizing) choice is to select the *integer matrices* $\{\boldsymbol{J}_0, \boldsymbol{h}_0\}$ such that:

$$\boldsymbol{J}_0 = \text{round}(\boldsymbol{F}) \quad ; \quad \boldsymbol{h}_0 = \text{round}(\boldsymbol{h}) \tag{5.115}$$

where for a scalar p, round(p) denotes the *nearest integer to* p, and for $\boldsymbol{P} = [p_{ij}]$ a matrix, then round$(\boldsymbol{P}) = [\text{round}(p_{ij})]$.

Example 5.4.1. Consider the third order digital filter with frequency response function $H(z)$ given by

$$\begin{aligned}H(z) &= \frac{(z-1)^3}{(z-0.8)[(z-0.9)^2 + (0.4)^2]} \\ &= \frac{z^3 - 3z^2 + 3z - 1}{z^3 - 2.6z^2 + 2.41z - 0.776} = 1 + \frac{-0.4z^2 + 0.59z - 0.224}{z^3 - 2.6z^2 + 2.41z - 0.776}\end{aligned}$$

Since the roots $\{\lambda_k\}$ of the characteristic equation

$$z^3 - 2.6z^2 + 2.41z - 0.776 = 0$$

are given by $\{\lambda_1 = 0.8, \lambda_2 = 0.9 + j0.4 = \overline{\lambda}_3\}$, this filter is asymptotically stable.

One possible state space realization $\{\boldsymbol{F}_1, \boldsymbol{g}_1, \boldsymbol{h}_1, d\}$ is given by

$$\boldsymbol{F}_1 = \begin{bmatrix} 0 & 0 & 0.776 \\ 1 & 0 & -2.41 \\ 0 & 1 & 2.6 \end{bmatrix} \quad ; \quad \boldsymbol{g}_1 = \begin{bmatrix} -0.224 \\ 0.59 \\ -0.4 \end{bmatrix} \quad ; \quad \boldsymbol{h}_1 = \begin{bmatrix} 0 \\ 0 \\ 1 \end{bmatrix} \quad ; \quad d = 1$$

and in this case, a good choice for $\{\boldsymbol{J}_0, \boldsymbol{h}_0\}$ is

$$\boldsymbol{J}_0 = \begin{bmatrix} 0 & 0 & 1 \\ 1 & 0 & \alpha_{23} \\ 0 & 1 & \alpha_{33} \end{bmatrix} \quad ; \quad \boldsymbol{h}_0 = \begin{bmatrix} 0 \\ 0 \\ 1 \end{bmatrix}$$

where $\alpha_{23} \in \{-2, -3\}$ and $\alpha_{33} \in \{2, 3\}$. (All possibilities should be checked to see which gives the best performance.)

If we now write

$$\frac{-0.4z^2 + 0.59z - 0.224}{z^3 - 2.6z^2 + 2.41z - 0.776} = \frac{\gamma_1}{z - 0.8} + \frac{\gamma_2 z + \gamma_3}{(z - 0.9)^2 + (0.4)^2}$$

where to 3 decimal places $\{\gamma_1 = -0.05, \gamma_2 = -0.35, \gamma_3 = 0.22\}$, then another state space representation $\{\boldsymbol{F}_2, \boldsymbol{g}_2, \boldsymbol{h}_2, d\}$ is given by

$$\boldsymbol{F}_2 = \begin{bmatrix} 0.8 & 0 & 0 \\ 0 & 0.9 & 0.4 \\ 0 & -0.4 & 0.9 \end{bmatrix} \quad ; \quad \boldsymbol{g}_2 = \begin{bmatrix} -0.05 \\ 0.238 \\ -0.35 \end{bmatrix} \quad ; \quad \boldsymbol{h}_2 = \begin{bmatrix} 1 \\ 0 \\ 1 \end{bmatrix} \quad ; \quad d = 1$$

and in this case, a good choice for $\{\boldsymbol{J}_0, \boldsymbol{h}_0\}$ is

$$\boldsymbol{J}_0 = \begin{bmatrix} 1 & 0 & 0 \\ 0 & 1 & 0 \\ 0 & 0 & 1 \end{bmatrix} \quad ; \quad \boldsymbol{h}_0 = \begin{bmatrix} 1 \\ 0 \\ 1 \end{bmatrix}$$

Error Bounds

An upper bound on the infinity norm $||e_y||_\infty$ of the output error can be expressed as the sum of three terms:

$$||e_y||_\infty \le L_{y\nabla u} + L_{yu} + L_{y\varepsilon} \tag{5.116}$$

where

$$L_{y\nabla u} \triangleq ||h_{y\nabla u}||_p \; ||\nabla u||_q \; ; \; L_{yu} \triangleq ||h_{yu}||_p \; ||u||_q \tag{5.117}$$
$$L_{y\varepsilon} \triangleq \sum_{j=1}^{n} ||h_{y\varepsilon_j}||_p \; ||\nabla \varepsilon_{Qj}||_q$$

and $\{p = 1, q = \infty\}$, $\{p = 2, q = 2\}$ or $\{p = \infty, q = 1\}$ where $\{h_{y\nabla u}, h_{yu}, h_{r\varepsilon_j}\}$ are the unit impulse responses of the various error systems as defined by (5.102), (5.107) and (5.113) respectively.

Once again, as indicated in (5.117), we have a choice in what norms we use to calculate any of the bounds. For example, when $n = 3$, one possible set of bounds is given by:

$$L_{y\nabla u} = ||h_{y\nabla u}||_\infty \; ||\nabla u||_1 \quad ; \quad L_{yu} = ||h_{yu}||_1 \; ||u||_\infty$$
$$L_{y\varepsilon} = ||h_{y\varepsilon_1}||_2 \; ||\nabla \varepsilon_{Q1}||_2 + ||h_{y\varepsilon_2}||_1 \; ||\nabla \varepsilon_{Q2}||_\infty + ||h_{y\varepsilon_3}||_\infty \; ||\nabla \varepsilon_{Q3}||_1$$

As we have already seen before, it is difficult to get exact expressions for both the 1-norm and infinity norm of an impulse response. However exact expressions for the 2-norm are available from Theorem 5.4.2, Theorem 5.4.3 and Theorem 5.4.4. The most useful bounds for $||e_y||_\infty$ in (5.116) are therefore given by $||e_y||_\infty \leq L$ where

$$L = ||h_{y\nabla u}||_1 ||\nabla u||_\infty + ||h_{yu}||_1 ||u||_\infty + (\sum_{j=1}^{n} ||h_{y\varepsilon_j}||_1) \; ||\varepsilon_Q||_\infty$$

or

$$L = ||h_{y\nabla u}||_2 ||\nabla u||_2 + ||h_{yu}||_2 ||u||_2 + (\sum_{j=1}^{n} ||h_{y\varepsilon_j}||_2) \; ||\varepsilon_Q||_2$$

If each state component $\hat{x}_j$ is rounded to a t_1 bit fractional representation, then in the absence of overflow, $||\varepsilon_Q||_\infty = 0.5 \times 2^{-t_1}$. Or, if a statistical model is used for the quantization error signal ε_Q, then from (5.51), $||\varepsilon_Q||_2 = 0.29 \times 2^{-t_1}$.

Example 5.4.2. Consider the ideal state space representation

$$x(k+1) = \begin{bmatrix} 0 & a_0 \\ 1 & a_1 \end{bmatrix} x(k) + \begin{bmatrix} b_0 \\ b_1 \end{bmatrix} u(k) \tag{5.118}$$
$$y(k) = [0 \quad 1] x(k)$$

of the ideal second order difference equation

$$y(k+2) = a_1 y(k+1) + a_0 y(k) + b_1 u(k+1) + b_0 u(k) \tag{5.119}$$

which has the frequency response function

$$H(z) = \frac{b_1 z + b_0}{z^2 - a_1 z - a_0}$$

A FWL state space realization which incorporates internal quantization error feedback is described by

$$\hat{\boldsymbol{x}}(k+1) = \begin{bmatrix} 0 & \hat{a}_0 \\ 1 & \hat{a}_1 \end{bmatrix} Q[\hat{\boldsymbol{x}}(k)] + \begin{bmatrix} \hat{b}_0 \\ \hat{b}_1 \end{bmatrix} \hat{u}(k) + \begin{bmatrix} 0 & \alpha_0 \\ \alpha_2 & \alpha_1 \end{bmatrix} \boldsymbol{\varepsilon}_Q(k)] \tag{5.120}$$

$$\hat{y}(k) = [0 \;\; 1] Q[\hat{\boldsymbol{x}}(k)] + \begin{bmatrix} h_1 & h_2 \end{bmatrix} \boldsymbol{\varepsilon}_Q(k)]$$

$$Q[\hat{\boldsymbol{x}}(k)] = \hat{\boldsymbol{x}}(k) - \boldsymbol{\varepsilon}_Q(k)$$

where $\{\alpha_2, \alpha_1, \alpha_0, h_1, h_2\}$ are integers.

Input Quantization Errors: In this case

$$E_{y\nabla u}(z) = \boldsymbol{h}^T (z\boldsymbol{I}_{\mathbf{n}} - \boldsymbol{F})^{-1} \boldsymbol{g} = H(z)$$

Coefficient Quantization Errors: In this case

$$\nabla \boldsymbol{F} = \begin{bmatrix} 0 & \nabla a_0 \\ 0 & \nabla a_1 \end{bmatrix} \;\; ; \;\; \nabla \boldsymbol{g} = \begin{bmatrix} \nabla b_0 \\ \nabla b_1 \end{bmatrix} \;\; ; \;\; \nabla \boldsymbol{h} = \begin{bmatrix} 0 \\ 0 \end{bmatrix} \;\; ; \;\; d = \nabla d = 0$$

and so from (5.106), we have that

$$\begin{aligned} E_{yu}(z) &= \hat{\boldsymbol{h}}^T (z\boldsymbol{I}_{\mathbf{n}} - \hat{\boldsymbol{F}})^{-1} \nabla \boldsymbol{g} + \boldsymbol{h}^T (z\boldsymbol{I}_{\mathbf{n}} - \hat{\boldsymbol{F}})^{-1} \nabla \boldsymbol{F} (z\boldsymbol{I}_{\mathbf{n}} - \boldsymbol{F})^{-1} \boldsymbol{g} \\ &= \frac{(\nabla b_1) z + \nabla b_0}{z^2 - \hat{a}_1 z - \hat{a}_0} + \frac{(b_1 z + b_0)[(\nabla a_1) z + \nabla a_0]}{(z^2 - \hat{a}_1 z - \hat{a}_0)(z^2 - a_1 z - a_0)} \end{aligned}$$

From Theorem 5.4.3

$$||h_{yu}||_2^2 = \boldsymbol{h}_{yu}^T \boldsymbol{K}_{yu} \boldsymbol{h}_{yu}$$

where $\boldsymbol{K}_{yu} = \boldsymbol{F}_{yu} \boldsymbol{K}_{yu} \boldsymbol{F}_{yu}^T + \boldsymbol{g}_{yu} \boldsymbol{g}_{yu}^T$, and

$$\boldsymbol{F}_{yu} = \begin{bmatrix} 0 & \hat{a}_0 & 0 & \nabla a_0 \\ 1 & \hat{a}_1 & 0 & \nabla a_1 \\ 0 & 0 & 0 & a_0 \\ 0 & 0 & 1 & a_1 \end{bmatrix} \;\; ; \;\; \boldsymbol{g}_{yu} = \begin{bmatrix} \nabla b_0 \\ \nabla b_1 \\ b_0 \\ b_1 \end{bmatrix} \;\; ; \;\; \boldsymbol{h}_{yu} = \begin{bmatrix} 0 \\ 1 \\ 0 \\ 0 \end{bmatrix}$$

Even though an analytical expression for $||h_{yu}||_2$ is not immediately available, numerical values may be readily obtained for specific values of the coefficients $\{a_i, b_j\}$ and their FWL approximations $\{\hat{a}_i, \hat{b}_j\}$.

Internal Quantization Errors: From (5.112)

$$E_{y\varepsilon_j}(z) = [-h_1 \;\; 1 - h_2] \boldsymbol{c}_j + [0 \;\; 1] \begin{bmatrix} z & -\hat{a}_0 \\ -1 & z - \hat{a}_1 \end{bmatrix}^{-1} \cdot \begin{bmatrix} 0 & \hat{a}_0 - \alpha_0 \\ 1 - \alpha_2 & \hat{a}_1 - \alpha_1 \end{bmatrix} \boldsymbol{c}_j$$

where

$$c_1^T = [1 \;\; 0] \;\; ; \;\; c_2^T = [0 \;\; 1]$$

In particular, when $\{h_1 = 0, h_2 = 1, \alpha_2 = 1\}$, we have

$$E_{y\varepsilon_1}(z) = 0 \;\; ; \;\; E_{y\varepsilon_2}(z) = \frac{(a_0 - \alpha_0) + (a_1 - \alpha_1)z}{z^2 - a_1 z - a_0}$$

Also from Theorem 5.4.4

$$||h_{y\varepsilon_j}||_2^2 = c_j^T Z c_j$$

where

$$Z = \begin{bmatrix} 0 & a_0 - \alpha_0 \\ 1 - \alpha_2 & a_1 - \alpha_1 \end{bmatrix}^T W \begin{bmatrix} 0 & a_0 - \alpha_0 \\ 1 - \alpha_2 & a_1 - \alpha_1 \end{bmatrix}$$
$$W = F^T W F + h h^T$$

It follows by inspection that a good choice is $\{\alpha_2 = 1, \alpha_0 = \text{round}(\hat{a}_0), \alpha_1 = \text{round}(\hat{a}_1)\}$. Then we have

$$W = \begin{bmatrix} w_{11} & w_{12} \\ w_{12} & w_{22} \end{bmatrix} \;\; ; \;\; Z = \begin{bmatrix} 0 & 0 \\ 0 & \delta_0 \end{bmatrix}$$

and so

$$h_{y\varepsilon_1} = 0 \;\; ; \;\; ||h_{y\varepsilon_2}||_2 = \sqrt{\delta_0}$$

where

$$w_0 \stackrel{\Delta}{=} w_{11} = w_{22} \;\; ; \;\; w_{12} = \left(\frac{a_1}{1 - a_0}\right) w_0$$
$$w_0 = \left(\frac{2a_0 a_1}{1 - a_0^2 - a_1^2}\right) w_{12} + \frac{1}{1 - a_0^2 - a_1^2}$$
$$\delta_0 = \left[(a_0 - \alpha_0)^2 + (a_1 - \alpha_1)^2 + \frac{2a_1(a_0 - \alpha_0)(a_1 - \alpha_1)}{1 - a_0}\right] w_0$$

5.4.2 Scaling

Consider again the ideal state space representation (5.96) of the nth order difference equation (5.92). In section 4.3.2, we showed that bounds L_j of the infinity norm $||x_j||_\infty$ of the component x_j of the state x for zero initial condition are given by

$$L_j = ||h_j||_1 \; ||u||_\infty \;\; ; \;\; L_j = ||h_j||_2 \; ||u||_2 \;\; ; \;\; L_j = ||h_j||_\infty \; ||u||_1$$

where

$$h_j(k) = \begin{cases} d & ; \ k = 0 \\ c_j^T F^{k-1} g & ; \ k \geq 1 \end{cases}$$

in which c_j is an n-vector defined in (5.111) whose components are all zero except for the jth component which is 1.

Now define

$$z(k) \triangleq Sx(k) \quad ; \quad S = \text{diag}\{L_1^{-1}, L_2^{-1}, \ldots, L_n^{-1}\}, \ L_j \neq 0 \tag{5.121}$$

so that $||x_j||_\infty \leq L_j$ for $1 \leq j \leq n$ implies $||z_j||_\infty \leq 1$ for $1 \leq j \leq n$. If (5.121) is substituted into (5.96), we have that

$$\begin{aligned} z(k+1) &= F_s z(k) + g_s u(k) \ ; \ z(k) \in \mathcal{R}^n \\ y(k) &= h_s^T z(k) + du(k) \end{aligned} \tag{5.122}$$

where

$$F_s \triangleq SFS^{-1} \quad ; \quad g_s \triangleq Sg \quad ; \quad h_s^T \triangleq h^T S^{-1} \tag{5.123}$$

This transformed state space representation $\{F_s, g_s, h_s, d\}$ in fact exists for *any* nonsingular matrix S, and not only for diagonal matrices as described in (5.121). Using Theorem 4.1.11, we have the following result.

Theorem 5.4.5. *Assuming zero initial conditions, upper bounds on the magnitude $||x_j||_\infty$ of the jth state component x_j of the state space representation (5.91) are given by*

$$||x_j||_\infty \leq ||h_j||_2 \ ||u||_2$$

where provided all eigenvalues λ_k of F are such that $|\lambda_k| < 1$, then

$$||h_j||_2 = \sqrt{c_j^T K c_j}$$

where the matrix K is the solution of the linear algebraic equation

$$K = FKF^T + gg^T$$

Moreover, after the invertible transformation

$$z(k) = Sx(k)$$

the jth component $z_j(k)$ of $z(k)$ in the transformed state space representation (5.122) is bounded according to

$$||z_j||_\infty \leq ||\tilde{h}_j||_2 \ ||u||_2$$

where

$$||\tilde{h}_j||_2 = \sqrt{c_j^T K_s c_j} \quad ; \quad K_s = SKS^T$$

where c_j is defined by (5.111).

We make the following observations:

- In the absence of scaling (i.e. $\boldsymbol{S} = \boldsymbol{I}$), we have $||\tilde{h}_j||_2 = ||h_j||_2$.
- The scaling matrix $\boldsymbol{S}$ can either *increase* or *decrease* the norm of the unit impulse response function between the filter input and an internal state component. For example, if $\boldsymbol{S} = \text{diag}\{\alpha_1, \alpha_2, \ \cdots \ , \alpha_n\}$ for some scalars $\alpha_j \neq 0$, then

$$||\tilde{h}_j||_2 = |\alpha_j| \ ||h_j||_2$$

and so the norm is *increased* for $|\alpha_j| > 1$ and *reduced* for $|\alpha_j| < 1$.

Adaptive Scaling

Define the invertible matrices $\{\boldsymbol{S}(k), \boldsymbol{D}(k)\}$, and the new state $\boldsymbol{z}(k)$ by

$$\boldsymbol{z}(k) \triangleq \boldsymbol{S}(k)\boldsymbol{x}(k) \quad ; \quad \boldsymbol{S}(k+1) \triangleq \boldsymbol{D}(k+1)\boldsymbol{S}(k) \tag{5.124}$$

Then it follows that the ideal state space representation (5.96) can be equivalently written in the form

$$\begin{aligned} y(k) &= \boldsymbol{h}^T \boldsymbol{S}^{-1}(k)\boldsymbol{z}(k) \\ \boldsymbol{p}(k) &= \boldsymbol{S}(k)\boldsymbol{F}\boldsymbol{S}^{-1}(k)\boldsymbol{z}(k) + \boldsymbol{S}(k)\boldsymbol{g}u(k) \\ \boldsymbol{z}(k+1) &= \boldsymbol{D}(k+1)\boldsymbol{p}(k) \end{aligned} \tag{5.125}$$

In particular, suppose

$$\boldsymbol{D}(k) = \text{diag}\{2^{-\beta_1(k)}, 2^{-\beta_2(k)}, \ldots, 2^{-\beta_n(k)}\} \tag{5.126}$$

is a diagonal matrix, and

$$\boldsymbol{S}(k) = \text{diag}\{2^{-\gamma_1(k)}, 2^{-\gamma_2(k)}, \ldots, 2^{-\gamma_n(k)}\} \tag{5.127}$$

is also diagonal. Also, for each component $p_j(k)$ of $\boldsymbol{p}(k)$ in (5.125), define the integer $\beta_j(k+1)$ such that:

$$2^{\beta_j(k+1)-1} \leq |p_j(k)| < 2^{\beta_j(k+1)} \tag{5.128}$$

Then it follows from (5.125) that the jth component $z_j(k+1)$ of $\boldsymbol{z}(k+1)$ is bounded according to

$$0.5 \leq |z_j(k+1)| < 1 \tag{5.129}$$

Furthermore, from (5.125) - (5.128), the integers $\{\gamma_j(k), \beta_j(k)\}$ are related by

$$\gamma_j(k+1) = \gamma_j(k) + \beta_j(k+1) \tag{5.130}$$

and if $\{f_{ij}\}$ denotes the (i,j)th component of the n-square matrix $\boldsymbol{F}$, then from (5.127), the (i,j)th component of the n-square matrix $\boldsymbol{D}(k)\boldsymbol{F}\boldsymbol{D}^{-1}(k)$ is given by $\{2^{\gamma_j(k)-\gamma_i(k)}f_{ij}\}$. We then have the following algorithm.

Adaptive State Space Scaling Algorithm: An adaptive scaling algorithm that is equivalent to the ideal state space representation (5.96) is provided as follows: Given $\{x_j(0)\}$, calculate $\{z_j(0),\gamma_j(0)\}$ such that

$$0.5 \le |z_j(0)| < 1 \quad ; \quad z_j(0) = 2^{-\gamma_j(0)}x_j(0)$$

and then determine $\{p_j(0)\}$ from

$$p_j(0) = f_{jj}z_j(0) + \sum_{i=1,i\neq j}^{n} 2^{\gamma_j(0)-\gamma_i(0)}f_{ij}z_i(0) + 2^{-\gamma_j(0)}g_ju(0)$$

Then for $k \ge 0$ and $1 \le j \le n$, given $\{z_j(k),u(k),\gamma_j(k)\}$:

- Calculate the integer $\beta_j(k+1)$ such that

$$2^{\beta_j(k+1)-1} \le |p_j(k)| \le 2^{\beta_j(k+1)}$$

- Update $\{z_j(k),\gamma_j(k)\}$ to $\{z_j(k+1),\gamma_j(k+1)\}$ by

$$z_j(k+1) = 2^{-\beta_j(k+1)}p_j(k) \quad ; \quad \gamma_j(k+1) = \gamma_j(k) + \beta_j(k+1)$$

- Update

$$\begin{aligned} p_j(k+1) &= f_{jj}z_j(k+1) + \sum_{i=1,i\neq j}^{n} 2^{\gamma_j(k+1)-\gamma_i(k+1)}f_{ij}z_i(k+1) \\ &\quad +2^{-\gamma_j(k+1)}g_ju(k+1) \\ y(k+1) &= \sum_{i=1}^{n} 2^{\gamma_i(k)}h_iz_i(k) + du(k) \end{aligned}$$

Effect of Scaling

Consider a constant scaling transformation

$$\boldsymbol{z}(k) \triangleq \boldsymbol{S}\boldsymbol{x}(k) \tag{5.131}$$

where $\boldsymbol{S}$ is *any* nonsingular matrix. Then the ideal state space representation (5.96) is transformed to the ideal state space representation (5.122). The corresponding FWL state space implementation of (5.122) is then given by

$$
\begin{aligned}
\hat{\boldsymbol{z}}(k+1) &= \hat{\boldsymbol{F}}_s Q[\hat{\boldsymbol{z}}(k)] + \boldsymbol{J}_0 \boldsymbol{\eta}_Q(k) + \hat{\boldsymbol{g}}_s \hat{u}(k) \\
\hat{y}(k) &= \hat{\boldsymbol{h}}_s^T Q[\hat{\boldsymbol{z}}(k)] + \boldsymbol{h}_0^T \boldsymbol{\eta}_Q(k) + \hat{d}\hat{u}(k)
\end{aligned} \tag{5.132}
$$

where the system matrices $\{\hat{\boldsymbol{F}}_s, \hat{\boldsymbol{g}}_s, \hat{\boldsymbol{h}}_s\}$ are the FWL approximations of $\{\boldsymbol{F}_s, \boldsymbol{g}_s, \boldsymbol{h}_s\}$, and all components of $\{\boldsymbol{J}_0, \boldsymbol{h}_0\}$ are *integers*. The components η_{Qj} of the internal quantization error vector $\boldsymbol{\eta}_Q$ are also given by

$$
\eta_{Qj}(k) \triangleq \hat{z}_j(k) - Q[\hat{z}_j(k)] \quad ; \quad 1 \le j \le n \tag{5.133}
$$

If we now define

$$
\boldsymbol{e}_z(k) \triangleq \boldsymbol{z}(k) - \hat{\boldsymbol{z}}(k) \quad ; \quad e_y(k) \triangleq y(k) - \hat{y}(k)
$$

it follows that

$$
\begin{aligned}
\boldsymbol{e}_z(k+1) &= \hat{\boldsymbol{F}}_s \boldsymbol{e}_z(k) + \hat{\boldsymbol{g}}_s \nabla u(k) + (\nabla \boldsymbol{F}_s)\boldsymbol{z}(k) \\
&\quad + (\nabla \boldsymbol{g}_s) u(k) + (\hat{\boldsymbol{F}}_s - \boldsymbol{J}_0)\boldsymbol{\eta}_Q(k) \\
e_y(k) &= \hat{\boldsymbol{h}}_s^T \boldsymbol{e}_z(k) + (\nabla \boldsymbol{h}_s)^T \boldsymbol{z}(k) + \hat{d}\nabla u(k) \\
&\quad + (\nabla d) u(k) + (\hat{\boldsymbol{h}}_s - \boldsymbol{h}_0)^T \boldsymbol{\eta}_Q(k)
\end{aligned}
$$

where

$$
\begin{aligned}
&\nabla \boldsymbol{F}_s \triangleq \boldsymbol{F}_s - \hat{\boldsymbol{F}}_s \quad ; \quad \nabla \boldsymbol{g}_s \triangleq \boldsymbol{g}_s - \hat{\boldsymbol{g}}_s \quad ; \quad \nabla \boldsymbol{h}_s \triangleq \boldsymbol{h}_s - \hat{\boldsymbol{h}}_s \\
&\nabla d \triangleq d - \hat{d} \quad ; \quad \nabla u \triangleq u - \hat{u}
\end{aligned}
$$

Let us now examine the effect on the output error e_y as it relates to the *internal quantization error*. In particular, if we assume

$$
\nabla \boldsymbol{F}_s = \boldsymbol{0} \quad ; \quad \nabla \boldsymbol{g}_s = \boldsymbol{0} \quad ; \quad \nabla \boldsymbol{h}_s = \boldsymbol{0} \quad ; \quad \nabla d = 0 \quad ; \quad \nabla u = 0
$$

the output error signal e_y then satisfies the equation

$$
\begin{aligned}
\boldsymbol{e}_z(k+1) &= \boldsymbol{F}_s \boldsymbol{e}_z(k) + (\boldsymbol{F}_s - \boldsymbol{J}_0)\boldsymbol{\eta}_Q(k) \\
e_y(k) &= \boldsymbol{h}_s^T \boldsymbol{e}_z(k) + (\boldsymbol{h}_s - \boldsymbol{h}_0)^T \boldsymbol{\eta}_Q(k)
\end{aligned} \tag{5.134}
$$

Alternatively, if we define $\tilde{\boldsymbol{e}}_z(k) \triangleq \boldsymbol{S}^{-1}\boldsymbol{e}_z(k)$, and substitute for $\{\boldsymbol{F}_s, \boldsymbol{g}_s, \boldsymbol{h}_s\}$ using (5.123), we conclude that the output error e_y can also be expressed in the form

$$
\begin{aligned}
\tilde{\boldsymbol{e}}_z(k+1) &= \boldsymbol{F}\tilde{\boldsymbol{e}}_z(k) + (\boldsymbol{F} - \tilde{\boldsymbol{J}}_0)\boldsymbol{S}^{-1}\boldsymbol{\eta}_Q(k) \\
e_y(k) &= \boldsymbol{h}^T \tilde{\boldsymbol{e}}_z(k) + (\boldsymbol{h} - \tilde{\boldsymbol{h}}_0)^T \boldsymbol{S}^{-1}\boldsymbol{\eta}_Q(k)
\end{aligned}
$$

where

$$\tilde{J}_0 \triangleq S^{-1} J_0 S \ ; \ \tilde{h}_0^T \triangleq h_0^T S$$

We have the following result.

Theorem 5.4.6. *Suppose that there are no input or coefficient quantization errors, and assume all eigenvalues λ_k of F have modulus less than unity. Then the 2-norm $||\tilde{h}_{y\eta_j}||_2$ of the unit impulse response function $\tilde{h}_{y\eta_j}$ between the internal quantization error signal component η_{Qj} defined in (5.133), and the FWL output signal e_y as a result of internal quantization is given by*

$$||\tilde{h}_{y\eta_j}||_2^2 = c_j^T (S^{-1})^T \tilde{Z} (S)^{-1} c_j$$

where

$$\begin{aligned} \tilde{Z} &= (h - \tilde{h}_0)(h - \tilde{h}_0)^T + (F - \tilde{J}_0) W (F - \tilde{J}_0)^T \\ W &= F W F^T + h h^T \\ \tilde{h}_0^T &= h_0^T S \ , \ \tilde{J}_0 = S^{-1} J_0 S \end{aligned}$$

where all components of $\{J_0, h_0\}$ are integers.

We make the following observations:

- In the absence of scaling (i.e. $S = I$), we have $\varepsilon_Q = \eta_Q$, and so (5.134) reduces to (5.109).
- The matrices $\{J_0, h_0\}$ in (5.109) and $\{J_0, h_0\}$ in (5.132) have all integer components. However the components of $\{\tilde{J}_0, \tilde{h}_0\}$ in general have *non integral* components. In particular, it is not the case that $\tilde{J}_0 = J_0$ for all S unless $J_0 = \mathbf{0_{nn}}$ or $J_0 = I_n$.
- If we assume the state components $\hat{x}_j(k)$ in (5.96) and that transformed state components $\hat{z}_j(k)$ in (5.132) are all rounded to the *same* fractional representation $[0, t_1]$, then: for $1 \leq j \leq n$

  $$|\varepsilon_{Qj}(k)| \leq 2^{-(t_1+1)} \ ; \ |\eta_{Qj}(k)| \leq 2^{-(t_1+1)}$$

  Consequently, the scaling matrix S in (5.131) can either *increase* or *decrease* the effect of the internal quantization error on the output error e_y. For example, if $S = \text{diag}\{\alpha_1, \alpha_2, \ \ldots \ , \alpha_n\}$ for some scalars $\alpha_j \neq 0$, then

  $$||\tilde{h}_{y\eta_j}||_2 = |\alpha_j^{-1}| \ ||h_{y\varepsilon_j}||_2$$

  and so the norm is *reduced* for $|\alpha_j| > 1$ and *increased* for $|\alpha_j| < 1$.

If one compares this last observation with the last observation made following the statement of Theorem 5.4.5, we conclude that using a diagonal scaling transformation with a scaling factor $|\alpha_j| > 1$ so as to *reduce* the norm $||\tilde{h}_{y\eta_j}||_2$ of the unit impulse response from the internal quantisation error on the state component z_j to the output error e_y will *increase* the norm $||\tilde{h}_j||_2$ of the unit impulse response from the input signal u to the state component z_j. In other words, a reduction in the output error as a result of internal quantization will result in a greater likelihood of overflow of the filter states.

This outcome while true for *diagonal* scaling transformation $\boldsymbol{S}$ is *not* necessarily true for *all* scaling transformations. There are different state space realizations of a given filter system function $H(z)$ which have the *same* scaling properties, but *different* FWL output error properties. This issue is pursued in the next section.

5.5 Filter Structures

Any linear time invariant digital filter defined by a system function $H(z)$ where

$$H(z) = \frac{\sum_{p=0}^{n} b_p z^p}{z^n - \sum_{p=0}^{n-1} a_p z^p} = d + \boldsymbol{h}^T (z\boldsymbol{I} - \boldsymbol{F})^{-1} \boldsymbol{g} \tag{5.135}$$

has an infinite number of state space representations. In particular, if $\{\boldsymbol{F}, \boldsymbol{g}, \boldsymbol{h}, d\}$ is one representation, then so to is $\{\boldsymbol{F}_s, \boldsymbol{g}_s, \boldsymbol{h}_s, d\}$ where

$$\boldsymbol{F}_s = \boldsymbol{S}\boldsymbol{F}\boldsymbol{S}^{-1} \;\; ; \;\; \boldsymbol{g}_s = \boldsymbol{S}\boldsymbol{g} \;\; ; \;\; \boldsymbol{h}_s^T = \boldsymbol{h}^T \boldsymbol{S}^{-1} \tag{5.136}$$

for *any* (and not necessarily diagonal) *nonsingular matrix* $\boldsymbol{S}$.

5.5.1 Error Performance Measure

Under the assumption of infinite precision arithmetic, all state space representations are *equivalent*. However when the filter coefficients and states are restricted to having a finite wordlength representation, then *different* state space representations provide *different* approximations to the ideal (i.e. infinite precision) behaviour. The effect of *finite coefficient wordlength* can be measured in terms of the error frequency function E_{yu} (or equivalently, the unit impulse response function h_{yu}) between the output error e_y and the input signal u. In the absence of coefficient errors, the effect of *input quantization errors* as described by the error frequency function $E_{y\nabla u}$ is *independent* of the internal structure of the filter, and depends only on the filter system function $H(z)$.

The effect of *internal quantization errors* can be measured by the error frequency function $E_{y\varepsilon_j}$ (or equivalently, the unit impulse response function

$h_{y\varepsilon_j}$) between the output error e_y and the internal quantization error signal ε_{Qj} which results from quantization of the jth state component x_j. However, unlike the case for coefficient quantization, the output error e_y as a result of internal quantization errors can not be directly calculated from the input u since the quantization error signals $\{\varepsilon_{Qj}\}$ must first be calculated. As a consequence, when comparing the effects of different filter structures with respect to the effect of internal quantization errors $\{\varepsilon_{Qj}\}$, it is more convenient to measure an 'average' effect in terms of the signal norms $\{||h_{y\varepsilon_j}||_2\}$.

Furthermore, since the *scaling* of a filter states also effect the FWL performance, a more appropriate measure of the effect of internal quantization error at the jth state on the output error e_y is given by the *product* $||h_j||_2\ ||h_{y\varepsilon_j}||_2$ where h_j is the unit impulse response between the input u and the jth state component x_j. Finally, since a filter consisting of n states will normally have n quantization error signals $\{\varepsilon_{Qj}\}$, an overall *FWL performance measure* J_{FWL} with respect to internal quantization errors is provided by the *average product*

$$J_{FWL} \triangleq \sqrt{\frac{1}{n}\sum_{j=1}^{n} ||h_j||_2^2} \cdot \sqrt{\frac{1}{n}\sum_{j=1}^{n} ||h_{y\varepsilon_j}||_2^2} \tag{5.137}$$

Statistical Model

If the input signal u is "sufficiently exciting", then it has been found for purposes of evaluating the FWL performance that the internal quantization error signals $\{\varepsilon_{Qj}\}$ can be usefully modelled using a *statistical model* for the error signal ε_Q. In particular, assuming *roundoff quantization* in which each state component $\hat{x}_j$ is rounded to $Q[\hat{x}_j]$ with a t_1 bit fractional representation, a useful approximation for ε_{Qj} is that of a zero mean uniformly distributed signal with

$$||\varepsilon_{Qj}||_2 = 0.29 \times 2^{-t_1} \quad ; \quad 1 \le j \le n$$

In this case, the 2-norm of the output error e_y is bounded by

$$||e_y||_2 \le (\sum_{j=1}^{n} ||h_{y\varepsilon_j}||_2)\ ||\varepsilon_{Qj}||_2$$

where equality is achieved when all n signals $\{\varepsilon_{Qj}\}$ are *independent.*

Also, if the input signal u is a zero mean white noise signal of unity variance, then the resulting variance of the state component x_j is given by $||h_j||_2^2$. This means that the larger the value of $||h_j||_2$, the greater the probability the state x_j will *overflow.* Since there are n states, the probability that any one state will overflow depends on $\sum_{j=1}^{n} ||h_j||_2^2$. Therefore from a *statistical point of view*, the FWL performance measure J_{FWL} in (5.137) provides a measure of the variance of the output error e_y as a result of internal quantization errors weighted by the probability that one of the filter states will overflow.

Effect of Scaling

After scaling a given state space representation $\{\boldsymbol{F}, \boldsymbol{g}, \boldsymbol{h}, d\}$ by the scaling transformation $\boldsymbol{S}$, the resulting state space representation $\{\boldsymbol{F}_s, \boldsymbol{g}_s, \boldsymbol{h}_s, d\}$ defined by (5.136) has the FWL performance measure $\tilde{J}_{FWL}$ given by

$$\tilde{J}_{FWL} = \sqrt{\frac{1}{n}\sum_{j=1}^{n} ||\tilde{h}_j||_2^2} \cdot \sqrt{\frac{1}{n}\sum_{j=1}^{n} ||\tilde{h}_{y\eta_j}||_2^2} \tag{5.138}$$

where $\tilde{h}_{y\eta_j}$ and $\tilde{h}_j$ are given by Theorem 5.4.6 and Theorem 5.4.5 respectively. We therefore have the following result.

Theorem 5.5.1. *Suppose there are no coefficient or input quantization errors.*

(a) Then

$$||h_j||_2^2 = \boldsymbol{c}_j^T \boldsymbol{K} \boldsymbol{c}_j \quad ; \quad ||h_{y_{F_j}}||_2^2 = \boldsymbol{c}_j^T \boldsymbol{Z} \boldsymbol{c}_j$$

where $\boldsymbol{c}_j$ is the unit vector defined by (5.111), and the FWL performance measure J_{FWL} in (5.137) of the state space realization

$$\begin{aligned} \hat{\boldsymbol{x}}(k+1) &= \boldsymbol{F} Q[\hat{\boldsymbol{x}}(k)] + \boldsymbol{J}_0 \boldsymbol{\varepsilon}_Q(k) + \boldsymbol{g} u(k) \\ Q[\hat{\boldsymbol{x}}(k)] &= \hat{\boldsymbol{x}}(k) - \boldsymbol{\varepsilon}_Q(k) \\ \hat{y}(k) &= \boldsymbol{h}^T Q[\hat{\boldsymbol{x}}(k)] + \boldsymbol{h}_0^T \boldsymbol{\varepsilon}_Q(k) \end{aligned} \tag{5.139}$$

where all components of $\{\boldsymbol{J}_0, \boldsymbol{h}_0\}$ are integers is given by

$$J_{FWL} = \frac{1}{n}\sqrt{tr[\boldsymbol{K}].tr[\boldsymbol{Z}]} \tag{5.140}$$

where

$$\boldsymbol{K} = \boldsymbol{F}\boldsymbol{K}\boldsymbol{F}^T + \boldsymbol{g}\boldsymbol{g}^T \tag{5.141}$$

and

$$\begin{aligned} \boldsymbol{Z} &= (\boldsymbol{h} - \boldsymbol{h}_0)(\boldsymbol{h} - \boldsymbol{h}_0)^T + (\boldsymbol{F} - \boldsymbol{J}_0)^T \boldsymbol{W} (\boldsymbol{F} - \boldsymbol{J}_0) \\ \boldsymbol{W} &= \boldsymbol{F}^T \boldsymbol{W} \boldsymbol{F} + \boldsymbol{h}\boldsymbol{h}^T \end{aligned} \tag{5.142}$$

(b) After applying an invertible scaling transformation $\boldsymbol{S}$, then

$$||\tilde{h}_j||_2^2 = \boldsymbol{c}_j^T \boldsymbol{S}\boldsymbol{K}\boldsymbol{S}^T \boldsymbol{c}_j \quad ; \quad ||h_{y\eta_j}||_2^2 = \boldsymbol{c}_j^T (\boldsymbol{S}^{-1})^T \tilde{\boldsymbol{Z}} \boldsymbol{S}^{-1} \boldsymbol{c}_j$$

and the FWL performance measure $\tilde{J}_{FWL}$ in (5.138) of the transformed state space representation

$$\begin{aligned}\hat{\boldsymbol{z}}(k+1) &= \boldsymbol{F}_s Q[\hat{\boldsymbol{z}}(k)] + \boldsymbol{J}_1 \boldsymbol{\eta}_Q(k) + \boldsymbol{g}_s u(k) \\ Q[\hat{\boldsymbol{z}}(k)] &= \hat{\boldsymbol{z}}(k) - \boldsymbol{\eta}_Q(k) \\ \hat{y}(k) &= \boldsymbol{h}_s^T Q[\hat{\boldsymbol{z}}(k)] + \boldsymbol{h}_1^T \boldsymbol{\eta}_Q(k)\end{aligned} \tag{5.143}$$

where $\{\boldsymbol{F}_s, \boldsymbol{g}_s, \boldsymbol{h}_s, d\}$ are defined by (5.136) and all components of $\{\boldsymbol{J}_1, \boldsymbol{h}_1\}$ are integers is given by

$$\tilde{J}_{FWL} = \frac{1}{n}\sqrt{tr[\boldsymbol{S}\boldsymbol{K}\boldsymbol{S}^T].tr[(\boldsymbol{S}^{-1})^T \tilde{\boldsymbol{Z}} \boldsymbol{S}^{-1}]} \tag{5.144}$$

where

$$\begin{aligned}\tilde{\boldsymbol{Z}} &= (\boldsymbol{h} - \tilde{\boldsymbol{h}}_1)(\boldsymbol{h} - \tilde{\boldsymbol{h}}_1)^T + (\boldsymbol{F} - \tilde{\boldsymbol{J}}_1)^T \boldsymbol{W} (\boldsymbol{F} - \tilde{\boldsymbol{J}}_1) \\ \tilde{\boldsymbol{h}}_1^T &= \boldsymbol{h}_1^T \boldsymbol{S} \ , \ \tilde{\boldsymbol{J}}_1 = \boldsymbol{S}^{-1} \boldsymbol{J}_1 \boldsymbol{S}\end{aligned} \tag{5.145}$$

Note that each state space realization (as defined by a transformation matrix $\boldsymbol{S}$) will have its own best choice for the integer components of $\{\boldsymbol{J}_1, \boldsymbol{h}_1\}$. Also, the components of $\{\tilde{\boldsymbol{J}}_1, \tilde{\boldsymbol{h}}_1\}$ which depend on $\{\boldsymbol{J}_1, \boldsymbol{h}_1\}$ via $\boldsymbol{S}$ are generally *not* integer valued. That is, $\{\tilde{\boldsymbol{J}}_1, \tilde{\boldsymbol{h}}_1\}$ can only be *indirectly* selected in the equation for $\tilde{\boldsymbol{Z}}$.

5.5.2 Low Complexity Structures

An nth linear filter $H(z)$ in (5.135) is in general defined by $2n$ coefficients $\{a_p, b_p\}$ which means that the computation of each value of the output signal y generally requires at least $2n$ multiplications. A state space realization $\{\boldsymbol{F}, \boldsymbol{g}, \boldsymbol{h}, d\}$ of $H(z)$ in (5.135) is defined by an n-square system matrix $\boldsymbol{F}$, an n-vector $\boldsymbol{g}$, an n-vector $\boldsymbol{h}$ and a scalar d. Therefore if all components of $\{\boldsymbol{F}, \boldsymbol{g}, \boldsymbol{h}, d\}$ are nonzero, then possibly as many as $n^2 + 2n + 1$ multiplications are needed to calculate each value of y.

The *complexity* of a state space structure can be described in terms of the total number of multiplications required. A structure whose multiplication count is the order of n^2 is said to have *high complexity*, while a structure whose multiplication count is linear in n is said to have *low complexity*. The lower the complexity of a structure, the less time it takes to compute a new value of the output signal y. While this property is important, so too is it important for the structure to have *low coefficient sensitivity* and *low sensitivity to internal quantization errors*. (Recall that the sensitivity to input errors is determined by the filter frequency function itself.)

Evaluation of FWL Performance

The FWL performance measure J_{FWL} which was defined in (5.137), and shown later to be equivalent to (5.140), provides a relevant and consistent measure with which to compare the sensitivity of two state space structures in terms of *internal quantization errors* irrespective of the relative values $\{||h_j||_2\}$ of the 2-norms of the unit impulse functions $\{h_j\}$ between the filter input signal u and the state signal component x_j. However since the likelihood of overflow of x_j is proportional to $||h_j||_2$, it is more usual to implement structures in which these 2-norms are approximately equal, or at least have the same order of magnitude.

While the digital filter is in operation, an 'occasional' overflow will normally not be significant whereas 'frequent' occurrences will result in unacceptable levels of signal distortion. The likelihood of signal overflow therefore needs proper consideration. Furthermore, the use of a floating point processor does not eliminate all such concerns. The ADC, the DAC and all the associated conditioning circuits which are associated with the digital filter need to be scaled to minimise the effect of a signal overflow. Even if a floating point processor is to be implemented, better performance can be achieved by improving the internal scaling characteristics of the digital filter.

During the filter operation, internal quantization errors appear as a 'noise like' signal on the filter output. The statistical model interpretation is therefore very relevant. The noise can be demonstrated using (say) MATLAB simulations. On other occasions when the filter input signal is not 'sufficiently exciting' such as when the input is constant or 'slowly' varying, the effect of internal quantization manifests itself in the form of 'limit cycles' on the filter output. Limit cycles are periodic signals which occur in recursive digital filters, and arise as a result of the nonlinear nature of the quantization characteristic. This text has given no consideration to this phenomenon. As the following example demonstrates, limit cycles can occur even when there is no input signal.

Example 5.5.1. Consider the initial condition response of a second order digital filter implementation in finite wordlength as described by the equations

$$\hat{\boldsymbol{x}}(k+1) = \begin{bmatrix} 0 & 1 \\ -0.8 & 0.9 \end{bmatrix} Q[\hat{\boldsymbol{x}}(k)]$$
$$\hat{y}(k) = [0 \quad 1]Q[\hat{\boldsymbol{x}}(k)$$

where $Q[\cdot]$ rounds each component of $\hat{\boldsymbol{x}}(k)$ to a t_1 bit fractional representation. Then when $\{x_1(0) = 0, \hat{x}_2(0) = 4 \times 2^{-t_1}\}$, the output signal is given by

$$2^{t_1}\hat{y}(k) = \{4, 4, 0, -3, -3, 0, 2, 2, 0, -2, -2, 0, 2, 2, \ldots\}$$

which exhibits a limit cycle $\{2, 2, 0, 0, -2, -2\}$ of amplitude equal to four quantization levels. That is, for the above realization, the magnitude can be reduced by increasing the wordlength t_1, but the limit cycle itself can not be eliminated. Other limit cycles can be shown to result from other initial conditions.

An important observation is that very often such limit cycles can be eliminated by using internal quantization error feedback. This outcome can be demonstrated in the filter implementation

$$\begin{aligned} \hat{\boldsymbol{x}}(k+1) &= \begin{bmatrix} 0 & 1 \\ -0.8 & 0.9 \end{bmatrix} Q[\hat{\boldsymbol{x}}(k)] + \begin{bmatrix} 1 & 1 \\ -1 & 1 \end{bmatrix} \varepsilon_Q(k) \\ Q[\hat{\boldsymbol{x}}(k)] &= \hat{\boldsymbol{x}}(k) - \varepsilon_Q(k) \\ \hat{y}(k) &= [0 \quad 1] Q[\hat{\boldsymbol{x}}(k)] \end{aligned}$$

or equivalently by the implementation

$$\begin{aligned} \hat{x}_1(k+1) &= \hat{x}_2(k) \\ \hat{x}_2(k+1) &= -\hat{x}_1(k) + \hat{x}_2(k) + 0.2Q[\hat{x}_1(k)] - 0.1Q[\hat{x}_2(k)] \\ \hat{y}(k) &= \hat{x}_2(k) \end{aligned}$$

feedback introduces 'dither' to the internal signals which has the effect and can be intuitively explained by the fact that internal quantization of 'quenching' the limit cycles.

It is generally true that a filter structure which exhibits low sensitivity to the effects of internal quantization errors also exhibits low sensitivity to the effects of coefficient quantization error. The advantage in the first instance of focusing on the effects of internal quantization errors is that there exists a rather simply analytical expression with which to measure the relative performance between different filter structures.

However given any particular filter structure, the effect of coefficient quantization can be readily evaluated using (say) MATLAB. In particular, if the $(n+1)$-square matrix $\boldsymbol{S}$ where

$$\boldsymbol{S} = \begin{bmatrix} \boldsymbol{F} & \boldsymbol{g} \\ \boldsymbol{h}^T & d \end{bmatrix}$$

represents the ideal filter $H(z) = d + \boldsymbol{c}^T(z\boldsymbol{I} - \boldsymbol{F})^{-1}\boldsymbol{g}$, then the MATLAB command:

$$\hat{\boldsymbol{S}} = 2^{-t_0} * round(2^{t_0} * \boldsymbol{S})$$

will represent all the coefficients $\{\boldsymbol{F}, \boldsymbol{g}, \boldsymbol{h}, d\}$ in $\boldsymbol{S}$ to then nearest t_0 bit fractional representatives in $\hat{\boldsymbol{S}}$. One can then directly compare the ideal frequency response function $H(z)$ to the approximation $\hat{H}(z) = \hat{d} + \hat{\boldsymbol{c}}^T(z\boldsymbol{I} - \hat{\boldsymbol{F}})^{-1}\hat{\boldsymbol{g}}$.

We now examine a number of low complexity structures in terms of such considerations.

Phase Variable Structure

The phase variable structure $\{\boldsymbol{F}, \boldsymbol{g}, \boldsymbol{h}, d\}$ for the digital filter (5.135) was first defined in Chapter 2 in (2.179) and (2.181). This structure has the *lowest complexity*, but generally has the *highest sensitive* to both coefficient quantization errors and input quantization errors. Therefore, except for low order cases, phase variable structures are generally *unsuitable* for implementation.

Example 5.5.2. Consider a narrowband 4th order Butterworth digital filter $H(z)$ with bandwidth 0.01π rads. Such a filter could also be used as the basis for designing a digital notch filter of bandwidth 0.02π rads. Furthermore, a digital filter with bandwidth ω_c rads would result from the transformation of an analog filter of bandwidth Ω_c rads/sec where $\omega_c = \Omega_c T$ where T is the sampling period (in secs), or $f_s = T^{-1}$ is the sampling rate (in Hz). Specifically, a digital bandwidth $\omega_c = 0.01\pi$ rads would result from the transformation of an analog filter of bandwidth $\Omega_c = 0.01\pi f_s$ rads/sec, or equivalently, a bandwidth $F_c = 0.005 f_s$ Hz. Possibilities are:

$$F_c = 1 \text{ Hz with } f_s = 200 \text{ Hz} \quad or \quad F_c = 100 \text{ Hz with } f_s = 20 \text{ kHz}$$

It follows from MATLAB using the command:

$$[B, A] = butter(4, 0.01)$$

that

$$H(z) = \frac{\beta(z+1)^4}{z^4 - a_3 z^3 - a_2 z^2 - a_1 z - a_0}$$

where $\beta = 5.8451 \times 10^{-8}$, and

$$a_3 = 3.9179 \; ; \; a_2 = -5.7571 \; ; \; a_1 = 3.7603 \; ; \; a_0 = -0.9212$$

One phase variable realization (see exercise 2.26) is given by

$$\boldsymbol{F} = \begin{bmatrix} 3.9179 & -5.7571 & 3.7603 & -0.9212 \\ 1 & 0 & 0 & 0 \\ 0 & 1 & 0 & 0 \\ 0 & 0 & 1 & 0 \end{bmatrix} \; ; \; \boldsymbol{g} = \begin{bmatrix} 1 \\ 0 \\ 0 \\ 0 \end{bmatrix}$$

$$\boldsymbol{h} = 10^{-6} \begin{bmatrix} 0.4628 \\ 0.0142 \\ 0.4536 \\ 0.0046 \end{bmatrix} \; ; \; d = 5.8451 \times 10^{-8}$$

which may be obtained using the MATLAB command:

$$[F, g, ht, d] = tf2ss(B, A) \ ; \ h = ht'$$

From Theorem 5.4.5, the 2-norm $||h_j||_2$ of the unit impulse response signal h_j is given by the square root of the jth diagonal component of the matrix $\boldsymbol{K}$ where $\boldsymbol{K} = \boldsymbol{F}\boldsymbol{K}\boldsymbol{F}^T + \boldsymbol{g}\boldsymbol{g}^T$ which in MATLAB is given by:

$$\boldsymbol{K} = dlyap(\boldsymbol{F}, \boldsymbol{g}\boldsymbol{g}^T)$$

It follows that all diagonal components of $\boldsymbol{K}$ are equal to 1.737×10^{10}. Therefore from Theorem 5.4.5, choosing a diagonal scaling transformation $\boldsymbol{S} = s\boldsymbol{I}_4$ where

$$s = \frac{1}{\sqrt{1.737 \times 10^{10}}} = 0.9230 \times 10^{-5}$$

will result in the scaled structure $\{\boldsymbol{F}_s, \boldsymbol{g}_s, \boldsymbol{h}_s, d\}$ where

$$\boldsymbol{F}_s = \boldsymbol{F} \ ; \ \boldsymbol{g}_s = s\boldsymbol{g} \ ; \ \boldsymbol{h}_s = s^{-1}\boldsymbol{h}$$

in which each 2-norm $||\tilde{h}_j||$ of the unit impulse responses $\tilde{h}_j$ between the filter input u and the transformed filter states is (approximately) given by $\{||h_j||_2 = 1; \ 1 \leq j \leq n\}$.

(a) *Coefficient sensitivity*: When all the filter coefficients $\{a_k\}$ are quantized to a t_1 bit fractional representation via the command:

$$\hat{a}_k = 2^{-t_1} \times \text{round}(2^{t_1} * a_k)$$

where round(p) rounds the real number p to the nearest integer, and the eigenvalues of $\hat{\boldsymbol{F}}$ recomputed, the results in Table 5.3 are obtained. The poor coefficient sensitivity characteristics of the phase variable representation are obvious. The ideal eigenvalues (to 4 dec. places) are achieved when 48 bits are used.

(b) *Sensitivity to Internal Quantization Errors*: (i) If no quantization error feedback is used (i.e. $\{\boldsymbol{J}_0 = \boldsymbol{0}_{44}, \boldsymbol{h}_0 = \boldsymbol{0}_4\}$, we have from Theorem 5.5.1 that

$$\text{tr}[\boldsymbol{K}] = 4.695 \times 10^{10} \ ; \ \text{tr}[\boldsymbol{W}] = 0.364$$

which can be found via the MATLAB commands:

$$\boldsymbol{K} = dlyap(\boldsymbol{F}, \boldsymbol{g}\boldsymbol{g}^T) \ ; \ \boldsymbol{W} = dlyap(\boldsymbol{F}^T, \boldsymbol{h}\boldsymbol{h}^T)$$

and gives the performance measure

$$J_{FWL} = \frac{1}{4}\sqrt{\text{tr}[\boldsymbol{K}].\text{tr}[\boldsymbol{W}]} = 3.27 \times 10^4$$

(ii) When quantization error feedback is used with

Table 5.3. Eigenvalues of quantized F_C

Wordlength	Eigenvalues
12	1.0000 (twice)
	$0.9591 \pm \jmath 0.0386$
16	$1.0205 \pm \jmath 0.0473$
	$0.9385 \pm \jmath 0.0442$
24	$0.9870 \pm \jmath 0.0290$
	$0.9719 \pm \jmath 0.0100$
48	$0.9876 \pm \jmath 0.0287$
	$0.9713 \pm \jmath 0.0117$

$$\boldsymbol{J}_0 = \text{round}(\boldsymbol{F}) = \begin{bmatrix} 4 & -6 & 4 & -1 \\ 1 & 0 & 0 & 0 \\ 0 & 1 & 0 & 0 \\ 0 & 0 & 1 & 0 \end{bmatrix} \; ; \; \boldsymbol{h}_0 = \boldsymbol{0}_4$$

we have $\text{tr}[\boldsymbol{Z}] = 0.0026$ which gives

$$J_{FWL} = 2.74 \times 10^3$$

which is approximately 10 times smaller than in (i).

(iii) Finally, when quantization error feedback is used with $\{\boldsymbol{J}_0 = \boldsymbol{I}_4 \; ; \; \boldsymbol{h}_0 = \boldsymbol{0}_4\}$ we have $\text{tr}[\boldsymbol{Z}] = 0.0015 \times 10^{-1}$ which gives the FWL performance measure

$$J_{FWL} = 6.61 \times 10^2$$

which is approximately 3 times smaller than in (ii). Note that $\boldsymbol{J}_0 = \text{round}(\boldsymbol{F})$ is not necessarily the best choice for the integer matrix $\boldsymbol{J}_0$.

Parallel (or Jordan) Structure

A realization $\{\boldsymbol{F}, \boldsymbol{g}, \boldsymbol{h}, d\}$ in which $\boldsymbol{F}$ is in *Jordan form* is said to be in *parallel form.* The Jordan form of a matrix is essentially unique up to a rearrangement of the position of the eigenvalues, although there is additional freedom available in the selection of components of the input vector $\boldsymbol{g}$, and output vector $\boldsymbol{h}$. However once the 2-norms $\{||h_j||_2\}$ are specified, the components of the input vector $\boldsymbol{g}$ are determined, and finally $H(z)$ then uniquely determines the output vector $\boldsymbol{h}$.

For example, consider a third order filter $H(z)$ as given by

$$H(z) = \frac{d_0(z-\mu_1)(z-\mu_2)(z-\mu_3)}{(z-\lambda_1)(z-\lambda_2)(z-\lambda_3)} \tag{5.146}$$

in which $\{\mu_k, \lambda_k\}$ are all real, and $\{\lambda_k\}$ are all distinct. A *parallel structure* is defined by

$$\boldsymbol{F} = \begin{bmatrix} \lambda_1 & 0 & 0 \\ 0 & \lambda_2 & 0 \\ 0 & 0 & \lambda_3 \end{bmatrix} \quad ; \quad \boldsymbol{g} = \begin{bmatrix} g_1 \\ g_2 \\ g_3 \end{bmatrix} \quad ; \quad \boldsymbol{h} = \begin{bmatrix} h_1 \\ h_2 \\ h_3 \end{bmatrix} \quad ; \quad d = d_0 \tag{5.147}$$

where

$$H(z) = d_0 + \sum_{k=1}^{3} \frac{r_k}{z-\lambda_k} \quad ; \quad r_k = g_k h_k \tag{5.148}$$

Given the coefficients $\{\mu_k\}$ in (5.146), the coefficients $\{r_k\}$ in (5.148) can be found by equating the expressions for $H(z)$. This can be shown to be equivalent to solving the algebraic system of equations which result after equating powers of z in the equation

$$\begin{aligned} r_1(z-\lambda_2)(z-\lambda_3) + r_2(z-\lambda_1)(z-\lambda_3) + r_3(z-\lambda_1)(z-\lambda_2) \\ +d_0(z-\lambda_1)(z-\lambda_2)(z-\lambda_3) = d_0(z-\mu_1)(z-\mu_2)(z-\mu_3) \end{aligned}$$

Selection of the components g_m of $\boldsymbol{g}$ can then be made based on scaling the 2-norms $\{||h_j||_2\}$, and finally the components h_m of the output vector $\boldsymbol{h}$ are given by $h_m = r_m g_m^{-1}$.

In this third order example, $H(z)$ in (5.146) is defined by 7 parameters, while the parallel form (5.147) requires 10 multiplications. An nth order filter $H(z)$ as given by

$$H(z) = d_0 \prod_{m=1}^{n} \left(\frac{z-\mu_m}{z-\lambda_m} \right) \tag{5.149}$$

is defined by $2n+1$ coefficients whereas the corresponding parallel state space structure requires $3n+1$ multiplications.

Even though the operation count changes a little when $\boldsymbol{F}$ has complex eigenvalues, the number of multiplications remains proportional to the filter order n, and consequently, all parallel state space structures have *low complexity.* A parallel structure is generally less sensitive than a phase variable structure to both coefficient quantization errors and internal quantization errors.

Example 5.5.3. Consider again the 4th order narrowband digital Butterworth filter defined in example 5.5.2. A parallel structure $\{\boldsymbol{F}, \boldsymbol{g}, \boldsymbol{h}, d\}$ is defined by

$$F = \begin{bmatrix} 0.9876 & 0.0287 & 0 & 0 \\ -0.0287 & 0.9876 & 0 & 0 \\ 0 & 0 & 0.9713 & 0.0117 \\ 0 & 0 & -0.0117 & 0.9713 \end{bmatrix} ; \; g = \begin{bmatrix} -0.0090 \\ -0.0200 \\ -0.0194 \\ 0.0484 \end{bmatrix}$$

$$h = \begin{bmatrix} 1 \\ 1 \\ 1 \\ 1 \end{bmatrix} ; \; d = 5.8451 \times 10^{-8}$$

After scaling by a diagonal scaling matrix

$$S = diag\{s_1, s_2, s_3, s_4\}$$

where

$$s_1 = 9.251 \;\; ; \;\; s_2 = 10.843 \;\; ; \;\; s_3 = 17.064 \;\; ; \;\; s_4 = 4.728$$

we have that the scaled state space structure $\{F_s = SFS^{-1}, g_s = Sg, h_s^T = h^T S^{-1}\}$ has unity 2-norms $\{||h_j||_2 = 1; \; 1 \le j \le 4\}$.

(a) *Coefficient Sensitivity*: The effect of coefficient sensitivity can be determined by restricting all coefficients to have a particular fractional wordlength representation, and then computing the frequency response characteristics of the resulting approximating filter. For example, when $\{F, g, h\}$ are rounded to a 12 bit fractional representation $\{\hat{F}, \hat{g}, \hat{h}\}$, we have

$$\hat{F} = \begin{bmatrix} 0.9875 & 0.0288 & 0 & 0 \\ -0.0288 & 0.9875 & 0 & 0 \\ 0 & 0 & 0.9712 & 0.0117 \\ 0 & 0 & -0.0117 & 0.9712 \end{bmatrix} ; \; \hat{g} = \begin{bmatrix} -0.0090 \\ -0.0200 \\ -0.0193 \\ 0.0483 \end{bmatrix}$$

$$\hat{h} = \begin{bmatrix} 1 \\ 1 \\ 1 \\ 1 \end{bmatrix} ; \; d = 5.8451 \times 10^{-8}$$

Whether the approximation is satisfactory or not will depend on the particular application. However the resulting bandwith, transition bandwidth and stopband attenuation would normally be important considerations.

(b) *Sensitivity to Internal Quantization Errors*: (i) When $\{J_0 = 0_{44}, h_0 = 0_4\}$, we have $\{\mathrm{tr}(K) = 0.0684, \mathrm{tr}(W) = 477.56\}$ which gives

$$J_{FWL} = 1.4285$$

(ii) When quantization error feedback is used with $\{J_0 = \mathrm{round}(F) = I_4 \; ; \; h_0 = 0_4\}$ we have $\mathrm{tr}(Z) = 0.4644$ which gives

$$J_{FWL} = 0.0445$$

which is about 30 times smaller than in (ii). This performance measure in turn is about 10,000 times smaller than that achieved using a phase variable structure.

Series (or Cascade) Form

A digital filter $H(z)$ which is realized in terms of sub-filters $H_m(z)$ in the form

$$H(z) = \prod_{m=1}^{P} H_m(z) \tag{5.150}$$

is said to be a *series* (or cascade) *structure*. In this structure, the overall filter input u is the input to sub-filter $H_1(z)$ and the overall output is the output of sub-filter $H_P(z)$, while internally, the output of the sub-filter $H_m(z)$ is the input to sub-filter $H_{m+1}(z)$. The sequential ordering of the sub-filters is a freedom that is available which can have an important effect on the sensitivity characteristics of the FWL filter implementation.

For example, a third order filter

$$H(z) = 0.008 \left(\frac{z+0.8}{z-0.9}\right)\left(\frac{z+0.7}{z-0.8}\right)\left(\frac{z+1}{z-0.6}\right)$$

has possible series representations in the form

$$H(z) = \prod_{m=1}^{n} H_m(z) \; ; \; H_m(z) = d_m \left(\frac{z-\mu_m}{z-\lambda_m}\right) \tag{5.151}$$

where

$$\mu_m \in \{-0.7, -0.8, -1\} \; ; \; \lambda_m \in \{0.6, 0.8, 0.9\} \; ; \; d_1 d_2 d_3 = 0.008$$

One possible representation is defined by

$$\{\lambda_1 = 0.9, \lambda_2 = 0.8, \lambda_3 = 0.6, \mu_1 = -0.8, \mu_2 = -0.7, \mu_3 = -1\}$$

while another representation is defined by

$$\{\lambda_1 = 0.9, \lambda_2 = 0.6, \lambda_3 = 0.8, \mu_1 = -0.7, \mu_2 = -1, \mu_3 = -0.8\}$$

Altogether, there are 6 different arrangements of the $\{\lambda_m\}$, and for every such arrangement, there are 6 different arrangements of the $\{\mu_m\}$ which gives a total of $N = 36$ different series representations. Both the pairing $\{\lambda_m, \mu_m\}$ and the positioning of the sub-filters can affect the FWL sensitivity characteristics of the overall filter.

An ideal state space representation of the third order filter $H(z)$ in (5.151) with input signal u and output signal y is then given by

$$\begin{aligned} x_1(k+1) &= \lambda_1 x_1(k) + \alpha_1 u(k) \\ y_1(k) &= \beta_1 x_1(k) + d_1 u(k) \\ x_2(k+1) &= \lambda_2 x_2(k) + \alpha_2 y_1(k) \\ y_2(k) &= \beta_2 x_1(k) + d_2 y_1(k) \\ x_3(k+1) &= \lambda_3 x_3(k) + \alpha_3 y_2(k) \\ y(k) &= \beta_3 x_1(k) + d_3 y_2(k) \end{aligned} \tag{5.152}$$

where

$$\alpha_m \beta_m = d_m(\lambda_m - \mu_m) \; ; \; m = 1, 2, 3$$

If $\boldsymbol{x}^T = [x_1 \;\; x_2 \;\; x_3]$, then these equations can be written in the form

$$\boldsymbol{x}(k+1) = \boldsymbol{F}\boldsymbol{x}(k) + \boldsymbol{g}u(k) \; ; \; y(k) = \boldsymbol{h}^T \boldsymbol{x}(k) + d_0 u(k)$$

in which

$$\boldsymbol{F} = \begin{bmatrix} \lambda_1 & 0 & 0 \\ f_{21} & \lambda_2 & 0 \\ f_{31} & f_{32} & \lambda_3 \end{bmatrix} \; ; \; \boldsymbol{g} = \begin{bmatrix} g_1 \\ g_2 \\ g_3 \end{bmatrix} \; ; \; \boldsymbol{h} = \begin{bmatrix} h_1 \\ h_2 \\ h_3 \end{bmatrix} \; ; \; d_0 = d_1 d_2 d_3 \tag{5.153}$$

where the state space filter coefficients $\{g_m, h_m, f_{ij}\}$ are precomputed from the filter parameters $\{\mu_m, d_m, \alpha_m, \beta_m\}$ by the equations

$$\begin{aligned} &f_{21} = \alpha_2 \beta_1 \; , \quad f_{31} = \alpha_3 d_2 \beta_1 \; , \quad f_{32} = \alpha_3 \beta_2 \\ &g_1 = \alpha_1 \; , \quad g_2 = \alpha_2 d_1 \; , \quad g_3 = \alpha_3 d_2 d_1 \\ &h_1 = d_3 d_2 d_1 \; , \quad h_2 = d_3 \beta_2 \; , \quad h_3 = \beta_3 \end{aligned} \tag{5.154}$$

The third order series state space structure (5.153) requires 13 multiplications to compute each new value of the output y even though $H(z)$ in (5.151) involves only the 7 coefficients $\{\lambda_m, \mu_m : m = 1, 2, 3\}$ and $d_0 = d_1 d_2 d_3$.

More generally, an nth order filter $H(z)$ in (5.149) which is defined by the $2n+1$ coefficients $\{\lambda_m, \mu_m \; : \; 1 \leq m \leq n\}$ and d_0 has a corresponding series state space structure $\{\boldsymbol{F}, \boldsymbol{g}, \boldsymbol{h}, d\}$ in which all components of $\boldsymbol{g}$ and $\boldsymbol{h}$, and all components of $\boldsymbol{F}$ which are on or below the diagonal, are nonzero. This means the resulting series state space structure requires a total of N_T multiplications where

$$N_T = 0.5n(n+1) + 2n + 1$$

Since this number is of the order of n^2, it would appear that a series structure has high complexity. However depending on how the structure is implemented, this need not be the case.

In particular, suppose the series structure for the filter (5.149) with input signal u and output signal y is actually implemented according to

$$\begin{aligned} x_m(k+1) &= \lambda_m x_m(k) + \alpha_m u_m(k) \; ; \; 1 \le m \le n \\ y_m(k) &= \beta_m x_m(k) + d_m u_m(k) \\ \alpha_m \beta_m d_m &= \lambda_m - \mu_m \; ; \; d_0 = \prod_{m=1}^{n} d_m \end{aligned} \tag{5.155}$$

where

$$\begin{aligned} u(k) &= u_1(k) \; ; \; y(k) = y_n(k) \\ u_{m+1}(k) &= y_m(k) \; ; \; 1 \le m \le n-1 \end{aligned} \tag{5.156}$$

Now at most 4 multiplications are required to both update each state component from $x_m(k)$ to $x_m(k+1)$, and calculate $y_m(k)$ from $x_m(k)$ and $u_m(k)$. Hence, overall the series structure (5.155) - (5.156) of an nth order filter (5.149) will only require a total of $N_T = 4n$ multiplications which therefore guarantees that this series structure has *low complexity*.

In the low complexity series implementation (5.155), (5.156), the "internal output signals" $\{y_m\}$ as well as the state components $\{x_m\}$ must be stored at each filter iteration which means that more memory is required. Also, since quantization of the internal signals $\{y_m\}$ will normally be required, the sensitivity of the low complexity series structure to internal quantization errors will generally be worse than the high complexity series state space filter implementation. On the other hand, the coefficient sensitivity characteristics of the low complexity series structure can be better than those of the high complexity series state space structure.

(a) *Scaling Characteristics*: The high complexity series state space structure (5.153) requires scaling of the state components in order to avoid overflow. For the third order filter, this can be achieved by means of a diagonal scaling matrix

$$\boldsymbol{S} = diag\{s_1, s_2, s_3\}$$

Then $\{\boldsymbol{F}_s = \boldsymbol{S}\boldsymbol{F}\boldsymbol{S}^{-1}, \boldsymbol{g}_s = \boldsymbol{S}\boldsymbol{g}, \boldsymbol{h}_s^T = \boldsymbol{h}^T\boldsymbol{S}^{-1}\}$ are given by

$$\boldsymbol{F} = \begin{bmatrix} \lambda_1 & 0 & 0 \\ \tilde{f}_{21} & \lambda_2 & 0 \\ \tilde{f}_{31} & \tilde{f}_{32} & \lambda_3 \end{bmatrix} \; ; \; \boldsymbol{g}_s = \begin{bmatrix} \tilde{g}_1 \\ \tilde{g}_2 \\ \tilde{g}_3 \end{bmatrix} \; ; \; \boldsymbol{h}_s = \begin{bmatrix} \tilde{h}_1 \\ \tilde{h}_2 \\ \tilde{h}_3 \end{bmatrix}$$

$$\tilde{f}_{21} = s_2 s_1^{-1} f_{21} \; ; \; \tilde{f}_{31} = s_3 s_1^{-1} f_{31} \; ; \; \tilde{f}_{32} = s_3 s_2^{-1} f_{32}$$

$$\tilde{g}_m = s_m g_m \; ; \; \tilde{h}_m = s_m^{-1} h_m$$

Implementation of the low complexity third order series structure (5.152) imposes additional scaling requirements. Specifically, since $\{y_1, y_2\}$ must be stored for subsequent calculations, $\{y_1, y_2\}$ as well as $\{x_1, x_2, x_3\}$ must all be scaled in order to avoid overflow. In particular, we see from (5.152) that:

- α_1 can be selected to scale the signal x_1. Then $\{\beta_1, d_1\}$ where $\beta_1 d_1^{-1} = \alpha_1^{-1}(\lambda_1 - \mu_1)$ can be used to scale y_1.
- Similarly, $\{\alpha_2, \beta_2, d_2\}$ can be used to scale $\{x_2, y_2\}$, and finally, $\{\alpha_3, \beta_3, d_3\}$ ca be used to scale x_3. (Note that since $d_0 = d_1 d_2 d_3$ is fixed, d_3 is determined once $\{d_1, d_2\}$ are selected, and so the output y cannot be scaled.)

(b) *Coefficient Sensitivity*: After quantizing f_{21} to $\hat{f}_{21}$ and $\{\alpha_2, \beta_1\}$ to $\{\hat{\alpha}_2, \hat{\beta}_1\}$ it will *not* generally be the case that

$$\hat{f}_{21} = \hat{\alpha}_2 . \hat{\beta}_1$$

Therefore the high complexity series state space structure (5.153) and the low complexity series structure (5.152) will have *different* coefficient sensitivity characteristics.

(c) *Sensitivity to Internal Quantization Errors*: A FWL implementation of the high complexity series state space structure (5.152) in the absence of input and coefficient errors and quantization error feedback is given by

$$\begin{aligned} \hat{\boldsymbol{x}}(k+1) &= \boldsymbol{F} Q[\hat{\boldsymbol{x}}(k)] + \boldsymbol{g} u(k) \\ \hat{y}(k) &= \boldsymbol{h}^T Q[\hat{\boldsymbol{x}}(k)] + d u(k) \end{aligned} \tag{5.157}$$

where $\{\boldsymbol{F}, \boldsymbol{g}, \boldsymbol{h}, d\}$ are given by (5.153). The corresponding FWL performance measure J_{FWL} can then be calculated as described in Theorem 5.5.1.

On the other hand, a FWL implementation of the low complexity series structure (5.152) in the absence of quantization error feedback is given by

$$\begin{aligned} \hat{x}_1(k+1) &= \lambda_1 Q[\hat{x}_1(k)] + \alpha_1 u(k) \\ \hat{y}_1(k) &= \beta_1 Q[\hat{x}_1(k)] + d_1 u(k) \\ \hat{x}_2(k+1) &= \lambda_2 Q[\hat{x}_2(k)] + \alpha_2 Q[\hat{y}_1(k)] \\ \hat{y}_2(k) &= \beta_2 Q[\hat{x}_2(k)] + d_1 Q[\hat{y}_1(k)] \\ \hat{x}_3(k+1) &= \lambda_3 Q[\hat{x}_3(k)] + \alpha_3 Q[\hat{y}_2(k)] \\ \hat{y}(k) &= \beta_3 Q[\hat{x}_3(k)] + d_3 Q[\hat{y}_2(k)] \end{aligned} \tag{5.158}$$

The four additional signal quantizations in (5.158) in which $\hat{y}_m(k)$ is quantized to $Q[\hat{y}_m(k)]$ means that the sensitivity characteristics with respect to internal quantization errors of the low complexity series structure are normally worse than those of the high complexity series state space implementation. The FWL implementation (5.158) can only be implemented according to (5.157) when $\{\alpha_m, d_m\}$ are integers.

Lattice Structure

An important structure for implementing a digital filter is a *lattice structure* which for an nth order filter is defined by

$$\begin{aligned} p_n(k) &= -\gamma_{n-1}x_{n-1}(k) - \gamma_n x_n(k) + u(k) \\ x_n(k+1) &= \gamma_{n-1}p_n(k) + x_{n-1}(k) \end{aligned} \tag{5.159}$$

and

$$\begin{aligned} p_m(k) &= -\gamma_{m-1}x_{m-1}(k) + p_{m+1}(k) \\ x_m(k+1) &= \gamma_{m-1}p_m(k) + x_{m-1}(k) \end{aligned} \tag{5.160}$$

for $m = n-1, n-2, \ \ldots \ , 2$, and

$$\begin{aligned} x_2(k) &= \gamma_1 p_2(k) + x_1(k) \\ x_1(k+1) &= -p_2(k) \\ y(k) &= \sum_{p=1}^{n} h_p x_p(k) + du(k) \end{aligned} \tag{5.161}$$

The corresponding nth order state space lattice structure $\{\boldsymbol{F}, \boldsymbol{g}, \boldsymbol{h}, d\}$ is then given by

$$\boldsymbol{F} = \begin{bmatrix} -\gamma_1 & -\gamma_2 & -\gamma_3 & \cdot\cdot & \cdot & -\gamma_n \\ 1-\gamma_1^2 & -\gamma_1\gamma_2 & -\gamma_1\gamma_3 & \cdot\cdot & \cdot & -\gamma_1\gamma_n \\ 0 & 1-\gamma_2^2 & -\gamma_2\gamma_3 & \cdot\cdot & \cdot & -\gamma_2\gamma_n \\ 0 & 0 & \cdot & \cdot\cdot & \cdot & \cdot \\ \cdot & \cdot & \cdot & \cdot\cdot & \cdot & \cdot \\ \cdot & \cdot & \cdot & \cdot\cdot & \cdot & \cdot \\ 0 & 0 & 0 & \cdot\cdot & 1-\gamma_{n-1}^2 & -\gamma_{n-1}\gamma_n \end{bmatrix} ; \ \boldsymbol{g} = \begin{bmatrix} 1 \\ \gamma_1 \\ \gamma_2 \\ \cdot \\ \cdot \\ \cdot \\ \cdot \\ \gamma_{n-1} \end{bmatrix}$$

$$\boldsymbol{h}^T = \begin{bmatrix} h_1 \ h_2 \ h_3 \cdots h_n \end{bmatrix} \tag{5.162}$$

It can be shown that the eigenvalues of the system matrix $\boldsymbol{F}$ has all eigenvalues with modulus less than unity (and hence that the corresponding digital filter $H(z)$ is asymptotically stable) if and only if $\{|\gamma_k| < 1; \ 1 \le k \le n\}$.

A calculation of each new value of the output signal y then requires a total of N_T multiplications where

$$N_T = N_{T_1} + N_{T_2} + N_{T_3}$$
$$N_{T_1} = 0.5n(n+1) + n - 1 \;\; ; \;\; N_{T_2} = n - 1 \;\; ; \;\; N_{T_3} = n$$

which corresponds to the fact that $\{\boldsymbol{F}, \boldsymbol{g}, \boldsymbol{h}\}$ involve N_{T_1}, N_{T_2} and N_{T_3} components respectively which are neither 0 nor 1. Hence since N_T is of the order of n^2, the state space lattice structure has *high complexity*.

In comparison, the implementation (5.159) - (5.161) requires only 4 multiplications to compute both $p_m(k)$ from $\{x_{m-1}(k), x_m(k)\}$ and $x_m(k+1)$ from $\{x_{m-1}(k), p_m(k)\}$. That is, overall only $5n+1$ multiplications are required in (5.159) - (5.161) to calculate each new output value $y(k)$ which means that this lattice structure implementation has *low complexity*.

5.5.3 Delay Replaced Structures

A FWL state space implementation with internal quantization error feedback is described by (5.96) in which all components of the n-square matrix $\boldsymbol{J}_0$ and the n-vector $\boldsymbol{h}_0$ are *integers*. In particular, suppose $\boldsymbol{J} = \boldsymbol{I}_\mathbf{n}$, $\boldsymbol{h}_0 = \mathbf{0}_\mathbf{n}$. Then in the absence of coefficient quantization error, the implementation is described by

$$\begin{aligned} \hat{\boldsymbol{x}}(k+1) &= \boldsymbol{F}Q[\hat{\boldsymbol{x}}(k)] + \varepsilon_Q(k) + \boldsymbol{g}u(k) \\ Q[\hat{\boldsymbol{x}}(k)] &= \hat{\boldsymbol{x}}(k) - \varepsilon_Q(k) \\ \hat{y}(k) &= \boldsymbol{h}^T Q[\hat{\boldsymbol{x}}(k)] + du(k) \end{aligned}$$

which can equivalently be written in the form

$$\begin{aligned} \hat{\boldsymbol{x}}(k+1) &= \hat{\boldsymbol{x}}(k) + \boldsymbol{F}_\delta Q[\hat{\boldsymbol{x}}(k)] + \boldsymbol{g}u(k) \\ Q[\hat{\boldsymbol{x}}(k)] &= \hat{\boldsymbol{x}}(k) - \varepsilon_Q(k) \\ \hat{y}(k) &= \boldsymbol{h}^T Q[\hat{\boldsymbol{x}}(k)] + du(k) \end{aligned} \tag{5.163}$$

where $\boldsymbol{F}_\delta \stackrel{\Delta}{=} \boldsymbol{F} - \boldsymbol{I}_\mathbf{n}$. In particular, since the filter frequency function $H(z)$ is given by

$$H(z) = d + \boldsymbol{h}^T(z\boldsymbol{I}_\mathbf{n} - \boldsymbol{F})^{-1}\boldsymbol{g}$$

it follows that $H(z-1) = d + \boldsymbol{h}^T(z\boldsymbol{I}_\mathbf{n} - \boldsymbol{F}_\delta)^{-1}\boldsymbol{g}$, That is

$$H(z) = H_\delta(z-1) \quad \text{or} \quad H_\delta(\mathcal{Z}) = H(\mathcal{Z}+1) \tag{5.164}$$

where

$$H_\delta(\mathcal{Z}) = d + \boldsymbol{h}^T(\mathcal{Z}\boldsymbol{I}_\mathbf{n} - \boldsymbol{F}_\delta)^{-1}\boldsymbol{g}$$

Since $H_\delta(\mathcal{Z}) = H(\mathcal{Z}+1)$, a state space structure of the form $\{\boldsymbol{F}_\delta, \boldsymbol{g}, \boldsymbol{h}, d\}$ is said to be a *delay replaced structure* with respect to the state space structure $\{\boldsymbol{F}, \boldsymbol{g}, \boldsymbol{h}, d\}$.

For example, suppose

$$H(z) = 1 + \frac{(-0.6)}{z - 0.8} + \frac{(0.2)}{z - 0.9}$$

Then

$$H_\delta(\mathcal{Z}) = H(\mathcal{Z}+1) = 1 + \frac{(-0.6)}{\mathcal{Z} + 0.2} + \frac{(0.2)}{\mathcal{Z} + 0.1}$$

which has a state space representation $\{\boldsymbol{F}_\delta, \boldsymbol{g}, \boldsymbol{h}, d\}$ where

$$\boldsymbol{F}_\delta = \begin{bmatrix} -0.2 & 0 \\ 0 & -0.1 \end{bmatrix} \ ; \ \boldsymbol{g} = \begin{bmatrix} -0.6 \\ 0.2 \end{bmatrix} \ ; \ \boldsymbol{h} = \begin{bmatrix} 1 \\ 1 \end{bmatrix} \ ; \ d = 1$$

Any state space structure, and particularly a low complexity state space structure, has a delay replaced equivalent. In particular, given $H(z)$, we first compute $H_\delta(\mathcal{Z})$ as defined in (5.164) and then construct a low complexity (such as phase variable, parallel, series or lattice) delay repleced version.

Delay Replaced Phase Variable Structure: Suppose

$$H(z) = d + \frac{b_2 z^2 + b_1 z + b_0}{z^3 - a_2 z^2 - a_1 z - a_0}$$

then

$$\begin{aligned} H_\delta(\mathcal{Z}) &= d + \frac{b_2(\mathcal{Z}+1)^2 + b_1(\mathcal{Z}+1) + b_0}{(\mathcal{Z}+1)^3 - a_2(\mathcal{Z}+1)^2 - a_1(\mathcal{Z}+1) - a_0} \\ &= d + \frac{\tilde{b}_2 \mathcal{Z}^2 + \tilde{b}_1 \mathcal{Z} + \tilde{b}_0}{\mathcal{Z}^3 - \tilde{a}_2 \mathcal{Z}^2 - \tilde{a}_1 \mathcal{Z} - \tilde{a}_0} \end{aligned}$$

where

$$\tilde{b}_2 = b_2 \ ; \ \tilde{b}_1 = 2b_2 + b_1 \ ; \ \tilde{b}_0 = b_2 + b_1 + b_0$$
$$\tilde{a}_2 = a_2 - 3 \ ; \ \tilde{a}_1 = a_1 + 2a_2 - 3 \ ; \ \tilde{a}_0 = a_0 + a_1 + a_2 - 1$$

A third order (unscaled) delay replaced phase variable structure $\{\boldsymbol{F}_\delta, \boldsymbol{g}, \boldsymbol{h}, d\}$ is therefore defined by

$$\boldsymbol{F}_\delta = \begin{bmatrix} 0 & 0 & \tilde{a}_0 \\ 1 & 0 & \tilde{a}_1 \\ 0 & 1 & \tilde{a}_2 \end{bmatrix} \ ; \ \boldsymbol{g} = \begin{bmatrix} \tilde{b}_0 \\ \tilde{b}_1 \\ \tilde{b}_2 \end{bmatrix} \ ; \ \boldsymbol{h} = \begin{bmatrix} 0 \\ 0 \\ 1 \end{bmatrix} \qquad (5.165)$$

A phase variable structure is known to be *very sensitive* to both coefficient quantization errors and internal quantization errors when the frequency response function $H(z)$ is narrowband. However the corresponding *delay replaced phase variable structure* (5.165) has low sensitivity to both coefficient and internal quantization errors. In fact, the sensitivity characteristics of all narrow band low complexity structures can be improved using a delay replaced structure. Low sensitivity delay replaced series, parallel and lattice structures may be similarly obtained.

5.5.4 Optimal Structures

Given any initial state space realization (5.139) of a filter $H(z)$, an *optimal FWL state space realization (5.143) with respect to internal quantization errors* is defined by a nonsingular matrix $\boldsymbol{S}$, an integer matrix $\boldsymbol{J}_0$, and an integer vector $\boldsymbol{h}_s$ which achieves the minimization:

$$\min_{\{\boldsymbol{S},\boldsymbol{J}_0,\boldsymbol{h}_s\}} \tilde{J}_{FWL}$$

where $\tilde{J}_{FWL}$ is given by (5.144), (5.145). In this section, we look at this optimization problem.

To begin, first observe that since the components of $\boldsymbol{h}_s$ and $\boldsymbol{J}_0$ must be *integers*, $\tilde{\boldsymbol{h}}$ and $\tilde{\boldsymbol{J}}_0$ are not arbitrary, but instead depend on $\boldsymbol{S}$. As a consequence, the matrix $\tilde{\boldsymbol{W}}_I$ in (5.144) as defined in (5.145) is itself generally dependent on the matrix $\boldsymbol{S}$ which means that there is no analytical solution for an optimal transformation $\boldsymbol{S}$. In addition, the fact that both $\boldsymbol{h}_s$ and $\boldsymbol{J}_0$ must have *integer* components means that even a numerical solution very difficult.

However there are two special cases when $\tilde{\boldsymbol{W}}_I$ is *independent* of $\boldsymbol{S}$ which then lead to analytical solutions for the optimal transformation $\boldsymbol{S}$. In particular, $\{\boldsymbol{h}_s = 0,\ \boldsymbol{J}_0 = 0\}$ implies $\{\boldsymbol{h} = 0,\ \boldsymbol{J} = 0\}$ for all $\boldsymbol{S}$ which in turn implies that for all $\boldsymbol{S}$:

$$\tilde{\boldsymbol{W}}_I = \boldsymbol{W}_0 \quad ; \quad \boldsymbol{h}_s = 0,\ \boldsymbol{J}_0 = 0 \tag{5.166}$$

Also, $\{\boldsymbol{h}_s = 0,\ \boldsymbol{J}_0 = \boldsymbol{I}_n\}$ implies $\{\boldsymbol{h} = 0,\ \boldsymbol{J} = \boldsymbol{I}_n\}$ for all $\boldsymbol{S}$ which in turn implies that for all $\boldsymbol{S}$:

$$\tilde{\boldsymbol{W}}_I = \boldsymbol{P}_0 \quad ; \quad \boldsymbol{h}_s = 0,\ \boldsymbol{J}_0 = \boldsymbol{I}_n \tag{5.167}$$

where

$$\boldsymbol{P}_0 = (\boldsymbol{I}_n - \boldsymbol{F}_0)^T \boldsymbol{W}_0 + \boldsymbol{W}_0(\boldsymbol{I}_n - \boldsymbol{F}_0) \tag{5.168}$$

We have the following result.

Theorem 5.5.2. *Consider the matrices* $\{\boldsymbol{K}_0, \boldsymbol{W}_0, \boldsymbol{P}_0\}$ *defined by (5.141), (5.142) and (5.168). Then:*

(i) All n *eigenvalues* $\{\sigma_k^2 > 0;\ 1 \le k \le n\}$ *of* $\boldsymbol{K}_0\boldsymbol{W}_0$ *are positive.*

(ii) Suppose $\{\boldsymbol{h}_s = 0,\ \boldsymbol{J}_0 = 0\}$. *Then with* $\sigma_k > 0$

$$\min_{\boldsymbol{S}} \tilde{J}_{FWL} = \min_{\boldsymbol{S}} \frac{1}{n}\sqrt{tr[\boldsymbol{S}\boldsymbol{K}_0\boldsymbol{S}^T].tr[(\boldsymbol{S}^{-1})^T\boldsymbol{W}_0\boldsymbol{S}^{-1}]}$$
$$= \frac{1}{n}\sum_{k=1}^{n}\sigma_k \tag{5.169}$$

(iii) All n eigenvalues $\{\varrho_k^2 > 0;\ 1 \leq k \leq n\}$ of $\boldsymbol{K}_0\boldsymbol{P}_0$ are positive.
(iv) Suppose $\{\boldsymbol{h}_s = \boldsymbol{0},\ \boldsymbol{J}_0 = \boldsymbol{I}_n\}$. Then with $\varrho_k > 0$

$$\min_{\boldsymbol{S}} \tilde{J}_{FWL} = \min_{\boldsymbol{S}} \frac{1}{n}\sqrt{tr[\boldsymbol{S}\boldsymbol{K}_0\boldsymbol{S}^T].tr[(\boldsymbol{S}^{-1})^T\boldsymbol{P}_0\boldsymbol{S}^{-1}]}$$
$$= \frac{1}{n}\sum_{k=1}^{n}\varrho_k \tag{5.170}$$

Proof via Singular Value Decomposition

In order to prove these results, we need to call on some key results in linear algebra. To begin, we recall that any real $n \times m$ matrix $\boldsymbol{M}$ of rank $p \leq (n, m)$ has a *Singular Value Decomposition*, or SVD, as given by

$$\boldsymbol{M} = \boldsymbol{U}\boldsymbol{\Sigma}\boldsymbol{V}^T \tag{5.171}$$

where $\{\boldsymbol{U}, \boldsymbol{V}\}$ are respectively n-square and m-square unitary matrices; that is

$$\boldsymbol{U}\boldsymbol{U}^T = \boldsymbol{I}_n \quad ; \quad \boldsymbol{V}\boldsymbol{V}^T = \boldsymbol{I}_m \tag{5.172}$$

which implies that

$$\boldsymbol{U}^{-1} = \boldsymbol{U}^T \quad ; \quad \boldsymbol{V}^{-1} = \boldsymbol{V}^T$$

and $\boldsymbol{\Sigma}$ is an $n \times m$ partitioned matrix of the form

$$\boldsymbol{\Sigma} = \begin{bmatrix} \boldsymbol{\Sigma}_1 & 0_{p,m-p} \\ 0_{n-p,p} & 0_{n-p,m-p} \end{bmatrix} \tag{5.173}$$

with the diagonal p-square matrix $\boldsymbol{\Sigma}_1$ given by

$$\boldsymbol{\Sigma}_1 = \mathrm{diag}\{s_1, s_2, \ \ldots\ , s_p\} \quad ; \quad s_k > 0 \text{ for all } k$$

The values $\{s_k > 0,\ 1 \leq k \leq p\}$ are called the *singular values* of $\boldsymbol{M}$.

Now, in order to prove part (i), define

$$\boldsymbol{K}_s \triangleq \boldsymbol{S}\boldsymbol{K}_0\boldsymbol{S}^T \quad ; \quad \boldsymbol{W}_s \triangleq (\boldsymbol{S}^{-1})^T\boldsymbol{W}_0\boldsymbol{S}^{-1} \tag{5.174}$$

Then

$$\boldsymbol{K}_s\boldsymbol{W}_s = \boldsymbol{S}(\boldsymbol{K}_0\boldsymbol{W}_0)\boldsymbol{S}^{-1}$$

and so $\boldsymbol{K}_s\boldsymbol{W}_s$ and $\boldsymbol{K}_0\boldsymbol{W}_0$ have the *same* eigenvalues for all nonsingular matrices $\boldsymbol{S}$.

Now, since $\boldsymbol{K}_0 = \boldsymbol{K}_0^T$ is positive definite, $\boldsymbol{K}_0$ has an SVD of the form

$$\boldsymbol{K}_0 = \boldsymbol{U}_0 \operatorname{diag}\{\sigma_{K1}, \sigma_{K2}, \ \ldots \ , \sigma_{Kn}\} \ \boldsymbol{U}_0^T \quad ; \quad \sigma_{Kj} > 0 \text{ for all } j$$

Then the *square root* $\boldsymbol{K}_0^{\frac{1}{2}}$ of $\boldsymbol{K}_0$ which is defined such that

$$\boldsymbol{K}_0^{\frac{1}{2}}.\boldsymbol{K}_0^{\frac{1}{2}} = \boldsymbol{K}_0$$

is given by

$$\boldsymbol{K}_0^{\frac{1}{2}} = \boldsymbol{U}_0 \operatorname{diag}\{\sqrt{\sigma_{K1}}, \sqrt{\sigma_{K2}}, \ \ldots \ , \sqrt{\sigma_{Kn}}\} \ \boldsymbol{U}_0^T$$

Since $\sigma_{Kj} > 0$ for all j, we can then define a matrix $\boldsymbol{S}_0$ by

$$\boldsymbol{S}_0 \triangleq \boldsymbol{K}_0^{-\frac{1}{2}} = \boldsymbol{U}_0 \operatorname{diag}\{\sqrt{\sigma_{K1}^{-1}}, \sqrt{\sigma_{K2}^{-1}}, \ \ldots \ , \sqrt{\sigma_{Kn}^{-1}}\} \ \boldsymbol{U}_0^T \tag{5.175}$$

It then follows that

$$\boldsymbol{K}_1 \triangleq \boldsymbol{S}_0\boldsymbol{K}_0\boldsymbol{S}_0^T = \boldsymbol{I}_n \quad ; \quad \boldsymbol{Z} \triangleq (\boldsymbol{S}_0^{-1})^T\boldsymbol{W}_0\boldsymbol{S}_0^{-1} = \boldsymbol{K}_0^{\frac{1}{2}}\boldsymbol{W}_0\boldsymbol{K}_0^{\frac{1}{2}} \tag{5.176}$$

which means that the eigenvalues of $\boldsymbol{K}_0\boldsymbol{W}_0$ are given by the eigenvalues of $\boldsymbol{K}_1\boldsymbol{Z} = \boldsymbol{Z}$. In particular, if the positive definite matrix $\boldsymbol{Z}$ is written in terms of its SVD according to

$$\boldsymbol{Z} = \boldsymbol{U}_1\boldsymbol{\Sigma}_0^2\boldsymbol{U}_1^T$$

where

$$\boldsymbol{\Sigma}_0 = \operatorname{diag}\{\sigma_1, \sigma_2, \ \ldots \ , \sigma_p\} \ ; \ \sigma_k > 0 \tag{5.177}$$

then $\{\sigma_k^2 > 0;\ 1 \le k \le n\}$ are the eigenvalues of $\boldsymbol{Z}$, and hence the eigenvalues of $\boldsymbol{K}_0\boldsymbol{W}_0$.

Now in order to prove part (ii), define

$$\boldsymbol{S}_1 \triangleq \boldsymbol{\Sigma}_0^{\frac{1}{2}}\boldsymbol{U}_1^T \tag{5.178}$$

Then

$$\boldsymbol{K}_2 \triangleq \boldsymbol{S}_1\boldsymbol{K}_1\boldsymbol{S}_1^T = \boldsymbol{\Sigma}_0^{\frac{1}{2}}\boldsymbol{U}_1^T.\boldsymbol{I}_n.\boldsymbol{U}_1\boldsymbol{\Sigma}_0^{\frac{1}{2}} = \boldsymbol{\Sigma}_0 \tag{5.179}$$

and

$$\boldsymbol{W}_2 \triangleq (\boldsymbol{S}_1^{-1})^T\boldsymbol{Z}\boldsymbol{S}_1^{-1} = \boldsymbol{\Sigma}_0^{-\frac{1}{2}}\boldsymbol{U}_1^T.\boldsymbol{U}_1\boldsymbol{\Sigma}_0^2\boldsymbol{U}_1^T.\boldsymbol{U}_1\boldsymbol{\Sigma}_0^{-\frac{1}{2}} = \boldsymbol{\Sigma}_0 \tag{5.180}$$

Hence from (5.176), (5.179) and (5.180)

$$\min_{\boldsymbol{S}} \sqrt{tr[\boldsymbol{S}\boldsymbol{K}_0\boldsymbol{S}^T].tr[(\boldsymbol{S}^{-1})^T\boldsymbol{W}_0\boldsymbol{S}^{-1}]}$$
$$= \min_{\boldsymbol{S}_2} \sqrt{tr[\boldsymbol{S}_2\boldsymbol{\Sigma}_0\boldsymbol{S}_2^T].tr[(\boldsymbol{S}_2^{-1})^T\boldsymbol{\Sigma}_0\boldsymbol{S}_2^{-1}]}$$

Once an optimal $\boldsymbol{S}_2$ is determined, the corresponding optimal value of $\boldsymbol{S}$ is given by

$$\boldsymbol{S} = \boldsymbol{S}_2\boldsymbol{S}_1\boldsymbol{S}_0 \tag{5.181}$$

where $\{\boldsymbol{S}_0, \boldsymbol{S}_1\}$ are given by (5.175) and (5.178) respectively.

Now from (5.177) define

$$\boldsymbol{P} = \boldsymbol{\Sigma}_0^{\frac{1}{2}}\boldsymbol{S}_2^T \ ; \ \boldsymbol{Q} = \boldsymbol{\Sigma}_0^{\frac{1}{2}}\boldsymbol{S}_2^{-1}$$

Then

$$\boldsymbol{P}^T\boldsymbol{P} = \boldsymbol{S}_2\boldsymbol{\Sigma}_0\boldsymbol{S}_2^T \ ; \ \boldsymbol{P}^T\boldsymbol{Q} = \boldsymbol{S}_2\boldsymbol{\Sigma}_0\boldsymbol{S}_2^{-1}$$
$$\boldsymbol{Q}^T\boldsymbol{Q} = (\boldsymbol{S}_2^{-1})^T\boldsymbol{\Sigma}_0(\boldsymbol{S}_2^{-1})$$

and

$$tr[\boldsymbol{P}^T\boldsymbol{Q}] = tr[\boldsymbol{\Sigma}_0]$$

It can be shown that

$$\langle \boldsymbol{P}, \boldsymbol{Q} \rangle \triangleq tr[\boldsymbol{P}^T\boldsymbol{Q}]$$

defines an inner product between two n-square matrices which then gives the induced matrix norm

$$||\boldsymbol{P}|| \triangleq \sqrt{tr[\boldsymbol{P}^T\boldsymbol{P}]}$$

Then the inner product inequality

$$(\langle \boldsymbol{P}, \boldsymbol{Q} \rangle)^2 \leq ||\boldsymbol{P}||^2 \ ||\boldsymbol{Q}||^2$$

implies that

$$tr[\boldsymbol{S}_2\boldsymbol{\Sigma}_0\boldsymbol{S}_2^T].tr[(\boldsymbol{S}_2^{-1})^T\boldsymbol{\Sigma}_0\boldsymbol{S}_2^{-1}] \geq tr^2[\boldsymbol{\Sigma}_0]$$

with equality if and only if

$$\boldsymbol{S}_2 = \lambda\boldsymbol{V}_2^T \tag{5.182}$$

for some real scalar λ and unitary matrix $\boldsymbol{V}_2$. Hence

$$\min_{\boldsymbol{S}_2} \sqrt{tr[\boldsymbol{S}_2\boldsymbol{\Sigma}_0\boldsymbol{S}_2^T].tr[(\boldsymbol{S}_2^{-1})^T\boldsymbol{\Sigma}_0\boldsymbol{S}_2^{-1}]} = \sum_{k=1}^{n}\sigma_k$$

independent of the value of λ in (5.182).

In order to prove part (iii), define

$$\boldsymbol{P}_s = (\boldsymbol{S}^{-1})^T \boldsymbol{P}_0 \boldsymbol{S}^{-1} \tag{5.183}$$

Then from (5.174), (5.183)

$$\boldsymbol{K}_s \boldsymbol{P}_s = \boldsymbol{S}(\boldsymbol{K}_0 \boldsymbol{P}_0)\boldsymbol{S}^{-1}$$

and so $\boldsymbol{K}_s\boldsymbol{P}_s$ and $\boldsymbol{K}_0\boldsymbol{P}_0$ have the same eigenvalues for all nonsingular matrices $\boldsymbol{S}$. The proof of the positiveness of the eigenvalues of $\boldsymbol{K}_0\boldsymbol{P}_0$ then follows in a similar way to the proof of the positiveness of the eigenvalues of $\boldsymbol{K}_0\boldsymbol{W}_0$ in part (i). The proof for part (iv) then follows in a similar way to the proof of part (ii).

Importance of Scaling Parameter λ

The value of λ in the choice of the (nonunique) optimal matrix $\boldsymbol{S}_2$ in (5.182) is important in the following way. First, note in part (ii) that

$$\boldsymbol{S}\boldsymbol{K}_0\boldsymbol{S}^T = \boldsymbol{S}_2\boldsymbol{\Sigma}_0\boldsymbol{S}_2^T \;\; ; \;\; (\boldsymbol{S}^{-1})^T\boldsymbol{W}_0\boldsymbol{S}^{-1} = (\boldsymbol{S}_2^{-1})^T\boldsymbol{\Sigma}_0\boldsymbol{S}_2^{-1} \tag{5.184}$$

and so

$$tr[\boldsymbol{S}\boldsymbol{K}_0\boldsymbol{S}^T] = tr[\boldsymbol{\Sigma}_0.\boldsymbol{S}_2^T\boldsymbol{S}_2] = \lambda^2 tr[\boldsymbol{\Sigma}_0] = \lambda^2.\sum_{k=1}^{n}\sigma_k$$

In other words, in an optimal state space structure with $\boldsymbol{S}_2$ given by (5.181), the value of λ determines the *scaling characteristics* according to

$$\sum_{j=1}^{n} ||\tilde{h}_j||_2^2 = tr[\boldsymbol{S}\boldsymbol{K}_0\boldsymbol{S}^T] = \lambda^2.\sum_{k=1}^{n}\sigma_k$$

More specifically, from (5.182) and (5.184) we have

$$\boldsymbol{S}\boldsymbol{K}_0\boldsymbol{S}^T = \lambda^2\boldsymbol{V}_2^T\boldsymbol{\Sigma}_0\boldsymbol{V}_2 \;\; ; \;\; (\boldsymbol{S}^{-1})^T\boldsymbol{W}_0\boldsymbol{S}^{-1} = \lambda^{-2}\boldsymbol{V}_2^T\boldsymbol{\Sigma}_0\boldsymbol{V}_2$$

where the diagonal matrix $\boldsymbol{\Sigma}_0$ is given by (5.177). We now make use of another result in linear algebra.

Specifically, given the positive definite matrix $\boldsymbol{\Sigma}_0$ in (5.177), there exists a unitary matrix $\boldsymbol{V}_2$ such that

$$\lambda^2\boldsymbol{V}_2^T\boldsymbol{\Sigma}_0\boldsymbol{V}_2 = \text{diag}\{\gamma_1^2, \gamma_2^2, \;\ldots\;, \gamma_n^2\}$$
$$\lambda^{-2}\boldsymbol{V}_2^T\boldsymbol{\Sigma}_0\boldsymbol{V}_2 = \lambda^{-4}\text{diag}\{\gamma_1^2, \gamma_2^2, \;\ldots\;, \gamma_n^2\}$$

where the parameters $\{\gamma_k > 0\}$ satisfy the equation

$$\sum_{k=1}^{n}\gamma_k = \lambda^2\sum_{k=1}^{n}\sigma_k \;\; (= \lambda^2 tr[\boldsymbol{\Sigma}_0])$$

From Theorem 5.5.1, this result then implies that there exists a particular optimal scaling transformation $\boldsymbol{S}$ such that: for $1 \le j \le n$

$$||\tilde{h}_j||_2 = \gamma_j \quad ; \quad ||\tilde{h}_{y\eta_j}||_2 = \left(\frac{\sum_{k=1}^{n} \sigma_k}{\sum_{k=1}^{n} \gamma_k}\right) \gamma_j$$

In particular, there exists a unitary matrix $\boldsymbol{V}_2$ such that

$$\lambda^2 \boldsymbol{V}_2^T \boldsymbol{\Sigma}_0 \boldsymbol{V}_2 = \gamma^2 \boldsymbol{I}_n \quad ; \quad \gamma > 0$$

where

$$n\gamma^2 = \lambda^2 \sum_{k=1}^{n} \sigma_k$$

That is, there exists an optimal transformation $\boldsymbol{S}$ such that: for all $1 \le j \le n$

$$||\tilde{h}_j||_2 = \gamma \quad ; \quad ||\tilde{h}_{y\eta_j}||_2 = \gamma^{-1} \left(\frac{\sum_{k=1}^{n} \sigma_k}{n}\right)$$

Similarly, in part (iv), there exists an optimal scaling transformation $\boldsymbol{S}$ such that: for $1 \le j \le n$

$$||\tilde{h}_j||_2 = \gamma_j \quad ; \quad ||\tilde{h}_{y\eta_j}||_2 = \left(\frac{\sum_{k=1}^{n} \varrho_k}{\sum_{k=1}^{n} \gamma_k}\right) \gamma_j$$

and in particular, an optimal $\boldsymbol{S}$ exists such that: for all $1 \le j \le n$

$$||\tilde{h}_j||_2 = \gamma \quad ; \quad ||\tilde{h}_{y\eta_j}||_2 = \gamma^{-1} \left(\frac{\sum_{k=1}^{n} \varrho_k}{n}\right)$$

Example 5.5.4. Consider again the 4th order narrowband digital Butterworth filter of bandwidth 0.01π defined in example 5.5.2.

(i) The square roots $\{\sigma_k\}$ of the eigenvalues $\{\sigma_k^2\}$ of $\boldsymbol{KW}$ are given by

$$\sigma_k \in \{1.2022, 0.6734, 0.1819, 0.0177\}$$

Hence from Theorem 5.5.2, the optimal performance measure J_{FWL} in the absence of quantization error feedback is given by

$$J_{FWL} = \frac{1}{4} \sum_{k=1}^{4} \sigma_k = 0.5188$$

As required, this value is less than the value $J_{FWL} = 1.4285$ which was obtained for the Jordan structure in example 5.5.3 when $\{\boldsymbol{J}_0 = \boldsymbol{0}, \boldsymbol{h}_0 = \boldsymbol{0}\}$.

(ii) The square roots $\{\varrho_k\}$ of the eigenvalues $\{\varrho_k^2\}$ of $\boldsymbol{KP}$ are given by

$$\varrho_k \in \{0.1413, 0.0219, 0.0097, 0.0012\}$$

Hence from Theorem 5.5.2, the optimal performance measure J_{FWL} with quantization error feedback coefficients $\{\boldsymbol{J}_0 = \boldsymbol{I}_4, \boldsymbol{h}_0 = \boldsymbol{0}\}$ is given by

$$J_{FWL} = \frac{1}{4}\sum_{k=1}^{4} \varrho_k = 0.0435$$

As required, this value is less than the value $J_{FWL} = 0.0445$ which was obtained for the Jordan structure in example 5.5.3 when $\{\boldsymbol{J}_0 = \boldsymbol{I}_4, \boldsymbol{h}_0 = \boldsymbol{0}\}$.

Summary

The implementation of a digital filter requires the evaluation of inner products. In *floating point* arithmetic, the evaluation of an inner product requires quantization (and hence arithmetic error) after every multiplication and every addition. On the other hand, *fixed point* arithmetic is able to evaluate an inner product exactly, and therefore only requires one quantization per inner product provided the accumulator has sufficient wordlength. The disadvantage of fixed point compared to floating point arithmetic is that overflow is more likely to occur unless appropriate *scaling* is introduced.

This Chapter for the most part was concerned with the FWL implementation of IIR filters in fixed point arithmetic. However filters designed based on these considerations will also provide appropriate structures with which to implement in floating point arithmetic. In particular, sensitivities that arise as a result of coefficient quantization apply in respect of both arithmetic formats. Furthermore, generally speaking, the effects of FWL arithmetic increase as either the filter order increases, or the filter bandwidth decreases. Low cost high accuracy real-time signal processing applications are best served by the strategies advanced.

Exercises

5.1 Suppose $\{w_1 = 0.375, w_2 = 0.875, w_3 = -0.75\}$

(i) Find the twos complement representation of each number w_j when

$$\text{(a)}\quad w_j \approx [2,4]\ ;\ \text{(b)}\quad w_j \approx [1,4]\ ;\ \text{(c)}\quad w_j \approx [1,3]$$

(ii) Find the representation for the summation $w = w_1 + w_2$ for each of the representations in (i) when overflow is detected and saturation arithmetic is implemented.

(iii) Repeat (ii) when overflow is not detected.

(iv) Repeat (ii) and (iii) for the summation $w = w_1 + w_2 + w_3$.

Comment on the results with respect to the overflow property of fixed point addition.

5.2 (i) Use Booths algorithm to obtain the 4 bit multiplication of the twos complement fixed point numbers -4 and 3.

(ii) Use the sign extension property to convert the twos complement fixed point numbers -4 and 3 to a 5 bit representation, and repeat (i).

5.3 (i) Find the twos complement fixed point representation of the following numbers: $\{w_1 = 0.21875, w_2 = -0.21875, w_3 = 0.3125\}$

(ii) Using truncation quantization, find the quantized numbers $Q[w_j]$ when:

$$\text{(a)}\quad Q[w_j] \approx [1,4]\ ;\ \text{(b)}\quad Q[w_j] \approx [1,3]$$

(iii) Repeat (ii) using roundoff (rounding up) quantization.

(iv) Repeat (ii) using roundoff (rounding down) quantization.

(v) Repeat (ii) using convergent roundoff quantization.

5.4 (i) Find the floating point representation $[m_\Delta exp]$ of the numbers $w_1 = 2.3125$ and $w_2 = -3.125$ when the mantissa m is a signed magnitude fraction and exp is a sign magnitude integer.

(ii) Repeat (i) when exp is an (unsigned) biased integer.

(iii) Find the resolution and the dynamic range of the floating point representation $[m_\Delta exp]$ when m is a signed magnitude fraction of 3 bits and exp is a sign magnitude integer of 2 bits.

(iv) Repeat (iii) when exp is an (unsigned) biased integer of 2 bits.

5.5 (i) Compute the addition of the two numbers: $\{w_1 = 2.3125\ ;\ w_2 = -3.125\}$ in floating point arithmetic when the mantissa has a $[1,3]$ signed magnitude representation, and the exponent has a [2,0] signed integer representation.

(ii) Repeat (i) when the exponent has a $[2,0]$ unsigned biased integer representation.

5.6 Consider the numbers $\{w_1 = 4.375 \text{ and } w_2 = -2.3125\}$.

(i) Obtain the closest representation $Q_{FP}[w_j] \approx [m_\Delta exp]$ for the numbers w_1 and w_2 when the mantissa is rounded to a 2 bit signed fraction.

(ii) Write $Q_{FP}[w_j]$ in the form

$$Q_{FP}[w_j] = w_j(1 - \delta_j)$$

and in each case, find δ_j.

(iii) Write the floating point product $w_1 w_2$ in the form

$$FL[w_1 w_2] = w_1 w_2 (1 - \delta_{12})$$

and determine δ_{12} when the product $w_1 w_2$ is quantized to a representation $FL[w_1 w_2]$ in which the mantissa is a signed 5 bit fraction, and the exponent is a signed 5 bit integer.

5.7 (i) Write down all the possible floating point numbers $[m_\Delta exp]$ in which the mantissa is a 3 bit signed fraction and the exponent is a 2 bit signed integer.

(ii) Repeat (i) when the exponent is a 2 bit unsigned biased integer.

5.8 (i) Find the response of the ideal first order digital filter

$$y(k+1) = -0.75y(k) + u(k)$$

when $y(0) = 0$ and $u(k) = 1$

(ii) Suppose the filter is implemented in floating point arithmetic in which the signal y, the coefficients and all products and summations have a $[m_\Delta exp]$ representation where m has a 4 bit signed fraction and exp is a 2 bit signed integer. Determine

$$\hat{y}(k+1) = FL[FL[0.25\hat{y}(k)] + FL[\hat{u}(k)]]$$

for $k = 1, 2, 3$ when $\hat{y}(0) = 0$, and $\hat{u}(k) = 1$ for all k.

5.9 Consider the fixed point implementation

$$\hat{y}(k+1) = -0.75Q[\hat{y}(k)] + \hat{u}(k)$$

of the ideal filter $y(k+1) = -0.75y(k) + u(k)$ when the coefficient and all signal values $\{\hat{y}(k), \hat{u}(k)\}$ have a [4,4] bit representation.

(i) Calculate $\hat{y}(k)$ and $\varepsilon_Q(k) \triangleq \hat{y}(k) - Q[\hat{y}(k)]$ for $k = 1, 2, 3, 4$ when $\hat{y}(0) = 0$ and $\hat{u}(k) = 1$ for all k using roundoff quantization with saturation arithmetic.

(ii) Repeat (i) using truncation quantization with saturation arithmetic.

(iii) What is the steady state response of $\hat{y}$ using: (a) roundoff quantization and (b) truncation quantization.

(iv) Repeat (i) - (iii) based on a $[3, 3]$ bit representation.

5.10 Suppose

$$\hat{z}(k+1) = Q[-0.75\hat{z}(k) + \hat{u}(k)]\ ;\ \hat{z}(0) = 0,\ \hat{u}(k) = 1$$

Calculate $\hat{z}(k)$ and

$$\varepsilon_Q(k) \stackrel{\Delta}{=} -0.75\hat{z}(k) + \hat{u}(k) - Q[-0.75\hat{z}(k) + \hat{u}(k)]$$

for $k = 1, 2, 3, 4$, and show $\hat{z}(k) = Q[\hat{y}(k)]$ where $\hat{y}$ is given by exercise 5.9.

5.11 Consider the ideal filter

$$y(k+1) = 0.8125y(k) + 0.25u(k)$$

and the FWL fixed point implementation

$$\hat{y}(k+1) = 0.8125Q[\hat{y}(k)] + 0.25\hat{u}(k)$$
$$\varepsilon_Q(k) = \hat{y}(k) - Q[\hat{y}(k)]$$

where $Q[\hat{y}(k)], \hat{u}(k) \approx [2, 3]$

(i) Obtain a difference equation in e_y where $e_y(k) \stackrel{\Delta}{=} y(k) - \hat{y}(k)$ when $u(k) = \hat{u}(k)$.

(ii) Compute the error system function $E_{yu}(z)$ between the input signal u and the output error signal e_y.

(iii) Obtain an upper bound for $||e_y||_\infty$ when $u(k) = 2\sin 4\pi k$.

5.12 (i) Find $\{d_0, \tilde{b}_0\}$ such that the two ideal first order filters

$$y(k+1) = a_0 y(k) + \tilde{b}_0 u(k)\ ;\ z(k) = y(k) + d_0 u(k)$$

and

$$z(k+1) = a_0 z(k) + b_0 u(k) + b_1 u(k+1)$$

are equivalent.

(ii) Suppose ideally $\{a_0 = 0.75, b_0 = 0.25, b_1 = 0.50\}$, and that each filter is realized in finite wordlength in which the coefficients have a [1,2] bit representation. Show that in terms of coefficient quantization, the two finite wordlength realizations are not equivalent.

(iii) For each filter, derive the coefficient error transfer function $E_{yu}(z)$ between the input signal u and the output error e_y.

5.13 Consider the finite wordlength realization

$$\hat{y}(k+1) = 0.75Q[\hat{y}(k)] + 0.25(u(k) + 0.50u(k+1)$$

of the ideal filter $y(k+1) = 0.75y(k) + 0.25u(k) + 0.50u(k+1)$.

(i) Find the equation for the output error $e_y \triangleq y - \hat{y}$ in terms of the quantization error signal $\varepsilon_Q \triangleq \hat{y} - Q[\hat{y}]$.

(ii) Suppose $Q[\hat{y}(k)]$ has a [4,4] bit representation and never overflows. Find an upper bound for $||e_y||_\infty$ as a result of roundoff quantization of $\hat{y}$.

(iii) Suppose a statistical model is used to model the quantization signal ε_Q. Find $||e_y||_2$ for: (a) roundoff quantization and (b) truncation quantization.

(iv) Suppose an internal quantization error feedback term $\alpha\varepsilon_Q(k)$ is introduced into the finite wordlength realization. Repeat the calculation in (ii) and (iii) in terms of α. What is the best choice of α in each case which minimizes $||e_y||_\infty$ and $||e_y||_2$.

5.14 (i) Ignoring all quantization errors, develop an adaptively scaled algorithm for the first order digital filter

$$y(k+1) = a_0 y(k) + b_0 u(k) + b_1 u(k+1) \ ; \ y(0) = 0$$

in terms of a state z such that $0.5 \le |z(k)| < 1, \ z(k) \ne 0$.

(ii) Suppose $\{a_0 = 0.9, b_0 = 0.1, c_0 = 0.3\}$. Calculate $\{z(k), y(k)\}$ for $1 \le k \le 4$ when $\{u(k) = 1, \ k \ge 0\}$.

(ii) Repeat (ii) when $y(k)$ is restricted such that $|y(k)| \le 3$.

(iii) Repeat (i) when only internal quantization errors are included in which $\{z(k), y(k)\}$ are rounded to a 3 bit fractional representation.

(iv) In (iii), find a difference equation for e_y in terms of the quantization error signal ε_Q.

5.15 Suppose h is the unit impulse response of the frequency response function $H(z)$ where

$$H(z) = \frac{a_1 z + a_0}{z^2 - a_1 z - a_0}$$

in which

$$z^2 - a_1 z - a_0 = (z - \lambda_1)(z - \lambda_1) \ ; \ |\lambda_1| < 1, \ |\lambda_1| < 1$$

(i) Show that for $\lambda_1 \ne \lambda_2$

$$||h||_1 \le \frac{1}{|\lambda_1 - \lambda_2|}\left\{\frac{|\lambda_1|^2}{1 - |\lambda_1|} + \frac{|\lambda_2|^2}{1 - |\lambda_2|}\right\}$$

$$||h||_\infty \le \frac{|\lambda_1|^2 + |\lambda_2|^2}{|\lambda_1 - \lambda_2|}$$

$$||h||_2 = \frac{1}{|\lambda_1 - \lambda_2|}\sqrt{\frac{\lambda_1^4}{1 - \lambda_1^2} + \frac{2\lambda_1^2\lambda_2^2}{1 - \lambda_1\lambda_2} + \frac{\lambda_2^4}{1 - \lambda_2^2}}$$

(ii) Suppose $\{\lambda_1 = \varrho + j\omega;\ \lambda_2 = \varrho - j\omega,\ \omega \neq 0\}$. Find an expression for $||h||_2$, and bounds on $||h||_1$, and $||h||_\infty$ in terms of $\{\varrho, \omega\}$.

5.16 Consider the second order FWL

$$\begin{aligned} \hat{y}(k+2) &= 1.25Q[\hat{y}(k+1) - 0.375Q[\hat{y}(k)] + u(k+2) + 2u(k+1) + u(k) \\ \varepsilon_Q(k) &= \hat{y}(k) - Q[\hat{y}(k)] \end{aligned}$$

where $Q[\hat{y}(k)] \approx [4,3]$ denotes roundoff quantization.

(i) Suppose $\{\hat{y}(-1) = \hat{y}(0) = 0;\ u(k) = 1, k \geq 0\}$. Calculate $\hat{y}(k)$ for $1 \leq k \leq 4$ and compare with the ideal case

$$y(k+2) = 1.25y(k+1) - 0.375y(k) + u(k+2) + 2u(k+1) + u(k)$$

(ii) Obtain the difference equation for $e_y \stackrel{\Delta}{=} y - \hat{y}$ in terms of ε_Q.

(iii) Calculate an upper bound for $||e_y||_\infty$ in terms of $||\varepsilon_Q||_\infty$.

(iv) Suppose a statistical model is used for ε_Q. Obtain an expression for $||e_y||_\infty$ in terms of $||\varepsilon_Q||_2$.

(v) Suppose the quantization error feedback terms $\alpha_1\varepsilon_Q(k+1) + \alpha_0\varepsilon_Q(k)$ is incorporated into the FWL implementation. Repeat (i) - (iv).

5.17 Consider a second order state space digital filter

$$\boldsymbol{x}(k+1) = \boldsymbol{F}\boldsymbol{x}(k) + \boldsymbol{g}u(k)\ ;\ y(k) = \boldsymbol{c}^T\boldsymbol{x}(k) + u(k)$$

where

$$\boldsymbol{F} = \begin{bmatrix} f_{11} & f_{12} \\ f_{21} & f_{22} \end{bmatrix}\ ;\ \boldsymbol{g} = \begin{bmatrix} g_1 \\ g_2 \end{bmatrix}\ ;\ \boldsymbol{c} = \begin{bmatrix} c_1 \\ c_2 \end{bmatrix}$$

(i) Show that the frequency response $H(z)$ from the input u to the output y is given by

$$H(z) = \frac{z^2 + b_1 z + b_0}{z^2 - a_1 z - a_0}$$

where

$$\begin{aligned} a_1 &= f_{11} + f_{22}\ ;\ a_0 = f_{12}f_{21} - f_{11}f_{22}\ ;\ b_1 = c_1 g_1 + c_2 g_2 - a_1 \\ b_0 &= (c_2 f_{21} - c_1 f_{22})g_1 + (c_1 f_{12} - c_2 f_{11})g_2 - a_0 \end{aligned}$$

(ii) Find all possible values for $f_{ij} \approx [2,1]$ such that both roots $\{\lambda_m; m = 1,2\}$ of the equation $\lambda^2 - a_1\lambda - a_0 = 0$ satisfy the condition $0 \leq \lambda_m < 1$.

(iii) Find all possible values for $g_1, g_2 \in \{0,1\}$ and $c_j \approx [2,1]$ such that both roots $\{r_m; m = 1,2\}$ of the equation $r^2 + b_1 r + b_0 = 0$ satify the condition $-1 \leq r_m \leq 0$.

5.18 Consider the second order digital filter

$$y(k+2) = 1.25y(k+1) - 0.375y(k) + u(k+2) + 2u(k+1) + u(k)$$

Define $z(k) \triangleq L_0^{-1} y(k)$
(i) Find L_0 in terms of $||u||_\infty$ such that $|z(k)| \leq 1$ when $y(0) = y(-1) = 0$.
(ii) Repeat (i) by expressing L_0 in terms of $||u||_2$.

5.19 Find the relationship between $\{\boldsymbol{F}_0, \boldsymbol{F}_1, \boldsymbol{F}, \boldsymbol{g}_0, \boldsymbol{g}_1, \boldsymbol{g}_2, \boldsymbol{c}, \boldsymbol{g}\}$ such that the two ideal state space filters

$$\begin{aligned} \boldsymbol{z}(k+2) &= \boldsymbol{F}_1 \boldsymbol{z}(k+1) + \boldsymbol{F}_0 \boldsymbol{z}(k) + \boldsymbol{g}_0 u(k) + \boldsymbol{g}_1 u(k+1) + \boldsymbol{g}_1 u(k+2) \\ y(k) &= \boldsymbol{c}^T \boldsymbol{z}(k) \end{aligned}$$

and

$$\boldsymbol{x}(k+1) = \boldsymbol{F}\boldsymbol{x}(k) + \boldsymbol{g}u(k) \; ; \; y(k) = \boldsymbol{c}^T \boldsymbol{x}(k) + du(k)$$

are equivalent with respect to the relationship between the input signal u and the output signal y.

5.20 An ideal second order state space filter is given by

$$\begin{aligned} x_1(k+1) &= 0.25x_1(k) - 0.5x_2(k) + u(k) \\ x_2(k+1) &= -0.5x_1(k) + 0.75x_2(k) - 2u(k) \\ y(k) &= x_1(k) - x_2(k) + 0.875u(k) \end{aligned}$$

(i) Suppose $Q[w] \approx [2,4]$. Explain how

$$\begin{aligned} \hat{x}_1(k+1) &= Q[0.25\hat{x}_1(k) + u(k)] - 0.5Q[\hat{x}_2(k)] \\ \hat{x}_2(k+1) &= -0.5Q[\hat{x}_1(k)] + Q[0.75\hat{x}_2(k) - 2\hat{u}(k)] \\ \hat{y}(k) &= \hat{x}_1(k) - \hat{x}_2(k) + Q[0.875\hat{u}(k)] \end{aligned}$$

is a valid FWL realization by defining the FWL wordlength representation of the variables $\{\hat{x}_j(k), \hat{u}(k), \hat{y}(k)\}$.
(ii) Find $\{\alpha_1, \alpha_2\}$ such that

$$\begin{aligned} z_1(k+1) &= 0.25z_1(k) - 0.5z_2(k) + u(k) + \alpha_1 u(k+1) \\ z_2(k+2) &= -0.5z_2(k) + 0.75z_2(k) - 2u(k) + \alpha_2 u(k+1) \\ y(k) &= z_1(k) - z_2(k) \end{aligned}$$

is also an equivalent filter representation.
(iii) If all coefficients must have a $[1,2]$ representation, what filter in (i) or (ii) best approximates the ideal filter. On what basis do you make your judgement ?

5.21 An ideal second order digital filter is given by

$$y(k+2) = \frac{1.8}{\sqrt{2}} y(k+1) - 0.81y(k) + u(k+2) + 2u(k+1) + u(k)$$

(i) Suppose all coefficients are restricted to a [2,6] bit representation, and $\hat{y}(k) \approx [2,6]$. Define an asymptotically stable FWL realization without quantization error feedback which expresses the FWL approximation $\hat{y}(k+2)$ directly in terms of $\hat{y}(k+1)$ and $\hat{y}(k)$.

(ii) Repeat (i) when a quantization error feedback term is included.

(iii) What is the best choice of the term in (ii) with respect to minimizing $||e_y||_\infty$ where $e_y = y - \hat{y}$ with respect to $||\hat{y} - Q[\hat{y}]||_\infty$.

5.22 Consider a FWL realization of the ideal state space representation

$$\boldsymbol{x}(k+1) = \begin{bmatrix} -0.9 & 0.8 \\ 0.36 & 0.72 \end{bmatrix} \boldsymbol{x}(k) + \begin{bmatrix} 1 \\ 0.9 \end{bmatrix} u(k)$$
$$y(k) = [-1.6 \;\; 2.1]\boldsymbol{x}(k) + 0.4u(k)$$

where all signals and all coefficients are restricted to a $[2,4]$ bit representation.

(i) Find the error frequency function $E_{y\nabla u}(z)$ between the input error $\nabla u = u - \hat{u}$ and the output error $e_y = y - \hat{y}$ as a result of only input errors.

(ii) Write down an expression for $||h_{y\nabla u}||_2$ where $h_{y\nabla u}$ is the unit impulse response of $E_{y\nabla u}(z)$, and hence obtain a bound for $||e_y||_\infty$.

(iii) Find the error frequency function $E_{yu}(z)$ between the input u and the output error e_y as a result of only coefficient errors.

(iv) Write down an expression for $||h_{yu}||_2$ where h_{yu} is the unit impulse response of $E_{yu}(z)$, and hence obtain a bound for $||e_y||_\infty$.

(v) Find the output error e_y in terms of only the internal quantization errors $\varepsilon_{Qj} = \hat{x}_j - Q[\hat{x}_j]$ on the FWL state components $\{\hat{x}_j; j = 1, 2\}$ in the presence of internal quantization error feedback.

(vi) Derive an expression for $||h_{y\varepsilon_j}||_2$ where $h_{y\varepsilon_j}$ is the unit impulse response between the internal quantization error ε_{Qj} and the output e_y.

5.23 Find the solution of $\boldsymbol{K} = \boldsymbol{K}^T$ of the equation $\boldsymbol{K} = \boldsymbol{F}\boldsymbol{K}\boldsymbol{F}^T + \boldsymbol{g}\boldsymbol{g}^T$ when:

(i) $\boldsymbol{F} = \begin{bmatrix} 0 & 1 \\ a_0 & a_1 \end{bmatrix}$; $\boldsymbol{g} = \begin{bmatrix} 0 \\ 1 \end{bmatrix}$

(ii) $\boldsymbol{F} = \begin{bmatrix} \lambda_1 & 0 \\ 0 & \lambda_2 \end{bmatrix}$ $(\lambda_1 \neq \lambda_2)$; $\boldsymbol{g} = \begin{bmatrix} 1 \\ 1 \end{bmatrix}$

(iii) $\boldsymbol{F} = \begin{bmatrix} \lambda & 1 \\ 0 & \lambda \end{bmatrix}$; $\boldsymbol{g} = \begin{bmatrix} 0 \\ 1 \end{bmatrix}$

(iv) $\boldsymbol{F} = \begin{bmatrix} \rho & \omega \\ -\omega & \rho \end{bmatrix}$ $(\omega \neq 0)$; $\boldsymbol{g} = \begin{bmatrix} 0 \\ 1 \end{bmatrix}$

5.24 Consider the function f defined by

$$f \triangleq (a_0 - \alpha_0)^2 + (a_1 - \alpha_1)^2 + \left(\frac{2a_1}{1-a_0}\right)(a_0 - \alpha_0)(a_1 - \alpha_1)$$

where $\{\alpha_0, \alpha_1\}$ are both integers. Find $\{\alpha_{0,min}, \alpha_{1,min}\}$ which minimize f when:

(a) $\{a_1 = 1.8, a_0 = 0.81\}$
(b) $\{a_1 = -1.8, a_0 = 0.81\}$
(c) $\{a_1 = 1.8cos\theta, a_0 = 0.81\}$
(d) $\{a_1 = 0.5, a_0 = 0.36\}$

Confirm that the choice $\{\alpha_{0,min} = round(a_0), \alpha_{1,min} = round(a_1)\}$ is not always optimal.

5.25 Consider the ideal state space representation

$$\boldsymbol{x}(k+1) = \begin{bmatrix} -0.90 & 0.80 \\ 0.36 & 0.72 \end{bmatrix} \boldsymbol{x}(k) + \begin{bmatrix} 1 \\ 0.9 \end{bmatrix} u(k)$$
$$y(k) = [-1.6 \;\; 2.1]\boldsymbol{x}(k) + 0.4u(k)$$

(i) Using MATLAB, compute the 2-norm $||h_j||_2$ for $j = 1, 2$ of the unit impulse response h_j between the input signal u and the state component x_j of the state $\boldsymbol{x}$.

(ii) Using MATLAB, compute the 2-norm $||h_{y\varepsilon_j}||_2$ for $j = 1, 2$ of the unit impulse response $h_{y\varepsilon_j}$ between the internal quantization error signal $\varepsilon_{Qj} = \hat{x}_j - Q[\hat{x}_j]$ and the output error $e_y = y - \hat{y}$ of the FWL realization.

$$\hat{\boldsymbol{x}}(k+1) = \begin{bmatrix} -0.90 & 0.80 \\ 0.36 & 0.72 \end{bmatrix} Q[\hat{\boldsymbol{x}}(k)] + \begin{bmatrix} 1 \\ 0.9 \end{bmatrix} u(k)$$
$$\hat{y}(k) = [-1.6 \;\; 2.1]Q[\hat{\boldsymbol{x}}(k)] + 0.4u(k)$$

(iii) Let

$$\boldsymbol{z}(k) = \boldsymbol{S}\boldsymbol{x}(k) \; ; \; \boldsymbol{S} = diag\{s_1, s_2\}$$

From (i), find $\{s_1, s_2\}$ such that

$$||\tilde{h}_j||_2 = 1 \; ; \; j = 1, 2$$

where $\tilde{h}_j$ is the unit impulse response between the input signal u and the state component z_j of the transformed state $\boldsymbol{z}$.

(iv) Using (ii), compute the 2-norm $||h_{y\eta_j}||_2$ for $j = 1, 2$ of the unit impulse response $h_{y\eta_j}$ between the internal quantization error signal $\varepsilon_{\eta j} = \hat{z}_j - Q[\hat{z}_j]$ and the output error $e_y = y - \hat{y}$ of the transformed FWL realization

$$\hat{z}(k+1) = S\begin{bmatrix} -0.90 & 0.80 \\ 0.36 & 0.72 \end{bmatrix} S^{-1}Q[\hat{z}(k)] + S\begin{bmatrix} 1 \\ 0.9 \end{bmatrix} u(k)$$

$$\hat{y}(k) = [-1.6 \quad 2.1]S^{-1}Q[\hat{z}(k)] + 0.4u(k)$$

(v) Repeat (ii) after including internal quantization error feedback.
(vi) Repeat (iv) after including internal quantization error feedback.

Index